Analysis of Microarray Data

Edited by
Frank Emmert-Streib and
Matthias Dehmer

Related Titles

J.E. Van Eyk, M.J. Dunn (Eds.)

Clinical Proteomics

From Diagnosis to Therapy

2008

ISBN 978-3-527-31637-3

S. Knudsen

Cancer Diagnostics with DNA Microarrays

2006

ISBN 978-0-471-78407-4

C.W. Sensen (Ed.)

Handbook of Genome Research

Genomics, Proteomics, Metabolomics, Bioinformatics, Ethical and Legal Issues

2005

ISBN 978-3-527-31348-8

E. Wit, J. McClure

Statistics for Microarrays

Design, Analysis and Inference

2004

ISBN 978-0-470-84993-4

M. Schena, S. Knudsen

Guide to Analysis of DNA Microarray Data, and Microarray Analysis Set

Second Edition

2004

ISBN 978-0-471-67853-3

Analysis of Microarray Data

A Network-Based Approach

Edited by
Frank Emmert-Streib and Matthias Dehmer

WILEY-VCH

WILEY-VCH Verlag GmbH & Co. KGaA

The Editors

Dr. Frank Emmert-Streib
University of Washington
Department of Biostatistics
University of Washington
Department of Genome Sciences
Seattle, WA 98195-5065
USA

Dr. Matthias Dehmer
Vienna University of Technology
Discrete Mathematics and Geometry
Wiedner Hauptstrasse 8–10
1040 Vienna
Austria

Library of Congress Card No.: applied for

British Library Cataloguing-in-Publication Data
A catalogue record for this book is available from the British Library.

Bibliographic information published by the Deutsche Nationalbibliothek
Die Deutsche Nationalbibliothek lists this publication in the Deutsche Nationalbibliografie; detailed bibliographic data are available in the Internet at http://dnb.d-nb.de

Printed in the Federal Republic of Germany
Printed on acid-free paper

Typesetting Thomson Digital, Noida, India
Printing Strauss GmbH, Mörlenbach
Book Binding Litges & Dopf GmbH, Heppenheim

ISBN: 978-3-527-31822-3

Contents

Analysis of Microarray Data: A Network-Based Approach. Edited by F. Emmert-Streib and M. Dehmer

ISBN: 978-3-527-31822-3

Preface

This book presents novel approaches for the statistical analysis of microarray data. All these methods share the common theme to be network based meaning that not single genes but groups of interacting genes are considered to be important. For this reason it is the first of its kind. This is in clear contrast to, e.g., traditional studies aiming to the detection of differentially expressed genes [7,10,11]. The advantage using network based statistical methods for the analysis of microarray data becomes apparent considering the fact that genes or products thereof within a cell form molecular networks [1,4,9]. Hence, network based methods are in some sense model based rather than ad hoc. Moreover, due to the fact that network based approaches consider per se groups of genes such approaches are at the very heart of systems biology [5,6] a derivative from systems theory [2,3] thematizing that a functional understanding of biological systems can only be gained by studying the system as a whole, which means practically one needs to study large enough parts to gain a functional understanding.

Network based analysis methods reflect our current knowledge about the molecular organization within and between cells unrevealed by an astonishing technological progress of high-throughput measurement devices nowadays used on a regular base in molecular biology. These high-throughput technologies allow to monitor the behavior of genes or products thereof on a genome-wide and, hence, systems rather than on a single gene level. Traditionally, molecular biology mainly deals with the description of macro molecules, especially DNA, RNA, proteins and interactions thereof concerning the regulation of biological processes like the cell cycle. In the past, experimental observations were restricted to, e.g., the interaction of single proteins or the expression or non-expression of single genes characterized by the mRNA concentration of the corresponding messenger ribosomes. These basic techniques have been developed into powerful assays that measure, e.g., the expression of all ~6000 genes in yeast. However, the high-throughput technology is not only a blessing but also a curse because the analysis of the obtained data is challenging due to the novelty and the characteristics of the data and their quantity requiring the development of efficient statistical and computational methods and algorithms. The purpose of this book is to provide some novel statistical methods to

Analysis of Microarray Data: A Network-Based Approach. Edited by F. Emmert-Streib and M. Dehmer

ISBN: 978-3-527-31822-3

further enhance our progress regarding a more systematic understanding of molecular biological processes. It is widely believed that a solid understanding of fundamental processes in molecular biology will have a strong impact for many related disciplines, above all medicine and, hence, public health. For example the development of diagnostic methods for complex diseases as well as methods for rational drug design could be two possible offsprings. Especially for complex diseases, e.g. cancer, insights and knowledge from molecular biology could be the key players for a thorough understanding of this disease because of the sophisticated mechanisms cancer adopts within an organism.

This book starts with an introductory chapter about the microarray technology and its underlying biology. This should equip people unfamiliar with this kind of technology with enough information to at least get a good starting point and an intuitive understanding for expression data analyzed in subsequent chapters. All following chapters present statistical methods for the analysis of microarray data. Methodologically, these chapters comprise a large area using, e.g., Bayesian networks, dynamical Bayesian networks, Boolean networks, structural equation models, graphical Gaussian models, regression models or the maximum entropy method to cluster genes, infer gene networks or modules, simulate gene regulatory networks or detect disease stages.

The present book is intended for researches in the interdisciplinary fields computational biology, biostatistics and systems biology comprising traditional fields as, e.g., biology, computer science, electrical engineering, mathematics, statistics or physics. During the initial phase in the preparation of this book we decided that the chapters of this book should be presented comprehensively accessible not only to researchers from this field but also to researchers from related fields or even students that passed already introductory courses. For this reason each chapter presents not only some novel results but also provides some background knowledge necessary to understand, e.g., the mathematical method or the biological problem under consideration. In research articles this background information is either completely omitted or the reader is referred to an original article. Hence, this book could also serve as textbook for, e.g., an interdisciplinary seminar for advanced students not only because of the comprehensiveness of the chapters but also because of its size allowing to fill a complete semester. In addition, the fact that each chapter is presented by different authors removes a possible bias regarding presented topics and the depth of the presentation and provides a more fair overview of state-of-the-art methods in this field.

We resisted to separate this book into several parts because this separation would have required the introduction of an evaluation schema either methodologically or phenomenologically. However, this separation would have been ad hoc and, hence, might be confusing for readers that misleadingly overestimate the presence of different parts. Instead, each chapter can be read independently from other chapters although the experienced scientist will immediately recognize that there are overlappings and common themes among chapters.

Many colleagues, whether consciously or unconsciously, have provides us with input, help and support before and during the formation of the present book. In particular we would like to thank Hamid Arabnia, Goekhan Bakir, Igor Bass, Danail Bonchev, Stefan Borgert, Chris Bausch, Mike Coleman, Malcolm Cook, Andrey A. Dobrynin, Michael Drmota, Tanja Gesell, Galina Glazko, Julio Vera Gonzalez, Martin Grabner, Earl Glynn, Arndt von Haeseler, Alexander Hartemink, Pengyu Hong, Petr Kovář, Tereza Kovářá, Jing Liu, Alexander Mehler, Thomas Millat, Max Mühlhäuser, Arcady Mushegian, Katja Rateitschak, William Reeves, Heinz Georg Schuster, Helmut Schwegler, Chris Seidel, Ricardo de Matos Simoes, Fred Sobik, Doru Stefanescu, John Storey, Dan Thomasset, Robert Tibshirani, Suzanne D. Vernon, Olaf Wolkenhauer, Wing H. Wong, Bohdan Zelinka, Dongxiao Zhu and apologize to all who have not been named mistakenly. We would like also to thank our editors Aiden Budd, Steffen Pauly, Andreas Sendtko and Frank Weinreich from Wiley-VCH who have been always available and helpful. Last but not least we would like to thank our families for support and encouragement during all that time.

Finally, we hope this book helps to spread out our enthusiasm and joy we have for this field and inspires people regarding their own practical or theoretical research problems.

Seattle and Vienna
January 2008

Frank Emmert-Streib
Matthias Dehmer

References

1 Barabsi, A.L. and Oltvai, Z.N. (2004) Network biology: Understanding the cell's functional organization. *Nature Reviews Genetics*, **5**, 101–113.

2 Bertalanffy, L. von (1950) An outline of general systems theory. *British Journal for the Philosophy of Science*, **1** (2).

3 Bertalanffy, L. von (1976) General System Theory: Foundations, Development, Applications, (revised edition). New York, George Braziller.

4 Jeong, H., Tombor, B., Albert, R., Oltvai, Z. and Barabsi, A.-L. (2000) The large-scale organization of metabolic networks. *Nature*, **407**, 651–655.

5 Kitano, H. (ed.) (2001) Foundations of Systems Biology. MIT Press.

6 Kriete, A. and Eils, R. (2005) Computational Systems Biology. Elsevier Academic Press.

7 Newton, M.A., Kendziorski, C.M., Richmond, C.S., Blattner, F.R. and Tsui, K.W. (2001) On differential variability of expression ratios: improving statistical inference about gene expression changes from microarray data. *Journal of Computational Biology*, **8**, 37–52.

8 Pan, W. (2002) A comparative review of statistical methods for discovering differentially expressed genes in replicated microarray experiments. *Bioinformatics*, **18**, 546–554.

9 Ravasz, E., Somera, A.L., Mongru, D.A., Oltvai, Z.N. and Barabsi, A.-L. (2002) Hierarchical organization of modularity in metabolic networks. *Science*, **297**, 1551–1555.

10 Storey, J.D. and Tibshirani, R. (2003) Statistical significance for genome-wide studies. *Proceedings of the National Academy of Sciences*, **100**, 9440–9445.

11 Tusher, V.G., Tibshirani, R. and Chu, G. (2001) Significance analysis of microarrays applied to the ionizing radiation response. *Proceedings of the National Academy of Sciences USA*, **98**, 5116–5121.

List of Contributors

Jason E. Aten
University of California Los Angeles
Department of Biomathematics
David Gaffen School of Medicine
AV-617 Center for Health Sciences
Box 951766
Los Angeles, CA 90095-1766
USA

Jayanth R. Banavar
Pennsylvania State University
Department of Physics
104 Davey Laboratory
University Park, PA 16802
USA

Bor-Sen Chen
National Tsing Hua University
Department of Electrical Engineering
101, Section 2, Kuang-Fu Road
Hsinchu, 30013
Taiwan

Yu Chen
Monsanto Company
800 North Lindbergh Blvd
St. Louis, MO 63167
USA

Marek Cieplak
Institute of Physics
Polish Academy of Sciences
Aleja Lotników 32/46
02-668 Warsaw
Poland

Matthias Dehmer
Vienna University of Technology
Discrete Mathematics and Geometry
Wiedner Hauptstrasse 8–10/104
1040 Vienna
Austria

Mary-Lee Dequéant
Stowers Institute for Medical Research
1000 East 50th Street
Kansas City, MO 64110
USA

Edward R. Dougherty
Texas A&M University
Department of Electrical and
Computer Engineering
College Station, TX 77843
USA

Analysis of Microarray Data: A Network-Based Approach. Edited by F. Emmert-Streib and M. Dehmer

ISBN: 978-3-527-31822-3

Frank Emmert-Streib
Stowers Institute for Medical Research
1000 East 50th Street
Kansas City, MO 64110
USA

Nina V. Fedoroff
Pennsylvania State University
Department of Biology and Huck
Institutes of the Life Sciences
University Park, PA 16802
USA

Anna Georgieva
Novartis Pharmaceuticals
Corporation
One Health Plaza
East Hanover, NJ 07901
USA

Marco Grzegorczyk
University of Edinburgh
School of Biological Sciences
and Biomathematics and Statistics
Scotland (BioSS)
The King's Buildings
Edinburgh EH9 3JZ
United Kingdom

Gabriel Helmlinger
Novartis Institutes for Biomedical
Research Inc.
400 Technology Square
Cambridge, MA 02139
USA

Yufei Huang
University of Texas at San Antonio
(UTSA)
Department of Electrical and
Compute Engineering
One UTSA Circle
San Antonio, TX 78249-0669
USA

Dirk Husmeier
University of Edinburgh
School of Biological Sciences
and Biomathematics and Statistics
Scotland (BioSS)
The King's Buildings
Edinburgh EH9 3JZ
United Kingdom

Inyoung Kim
Virginia Tech
Department of Statistics
406-A Hutcheson Hall
Blacksburg, VA 24061
USA

Timothy Lezon
Pennsylvania State University
Department of Physics
104 Davey Laboratory
University Park, PA 16802
USA

Hongzhe Li
University of Pennsylvania
Department of Biostatistics
and Epidemiology
920 Blockley Hall
423 Guardian Drive
Philadelphia, PA 19104-6021
USA

Hua Li
Stowers Institute for Medical Research
1000 East 50th Street
Kansas City, MO 64110
USA

Wen-Hsiung Li
University of Chicago
Department of Evolution and Ecology
1101 East 57th Street
Chicago, IL 60637
USA

Amos Maritan
Universita di Padova
Dipartimento di Fisica "G. Galilei"
and INFN
Via Marzolo 8
35131 Padova
Italy

Herbert Pang
Yale University School of Medicine
Department of Epidemiology and
Public Health
60 College Street
New Haven, CT 06520
USA

Andre S. Ribeiro
University of Calgary
Institute for Biocomplexity and
Informatics
Department of Physics and
Astronomy
Calgary
Canada
and
Center for Computational Physics
University of Coimbra
3004-516 Coimbra
Portugal

Muriel Saulnier
Novartis Pharmaceuticals
Corporation
One Health Plaza
East Hanover, NJ 07901
USA

Christopher Seidel
Stowers Institute for Medical
Research
1000 East 50th Street
Kansas City, MO 64110
USA

Adriano V. Werhli
Pontifical Catholic University
of Rio Grande do Sul
Department of Computing Science
Av. Ipiranga, 6681
Porto Alegre/RS
Brazil

Wei-Sheng Wu
National Tsing Hua University
Department of Electrical Engineering
101, Section 2, Kuang-Fu Road
Hsinchu, 30013
Taiwan

Momiao Xiong
University of Texas School of
Public Health
Division of Biostatistics
1200 Herman Pressler Street
Houston, TX 77030
USA

Dong Xu
University of Missouri-Columbia
Department of Computer Science
201 Engineering Building West
Columbia, MO 65211
USA

Jing Yu
Novartis Institutes for Biomedical
Research Inc.
400 Technology Square
Cambridge, MA 02139
USA

Hongyu Zhao
Yale University School of Medicine
Department of Epidemiology
and Public Health
Department of Genetics
60 College Street
New Haven, CT 06520
USA

Dongxiao Zhu
Stowers Institute for Medical Research
1000 East 50th Street
Kansas City, MO 64110
USA
and
University of Kansas Medical Center
Department of Biostatistics
3901 Rainbow Boulevard
Kansas City, MO 66160
USA

1 Introduction to DNA Microarrays

Chris Seidel

1.1 Introduction

All living organisms contain DNA, a molecule that encodes all the information required for the development and functioning of an organism. Finding and deciphering the information encoded in DNA, and understanding how such a simple molecule can give rise to the amazing biological diversity of life, is a goal shared in some way by all life scientists. Microarrays provide an unprecedented view into the biology of DNA, and thus a rich way to examine living systems. DNA is a physical molecule that is able to encode information in a linear structure. Cells express information from different parts of this structure in a context-dependent fashion. DNA encodes for genes, and regulatory elements control whether genes are on or off. For instance, all the cells of the human body contain the same DNA, yet there are hundreds of different types of cells, each expressing a unique configuration of genes from the DNA. In this regard, DNA could be described as existing in some number of states. Microarrays are a tool used to read the states of DNA.

Microarrays have had a transforming effect on the biological sciences. In the past, biologists had to work very hard to generate small amounts of data that could be used to explore a hypothesis with one observation at a time. With the advent of microarrays, individual experiments generate thousands of data points or observations. This turns the experiment from a hypothesis-driven endeavor to a hypothesis-generating endeavor because every experiment sheds light across an entire terrain of gene expression, letting relevant genes reveal themselves, often in surprising ways. Concomitantly, what used to be primarily wet science becomes in large part information science. The highly parallel nature of microarrays that are used to make biological observations signifies that most experiments generate more information than the experimenter could possibly interpret. Indeed, from a statistical point of view, every gene measured on a microarray is an independent variable in a highly parallel experiment. The number of hypotheses to which the data may or may not lend support cannot be known in advance. To take advantage of the excess

Analysis of Microarray Data: A Network-Based Approach. Edited by F. Emmert-Streib and M. Dehmer

ISBN: 978-3-527-31822-3

information in microarray data, repositories have been set up in which people can deposit their experiments, thus making them available to a wide community of researchers with questions to explore (Gene Expression Ominbus GEO, ArrayExpress).

1.1.1
The Genome is an Information Scaffold

Microarrays measure events in the genome. An event may be the transcription of a gene, the binding of a protein to a segment of the DNA, the presence or absence of a mutation, a change in the copy number of a locus, a change in the methylation state of the DNA, or any of a number of states or activities that are associated with DNA or RNA molecules. As a genomic readout, microarrays identify where these events occur.

The idea that one can accurately describe the genome, let alone measure its activity in a comprehensive way, is a relatively novel concept. Several factors have led to the recent enhancement and blending of molecular biology into a field called genomics. The first is genome-sequencing projects. Today, sequencing a genome is considered a routine activity. However, in the late 1980s when sequencing the human genome was first suggested as a serious endeavor, the community was divided. Given the sequencing technology available at the time, the project looked as if it would consume colossal resources over a long time frame that many thought could be put to better use on more practical projects. However, visionaries were banking on two precepts: once given the mandate, the technology would transform itself and new sequencing methods would be invented that would increase the rate of sequence accumulation. The second aspect is that the finished project, full genome sequences, would be a public gold mine of a resource that would pay off for all biologists. Both of these assumptions have come to fruition. Genome sequences accumulate at rates few imagined possible. Biologists can expect the sequence of their model organism to exist in GenBank or to be in someone's sequencing pipeline. More important, having a map of the full genomic sequence of an organism has transformed the way biology is studied.

DNA gives rise to the organism and so is a scaffold for information. The genomic map is like a landscape of code, openly visible to all and for anyone to figure out. Through experimentation, often involving microarrays, DNA is annotated with functional information. In addition, the large-scale sequencing effort served as a kind of space program for biology, whereby the genome was a new frontier [1]. It made possible previously unforeseen possibilities and conceptually paved the way for a host of parallel analysis methods. The unveiling of a unified map begged the creation of microarrays, as well as other large genome-sized projects, such as the systematic deletion of every yeast gene [2], the systematic fusion of every yeast promoter to a reporter gene [3], and many other similar projects [4]. As the invention of the telescope changed how we view the universe, microarrays have changed the way we view the genome.

1.1.2
Gene Expression is Detected by Hybridization

The purpose of a microarray is to examine expression of multiple genes simultaneously in response to some biological perturbation. More generally, a microarray serves to interrogate the concentrations of molecules in a complex mixture and thus can serve as a powerful analytical tool for many kinds of experiments. To understand how this occurs, it may be useful to review the structure of DNA and examine how the unique structure of this molecule plays a role in identifying itself.

Although DNA is remarkably informationally complex, the general structure of the molecule is really quite simple. DNA is made up of four chemical building blocks called bases: adenine, cytosine, guanosine, and thymidine (abbreviated respectively as A, C, G, or T). As individual subunits these building blocks are also referred to as nucleotides. A strand of DNA consists of a sugar phosphate backbone to which these bases are covalently linked such that they form a series. Because these four bases can form sequences, it is possible to use them to encode information based on their patterns of occurrence. Indeed, from an information point of view, DNA has a potential data density of 145 million bits per inch and has been considered as a substrate for computation whereby the sequences are referred to as software [5,6].

Like strings of text in a book, the sequences that make up a strand of DNA have directionality such that information can be encoded in a given direction. The amount of DNA, and thus the amount of sequence, varies from organism to organism. For instance, the microorganism *Escherichia coli* has 4.5 million bases of sequence, whereas human cells have about 3 billion bases. Exactly how much biological information is encoded in these sequences is unknown, representing one of the deepest mysteries of biology, but microarrays provide a way to gain clues.

Cellular DNA most often consists not just of one strand but of two strands antiparallel to each other. The two strands are hydrogen bonded together by interactions between the bases, forming a structure in the cell. The structure is helical, similar to a spiral staircase in which the bases are attached to each side and interact in a plane to form the steps of the staircase. Besides the hydrogen bonds between the bases of opposite strands, the overlapping and proximity of the bases to each other lead to a second kind of noncovalent force called a stacking interaction that contributes to the stability of the double-stranded structure.

The bases of one strand interact with the bases of the other strand according to a set of pairing rules, such that A pairs with T and C pairs with G. Thus, if one knows the sequence of one strand, by definition, one then knows the sequence of the opposite strand. This property has profound consequences in the study of biology. It is also what the cell uses to replicate itself. As the interaction between the bases is noncovalent, consisting only of hydrogen bonds, the strands can essentially be melted apart and separated, thus opening the way for a copying mechanism to read each single strand and re-create the second complementary strand for each half of the pair, resulting in a new double-stranded molecule for each cell. This is also the mechanism by which cells express genes. The strands are opened by the gene expression machinery so that some number of RNA copies of a gene can be synthesized. The RNA transcript has the same sequence as the gene with the exception that uracil (U) replaces T, though the hybridization pairing rules remain the same (U and T can both pair with A).

This property of complementarity is also what is used for measuring gene expression on microarrays. Just as energy can melt strands apart and separate them into single molecules, the process is reversible such that single strands that are complementary to each other can come together and reanneal to form a double-stranded complex [109]. This process is called hybridization and is the basis for many assays or experiments in molecular biology. In the cell, hybridization is at the center of several biological processes, whereas in the lab complementarity is identity and thus hybridization is at the center of many *in vitro* reactions and analytical techniques. The molecules can come from completely different sources, but if they match, they will hybridize.

1.1.2.1 **Hybridization is Used to Measure Gene Expression**

Hybridization-based methods for measuring nucleic acid sequences have been used as a standard technique for decades [7]. The most widely accepted and understood standard for measuring gene expression is a hybridization-based assay called the Northern blot [8]. Microarrays are derived from blotting assays. Thus, it is useful to understand the basics of how they work and explain the origin of some of the terminology. Northern blots are also a common method for verifying the results from a microarray study.

The purpose of a Northern blot is to measure the size and abundance of RNA transcribed from a gene. To compare the expression of a gene under various conditions, extracts of cellular RNA are prepared from different batches of cells or tissues and fractionated by size on agarose gels prior to being transferred to a nylon membrane. The membrane binds nucleic acid quantitatively and preserves the pattern of RNA from the gel. The result is a porous membrane with RNA from different samples in discrete regions of the membrane. To quantify the expression level of a gene of interest in each sample, a radioactive DNA probe complementary to the gene of interest is prepared. Upon applying the probe to the membrane, hybridization will occur between the probe and the messenger ribonucleic acid (mRNA) from the gene of interest. In this way, both the location of the RNA within the fractionated sample and the relative abundance will be revealed. The amount of radioactivity captured by each sample on the membrane is a readout of the relative expression level of the gene of interest between the two samples.

In this scenario, the mRNA sample on the membrane is a complex mixture of molecules of unknown concentrations. Somewhere in the discrete region occupied by the sample RNA is a target transcript, which will be revealed when the probe hybridizes to it. Thus, the probe is known, and the target is what is revealed by hybridization. Although this process is robust, it is limited to questions of one target at a time. To measure a second gene, the membrane must be washed under conditions that disrupt hybridization and strip the membrane of radioactivity, so that it can be reprobed for another target.

1.1.2.2 Microarrays Provide a New Twist to an Old Technique

Microarrays provide a way to measure many genes at once by reversing and parallelizing this process. Instead of labeling what is known and probing a complex mixture of unknowns to highlight a single target, microarray methods label the complex mixture that is in a solution and utilize a two-dimensional surface of known molecules or probes in discrete locations, as a readout. In this context, the cellular RNA mixture is labeled, creating a labeled complex mixture of unknowns. Complementarity between target molecules in the complex mixture and probes arrayed on the solid surface will result in annealing and hybridization, thus capturing the labeled molecules on the surface. Unhybridized molecules are washed away prior to quantification.

Although this technique has been used in the past with membranes to create macroarrays, usually for screening libraries of clones [9], including identification of differentially expressed genes [10], the success of microarrays has to do with miniaturization. The interaction between a probe and its target is a binding assay. As many genes are expressed at low levels, sensitivity is often an issue. In a finding most would consider counterintuitive, a series of studies on binding assays in the mid-1980s found that decreasing the size of a binding target does not reduce the sensitivity or accuracy of an assay and actually leads to an increase in sensitivity [11,12]. Thus, miniaturization of the assay is possible. As the size of a spot decreases, the fractional occupancy of the immobilized probe is dependent only on the concentration of the corresponding target molecule in solution. This is the principle by which microarrays operate.

Other developments have contributed to efficient parallelization through miniaturization. The use of glass substrates instead of porous filters as a medium for capturing nucleic acid in an array format means that small hybridization volumes can be used. The rate of hybridization is concentration dependent. Thus, smaller volumes mean higher rates of hybridization and increased sensitivity. In addition, solid glass surfaces have low inherent fluorescence, allowing the use of fluorescent dyes for labeling samples, and along with methods adapted from confocal microscopy, they allow for efficient quantification of a highly parallel miniaturized assay.

Several groups developed and applied these concepts, combining miniaturization, robots, and genomes to develop new assays for the measurement of gene expression in parallel [13,14].

1.2 Types of Arrays

There are predominantly three kinds of microarray technologies in widespread use among most laboratories: spotted microarrays consisting of presynthesized oligos or PCR products robotically deposited onto a surface, Affymetrix GeneChips composed of relatively short oligonucleotides synthesized on a chip surface, and other *in situ* synthesis platforms such as arrays made by Agilent and NimbleGen. Although each technology effectively serves as a genomic readout, each has unique characteristics that offer advantages or disadvantages in a given context. Parallel forms of measuring DNA and RNA will continue to change and evolve; however, these three platforms are currently the most ubiquitous.

1.2.1 Spotted Microarrays

Spotted microarrays were the first widely available array platform and continue to enjoy broad use. Originating in the laboratory of Pat Brown, they consist of glass microscope slides onto which libraries of PCR products or long oligonucleotides are printed using a robot equipped with nibs capable of wicking up DNA from microtiter plates and depositing it onto the glass surface with micron precision [13,15]. Since their inception, demand for microarrays has exceeded availability. Because the Brown laboratory expended effort in every aspect of distributing the technology, including plans to build the robot and all protocols required for array manufacture and use, many academic laboratories invest resources into producing these arrays locally. This includes building or purchasing a robot, as well as performing PCR or oligo design and synthesis to create probes for spotting onto glass. The basic principle by which the arrays function is fairly simple, and all the reagents required are available to most researchers with some initial investment. However, apart from praising the benefits of putting technology into the hands of researchers, the reason for highlighting this aspect of spotted arrays is to point out the nonuniform nature of spotted microarrays. Because there is not one manufacturer, one source of materials, or a uniform method of production, variability exists among batches of microarrays and must be considered when planning experiments or when comparing experiments from different array sources.

Spotted microarrays are primarily a comparative technology. They are used to examine relative concentrations of targets between two samples. Complex samples to be compared are labeled with uniquely colored fluorescent tags before being mixed together and allowed to compete for hybridization to the microarray spots (see [16,17] for review). In this way, differences between the samples are observed on a per spot basis because the fractional occupancy of the spot hybridized by each sample reflects the relative concentration of that gene or target in the original complex mixture. Thus, for any probe on the microarray, one gets a readout of the relative concentrations of the target between the two input samples. For this reason, spotted microarrays are often called two-color or two-sample arrays.

1.2.2 Affymetrix GeneChips

Affymetrix GeneChips are the most ubiquitous and long-standing commercial array platform in use. The arrays consist of 25-mer oligonucleotides synthesized *in situ* on the surface of a glass chip. A photolithography mask, similar to that used to construct semiconductor chips, is used to control light-directed DNA synthesis chemistry such that oligo sequences are built up one nucleotide at a time at defined locations on a solid substrate or glass chip [18,19]. Current chips contain 6.5 million unique probes in an area of 1.28 cm^2. The highly precise nature of the lithographic method allows the construction of compact matrices of square patches of probes.

Instead of using a single sequence to probe expression of each gene, as would be common for a spotted array, Affymetrix employs a set of probes to measure expression of a gene. Probe sets contain two types of probes to measure the gene of interest, perfect match (PM) and mismatch (MM) probes. Perfect match probes are chosen to match the gene exactly and are designed against an exemplar sequence representing the gene. Although each probe is unique, probes may occasionally overlap. Mismatch probes are identical to the perfect match probes except that they contain a single base mismatch in the center of the probe. A single mismatch in a short sequence such as a 25-mer is very disruptive to hybridization. The purpose of the mismatch probe is to serve as a negative control for background hybridization. A typical probe set contains 11 perfect match probes and 11 mismatch probes. The positioning of probes for a single gene on the array is chosen by a random process to protect against local hybridization artifacts that could otherwise affect all the probes for a gene if they were clustered together. As most spotted arrays use only one probe per gene, local hybridization artifacts can be a problem.

To read the expression level of a gene, the perfect match and mismatch probe readings must be summarized. Although Affymetrix has a standard method for summarizing 22 readouts to obtain a single number for gene expression [20], many approaches are available [21].

Affymetrix GeneChips are single sample microarrays (also known as one color or one channel). These arrays measure the relative abundance of every gene in a single sample. In this way, one can examine whether one gene is expressed at a higher or lower level than some other gene in the same sample. If samples are to be compared, a separate chip must be performed for each sample, and the data adjusted by scaling or normalization before comparison.

1.2.2.1 Other *In Situ* Synthesis Platforms

Apart from Affymetrix, two alternative *in situ* synthesis methods exist by which oligonucleotides are built up one nucleotide at a time in successive steps to create probes of length 25–60 nucleotides long [108]. These methods are almost exclusively commercial and different companies take different approaches. Although Affymetrix uses a mask-based photolithographic process to control light-directed DNA synthesis, an alternative method employed by NimbleGen makes use of small rotating mirrors to control light and accomplish a similar task [22]. This approach is called Maskless Photolithography, and uses technology developed by Texas Instruments for projection televisions in which arrays of digitally controlled micromirrors can be used to direct light. In combination with light activated chemistry, light of the appropriate intensity and wavelength can be actuated in patterns required to build up any series of nucleotides into an oligonucleotide on a solid surface [23,24].

The NimbleGen approach has two great advantages over the method worked out by Affymetrix. The first is that it does not require a mask. To build an array of different oligonucleotides of length N requires a series of $4N$ synthesis steps. Thus, to build a library of unique 25-mers on a surface requires 100 chemical synthesis steps. For Affymetrix, a unique photolithography mask is required to control the chemistry at each step. These masks are expensive to construct; thus, the arrays are very costly. In addition, once a set of masks is constructed, it describes only a single array design. Changing the design requires a whole new set of masks. However, changing a pattern of micromirrors under electronic control is very easy; thus, each array produced by NimbleGen can have a different design.

The second alternative *in situ* synthesis approach to array construction uses traditional oligo synthesis chemistry, but the method of controlling base addition is novel. Ink-jet technology, developed by Hewlett Packard for consumer printers, has been adapted to control the liquid precursors of DNA synthesis. Agilent, a spin-off of Hewlett Packard, uses this technology to synthesize 60-mer oligos on glass slides [22,25]. Like ink-jet printing itself, this technology is very flexible. Every array can be customized and thus possess unique content.

The flexibility to change the design of an array easily is both a blessing and a curse. The positive aspect is that one can easily change the array design to explore the genome or expression space as required by the experiment. The negative aspect is that data analysis becomes more cumbersome, in general, because one can easily change the probes used to represent a gene from array to array, as well as the content of the array from experiment to experiment. When everyone is using the same chip, as in chips mass produced by Affymetrix, comparisons between data sets are fairly easy. If every chip is unique, comparison between data sets becomes difficult.

Another positive aspect of arrays created using *in situ* synthesis methods is that they do not depend on libraries of clones or molecules created elsewhere. Instead, the content is freshly created with each array. This is good because with spotted arrays one never knows the history of the library, or how many times it has been used to create arrays. However, quality control for *in situ* synthesized arrays remains obscure.

1.2.2.2 Uses of Microarrays

An example of the power of microarrays can be seen in Figure 1.1. Four sets of cells purified in triplicate by the abundance of two cell surface markers exhibit different differentiation fates yet are otherwise indistinguishable. However, microarray analysis and simple clustering of differentially expressed genes reveal previously unknown differences.

As a genomic readout, microarrays can serve many purposes, and novel applications continue to emerge. A common application of microarrays has been the measurement of gene expression, from characterizing cells and processes [26–28] to clinical applications such as tumor classification [29,30]. Another very common use of microarrays is in genotyping and the measurement of genetic variation [31,32].

As microarrays are capable of quantifying complex mixtures of DNA or RNA fragments derived from a variety of experiments, they have been used for a number of purposes. For instance, RNA interacts with many proteins in the cell and is a part

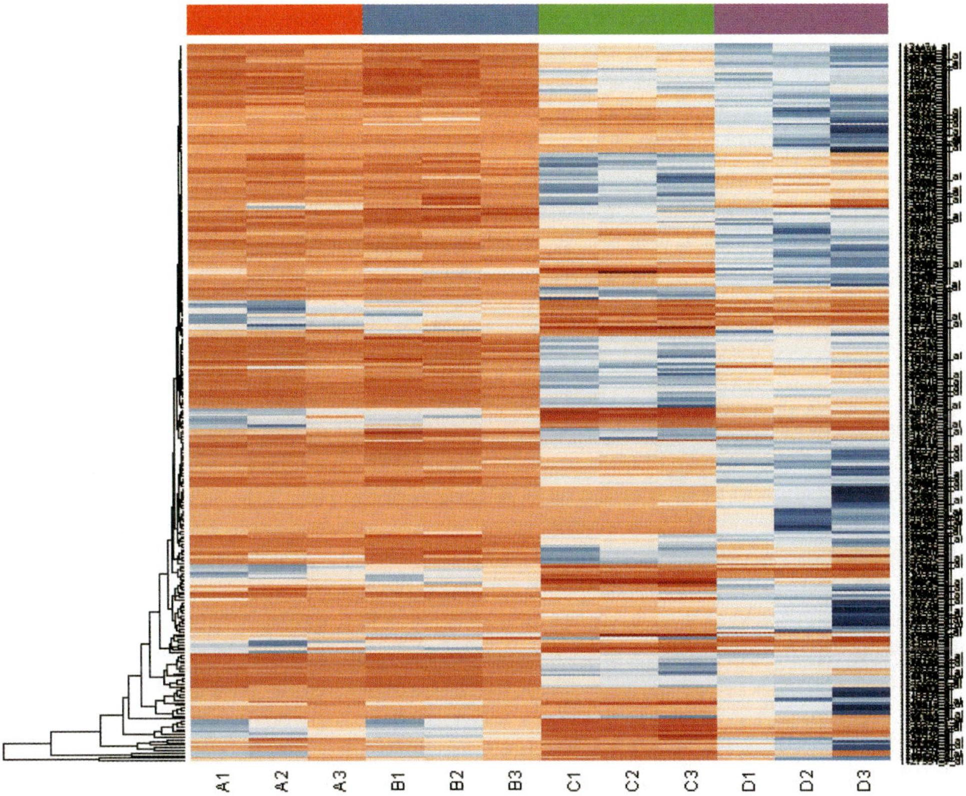

Figure 1.1 Simple clustering to find cell types. Four sets of cells cytometrically sorted in triplicate based on two surface markers are assayed by expression microarray. Clustering of differentially expressed genes defines states of gene expression for each class.

of many molecular complexes, which have traditionally been difficult to characterize. Microarrays have been used to identify the RNA components of various complexes, shedding light on biological mechanisms of RNA translation and transport [33,34]. Recently identified complexes of protein and RNA, called P-bodies, are thought to be involved in gene expression by regulating mRNA in the cytoplasm [35]. Microarrays could be used to monitor and characterize the trafficking of cellular RNA through this complex. Changes in DNA copy number at various loci have been implicated in tumorigenesis and cancer. Using comparative genomic hybridization, microarrays have been used to examine aneuploidy and changes in loci copy number in a variety of cell types [36,37]. Microarrays have been used to examine the progress of replication forks as they copy the genome [38], as well as for genome-wide screens of RNA modifying enzymes [39]. The full range of applications is too numerous to mention, improvements and adaptations are continually being made, and one must continually review the literature (see [40] for review).

However, a rapidly burgeoning area of microarray research over the past few years involves the use of microarrays to identify functional elements in the genome. Expression of a gene in the form of an RNA transcript is but one small slice of the biology of a gene. A fundamental aspect of gene expression currently being explored by microarrays is the revelation of control elements in the genome that are responsible for turning genes on and off. Every gene is under the control of a regulatory code. However, this code is largely unknown. The code is composed of relatively short sequences termed *cis*-acting regulatory sites. These sites are bound by transcription factors, proteins that are responsible for activating or repressing the transcription of a gene. A majority of these sites, and combinations in which they occur, are not known or understood. However, it is possible to map the locations where proteins interact with the genome using a technique called chromatin immunoprecipitation (see [41–43] for review). This technique has been used to map the locations where specific transcription factors bind to the genome under a specific set of conditions [44], as well as to map proteins that interact with the genome on a larger scale, such as cohesins [45] and histones [46].

The general idea is to map protein/nucleic acid interactions by utilizing the specificity of antibodies for a protein of interest. Armed with an antibody against a specific protein, one can target and purify the protein as well as the associated nucleic acid. After purification the nucleic acid is removed, labeled, and hybridized to the microarray to identify which regions of the genome were associated with the protein in the cell. The identification of regulatory elements in this way has had profound implications on our understanding of gene regulatory circuitry [47,48], and provides data for a system's biological approach to understanding the cell.

The ability of chromatin IP experiments to detect functional elements such as protein binding sites in the genome depends on the resolution of probes on the microarray. Early ChIP-chip experiments in yeast used PCR products representing each gene and each intergenic region as a single DNA segment. Thus, the resolution of the technique was limited by the size of the DNA fragment.

However, advances in probe technology, such as the use of long oligos (60–90 bases) for spotted arrays and the ability to create higher and higher density *in situ* synthesized arrays (25–60mers), have allowed for the creation of tiling arrays. In a tiling array, probes are laid down across the genome, either overlapping or at some degree of spacing. Experiments with these arrays have yielded a number of amazing findings, including the ability to view single nucleosome resolution in the genome [49].

Tiling arrays have also had another benefit. High-density arrays consisting of probes that cover both strands of DNA have been used to show that transcription in the genome is ubiquitous [50,51]. The majority of nucleotides that make up the genome are part of a primary transcript, and the amount of antisense transcription is much higher than previously appreciated. In this regard, microarrays serve to directly identify functional components of the genome.

Taken together, the data being collected by tiling arrays for both protein/DNA interactions and identification of novel transcripts in humans are being systematically and jointly analyzed as part of a large consortium termed as the ENCODE

Project with the aim of compiling a comprehensive encyclopedia of DNA elements [52,53]. Microarrays are a fundamental aspect of this effort.

Although tiling arrays are yielding a wealth of productive information, they are still relatively nonstandard both in terms of design and analysis. Depending on the question being asked, a number of analysis and normalization strategies have been developed.

1.3
Array Content

1.3.1
ESTs Are the First View

Microarrays are designed to measure active parts of the genome. The most obviously functional parts of the genome are genes. These are also the parts where sequencing efforts were initially focused and used to start building sequence-based maps of the genome. Organisms express the genome by copying parts of the DNA into a form called mRNA, which is easy to isolate and obtain sequence from. Thus, the first sequenced parts of a genome that have traditionally been available are usually the expressed parts. RNA isolated from cells is converted to DNA and sequenced; however, due to certain technical aspects of the enzymes involved, it is often difficult to get a complete sequence for each RNA molecule. Thus, the resulting sequences are termed as expressed sequence tags (ESTs) because they represent short reads of an expressed sequence. It is through the collection and assembly of ESTs that we get our first view of the genome, and from which gene models are built [54]. Even after genomes are completely sequenced through systematic conventional means, ESTs are mapped back onto the genome to define or indicate which parts of the genome are expressed as genes. Microarrays used to measure gene expression have traditionally been constructed with probes designed against assemblies of ESTs.

The human body contains hundreds of cell types. Consider for a moment how even a single tissue such as skin can be different depending on where from the body it is isolated. As all the tissues and cell types arise from the same DNA, each has a unique gene expression program. As mentioned above, RNA is easy to isolate and convert to DNA. The RNA isolated from cells or tissues represents the result of gene expression occurring in that cell or tissue at the time of isolation. Once RNA has been converted to DNA, techniques exist for cloning the DNA fragments into plasmids or other vectors such that each molecule can be treated as a separate isolate, and thus can be amplified or manipulated at will. The process of converting a complex population of molecules into a collection of individually manipulatable molecules is called library construction. Many microarrays are constructed by individually amplifying the DNA segments represented by a library and printing them onto glass. In this way, one can study the activity of genes expressed in a certain tissue by performing array analysis using arrays constructed from a library of genes isolated from that tissue [55]. In some cases, libraries are constructed from organisms that have not been sequenced. For instance, microarray studies of the human pathogen Histoplasma began before a sequence of the genome was known. Libraries of Histoplasma genes were constructed and used to create microarrays that led to the identity of genes active during infection [56]. In this situation, the identity of the spots on the array is often not known at the outset. Through array analysis, genes that appear to change under certain conditions can then be sequenced to get a foothold into the biology of the organism.

However, another way in which ESTs are used to create microarrays is through the use of a rational design process to create and design microarray probes from assemblies of ESTs. As also mentioned above, ESTs are assembled to create gene models *in silico*. Thus, with a collection of gene models, it is possible to design complementary DNA sequences that can be synthetically produced for microarray construction and that can be used to identify the expression of the target gene uniquely. Affymetrix arrays, since they are produced *in situ*, have always been based on a rational design process.

1.3.1.1 **Probe Design**

The ability of a microarray to report on the identity of a hybridization target depends on the probes attached to the microarray surface. Spotted arrays have traditionally been produced with experimentally derived libraries or with PCR products representing known gene segments [57]. Many genes share some degree of sequence identity; thus, cross-hybridization of labeled transcripts can occur between them, making it difficult to interpret results. However, most genes can be uniquely distinguished using optimized probes, which can be accomplished using synthetic DNA oligonucleotides. The wealth of sequence information available and the realization that long oligonucleotides could be substituted for PCR products with little modification to existing array production protocols [58] have changed the content of most spotted arrays to rationally designed probes.

Although PCR products serve as effective array reagents, because they are easy to produce and they stick well to glass without advanced chemistry, they can also be a source of error and variability. For instance, early attempts to create yeast genome microarrays representing more than 6000 yeast genes involved weeks and months of effort to perform 6000 PCR reactions with a typical success rate of 85%. The resulting library represented a collection of printable DNA fragments at various concentrations and of varying purity. This kind of variation differs from amplification to amplification and from group to group. Systematic errors such as erroneous plate mappings are not uncommon. Some widely used libraries have well-documented problems of misidentified or contaminated clones [59].

The length of a PCR product makes it a sensitive hybridization reagent, as after a certain length DNA molecules converge on a general melting temperature. However, length also serves a disadvantage because long pieces of DNA are likely to have a region of homology or similarity to more than one place in a genome. Experiments with differing lengths of long oligonucleotides found that long oligos were capable of sticking to glass, and exhibited sensitivity similar to or better than PCR products in a spotted array format [60]. A number of groups have evaluated the use of long oligos as array reagents [61,62].

Long oligos are synthetically produced array reagents, and thus are amenable to bioinformatic optimization. General considerations when designing probes for measuring gene expression are the uniqueness in the genome, temperature of hybridization, secondary structure, and proximity to the 3′ end of a gene [63,64]. To find probes targeted to measure the expression of a single gene, design targets, usually consisting of cDNA sequences or unigene clusters, are screened against a genomic background to find unique regions from which probes can be designed. The oligos are chosen to have similar melting temperatures, so that they behave similarly under a single hybridization condition on the array. They are also chosen to be free of secondary structures that would interfere with hybridization. Long oligo probes for gene expression are also chosen with a bias toward the 3′ end of a gene. The labeling procedure for most gene expression experiments in eukaryotes utilizes the polyA tail at the end of the transcript resulting in a 3′-end bias, so the probe is chosen to be near the 3′ end to have a higher chance of detecting the labeled transcript. Although these criteria have made for good quality data collection on gene expression, it is likely that design criteria can be improved further [65].

Advances in oligo synthesis technology have lowered costs significantly such that design and production of a genome-sized reagent is within reach of an organized group of researchers. A variety of long oligo design programs have become available. An additional advantage of printing spotted arrays with long oligonucleotides is the practical matter that the oligos are synthesized by machine; thus, many of the pitfalls associated with PCR products are avoided. For example, every oligo is guaranteed to exist, plate location errors are avoided, and the concentration of oligos is made uniform for printing.

Since Affymetrix arrays are a commercial reagent, oligo design for the arrays is not usually in the hands of researchers. However, some of the principles are worth noting both for issues of data analysis and for design considerations when using other *in situ* synthesis platforms. In early work, Affymetrix performed a careful examination of characteristics required for the design of probes from which hybridization signals could be used to determine relative expression level of a gene from a complex sample. An array consisting of 16 000 probes was used to examine the expression of just 12 genes [14]. By making hundreds of probes per gene and then examining the characteristics of the probes that most accurately reported the expression level of the gene, a set of heuristic probe design criteria was derived. These results were then applied to larger numbers of genes. In addition, a minimum number of probes required was determined, leading to a current set size of 11–20 probes per gene.

Tiling arrays have been designed both with and without regard to bioinformatic optimization. The presence of repeat regions and elements with sequence similarity in multiple places is likely to cause problems and should be taken into account in some way. Fortunately, such issues are being examined and addressed, and tools are beginning to emerge [66].

The importance of good probe design has to do with hybridization. Hybridization signal is used to infer the expression level of a gene. By comparing hybridization signals between different genes, quantitative statements can be made about the expression levels of those genes. However, the signal derived from hybridization is affected by many factors, and unless those factors are identical from one gene to the next on a chip, our inferences about gene levels may be incorrect. In addition, the hybridization characteristics of a DNA probe are unique and sequence dependent. Each type of array that utilizes hybridization as a readout suffers from the effects of hybridization differences between DNA sequences representing genes, and measures are taken to minimize them. Affymetrix and NimbleGen address this problem by using multiple probes to represent a gene.

For spotted arrays the problem of hybridization differences between probes is less of an issue because comparisons are not made on a gene-by-gene basis within a sample. Instead, spotted arrays are used for comparing gene expression differences between two samples. The samples are uniquely labeled and then mixed together prior to array hybridization. Thus, for any given gene, what is observed on the array is not an absolute hybridization signal for a gene, but rather a ratio of hybridization signals for a gene – one from each sample. Whatever the characteristics of the hybridization probe on the chip surface, uniquely labeled complements from each sample are competing for hybridization to the spot on the array surface, and the ratio of hybridization is the primary quantity of interest.

1.4 Normalization and Scaling

Microarray experiments, whether utilizing one-channel or two-channel technology, are comparative experiments involving populations of measurements, with the end goal being to compare abundance of targets in complex populations. Most of the time, targets are compared between populations. To compare two populations to find meaningful differences, one must make assumptions about the populations. As such, the set of fluorescent measurements representing a population must undergo a process of normalization so that the parts that have changed can be distinguished from the parts that have not changed.

In the generation and collection of microarray data, many factors can give rise to systematic differences in the population. For instance, samples to be compared are not always labeled with the same efficiency. Purification steps for each sample often have differing yields. Samples to be compared on an array are not always mixed in equal proportions prior to hybridization (either by mass or by moles of dye). The dyes themselves have differing rates of decay from atmospheric elements such as ozone [67], as well as photo bleaching during the scanning of the array. Thus, if the labeled transcripts from one sample are uniformly less bright than those of the other sample, the uniform nature of such an effect allows for correction by adjusting the brightness of the dim sample by a correction factor. One goal of normalization is to remove these systematic biases [68].

Embedded within the goal of normalization is an assumption about gene expression and how large populations of gene expression measurements should be compared. For two-color experiments to be compared on one chip, the assumption is that in response to some factor – whether it is genotype – or treatment, the majority of genes will remain unchanged, and some small fraction of genes (e.g., 2%) involved in pathways responsive to the effect being studied will be regulated up or down in roughly equal proportion [69]. According to this assumption, the average signal coming from each population as measured on the array should be equal, whereas small numbers of specific genes will change their levels and show ratios of expression significantly different from the overall population. Of course, this assumption is not always true.

Early normalization methods consisted of calculating a median of each population after excluding the upper and lower ends of the distribution of the data, and then adjusting one population through multiplication of a constant to bring the means of the populations to equality. However, in plotting one population against the other, one often notices that the populations too do not differ by a single constant across their range; that is, an intensity dependence is often observed for differences between populations. This is illustrated in Figure 1.2a. A common method of viewing two-color microarray data is to examine the ratio and the intensity together on a microarray plot [70]. In this kind of plot, the ratio of colors (R for red and G for green) is represented on the y-axis by M and is calculated as $M = \log_2(R/G)$, and the combined intensity from each fluorescent dye is represented on the x-axis by A, which is calculated as $A = (1/2)\ (\log_2(R) + \log_2(G))$. Figure 1.2a illustrates raw data from a pair of wild-type (wt) and mutant samples hybridized on a two-color array. The observed skew in the ratio is not a function of the difference between wt and mutant gene expression, but rather comes from a systematic difference arising in the process from the microarray experiment. Such plots are not uncommon and can

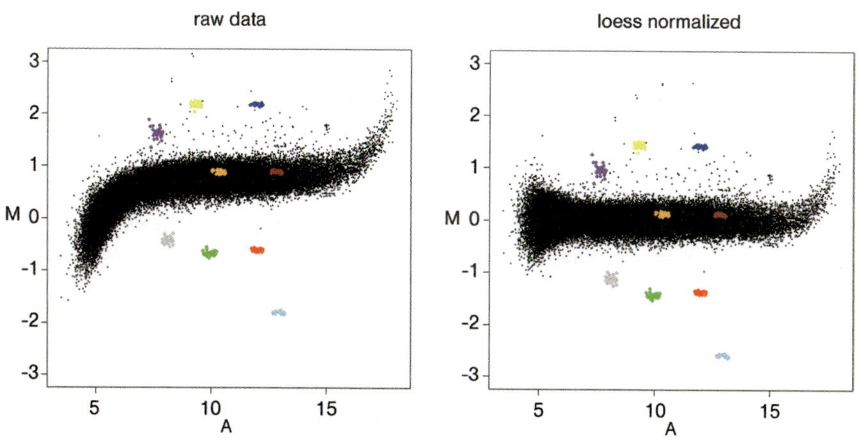

Figure 1.2 Lowess normalization of data. MA plot of a single-spotted array of expression data from a wt versus a mutant cell. Systematic skew in the data is corrected by lowess normalization. Control targets mixed at known ratios are indicated by color. (a) Left panel: unnormalized data. (b) Right panel: normalized data.

be seen even from single samples that are split in two, separately labeled and hybridized.

To alleviate this effect, many methods of normalization have been developed. The most popular is a lowess normalization method utilizing local regression to fit each population in an intensity-dependent fashion [69]. Figure 1.2b illustrates the same data as in Figure 1.2a after lowess normalization. Much of the skew from the original data set has been removed. A variety of methods have been developed for the normalization of two-color array data [71–75].

The methods above assume that the populations to be normalized are roughly equally distributed, and that the number of genes differentially expressed is small and the direction of expression is symmetric. This assumption may not be valid in some cases. For instance, small custom arrays consisting of only a subset of genes can invalidate the assumption that the effect being studied is evenly and symmetrically distributed across the gene set. However, in simulated experiments lowess normalization has been found to be robust even when 20% of the genes show differential expression in just one direction [76]. Chromatin immunoprecipitation experiments represent another area where normalization methods for the populations being compared may require special consideration [42]. In this regard, the relatively new emergence of tiling arrays has prompted the development of new methods for normalization [77]. Current methods do not take sequence composition into account, yet is clear that sequence effects will contribute to hybridization signals and should be accounted for.

Affymetrix GeneChips use hybridization intensities of single samples as a readout of gene expression. Since many factors unrelated to gene expression can affect the hybridization properties of a probe, each gene is represented not by one probe (like most other types of arrays) but by a population of probes. Summarizing a readout of several probes into a single value for gene expression adds a layer of complexity to data analysis because there are several ways probe sets can be polled and opinions differ on which method is best. Indeed, dozens of methods have been developed [21].

In dealing with Affymetrix arrays, there are two basic steps involved in data analysis. The first is summarizing the probe sets representing a gene into a readout of expression for that gene. If gene expression between two samples is to be compared, the results from each chip must be scaled to each other in a second step to account for differences in labeling and scanning of the samples. There are numerous different approaches for carrying out these steps. As of 2006, more than 30 methods had been identified [21].

The most popular methods are MAS5 [20], RMA (Robust Multichip Average) [78], and dCHIP [79]. Some methods such as GCRMA (GC Robust Multichip Average) take probe sequence into account [80]. No method is clearly the best [81], each has trade-offs and makes different assumptions about the data. The Affymetrix algorithm (MAS5) includes a background subtraction that considers both the perfect match and mismatch probes, whereas some algorithms such as RMA and GLA (Generalized Logarithm Average) [82] do not consider data from the MM probes. In a small data set with many values close to background, such differences may lead to

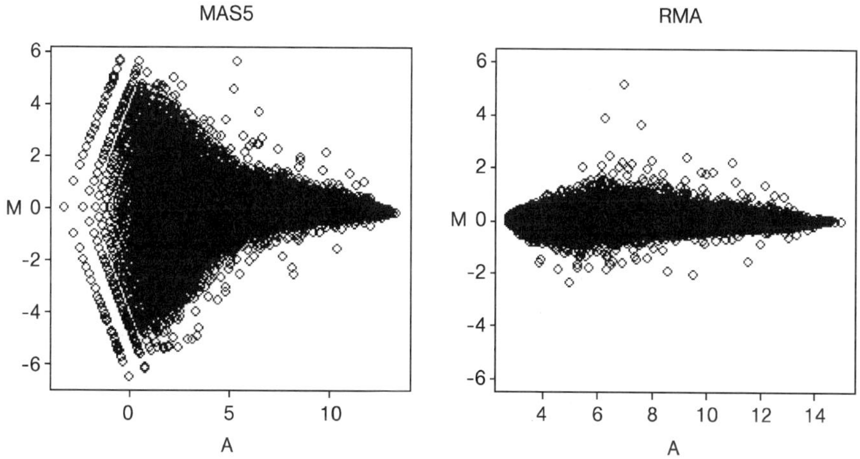

Figure 1.3 MAS5 versus RMA comparison of two data sets. Two Affymetrix expression data sets on amplified mouse cell RNA comparing a wt and mutant mouse using either MAS5- or RMA-treated data. Replicates should be performed as small data sets can exhibit large differences with different data treatments.

very different views of the data. Figure 1.3 shows MA plots of a pair of arrays summarized with MAS5 or RMA. An unsophisticated user taking only ratios into account might be misled. However, such a difference should be seen as a cautionary tail that replicates should be performed, as the two methods would otherwise have more agreement.

Control data sets have been generated and analyzed for the comparison and validation of different analysis methods [83], and then analyzed by others with different conclusions [84]. Validated data sets will be an important component in deciphering which method to apply to the analysis of raw microarray data. It is important to note the existence of different approaches, and realize trade-offs of various methods. However, most agree that the biology of the system is usually the noisiest component in the analysis. Regardless of the analysis method, microarrays are sensitive, and it is important to be a good experimentalist.

Microarrays have a difficult job to do. Expression of RNA in the cell can vary over several orders of magnitude. It is estimated that the majority of mRNA extracted from the cell contains transcripts from a small minority of genes [85]. The majority of transcripts are considered rare and expressed at very low levels [86,87]. Thus, it is not surprising that a typical Affymetrix GeneChip hybridized with a mammalian sample and analyzed with Affymetrix software reports that half of the probes are considered "absent" such that the signal was too low to be reliably detected. In addition, many protocols for array hybridization are suboptimal. Historically, quantitative PCR and Northern blots represent a standard method for measuring gene expression, and are used as a way to verify microarray results. It is common when performing Northern blot analysis, or qPCR, to optimize the probe and hybridization conditions. In a microarray experiment, all the probes are different, yet they all experience identical hybridization conditions. Few experimentalists take time to optimize one of the most important steps: hybridization and wash conditions. Most spotted array protocols call for suboptimal hybridization times [88]. In addition, the specificity of hybridization is established in the washing steps when nonhybridized material is washed away. Unlike Northern blotting, few optimize this step, yet doing so would decrease noise from background hybridization, and increase the reproducibility of expression values [89].

Despite the caveats of different analysis methods, and the general difficulty of measuring gene expression, comparisons of different kinds of microarrays and across different laboratories show that the technology is robust [90–94], though issues in lack of concordance have been observed [95,96]. Most would agree that the majority of noise is associated with biological variability rather than with microarray measurement platform. Often, the question arises as to which microarray platform to use in an experiment. In terms of sensitivity, specificity, and reproducibility, a large comprehensive study recently indicated that both single and double sample platforms are equivalent [97].

1.4.1 Be Unbiased, Be Complete

Microarrays afford a new luxury that changes the way science is done. Just a short time ago, the measurement of gene expression was performed one or two genes at a time, and almost always with a hypothesis about the gene in mind. These days, however, this would be considered a biased approach. Since microarrays confer the ability to measure virtually all the genes, one can take an unbiased approach to a problem by examining all genes and identifying those genes or elements that exhibit change under some condition. Thus, the idea that one can be unbiased and complete when looking for a gene expression response emerges.

However, even here technological advances show us how biased we still are. Because microarrays will measure only what they are designed to measure, if they contain only sequences corresponding to known genes, they will miss novel biological responses that occur in unexpected places.

For example, a genomic probe set from Operon Technologies has been designed to perform whole genome analysis of gene expression in yeast, but since it consists of 70-mer oligonucleotides for 6300 known and predicted genes, the probes represent only 1.8% of the genome when taking both strands into account. Given the ubiquity of genomic transcription recently observed on tiling arrays, we can see that there are many transcriptional events that will not be detected.

1.4.2 Sequence Counts

There are two ways to identify a nucleic acid sequence: match it to something that is known or sequence it directly. Microarrays identify DNA by hybridizing it to a known complement. However, new advances in sequencing technologies allow large populations of molecules to be identified by direct sequencing.

As mentioned earlier, advances in sequencing technology seeded a paradigm shift by creating an appetite for parallel gene expression. The switch from serial to parallel has extended into other areas such as protein arrays [98], cellular arrays [99], and phenotype arrays [100]. Recent advances in sequencing technology are poised to take over for microarrays and change the way we measure gene expression, or quantify nucleic acids in complex mixtures, yet again. Bead-based sequencing technologies allow the collection of hundreds of millions of sequence reads on individual samples [101–103]. Thus, rather than label fragments and hybridize them to an array of probes, one can simply gather sequences and count the occurrence of tags as a readout of expression.

This technology fundamentally changes what is measured in a gene expression experiment. It does not depend on gene models or building a probe, rather it counts the nucleotides directly, thus directly reading the results of gene expression. This opens the door to a long sought goal of understanding splice variation in the genome. It also sidesteps many of the technical issues behind hybridization-based methods such as designing good probes or carefully controlling hybridization conditions. Much like SAGE (Serial Analysis of Gene Expression) techniques that provide a more digital view of transcription [104] and have nucleotide level resolution, high volume sequencing techniques will do the same but are much more efficient.

As mentioned previously, the highly parallel nature of microarray data means that more is measured in a typical experiment than may be realized by the experimenter. To keep data available as a resource for mining, standards have been created for the capture and description of microarray data [105], and public repositories such as GEO [106] and ArrayExpress [107] have sprung up to archive data for public mining. The bottleneck is no longer in the generation of data, but in our ability to make sense of it.

References

1 Lander, E.S. (1996) The new genomics: global views of biology. *Science*, **274** (5287), 536–539.

2 Winzeler, E.A., Shoemaker, D.D., Astromoff, A., Liang, H., Anderson, K., Andre, B., Bangham, R., Benito, R., Boeke, J.D., Bussey, H., Chu, A.M., Connelly, C., Davis, K., Dietrich, F., Dow, S.W., El Bakkoury, M., Foury, F., Friend, S.H., Gentalen, E., Giaever, G., Hegemann, J.H., Jones, T., Laub, M., Liao, H., Liebundguth, N., Lockhart, D.J., Lucau-Danila, A., Lussier, M., M'Rabet, N., Menard, P., Mittmann, M., Pai, C., Rebischung, C., Revuelta, J.L., Riles, L., Roberts, C.J., Ross-MacDonald, P., Scherens, B., Snyder, M., Sookhai-Mahadeo, S., Storms, R.K., Véronneau, S., Voet, M., Volckaert, G., Ward, T.R., Wysocki, R., Yen, G.S., Yu, K., Zimmermann, K., Philippsen, P., Johnston, M. and Davis, R.W. (1999) Functional characterization of the *S. cerevisiae* genome by gene deletion and

parallel analysis. *Science*, **285** (5429), 901–906.

3 Dimster-Denk, D., Rine, J., Phillips, J., Scherer, S., Cundiff, P., DeBord, K., Gilliland, D., Hickman, S., Jarvis, A., Tong, L. and Ashby, M. (1999) Comprehensive evaluation of isoprenoid biosynthesis regulation in *Saccharomyces cerevisiae* utilizing the Genome Reporter Matrix. *Journal of Liposome Research*, **40** (5), 850–860.

4 Brenner, C. (2000) A cultivated taste for yeast. *Genome Biology*, **1** (1), 1031–1034.

5 Adleman, L.M. (1994) Molecular computation of solutions to combinatorial problems. *Science*, **266** (5187), 1021–1024.

6 Amos, M. (2005) *Theoretical and Experimental DNA Computation*, Springer.

7 Gillespie, D. and Spiegelman, S. (1965) A quantitative assay for DNA–RNA hybrids with DNA immobilized on a membrane. *Journal of Molecular Biology*, **12** (3), 829–842.

8 Allwine, J.C., Kemp, D.J. and Stark, G.R. (1977) Method for detection of specific RNAs in agarose gels by transfer to diazobenzylmethyl-paper and hybridization with DNA probes. *Proceedings of the National Academy of Sciences of the United States of America*, **74**, 5350–5354.

9 Gergen, J.P., Stern, R.H. and Wensink, P.C. (1979) Filter replicas and permanent collections of recombinant DNA plasmids. *Nucleic Acids Research*, **7** (8), 2115–2136.

10 Riggleman, R.C., Fristensky, B. and Hadwiger, L.A. (1985) The disease resistance response in pea is associated with increased levels of specific mRNAs. *Journal Plant Molecular Biology*, **4** (2–3), 81–86.

11 Ekins, R.P. and Chu, F.W. (1991) Multianalyte microspot immunoassay – microanalytical "compact disk" of the future. *Clinical Chemistry*, **37** (11), 1955–1967.

12 Ekins, R.P. (1998) Ligand assays: from electrophoresis to miniaturized microarrays. *Clinical Chemistry*, **44** (9), 2015–2030.

13 Shalon, T.D. (1995) DNA Microarrays: a new tool for genetic analysis.Stanford University. Ph.D. thesis.

14 Lockhart, D.J., Dong, H., Byrne, M.C., Follettie, M.T., Gallo, M.V., Chee, M.S., Mittmann, M., Wang, C., Kobayashi, M., Horton, H. and Brown, E.L. (1996) Expression monitoring by hybridization to high-density oligonucleotide arrays. *Nature Biotechnology*, **14** (13), 1675–1680.

15 Schena, M., Shalon, D., Davis, R.W. and Brown, P.O. (1995) Quantitative monitoring of gene expression patterns with a complementary DNA microarray. *Science*, **270** (5235), 467–470.

16 Duggan, D.J., Bittner, M., Chen, Y., Meltzer, P. and Trent, J.M. (1999) Expression profiling using cDNA microarrays. *Nature Genetics*, **21** (1 Suppl.), 10–14.

17 Brown, P.O. and Botstein, D. (1999) Exploring the new world of the genome with DNA microarrays. *Nature Genetics*, **21** (1 Suppl.), 33–37. Review.

18 Fodor, S.P., Read, J.L., Pirrung, M.C., Stryer, L., Lu, A.T. and Solas, D. (1991) Light-directed, spatially addressable parallel chemical synthesis. *Science*, **251** (4995), 767–773.

19 Pease, A.C., Solas, D., Sullivan, E.J., Cronin, M.T., Holmes, C.P. and Fodor, S.P. (1994) Light-generated oligonucleotide arrays for rapid DNA sequence analysis. *Proceedings of the National Academy of Sciences of the United States of America*, **91** (11), 5022–5026.

20 Affymetrix((2002) Statistical Algorithms description document. Technical Report.

21 Irizarry, R.A., Wu, Z. and Jaffee, H.A. (2006) Comparison of Affymetrix

GeneChip expression measures. *Bioinformatics*, **22** (7), 789–794.

22 Blanchard, A.P. and Friend, S.H. (1999) Cheap DNA arrays – it's not all smoke and mirrors. *Nature Biotechnology*, **17** (10), 953.

23 Singh-Gasson, S., Green, R.D., Yue, Y., Nelson, C., Blattner, F., Sussman, M.R. and Cerrina, F. (1999) Maskless fabrication of light-directed oligonucleotide microarrays using a digital micromirror array. *Nature Biotechnology*, **17** (10), 974–978.

24 Nuwaysir, E.F., Huang, W., Albert, T.J., Singh, J., Nuwaysir, K., Pitas, A., Richmond, T., Gorski, T., Berg, J.P., Ballin, J., McCormick, M., Norton, J., Pollock, T., Sumwalt, T., Butcher, L., Porter, D., Molla, M., Hall, C., Blattner, F., Sussman, M.R., Wallace, R.L., Cerrina, F. and Green, R.D. (2002) Gene expression analysis using oligonucleotide arrays produced by maskless photolithography. *Genome Research*, **12** (11), 1749–1755.

25 Hughes, T.R., Mao, M., Jones, A.R., Burchard, J., Marton, M.J., Shannon, K.W., Lefkowitz, S.M., Ziman, M., Schelter, J.M., Meyer, M.R., Kobayashi, S., Davis, C., Dai, H., He, Y.D., Stephaniants, S.B., Cavet, G., Walker, W.L., West, A., Coffey, E., Shoemaker, D.D., Stoughton, R., Blanchard, A.P., Friend, S.H. and Linsley, P.S. (2001) Expression profiling using microarrays fabricated by an ink-jet oligonucleotide synthesizer. *Nature Biotechnology*, **19** (4), 342–347.

26 DeRisi, J., Penland, L., Brown, P.O., Bittner, M.L., Meltzer, P.S., Ray, M., Chen, Y., Su, Y.A. and Trent, J.M. (1996) Use of a cDNA microarray to analyse gene expression patterns in human cancer. *Nature Genetics*, **14** (4), 457–460.

27 DeRisi, J.L., Iyer, V.R. and Brown, P.O. (1997) Exploring the metabolic and genetic control of gene expression on a genomic scale. *Science*, **278** (5338), 680–686.

28 Hughes, T.R., Marton, M.J., Jones, A.R., Roberts, C.J., Stoughton, R., Armour, C.D., Bennett, H.A., Coffey, E., Dai, H., He, Y.D., Kidd, M.J., King, A.M., Meyer, M.R., Slade, D., Lum, P.Y., Stepaniants, S.B., Shoemaker, D.D., Gachotte, D., Chakraburtty, K., Simon, J., Bard, M. and Friend, S.H. (2000) Functional discovery via a compendium of expression profiles. *Cell*, **102** (1), 109–126.

29 Golub, T.R., Slonim, D.K., Tamayo, P., Huard, C., Gaasenbeek, M., Mesirov, J.P., Coller, H., Loh, M.L., Downing, J.R., Caligiuri, M.A., Bloomfield, C.D. and Lander, E.S. (1999) Molecular classification of cancer: class discovery and class prediction by gene expression monitoring. *Science*, **286** (5439), 531–537.

30 Alizadeh, A.A., Eisen, M.B., Davis, R.E., Ma, C., Lossos, I.S., Rosenwald, A., Boldrick, J.C., Sabet, H., Tran, T., Yu, X., Powell, J.I., Yang, L., Marti, G.E., Moore, T., Hudson, J. Jr, Lu, L., Lewis, D.B., Tibshirani, R., Sherlock, G., Chan, W.C., Greiner, T.C., Weisenburger, D.D., Armitage, J.O., Warnke, R., Levy, R., Wilson, W., Grever, M.R., Byrd, J.C., Botstein, D., Brown, P.O. and Staudt, L.M. (2000) Distinct types of diffuse large B-cell lymphoma identified by gene expression profiling. *Nature*, **403** (6769), 503–511.

31 Winzeler, E.A., Richards, D.R., Conway, A.R., Goldstein, A.L., Kalman, S., McCullough, M.J., McCusker, J.H., Stevens, D.A., Wodicka, L., Lockhart, D.J. and Davis, R.W. (1998) Direct allelic variation scanning of the yeast genome. *Science*, **281** (5380), 1194–1197.

32 Mägi, R., Pfeufer, A., Nelis, M., Montpetit, A., Metspalu, A. and Remm, M. (2007) Evaluating the performance of commercial whole-genome marker sets for capturing common genetic variation. *BMC Genomics*, **118**, 159.

33 Johannes, G., Carter, M.S., Eisen, M.B., Brown, P.O. and Sarnow, P. (1999)

Identification of eukaryotic mRNAs that are translated at reduced cap binding complex eIF4F concentrations using a cDNA microarray. *Proceedings of the National Academy of Sciences of the United States of America*, **96** (23), 13118–131123.

34 Takizawa, P.A., DeRisi, J.L., Wilhelm, J.E. and Vale, R.D. (2000) Plasma membrane compartmentalization in yeast by messenger RNA transport and a septin diffusion barrier. *Science*, **290** (5490), 341–344.

35 Parker, R. and Sheth, U. (2007) P bodies and the control of mRNA translation and degradation. *Molecules and Cells*, **25** (5), 635–646.

36 Pollack, J.R., Perou, C.M., Alizadeh, A.A., Eisen, M.B., Pergamenschikov, A., Williams, C.F., Jeffrey, S.S., Botstein, D. and Brown, P.O. (1999) Genome-wide analysis of DNA copy-number changes using cDNA microarrays. *Nature Genetics*, **23** (1), 41–46.

37 Shadeo, A. and Lam, W.L. (2006) Comprehensive copy number profiles of breast cancer cell model genomes. *Breast Cancer Research*, **8** (1), R9.

38 Khodursky, A.B., Peter, B.J., Schmid, M.B., DeRisi, J., Botstein, D., Brown, P.O. and Cozzarelli, N.R. (2000) Analysis of topoisomerase function in bacterial replication fork movement: use of DNA microarrays. *Proceedings of the National Academy of Sciences of the United States of America*, **97** (17), 9419–9424.

39 Hiley, S.L., Jackman, J., Babak, T., Trochesset, M., Morris, Q.D., Phizicky, E. and Hughes, T.R. (2005) Detection and discovery of RNA modifications using microarrays. *Nucleic Acids Research*, **733** (1), e2.

40 Hoheisel, J.D. (2006) Microarray technology: beyond transcript profiling and genotype analysis. *Nature Reviews Genetics*, **7** (3), 200–210.

41 Hanlon, S.E. and Lieb, J.D. (2004) Progress and challenges in profiling the dynamics of chromatin and transcription factor binding with DNA microarrays. *Current Opinion in Genetics and Development*, **14** (6), 697–705.

42 Buck, M.J. and Lieb, J.D. (2004) ChIP-chip: considerations for the design, analysis, and application of genome-wide chromatin immunoprecipitation experiments. *Genomics*, **83** (3), 349–360.

43 Wu, J., Smith, L.T., Plass, C. and Huang, T.H. (2006) ChIP-chip comes of age for genome-wide functional analysis. *Cancer Research*, **66** (14), 6899–6902.

44 Ren, B., Robert, F., Wyrick, J.J., Aparicio, O., Jennings, E.G., Simon, I., Zeitlinger, J., Schreiber, J., Hannett, N., Kanin, E., Volkert, T.L., Wilson, C.J., Bell, S.P. and Young, R.A. (2000) Genome-wide location and function of DNA binding proteins. *Science*, **290** (5500), 2306–2309.

45 Glynn, E.F., Megee, P.C., Yu, H.G., Mistrot, C., Unal, E., Koshland, D.E., DeRisi, J.L. and Gerton, J.L. (2004) Genome-wide mapping of the cohesin complex in the yeast *Saccharomyces cerevisiae*. *PLoS Biology*, **2** (9), E259.

46 Schübeler, D., MacAlpine, D.M., Scalzo, D., Wirbelauer, C., Kooperberg, C., van Leeuwen, F., Gottschling, D.E., O'Neill, L.P., Turner, B.M., Delrow, J., Bell, S.P. and Groudine, M. (2004) The histone modification pattern of active genes revealed through genome-wide chromatin analysis of a higher eukaryote. *Genes and Development*, **18** (11), 1263–1271.

47 Lee, T.I., Rinaldi, N.J., Robert, F., Odom, D.T., Bar-Joseph, Z., Gerber, G.K., Hannett, N.M., Harbison, C.T., Thompson, C.M., Simon, I., Zeitlinger, J., Jennings, E.G., Murray, H.L., Gordon, D.B., Ren, B., Wyrick, J.J., Tagne, J.B., Volkert, T.L., Fraenkel, E., Gifford, D.K. and Young, R.A. (2002) Transcriptional regulatory networks in *Saccharomyces cerevisiae*. *Science*, **298** (5594), 799–804.

48 Boyer, L.A., Lee, T.I., Cole, M.F., Johnstone, S.E., Levine, S.S., Zucker, J.P., Guenther, M.G., Kumar, R.M., Murray, H.L., Jenner, R.G., Gifford,

D.K., Melton, D.A., Jaenisch, R. and Young, R.A. (2005) Core transcriptional regulatory circuitry in human embryonic stem cells. *Cell*, **122** (6), 947–956.

49 Liu, C.L., Kaplan, T., Kim, M., Buratowski, S., Schreiber, S.L., Friedman, N. and Rando, O.J. (2005) Single-nucleosome mapping of histone modifications in *S. cerevisiae*. *PLoS Biology*, **3** (10), e328.

50 Johnson, J.M., Edwards, S., Shoemaker, D. and Schadt, E.E. (2005) Dark matter in the genome: evidence of widespread transcription detected by microarray tiling experiments. *Trends in Genetics*, **21** (2), 93–102. Review.

51 David, L., Huber, W., Granovskaia, M., Toedling, J., Palm, C.J., Bofkin, L., Jones, T., Davis, R.W. and Steinmetz, L.M. (2006) A high-resolution map of transcription in the yeast genome. *Proceedings of the National Academy of Sciences of the United States of America*, **103** (14), 5320–5325.

52 ENCODE Project Consortium,((2004) The ENCODE (ENCyclopedia of DNA Elements) Project. *Science*, **306**, 636–640.

53 ENCODE Project Consortium,((2007) Identification and analysis of functional elements in 1% of the human genome by the ENCODE pilot project. *Nature*, **447**, 799–816.

54 Davies, K. (1993) The EST express gathers speed. *Nature*, **364**, 554.

55 Altmann, C.R., Bell, E., Sczyrba, A., Pun, J., Bekiranov, S., Gaasterland, T. and Brivanlou, A.H. (2001) Microarray-based analysis of early development in *Xenopus laevis*. *Developmental Biology*, **236** (1), 64–75.

56 Hwang, L., Hocking-Murray, D., Bahrami, A.K., Andersson, M., Rine, J. and Sil, A. (2003) Identifying phase-specific genes in the fungal pathogen *Histoplasma capsulatum* using a genomic shotgun microarray. *Molecular Biology of the Cell*, **14** (6), 2314–2326.

57 Tomiuk, S. and Hofmann, K. (2001) Microarray probe selection strategies. *Briefings in Bioinformatics*, **2** (4), 329–340.

58 Chen, M., ten Bosch, J., Beckman, K., Saljoughi, S., Seidel, C., Tuason, N., Larka, L., Lam, H., Sinibaldi, R. and Saul, R. (1999) Covalent attachment of sequence optimized PCR products and oligos for DNA microarrays. Abstract TIGR Conference.

59 Halgren, R.G., Fielden, M.R., Fong, C.J. and Zacharewski, T.R. (2001) Assessment of clone identity and sequence fidelity for 1189 IMAGE cDNA clones. *Nucleic Acids Research*, **29** (2), 582–588.

60 Sinibaldi, R., O'Connell, C., Seidel, C. and Rodriguez, H. (2001) Gene expression analysis on medium-density oligonucleotide arrays. *Methods in Molecular Biology*, **170**, 211–222.

61 Kane, M.D., Jatkoe, T.A., Stumpf, C.R., Lu, J., Thomas, J.D. and Madore, S.J. (2000) Assessment of the sensitivity and specificity of oligonucleotide (50 mer) microarrays. *Nucleic Acids Research*, **28** (22), 4552–4557.

62 Chou, C.C., Chen, C.H., Lee, T.T. and Peck, K. (2004) Optimization of probe length and the number of probes per gene for optimal microarray analysis of gene expression. *Nucleic Acids Research*, **12** (32), e99.

63 Bozdech, Z., Zhu, J., Joachimiak, M.P., Cohen, F.E., Pulliam, B. and DeRisi, J.L. (2003) Expression profiling of the schizont and trophozoite stages of *Plasmodium falciparum* with a long-oligonucleotide microarray. *Genome Biology*, **4** (2), R9.

64 Kreil, D.P., Russell, R.R. and Russell, S. (2006) Microarray oligonucleotide probes. *Methods in Enzymology*, **410**, 73–98.

65 Pozhitkov, A.E., Tautz, D. and Noble, P.A. (2007) Oligonucleotide microarrays: widely applied poorly understood.

Briefings in Functional Genomics and Proteomics, **6** (2), 141–148.

66 Graf, S., Nielsen, F.G., Kurtz, S., Huynen, M.A., Birney, E., Stunnenberg, H. and Flicek, P. (2007) Optimized design and assessment of whole genome tiling arrays. *Bioinformatics*, **23** (13), i195–i204.

67 Fare, T.L., Coffey, E.M., Dai, H., He, Y.D., Kessler, D.A., Kilian, K.A., Koch, J.E., LeProust, E., Marton, M.J., Meyer, M.R., Stoughton, R.B., Tokiwa, G.Y. and Wang, Y. (2003) Effects of atmospheric ozone on microarray data quality. *Analytical Chemistry*, **75** (17), 4672–4675.

68 Quackenbush, J. (2002) Microarray data normalization and transformation. *Nature Genetics*, **32** (Suppl.), 496–501. Review.

69 Yang, Y.H., Dudoit, S., Luu, P., Lin, D.M., Peng, V., Ngai, J. and Speed, T.P. (2002) Normalization for cDNA microarray data: a robust composite method addressing single and multiple slide systematic variation. *Nucleic Acids Research*, **30** (4), e15.

70 Dudoit, S., Yang, Y.H., Callow, M. and Speed, T. (2002) Statistical methods for identifying genes with differential expression in replicated cDNA experiments. *Statistica Sinica*, **12**, 111–139.

71 Dabney, A.R. and Storey, J.D. (2007) Normalization of two-channel microarrays accounting for experimental design and intensity-dependent relationships. *Genome Biology*, **8** (3), R44.

72 Wilson, D.L., Buckley, M.J., Helliwell, C.A. and Wilson, I.W. (2003) New normalization methods for cDNA microarray data. *Bioinformatics*, **19** (11), 1325–1332.

73 Baird, D., Johnstone, P. and Wilson, T. (2004) Normalization of microarray data using a spatial mixed model analysis which includes splines. *Bioinformatics*, **20** (17), 3196–3205.

74 Wit, E. and McClure, J. (2004) Normalization in *Statistics for Microarrays: Design, Analysis and Inference*, John Wiley & Sons, Ltd., Chapter 4.

75 Tseng, G.C., Oh, M.K., Rohlin, L., Liao, J.C. and Wong, W.H. (2001) Issues in cDNA microarray analysis: quality filtering channel normalization, models of variations and assessment of gene effects. *Nucleic Acids Research*, **29** (12), 2549–2557.

76 Oshlack, A., Emslie, D., Corcoran, L.M. and Smyth, G.K. (2007) Normalization of boutique two-color microarrays with a high proportion of differentially expressed probes. *Genome Biology*, **8** (1), R2.

77 Royce, T.E., Rozowsky, J.S. and Gerstein, M.B. (2007) Assessing the need for sequence-based normalization in tiling microarray experiments. *Bioinformatics*, **23** (8), 988–997.

78 Irizarry, R.A., Hobbs, B., Collin, F., Beazer-Barclay, Y.D., Antonellis, K.J., Scherf, U. and Speed, T.P. (2003) Exploration normalization, and summaries of high density oligonucleotide array probe level data. *Biostatistics*, **4** (2), 249–264.

79 Li, C. and Wong, W.H. (2001) Model-based analysis of oligonucleotide arrays: expression index computation and outlier detection. *Proceedings of the National Academy of Sciences of the United States of America*, **98** (1), 31–36.

80 Wu, Z., Irizarry, R.A., Gentleman, R., Francisco, M.M. and Spencer, F. (2004) A Model Based Background Adjustment for Oligonucleotide Expression Arrays. Johns Hopkins University Department of Biostatistics Working Papers. Working Paper 1.

81 Qin, L.X., Beyer, R.P., Hudson, F.N., Linford, N.J., Morris, D.E. and Kerr, K.F. (2006) Evaluation of methods for oligonucleotide array data via quantitative real-time PCR. *BMC Bioinformatics*, **177**, 23.

82 Zhou, L. and Rocke, D.M. (2005) An expression index for Affymetrix GeneChips based on the generalized logarithm. *Bioinformatics*, **21** (21), 3983–3989.

83 Choe, S.E., Boutros, M., Michelson, A.M., Church, G.M. and Halfon, M.S. (2005) Preferred analysis methods for Affymetrix GeneChips revealed by a wholly defined control dataset. *Genome Biology*, **6** (2), R16.

84 Dabney, A.R. and Storey, J.D. (2006) A reanalysis of a published Affymetrix GeneChip control dataset. *Genome Biology*, **7** (3), 401.

85 Holland, M.J. (2002) Transcript abundance in yeast varies over six orders of magnitude. *Journal of Biological Chemistry*, **277** (17), 14363–14366.

86 Axel, R., Feigelson, P. and Schutz, G. (1976) Analysis of the complexity and diversity of mRNA from chicken liver and oviduct. *Cell*, **7** (2), 247–254.

87 Bishop, J.O., Morton, J.G., Rosbash, M. and Richardson, M. (1974) Three abundance classes in HeLa cell messenger RNA. *Nature*, **250** (463), 199–204.

88 Sartor, M., Schwanekamp, J., Halbleib, D., Mohamed, I., Karyala, S., Medvedovic, M. and Tomlinson, C.R. (2004) Microarray results improve significantly as hybridization approaches equilibrium. *Biotechniques*, **36** (5), 790–796.

89 Korkola, J.E., Estep, A.L., Pejavar, S., DeVries, S., Jensen, R. and Waldman, F.M. (2003) Optimizing stringency for expression microarrays. *Biotechniques*, **35** (4), 828–835.

90 Wang, Y., Barbacioru, C., Hyland, F., Xiao, W., Hunkapiller, K.L., Blake, J., Chan, F., Gonzalez, C., Zhang, L. and Samaha, R.R. (2006) Large scale real-time PCR validation on gene expression measurements from two commercial long-oligonucleotide microarrays. *BMC Genomics*, **7**, 59.

91 de Reynies, A., Geromin, D., Cayuela, J.M., Petel, F., Dessen, P., Sigaux, F. and Rickman, D.S. (2006) Comparison of the latest commercial short and long oligonucleotide microarray technologies. *BMC Genomics*, **7**, 51.

92 Petersen, D., Chandramouli, G.V., Geoghegan, J., Hilburn, J., Paarlberg, J., Kim, C.H., Munroe, D., Gangi, L., Han, J., Puri, R., Staudt, L., Weinstein, J., Barrett, J.C., Green, J. and Kawasaki, E.S. (2005) Three microarray platforms: an analysis of their concordance in profiling gene expression. *BMC Genomics*, **56** (1), 63.

93 Yauk, C.L., Berndt, M.L., Williams, A. and Douglas, G.R. (2004) Comprehensive comparison of six microarray technologies. *Nucleic Acids Research*, **32** (15), e124.

94 Draghici, S., Khatri, P., Eklund, A.C. and Szallasi, Z. (2005) Reliability and reproducibility issues in DNA microarray measurements. *Trends in Genetics*, **22** (2), 101–109. Review.

95 Kothapalli, R., Yoder, S.J., Mane, S. and Loughran, T.P. (2002) Microarray results: how accurate are they? *BMC Bioinformatics*, **233**, 22.

96 Tan, P.K., Downey, T.J., Spitznagel, E.L., Jr, Xu, P., Fu, D., Dimitrov, D.S., Lempicki, R.A., Raaka, B.M. and Cam, M.C. (2003) Evaluation of gene expression measurements from commercial microarray platforms. *Nucleic Acids Research*, **31** (19), 5676–5684.

97 Patterson, T.A., Lobenhofer, E.K., Fulmer-Smentek, S.B., Collins, P.J., Chu, T.M., Bao, W., Fang, H., Kawasaki, E.S., Hager, J., Tikhonova, I.R., Walker, S.J., Zhang, L., Hurban, P., de Longueville, F., Fuscoe, J.C., Tong, W., Shi, L. and Wolfinger, R.D. (2006) Performance comparison of one-color and two-color platforms within the MicroArray Quality Control (MAQC) project. *Nature Biotechnology*, **24** (9), 1140–1150.

98 Haab, B.B., Dunham, M.J. Brown, P.O., (2001) Protein microarrays for highly parallel detection and quantitation of specific proteins and antibodies in complex solutions. *Genome Biology*, **2** (2), 1–13.

99 Wheeler, D.B., Carpenter, A.E. and Sabatini, D.M. (2005) Cell microarrays and RNA interference chip away at gene function. *Nature Genetics*, **37** (Suppl.), S25–S30. Review.

100 Homann, O.R., Cai, H., Becker, J.M. and Lindquist, S.L. (2005) Harnessing natural diversity to probe metabolic pathways. *PLoS Genetics*, **1** (6), e80.

101 Margulies, M., Egholm, M., Altman, W.E., Attiya, S., Bader, J.S., Bemben, L.A., Berka, J., Braverman, M.S., Chen, Y.J., Chen, Z., Dewell, S.B., Du, L., Fierro, J.M., Gomes, X.V., Godwin, B.C., He, W., Helgesen, S., Ho, C.H., Irzyk, G.P., Jando, S.C., Alenquer, M.L., Jarvie, T.P., Jirage, K.B., Kim, J.B., Knight, J.R., Lanza, J.R., Leamon, J.H., Lefkowitz, S.M., Lei, M., Li, J., Lohman, K.L., Lu, H., Makhijani, V.B., McDade, K.E., McKenna, M.P., Myers, E.W., Nickerson, E., Nobile, J.R., Plant, R., Puc, B.P., Ronan, M.T., Roth, G.T., Sarkis, G.J., Simons, J.F., Simpson, J.W., Srinivasan, M., Tartaro, K.R., Tomasz, A., Vogt, K.A., Volkmer, G.A., Wang, S.H., Wang, Y., Weiner, M.P., Yu, P., Begley, R.F. and Rothberg, J.M. (2005) Genome sequencing in microfabricated high-density picolitre reactors. *Nature*, **437** (7057), 376–380.

102 Barski, A., Cuddapah, S., Cui, K., Roh, T.Y., Schones, D.E., Wang, Z., Wei, G., Chepelev, I. and Zhao, K. (2007) High-resolution profiling of histone methylations in the human genome. *Cell*, **129** (4), 823–837.

103 Johnson, D.S., Mortazavi, A., Myers, R.M. and Wold, B. (2007) Genome-wide mapping of *in vivo* protein–DNA interactions. *Science*, **316** (5830), 1497–1502.

104 Velculescu, V.E., Zhang, L., Zhou, W., Vogelstein, J., Basrai, M.A., Bassett, D.E., Jr, Hieter, P., Vogelstein, B. and Kinzler, K.W. (1997) Characterization of the yeast transcriptome. *Cell*, **88** (2), 243–251.

105 MGED Society (2002) Microarray standards at last. *Nature*, **419**, 323.

106 Edgar, R., Domrachev, M. and Lash, A.E. (2002) Gene Expression Omnibus: NCBI gene expression and hybridization array data repository. *Nucleic Acids Research*, **30** (1), 207–210.

107 Parkinson, H., Sarkans, U., Shojatalab, M., Abeygunawardena, N., Contrino, S., Coulson, R., Farne, A. and Lara, G.G., Holloway, E., Kapushesky, M. Lilja, P., Mukherjee, G., Oezcimen, A., Rayner, T., Rocca-Serra, P., Sharma, A., Sansone, S. and, Brazma, A. (2005) ArrayExpress – a public repository for microarray gene expression data at the EBI. *Nucleic Acids Research*, **33**,D553–D555.

108 Blanchard, A.P., Kaiser, R.J. and Hood, L.E. (1996) High-density oligonucleotide arrays. *Biosensors and Bioelectronics*, **6/7**, 687–690.

109 Marmur, J. and Doty, P. (1961) Thermal renaturation of deoxyribonucleic acids. *Journal of Molecular Biology*, **3**, 585–594.

2
Comparative Analysis of Clustering Methods for Microarray Data

Dongxiao Zhu, Mary-Lee Dequeéant, and Hua Li

2.1
Introduction

Hierarchical clustering is one of the approaches used most frequently to uncover the gene expression patterns and pathways from microarray data. The nested structure of hierarchical clustering solution appeals to biologists raised on the tree metaphor (even though, evolutionary information is at best implicit in the transcription readouts) and, more importantly, offers several algorithmically simple ways of partitioning the data set into clusters at different resolution levels. The hierarchical clustering methods can be classified into bottom-up (agglomerative) or top-down (divisive) approaches. The bottom-up algorithms proceed in a stepwise fashion, so that at each step two closest objects are joined until all objects are in a single cluster; here and in other approaches, "closeness" is determined by a specific similarity/distance measure (that should be defined for the distances between individual entities, as well as for distances between clusters), examples of which are given below. The top-down algorithms also proceed step by step, dividing objects into groups until they are nondivisible according to certain criteria, which amount to distance/similarity measures, too.

The bottom-up methods are usually better at identifying (perhaps large number of) small clusters, while the top-down methods are more successful in identifying a few large clusters. Chipman and Tibshirani recently introduced a hybrid hierarchical clustering method, which seeks to combine this differential sensitivity of bottom-up and top-down methods to different cluster sizes [1]. They modified top-down procedures with information gained from a preliminary bottom-up clustering facilitated by introducing a new concept, "mutual cluster." It is defined as a group of objects collectively closer, as measured by a distance measure, to each other than to any other object [1]. The claimed combination of advantages over top-down and bottom-up approaches is mainly demonstrated through simulations using synthetic data and examining internal criteria, such as tightness and separation of clusters, using real-life data. Applications of this approach to biological discovery, to the best of our knowledge, have not been published yet.

Analysis of Microarray Data: A Network-Based Approach. Edited by F. Emmert-Streib and M. Dehmer

ISBN: 978-3-527-31822-3

In this chapter, we review several aspects of clustering procedures. First, we discuss some of the popular distance measures used for computing distance or similarity between gene vectors, that is, strings of numbers associated with genes, such as gene expression profiles. Second, we discuss the networks that can be built and point out the notion of the shortest path between genes in the network, first proposed by Zhou *et al.* [2]. Finally, we show how this and related network-derived information can be used to augment the distance measure between vectors, which results in improved hierarchical clustering solutions.

2.2
Measuring Distance Between Genes or Clusters

In the agglomerative clustering merging process, the choice of distance measure between genes and between clusters may significantly affect the results. There are no good guidelines on how to properly select a distance measure, but many of them have been proposed.

For measuring distance between genes, assume X and Y are a pair of genes for which sample expression data were collected across some conditions, where i represents the condition index. Some of the more familiar methods to calculate pairwise distance matrix are as follows:

- Euclidean distance between X and Y is [3]

$$||X-Y|| = \sqrt{\sum_{i=1}^{p}(x_i-y_i)^2},$$

or its squared form

$$||X-Y||^2 = \sum_{i=1}^{p}(x_i-y_i)^2$$

or Manhattan distance

$$d_{XY} = \sum_{i=1}^{p}|x_i-y_i|.$$

- Scale-free correlation between X and Y is

$$d_{XY} = 1-\rho_{XY}$$

or

$$d_{XY} = 1-\mathrm{abs}(\rho_{XY}),$$

where ρ is either Pearson correlation coefficient, or Kendall or Spearman rank-based correlation coefficient. Pearson correlation coefficient is suitable for capturing linear correlation between a pair of gene expression profiles, and the rank-based methods are good for capturing both linear and nonlinear correlations, albeit with reduced power when the pairwise gene expression correlations are truly linear. Both linear and nonlinear pairwise gene expression correlations may exist in gene expression data, and Pearson correlation may be used concurrently with rank-based correlations [4]. In general, the question of dependencies in gene expression program is not well studied [4].

- Mutual information in the discrete case,

$$I(X, Y) = \sum_{y \in Y} \sum_{x \in X} p(x, y) \log \frac{p(x, y)}{p(x)\, p(y)},$$

where $p(x, y)$ is the joint distribution of X and Y, and $p(x)$ and $p(x)$ are the marginal probability distributions of X and Y, respectively; in continuous case,

$$I(X, Y) = \int_Y \int_X p(x, y) \log \frac{p(x, y)}{p(x)\, p(y)}.$$

Intuitively, mutual information measures the information that X and Y share: it measures to what extent the uncertainty of one variable can be reduced by knowing the other variable. Note that mutual information and correlation measures converge for binary data [5].

- Covariance and coherence, especially for measuring dependency between X and Y when each of them is serially correlated over time; for example, the observation of X at time t, x_t, depends in some way on the past values $x_{t-1}, x_{t-2}, \ldots$. The two measures are to be used in time domain and frequency domain, respectively, where the latter is usually transformed from the former using Fourier transform. We describe this measure in more detail here because it is less comprehensively introduced elsewhere.
For mathematical convenience, we first assume that the time-series data are *weakly stationary*. It imposes conditions only on the first two moments of a time series, that is, mean and (autoco)variance. These conditions are [6] as follows:
  – The mean value function $\mu = E(x_t)$ is constant and does not depend on time t.
  – The covariance function γ depends only on the *lag* h of the two time points, that is, $\gamma(h) = E[(x_{t+h} - \mu)(x_t - \mu)]$.
(Weakly) stationary time series is viewed as time invariant, meaning that although the measurement may change over time, the variability does not change. The methods assume that the dynamic properties of the time series remain constant over time [7].
For joint stationary series, x_t and y_t, in the time domain, their covariance function γ is defined as $\gamma_{xy}(h) = E[(x_{t+h} - \mu_x)(y_t - \mu_y)]$, where h represents the time shift (lag),

μ_x and μ_y represent the mean of each series, and E is the statistical expectation. In the frequency domain, we define the cross-spectrum as the Fourier transform of the covariance function $\gamma_{xy}(h)$ [6]:

$$f_{xy}(\nu) = \sum_{h=-\infty}^{\infty} \gamma_{xy}(h) e^{-2\pi i \nu h}, \quad -1/2 \leq \nu \leq 1/2. \tag{2.1}$$

For any stationary process for which the autocovariance function $\gamma(h)$ satisfies $\sum_{h=-\infty}^{\infty} |\gamma(h)| < \infty$, we also define the spectral density of series x_t or y_t as

$$f(\nu) = \sum_{h=-\infty}^{\infty} \gamma(h) e^{-2\pi i \nu h}. \tag{2.2}$$

The spectrum of marginal dependency of the two series at a number of frequencies can be quantified using (squared) *coherence* function, defined as [6]

$$\rho_{xy}^2(\nu) = \frac{|f_{xy}(\nu)|^2}{f_x(\nu) f_y(\nu)}, \tag{2.3}$$

where $|f_{xy}(\nu)|^2$ is the cross-spectrum, and f_x and f_y are spectrum densities. The squared coherency is analogous to conventional squared correlation.

- Shortest path represents a different class of measure, proposed only recently in the context of gene expression analysis [2,8]. This measure is derived in two steps. First, pairwise distances between gene vectors are defined, using any of the measures, such as those listed above (in the original study, Pearson correlation was used), and a gene coexpression network is formed, in which edges represent all gene pairs such that the correlation of their expression vectors exceeds certain cutoff. Second, the so-called shortest path is measured for each pair of genes. This helps to quantitatively express the situation when expression vectors of two genes are not strongly correlated to each other, but they both exhibit much stronger correlation with the third gene, or a set of genes (Figure 2.1).

The network defined by pairwise distances between gene expression vectors and then redefined using the shortest path distance, has many interesting properties, many of which are not well studied. When it comes to the identification of clusters of functionally connected genes, however, such a nonhierarchical network is not well tractable from the computational point of view: optimal partition of an arbitrary graph is known to be a NP-hard problem (but see recent work [9] for novel practical approaches to a related problem of network modularity). Therefore, it would be advantageous to incorporate the information about indirect but significant relationships between genes, thrown up by the analysis of the shortest paths in the network, into the hierarchical clustering of gene expression vectors. In this chapter, we expand our earlier proposal [8] to do exactly this. Network connection

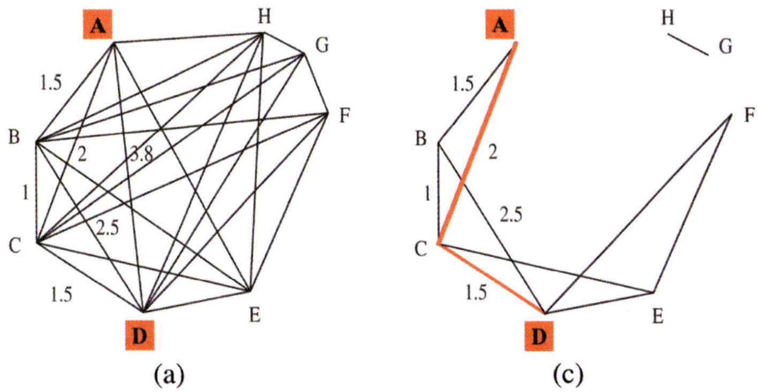

A	0			
B	1.5	0		
C	2	1	0	
D	3.8	2.5	1.5	0

(b)

A	0			
B	1.5	0		
C	2	1	0	
D	3.5	2.5	1.5	0

(d)

Figure 2.1 Network models and distance matrices for traditional clustering (a, b) and network constrained clustering (c, d). Figure 2.3c is obtained by removing some edges of weak correlations (long distances), for example, distance longer than 3. The distance between two genes is a decreasing function of their correlation (see Eq. 2.15). (a) Fully connected network: any two genes interact with each other directly in the network (connected). (b) Part of the distance matrix for the network model (a). (c) Partially connected network; it assumes only two genes with high correlation (e.g., 0.6) directly interacting with each other (connected). Gray edges represent the shortest path from A to D. (d) Part of the distance matrix for the network model (c).

patterns revealed by the shortest path analysis are called network constraint in the rest of this chapter (Figure 2.1).

Several methods of measuring distance between clusters are widely known, including

- Single linkage: Assume there are I and J objects in the two clusters, the method calculates $\min_{i,j} d(i, j)$, $i = 1, \ldots, I$ and $j = 1, \ldots, J$, which represents the minimal pairwise distance between objects of the two clusters.
- Complete linkage: The method calculates $\max_{i,j} d(i, j)$.
- Mean linkage: The method calculates the $\text{mean}_{i,j} d(i, j)$.

One of the ultimate goals of all gene-clustering algorithms is to discover the gene pathways underlying each biological process. Genes that are lying in the same pathway are often activated or depressed *simultaneously* or *sequentially* upon receiving stimuli. The biological signal is typically transmitted through shared biochemical intermediates, through intermediate gene interactions that involve physical binding or chemical modifications (phosphorylation, acetylation, methylation, etc.)

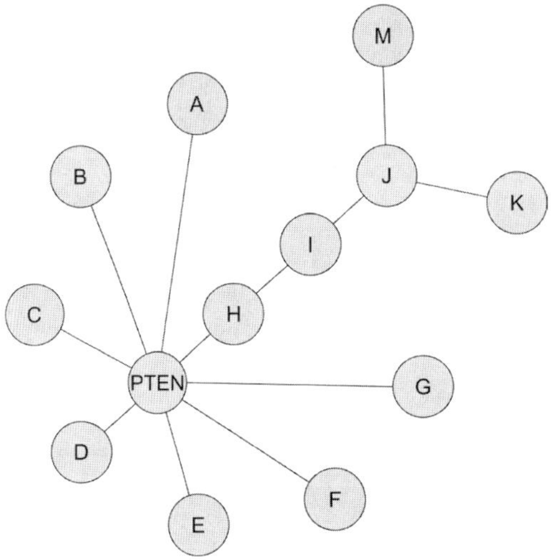

Figure 2.2 A representative signaling pathway (scheme is based on PTEN signaling involved in maintaining hematopoietic stem cells (insert references). Genes A, B, C, D, E, and so on are *simultaneously* regulated, while H, I, J, K, and so on are *sequentially* regulated.

of the gene products, or by other means. The simultaneous and sequential gene activation or repression is delineated by network connection patterns.

The small value of the shortest path distance in the nonhierarchical network is the network constraint that can be used for augmenting the distance matrix used in hierarchical clustering. Such a clustering solution may be suitable for discovering part of the signaling pathways where all genes are simultaneously activated or repressed assuming proper distance measures were used (Figure 2.2). For the majority of genes lying in the same signaling pathway that are repressed or activated sequentially, the network constraint must be taken into account for the clustering algorithm to discover the whole pathways. In this chapter, we introduce a new clustering technique taking account of the network constraint, termed "network constrained (NC) clustering." Our approach is to impose network connection patterns as realistic network constraints to improve the hierarchical clustering technique.

We first review the existing ways of inferring network constraint/network structure from high-throughput data. Some of the more popular approaches to this problem include Boolean network [10–12], coexpression network [4], Bayesian network [13], and co-occurrence network [14,15]. Deterministic Boolean network methods discretize high-throughput data into binary values, sometimes also called outcomes. In the context of gene expression, binary values may correspond, for example, to "gene turned on/turned off," or "gene expression changed/did not change." The algorithm detects the presence/absence of the network edges using a logic rule (Figure 2.1A). The method is easy to implement but the deterministic discretization oversimplifies the complicated modes of gene regulation: on the one hand, it loses information about expression levels; on the other hand, it results in uncontrolled error rate. Coexpression network methods infer network edges in a probabilistic manner through simultaneous statistical hypothesis tests with significance adjusted by false discovery rate (FDR) analysis [4]. This method is scalable to fairly large network construction (Figure 2.1B). Bayesian network methods strive for searching the most likely network topology given data, measured by maximal likelihood score or Bayesian score (Figure 2.1C). The co-occurrence network methods focus on finding the most likely network topology from co-occurrence data [14,15]. The network topology may be cast in the first-order Markov chain framework, in which the transition matrix **A** fully defines the network topology. **A** is estimated using expectation-maximization (EM) algorithm. In each EM iteration, the observed co-occurrence data are *incomplete* data (e.g., see Figure 2.3D), and the true order of these co-occurred data is *augmented* data. Incomplete data combined with augmented data constitute *complete* data. With the current estimate of $\hat{\mathbf{A}}$, one can augment the

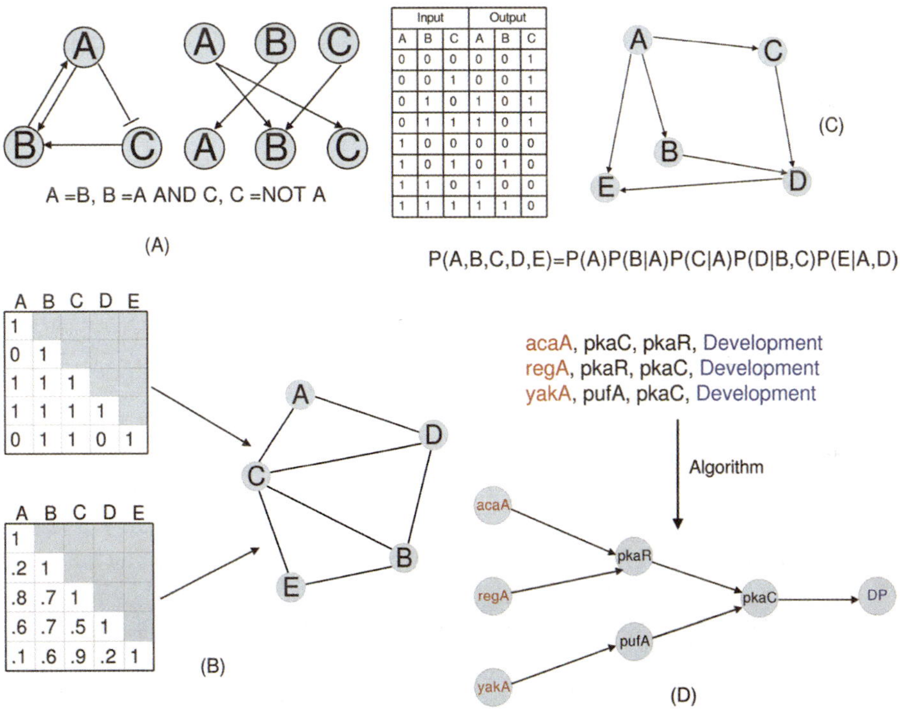

Figure 2.3 Four network reconstruction algorithms. Edges (arcs) correspond to hypothetical biological relationships. Nodes correspond to biomolecules such as genes and proteins. (A) Boolean network. In this model, gene expression is crudely approximated to take only two levels: ON and OFF. The expression level (state) of each gene is related to the expression states of some other genes, using logical rules [10]. (B) Coexpression network. Edges correspond to hypothetical functional relationships. (C) Bayesian network. Arcs correspond to inferred (statistical) causal relationships. (D) Co-occurrence network. Signaling transduction networks are inferred from text data. The inputs are lists of pathway components with known terminals (red and blue).

data and update $\hat{\mathbf{A}}$ based on complete data. The process iterates until it converges to a locally optimized **A**. The Bayesian networks are able to infer (statistical) causal networks but with exponentially increasing computational burden. The co-occurrence networks enjoy most favorable features of other methods but the input (see Figure 2.3D) requires painstaking manual processing and therefore is difficult to extend to larger networks.

Here we assume the true network constraint has been reasonably well approximated by some of the methods discussed above in the form of hybrid distance matrix consisting of both direct pairwise distances and shortest path distances (see Figure 2.1). Genes along the shortest path tend to be lying in the same signaling pathway or share the similar function [2]. Thus, the network constrained clustering is based on the idea that distance between adjacent genes in the network is estimated the same way as in other clustering methods, and distance between other pairs of genes is estimated by the length of shortest path connecting them (Figure 2.1). Since the latter depends on the network backbone, we elaborate on a number of frequently applied network models before describing the network constrained clustering.

2.3 Network Models

2.3.1 Boolean Network

A Boolean network $G(V, F)$ is defined by a set of nodes corresponding to genes $V = x_1, \ldots, x_n$ and a list of Boolean functions $F = (f_1, \ldots, f_n)$. The state of a node is completely determined by the binary values of other nodes at time t through underlying logical Boolean functions, for example, AND, OR, NOT, NOR, and NAND. The model is represented in the form of directed graph. Each x_i represents the state (expression) of gene i, where $x_i = 1$ is translated as an expression of gene i while $x_i = 0$ means otherwise. The list of Boolean functions F defines the rules of regulatory relationships between genes. That is, any given gene transforms its inputs (regulatory factors that bind to it) into an output, which is the state or expression of the gene itself; refer to Figure 2.3a for example. The underlying assumption of the Boolean network mode is that all genes' states update synchronously in accordance with the functions assigned to them and this process is iterated. The artificial synchrony simplifies computation while preserving the qualitative, generic properties of global network dynamics [16,17].

2.3.2 Coexpression Network

A gene coexpression network is defined by a set of nodes corresponding to genes and a list of edges corresponding to coexpression. Based on the assumption that coexpressed genes tend to be coregulated and possibly have the similar function, the coexpression network may represent hypothetical gene coregulation relationships that provide foundations to detect functional modules. One key problem for reconstructing gene coexpression network is to detect truly coexpressed gene pairs from genomic gene expression data. Assuming Γ is the measure of strength of coexpression between a pair of gene expression profiles, cormin is the cutoff for declaring coexpression or not and G is the number of genes to be used to construct coexpression networks, we need to simultaneously test $\Lambda = \binom{G}{2}$ pairs of two-sided hypotheses:

$$H_0 : \Gamma_{g_i,g_j} \leq \text{cormin} \quad \text{versus} \quad H_\alpha : \Gamma_{g_i,g_j} > \text{cormin}, \quad \text{for} \quad g_i \neq g_j \\ \text{and} \quad g_i, g_j \in (1, 2, \ldots, G), \tag{2.4}$$

where cormin is a minimum acceptable strength (MAS) of correlation.

The sample correlation coefficient $\hat{\Gamma}$ ($\hat{\rho}$ or $\hat{\tau}$) is used as a decision statistic to decide on pairwise dependency between two genes in the network. One key aspect of the coexpression network reconstruction is the selection of network models that differs in estimating Γ. Popular methods include relevance network model in which marginal correlation is estimated and Gaussian graphic model (GGM) in which partial correlation is estimated. The difference between the two methods, in the view of the path analysis, lies in the fact that the marginal correlation quantifies a composite correlation between a gene pair taking into account effects of all other genes in the network, while the partial correlation measures the direct correlation excluding all other genes in the network. We next discuss the statistical test of correlation using marginal correlation as example described in [8]; the situation remains similar for partial correlation but with a reduced degree of freedom [18].

For N realizations of any pair of gene responses, $\{x_{gi(n)}, x_{gj(n)}\}_{n=1}^{N}$, we first calculate $\hat{\tau}$ or $\hat{\rho}$. For large N, the per comparison error rate (PCER) p-values for ρ or τ are

$$p_{\rho_{i,j}} = 2\left(1 - \Phi\left(\frac{\tanh^{-1}(\hat{\rho}_{i,j})}{(N-3)^{-1/2}}\right)\right), \tag{2.5}$$

$$p_{\tau_{i,j}} = 2\left(1 - \Phi\left(\frac{K}{N(N-1)(2N+5)/18^{1/2}}\right)\right), \tag{2.6}$$

where Φ is the cumulative density function of a standard Gaussian random variable and $K = \Sigma\Sigma_{1 \leq n \leq m \leq N} K_{nm}$. The above expressions are based on asymptotic Gaussian approximations to $\hat{\rho}_{i,j}$ [19] and to $\hat{\tau}_{i,j}$ [20].

The PCER p-value refers to the probability of type I error rate incurred in testing a single pair of hypothesis for a single pair of genes g_i, g_j. When considering the Λ multiple hypotheses for all possible pairs, as in previous studies, we adopt the FDR to control statistical significance of the selected gene pair correlations in our screening procedure [21]. The procedure guarantees that the false discovery rate associated with testing the hypotheses [1] does not exceed α.

Select a level α of FDR and a level cormin of MAS significance levels. We use a modified version of the two-stage screening procedure applied to gene screening [22]. This procedure consists of the following:

- Stage I: Test the simple null hypothesis.

$$H_0 : \Gamma_{g_i,g_j} = 0 \,\text{versus}\, H_\alpha : \Gamma_{g_i,g_j} \neq 0$$

  at FDR level. The step-down procedure of Benjamini and Hochberg [23] is used.

- Stage II: Suppose Λ_1 pairs of genes pass stage I procedure. In stage II, we first construct asymptotic PCER confidence intervals (PCER-CIs): $\hat{I}^{\lambda}(\alpha)$ for each Γ (ρ or τ) in subset $\mathcal{G}_1$, and convert into FDR confidence intervals (FDR-CIs): $I^g(\Lambda_1/\Lambda)$ [24]. A gene pair in subset $\mathcal{G}_1$ is declared to be both statistically and biologically significant if its FDR-CI does not intersect the MAS interval [−cormin, cormin].

A relevance network composed of these screened gene pairs can then be constructed with simultaneously controlled statistical and biological significance.

2.3.3
Bayesian Network

In Bayesian network, genes are abstracted as random variable X_i, $i = 1, \ldots, n$, and the joint distribution of these genes is represented as the product of conditional probabilities on their parents **U** (Equation 2.7):

$$P(X_1, \ldots, X_n) = \prod_i P(X_i|\mathbf{U}). \tag{2.7}$$

The graphical representation is given by a directed graph where we put arcs from X_i's parents to X_i. For example, the joint distribution of genes B and C is equivalent to the product of the conditional distribution of B and C given A, that is, $P(B, C) = P(B|A)P(C|A)$ (Figure 2.3c). Note that the joint distribution of all genes in the network depends on the network connection patterns (or topology). It follows that the likelihood of observing a gene expression data depends also on network topologies. The practical interest is often to find the network topology that maximizes the likelihood of the data. If the graph is cyclic, the product decomposition of Equation 2.7 is guaranteed to be a coherent probability distribution [13]. Many Bayesian network applications involve, more or less, enumerating all possible network topology and selecting one of the maximum likelihood as specified in Equation 2.7. Bayesian networks provide a flexible framework for inferring causal relationships among genes. However, the computational burden increases exponentially as the size of network increases. Therefore, Bayesian network, at the current stage, is only suitable for constructing very small networks unless sophisticated biological constraint is imposed to significantly reduce the search space [25].

Another limitation is that statistical causal relationships inferred using Bayesian networks are not necessarily biologically causal ones; instead, they serve as a starting point for generating a novel biological hypothesis.

2.3.4
Co-Occurrence Network

Co-occurrence network model reconstructs signaling networks from co-occurrence data that contain information on which genes are in which pathways without knowing the order and mode of their interaction. We describe a recently developed algorithm for constructing co-occurrence network in more detail [14,15].

The true network that we are trying to elucidate consists of an ensemble of signaling pathways, which can be viewed as a collection $Y = y^{(1)}, y^{(2)}, \ldots, y^{(T)}$ of T independent samples of the first-order Markov chain. The parameters of the Markov chain are initial state distribution π and a transition matrix **A**. **A** uniquely defines the network topology. To recover the signaling network topology from Y, we treat the corresponding unobserved permutations $\tau^{(1)}, \tau^{(2)}, \ldots, \tau^{(T)}$ as hidden variables and describe an EM algorithm for computing maximum likelihood estimates of the Markov chain parameters. We would like to compute maximum likelihood estimates of the Markov chain transition matrix **A** and the initial state distribution π from a collection of unordered co-occurrences $Y = y^{(1)}, y^{(2)}, \ldots, y^{(T)}$.

In practice, we do not know the corresponding permutations, so we treat them as hidden variables and derive an EM algorithm, assuming each permutation as being drawn from an equally likely distribution on all permutations of the appropriate length; that is, if $x^{(m)}$ corresponds to a path of N_m elements, then $r_{(m)}$ is modeled as a random permutation matrix drawn uniformly from the collection of all permutations of N_m elements, denoted by Ψ_{N_m}. The EM algorithm alternates between the expectation or E-step, which amounts to estimating expected permutations for each path conditioned on the current parameter estimates, and the maximization or M-step, where the parameter estimates are updated based on the expected permutations computed in the E-step.

More precisely, in the E-step we compute sufficient statistics

$$\bar{\alpha}^{(m)}_{t',t'} = E\left[\sum_{t=2}^{N_m} r^{(m)}_{t',t'} r^{(m)}_{t-1,t''} | x^{(m)}, \hat{A}, \hat{\pi}\right], \tag{2.8}$$

$$\bar{\alpha}^{(m)}_{t',t'} = \frac{\sum_{r \in \psi N_m} r_{t,t'} r_{t-1,t''} P[x^{(m)} | r, \hat{A}, \hat{\pi}]}{\sum_{r \in \psi N_m} P[x^{(m)} | r, \hat{A}, \hat{\pi}]}, \tag{2.9}$$

and

$$\bar{r}^{(m)}_{1,t'} = E\left[r^{(m)}_{t',t'} | x^{(m)}, \hat{A}, \hat{\pi}\right], \tag{2.10}$$

$$\bar{r}_{1,t'}^{(m)} = \frac{\sum_{r \in \psi N_m} r_{1,t'} P[x^{(m)}|r, \hat{A}, \hat{\pi}]}{\sum_{r \in \psi N_m} P[x^{(m)}|r, \hat{A}, \hat{\pi}]}, \tag{2.11}$$

where each term $P[x^{(m)}|r, \hat{A}, \hat{\pi}]$ is easily computed after using r to unshuffle $x^{(m)}$:

$$P[x^{(m)}|r, \hat{A}, \hat{\pi}] = P[y^{(m)}|\tau, \hat{A}, \hat{\pi}] = \hat{\pi}_{y_{\tau_1}^{(m)}} \prod_{t=2}^{N_m} \hat{A}_{y_{\tau_{t-1}}^{(m)}, y_{\tau_t}^{(m)}}. \tag{2.12}$$

Closed form expressions for the M-step updates are then given by

$$\left(\hat{A}_{i,j}\right)_{\text{new}} = \frac{\sum_{m=1}^{T} \sum_{t',t''=1}^{N_m} \bar{\alpha}_{t',t''}^{(m)} x_{t'',i}^{(m)} x_{t',j}^{(m)}}{\sum_{j=1}^{|S|} \sum_{m=1}^{T} \sum_{t',t''=1}^{N_m} \bar{\alpha}_{t',t''}^{(m)} x_{t'',i}^{(m)} x_{t',j}^{(m)}} \tag{2.13}$$

and

$$\left(\bar{\pi}_i\right)_{\text{new}} = \frac{\sum_{m=1}^{T} \sum_{t'=1}^{N_m} \bar{r}_{1,t'}^{(m)} x_{t',i}^{(m)}}{\sum_{i=1}^{|S|} \sum_{m=1}^{T} \sum_{t'=1}^{N_m} \bar{r}_{1,t'}^{(m)} x_{t',i}^{(m)}}. \tag{2.14}$$

This algorithm can also be modified to incorporate biological prior knowledge about partially known epistatic relationships. Instead of assembling networks from pairwise gene relationships or searching for the most likely network topology, the co-occurrence networks provide attractive alternatives to take advantages of another source of information of pathway composition.

2.4 Network Constrained Clustering Method

2.4.1 Extract the Giant Connected Component

From reconstructed networks, we proceed to present a novel gene-clustering approach that takes account of network constraint. While the network constrained clustering algorithm can be applied to any kind of network, we will draw all examples of our approach from the analysis of gene expression. Many networks contain several connected components (CC). Only gene pairs that are in the same CC of the network have finite distances and thus can be clustered. The largest connected component called giant connected component (GCC) is usually important for both biological function and network topology [26] (see Figure 2.1c, genes A, B, C, D, E, and F form a GCC). The GCC of an undirected graph $G = (V, E)$, where V is the set of all vertices and E is the set of all edges, is the maximal set of vertices $U \subset V$ such that every pair of vertices u and v in U are reachable from each other. Our network constrained clustering is done on the GCC, and using three different examples we show later that almost all important genes are included in the GCC. The standard depth first search (DFS) algorithm [27] was used to extract the GCC.

2.4.2
Compute "Network Constrained Distance Matrix"

Let $\hat{\Gamma}_{ij}$ be the sample correlation coefficient between gene i and j, for example, estimated from a gene microarray data by Pearson or Kendall correlation statistic. Let w_{ij} be the weight of the edge between gene i and gene j. Similar to Zhou *et al.* [2], w_{ij} is defined as

$$w_{ij} = (1-\text{abs}(\hat{\Gamma}_{ij}))^p. \tag{2.15}$$

The integer p is an exponential tuning parameter used to enhance the differences between low and high correlations. We define the matrix $W=[w_{ij}]$ as the "traditional distance matrix" (e.g., Figure 2.1b).

We use the standard Floyd–Warshall algorithm to search among all pairs for the shortest paths within the GCC. Let $d_{ij}^{(k)}$ be the length of a shortest path from vertex i to vertex j such that all intermediate vertices on the path (if any) are in set $\{1, 2, \ldots, k\}$. When $k=0$, there is no intermediate vertex between vertices i and j, and we define $d_{ij}^{(0)} = w_{ij}$. A recursive definition of $d_{ij}^{(k)}$ is given by [27] as

$$d_{ij}^{(k)} = \begin{cases} w_{ij} & \text{if } k=0, \\ \min(d_{ij}^{(k-1)}, d_{ik}^{(k-1)} + d_{kj}^{(k-1)}) & \text{if } k \geq 1, \end{cases} \tag{2.16}$$

where $d_{ij}^{(k-1)}$ is the length of shortest path when k is not a vertex on the path, and $d_{ik}^{(k-1)} + d_{kj}^{(k-1)}$ is the length when k is a vertex on the path. We define the matrix $D=[d_{ij}]$ as the "network constrained distance matrix" (e.g., Figure 2.1d). It can be used as an input to many distance matrix based clustering softwares such as hierarchical clustering [28] and K-medoids [29]. The calculation of matrix D can be easily extended to higher eukaryotes since the algorithm runs in polynomial time, that is, $O(V^3 + V + E)$.

2.5
Network Constrained Clustering Results

We compare the results of the network constrained method (NC clustering) to the more traditional methods (TD clustering) using three examples. In the first example, we attempt to recover the well-known galactose metabolism pathway in yeast *S. cerevisiae*. The second example compares the ability of two classes of methods to enrich for biologically relevant genes in the transcriptome of mouse photoreceptor cells. The third example compares the error rate in different procedures of clustering of genes involved in the transcriptional program of mouse somitogenesis.

2.5.1
Yeast Galactose Metabolism Pathway

We empirically evaluated the performance of the NC clustering approach by applying it to a relatively well-studied yeast galactose metabolism pathway and comparing it with the TD clustering approaches. We used a subset of 997 genes that have been identified by Ideker *et al.* [30]. In their experiment, nine genes whose gene products are known to be lying in the pathway were knocked out, one at a time, and cells were grown in galactose or glucose media before their gene expression under each perturbation was profiled. Genes having a likelihood statistic $\lambda \leq 45$ were selected as differentially expressed, whose mRNA levels differed significantly from the reference strain under one or more perturbations.

We measured the pairwise gene correlations using both Pearson and Kendall correlation coefficients, and applied a two-stage algorithm to screen gene pairs with $\mathrm{FDR} \leq 5\%$ and $\mathrm{MAS} = 0.6$ [4]. The resulting network is a mixed network within which edges are discovered with one or both of statistics. Our network construction algorithm and the screening criteria ensure false discovery of no more than 5% edges having strength of association greater than 0.6 [4].

We extracted the GCC from the coexpression network using a DFS type algorithm (see Section 2.4). The GCC contains 772 genes within which almost all known structural genes in the pathway are included. This confirms the notion that GCC of the network has topological as well as functional significance [26]. The network constrained distance matrix for GCC was computed according to Equations 2.13 and 2.16 using GCC selected genes (see Section 2.4) while the distance matrix for the TD clustering was computed according to Equation 2.15.

The yeast galactose metabolism pathway consists of at least three types of genes including transporter genes such as GAL2, HXT1-10 (the role of other HXT genes is not clear), enzyme genes (GAL1, GAL7, GAL10, etc.), and transcription factor genes (GAL3, GAL4, GAL80, etc.) [31]. Transcription factor genes are not discoverable from this microarray experiment, as their expression levels peak prior to the maxima of the regulated genes; and the sample was taken at the time corresponding to that later time. We sought to compare our NC clustering approach with the TD clustering approach through rediscovering the 14 important transporter genes and enzymes of the yeast galactose metabolism pathway.

We used agglomerative hierarchical clustering (implemented in R function `hclust`) as the comparison platform. Similarly, other platforms such as *K*-medoids can be used interchangeably. For calculating distance between clusters, we used "complete" method in which the longest geodesics between genes in the two clusters are used as distance between clusters.

Figure 2.4 shows the traditional clustering approach using all 997 genes and Figure 2.5 shows the traditional clustering approach using the GCC selected 772 genes. In both cases, the 14 structural genes are separated into three subclusters

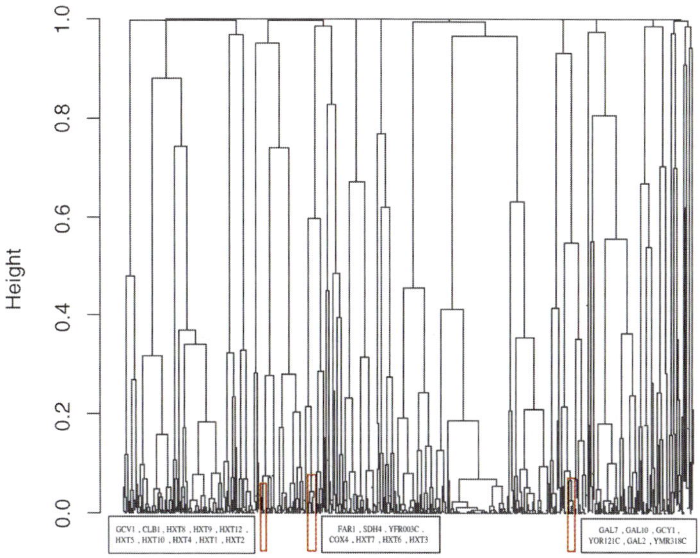

Figure 2.4 Traditional clustering: agglomerative hierarchical clustering using all 997 differentially expressed genes. The 14 structural genes are separated into three clusters (red rectangular).

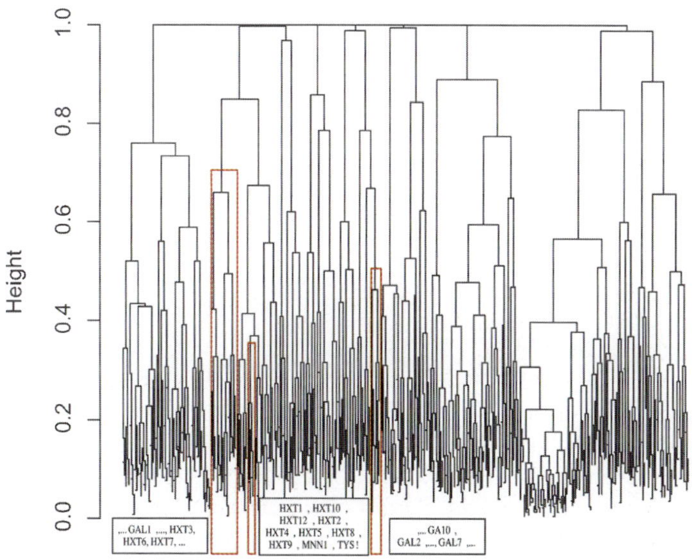

Figure 2.5 Traditional clustering: agglomerative hierarchical clustering using the GCC selected 772 genes. The 14 structural genes are separated into three clusters (red rectangular). Dots indicating incomplete clusters are shown due to space limitation.

(Figures 2.4 and 2.5). In Figure 2.4, all GAL genes are nicely grouped in a cluster, but not HXT genes, while in Figure 2.5, all HXT genes are grouped into one cluster, but the algorithm failed to combine GAL gene clusters with HXT gene clusters. Figures 2.4 and 2.5 show that the GCC gene selection process has some desirable effects on clustering by removing a few unrelated genes that are not relevant to the biological pathway; however, the process does not improve clustering performance significantly.

We hypothesize that the undesirable separation of genes in the pathway is due to the gene expression dissimilarity between subclusters and gene expression similarity within subclusters. To test this hypothesis, we plotted the correlation matrix of 14 genes in the structural module and did hierarchical clustering (Figure 2.6). The color intensities in Figure 2.6 correspond to the levels of correlations (increasing correlations are represented from yellow to red). It is evident from Figure 2.6 that expression correlations within GAL genes and HXT genes are much higher than the correlations between the two groups. This explains the separation of these two gene clusters in the traditional clustering dendrogram (Figures 2.4 and 2.5). Among the HXT gene clusters, HXT3, HXT6, and HXT7 are highly correlated (red–dark zone in Figure 2.6). It explains the actual separation of these three genes from the remaining HXT genes shown in the clustering dendrogram (Figure 2.4). Figures 2.4 to 2.6 show

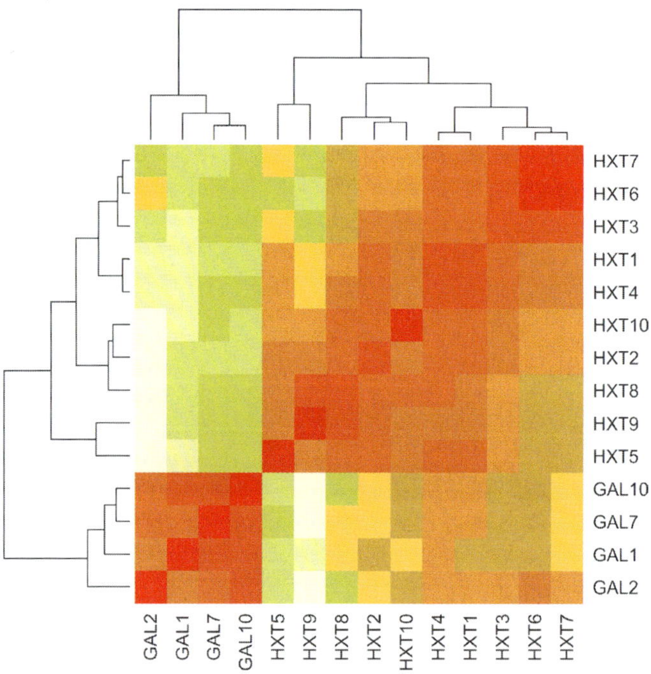

Figure 2.6 Correlation matrix of 14 structural genes with clustering dendrogram. Yellow to red corresponds to correlations from low to high.

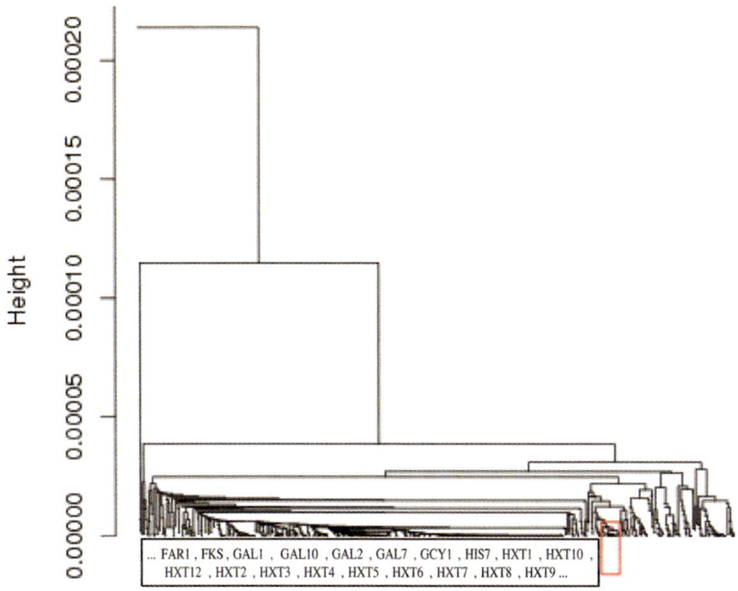

Figure 2.7 Network constrained clustering: agglomerative hierarchical clustering using network constrained distance matrix calculated from relevance network (Equation 2.5).

that traditional clustering methods failed to group functionally related genes with dissimilar expression profiles (low correlations) into one cluster.

Figure 2.7 presents network constrained clustering of 772 genes selected by the GCC extraction. All 14 structural genes that failed to cluster together by the traditional approach (Figure 2.4) are grouped into a single cluster by the network constrained clustering approach. As has been shown, the GCC selection process contributes only moderately to the apparent success. The employment of the network constrained distance matrix may play an essential role in the significant performance improvement.

2.5.2
Retinal Gene Expression Data

The aim of gene expression analysis in mouse photoreceptors is to study retinal development and to discover unknown genes involved in this developmental pathway. The retina data represent a total of 45 101 probe set expression profiles over five time points, measured in both wild-type mice and Nrl knockout line [32], in which the Maf family transcription factor, key regulator of photoreceptor differentiation in mammals [33], is inactivated. The data are deposited in NCBI Gene Expression Ominibus (GEO) with accession number GSE4051.

The data were preprocessed using "rma" method [34] and subjected to an initial screening using a two-stage screening method proposed by Hero *et al.* [22]. The top

1000 genes ranked by FDR and fold change are kept for further analysis. We constructed a coexpression network similar to the yeast analysis (FDR $\leq$ 5% and MAS = 0.6). A GCC of size 790 genes was extracted (see Section 2.4). These 790 genes were used in NC clustering according to Equations 2.15 and 2.16 and in all 1000 genes were used in traditional clustering according to Equation 2.15 only.

We used Gene Ontology (GO) [35] annotation as the objective criterion to compare the two clustering approaches. GO is a set of standard hierarchical vocabularies to describe the *biological process, molecular function,* and *cellular component* of genes. It is conveniently represented as a graph where nodes represent standard vocabularies and edges represent the relationship (either "is a" or "part of") between vocabularies. A child node is the more specific vocabulary than its parent node(s). A list of probe sets from either NC clustering or TD clustering can be mapped to a GO graph (e.g., *biological process* graph), while the appearance counts of all nodes of the GO graph as well as their *p*-values of chi-square statistics can be calculated. The most significant node(s) (corresponding to the smallest *p*-value(s) usually describe(s) the biological functions of the probe set list. Specifically, all genes with GO annotation "visual perception (GO:0007601)" are expected to belong to photoreceptor differentiation pathway.

We thoroughly compared the two clustering results with respect to three criteria (appearance counts; separation; and *p*-values of the GO category: visual perception) at each cluster number ranging from 1 to 20. The first two criteria measure stability of the "visual perception" cluster as a function of cluster numbers, and the third criterion measures the enrichment of the interested GO vocabulary as a function of cluster numbers. We end our analyses at cluster number 20 because it results in many tiny clusters of size 4 or less, indicating that further division is unnecessary.

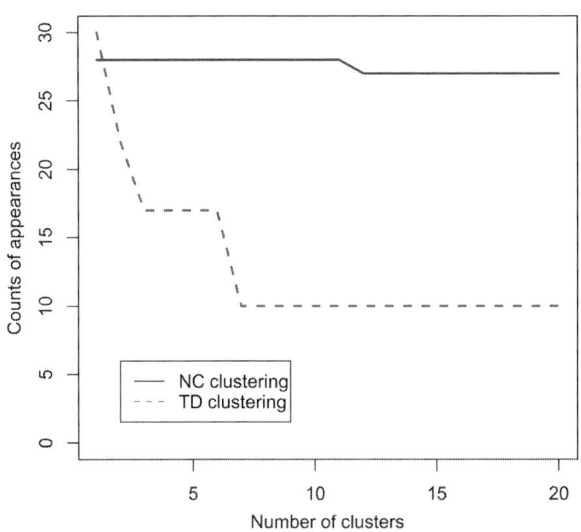

Figure 2.8 Clustering comparison – GO vocabulary counts.

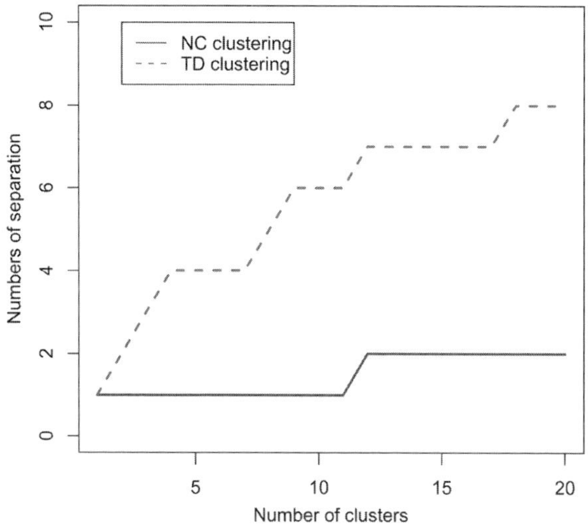

Figure 2.9 Clustering comparison – GO vocabulary separation.

Figures 2.8 and 2.9 demonstrate that the "visual perception" cluster acquired by NC clustering is quite stable in terms of counts and separation over different cluster numbers, but not the cluster that is acquired by TD clustering. Figure 2.10 demonstrates that the interested GO vocabulary "visual perception" is enriched much more in the NC cluster than in the TD cluster. In Figure 2.8, the initial (cluster number = 1) count difference (28 versus 30) is due to the GCC selection.

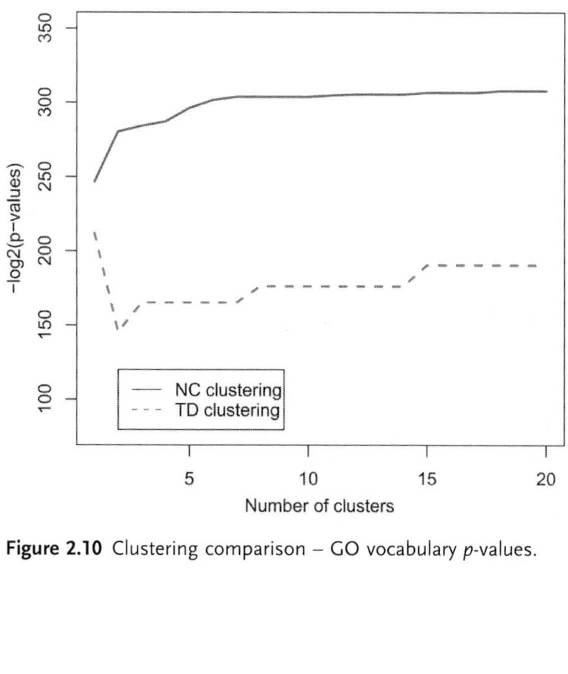

Figure 2.10 Clustering comparison – GO vocabulary *p*-values.

2.5.3 Mouse Segmentation Clock Data

In our third example, we compared the impact of the clustering method on the error rate of a list of experimentally validated cyclic genes regulated by the segmentation clock, a molecular oscillator acting during somitogenesis [36]. A set of periodic processes linked to the formation of the vertebrate embryo segments (somites), the segmentation clock gives rise to the segments in the adult body plan of a vertebrate animal. The segmentation clock drives the periodic expression of a number of "cyclic" genes. The time series consists of 17 time points covering one cycle of the mouse segmentation clock and follows the expression profiles of 22 690 probe sets. Microrarray data are available at ArrayExpress at www.ebi.ac.uk/arrayexpress/ under accession number E-TABM-163. Recent large-scale discovery of periodic genes suggests that there are at least 30 such genes and a significant portion belongs to three signaling pathways: Notch, FGF/MAPK, and Wnt [36]. Characterization of individual transcripts by RT-PCR and whole mount *in situ* hybridization confirmed cyclic behavior of 27 genes in that list. When 27 expression profiles were ordered by time at which gene expression reaches its maximum, two main clusters of genes cycling in the opposite phases were found: one including genes from the Notch and FGF/MAPK pathways, and the other including genes from the Wnt pathway. Here, we compared NC clustering methods using marginal and partial correlation distance measures with regular hierarchical clustering methods using correlation and Euclidean distance measures.

The raw data were normalized using gcrma method and the extent of periodicity was determined as previously described [36]. The clustering solutions for the top-scoring ("most periodic") 687 probe sets were studied; this was the smallest set of probes, which included all 27 validated probe sets. In the NC clustering, since the *p*-values from marginal correlation and partial correlation are not directly comparable, we constructed networks of the same number of edges; that is, we kept 0.5% of edges in the network. In each round of clustering, we defined the cluster including most the validated genes as cyclic cluster, and the clustering error rate was calculated as the number of genes falling out of the cyclic cluster divided by the total number of genes in the cluster. As the correlation-based distance measure takes absolute values into account, all 27 expression profiles cycling in the same or the opposite phase are expected to cluster together. Figure 2.11 shows the clustering error rates over different cluster numbers for each method. Overall, the NC clustering (red and navy blue) achieves a lower clustering error rate than its competitors (yellow and navy blue). In particular, the clustering error rate of the NC clustering using marginal correlation distance measure achieved a low clustering error rate throughout a wide range of cluster numbers (red). Since the true cluster number will often fall in the range between 3 and 10, we conclude that the NC clustering outperforms the TD clustering methods for clustering cyclic genes using this data set. Note that the NC clustering behaves very much like it did in the previous example; that is, the stability of cyclic cluster in terms of clustering error rate is quite robust as the number of clusters increases.

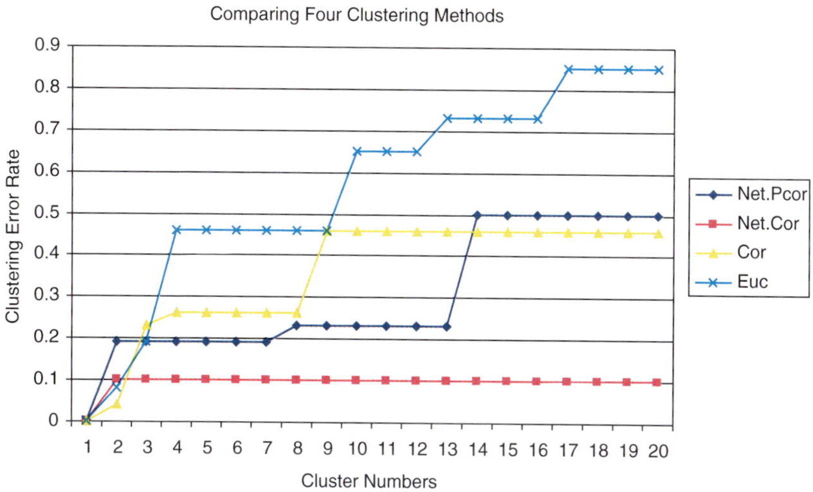

Figure 2.11 Comparing the error rate of four clustering methods. The error rate is defined in terms of 27-gene validation set. "Net.Cor" and "Net.Pcor" stand for the network constrained clustering using marginal correlation and partial correlation, respectively; "Cor" and "Euc" stand for the regular clustering using marginal correlation and Euclidean distance, respectively.

2.6 Discussion and Conclusion

In this chapter, we presented a new method of hierarchical clustering that we call network constrained clustering. We applied this method to analyze gene expression, though it may also be suitable for the analysis of other multidimensional numeric data sets. Data analysis examples demonstrate that the method aggregates functionally similar genes into a single cluster, expanding the definition of the coexpression of functionally related genes.

Genes, their products, and many other molecules in living organisms exist in a complicated network of tight regulation and interaction. There is a considerable interest among scientists in inferring the gene network from high-throughput data, which holds the key to the understanding of the complexity of biological systems. Under the framework of a statistical hypothesis test, the null network topology model may be fully connected, meaning that all pairs of genes have direct relationships; for example, coregulation, interaction, chemical modification, and so on. However, the null network model does neither reflect the biological reality nor conform to the rules of parsimony in life. In the real world, many biological networks are found to be only partially connected and very sparse. For example, in metabolic networks of the selected single cell organisms, in which edges are biochemical reactions and nodes are metabolites, the "concentration" (defined as the ratio of the total number of network edges over the maximal allowable number of edges) of the edges is estimated to be less than 1% [37].

Many current data analysis schemes implicitly assume an unconstrained model of the cellular networks. Recently, however, network constraints have been imposed to identify differentially expressed genes. For example, Morrison *et al.* adjusted the gene rank obtained from regular statistical tests using the network structure inferred from gene annotations (gene ontology) or expression profile correlations [38]. Thus, the original gene rank was altered by the corresponding network connectivity that can be treated as a network constraint. This approach reveals additional functionally important genes with weak differential expression. We define the single gene approach as "network constrained screening of differentially expressed genes." However, there are relatively few studies on imposing multigene network constraints to analyze high-throughput data analysis. In this chapter, we proposed a generalized multigene network constraint using clustering and signaling pathway reconstruction. Our method shows some promise in the future research on network constrained high-throughput data analysis.

References

1 Chipman, H. and Tibshirani, R. (2006) Hybrid hierarchical clustering with applications to microarray data. *Biostatistics*, **7**, 286–301.

2 Zhou, X.J., Kao, M. and Wong, W.H. (2002) Transitive functional annotation by shortest path analysis of gene expression data. *Proceedings of the National Academy of Sciences of the United States of America*, **99**, 12783–12788.

3 Speed, T. (ed.) (2003) *Statistical Analysis of Gene Expression Microarray Data*, Chapman & Hall/CRC Press, Boca Raton, FL.

4 Zhu, D., Hero, A.O., Qin, Z.S. and Swaroop, A. (2005) High throughput screening of co-expressed gene pairs with controlled false discovery rate (FDR) and minimum acceptable strength (MAS). *Journal of Computational Biology*, **12** (7), 1029–1045.

5 Li, W.T. (1990) Mutual information functions versus correlation functions. *Journal of Statistical Physics*, **60** (5–6), 823–837.

6 Shumway, R.H. and Stoffer, D.S. (2000) *Time Series Analysis and Its Applications*, Springer-Verlag, New York.

7 Butte, A.J., Bao, L., Reis, B.Y., Watkins, T.W. and Kohane, I.S. (2001) Comparing the similarity of time series gene expression using signal processing metrics. *Journal of Biomedical Informatics*, **34**, 396–405.

8 Zhu, D., Hero, A.O., Cheng, H., Khanna, R. and Swaroop, A. (2005) Network constrained clustering for gene microarray data. *Bioinformatics*, **21** (21), 4014–4020.

9 Newman, M.E.J. (2006) Modularity and community structure in networks. *Proceedings of the National Academy of Sciences of the United States of America*, **103**, 8577–8582.

10 Akutsu, T., Miyano, S. and Kuhara, S. (1999) Identification of genetic networks from a small network of gene expression patterns under the Boolean network model. *Pacific Symposium on Biocomputing*, **4**, 17–28.

11 Shmulevich, I. and Zhang, W. (2002) Binary analysis and optimization based normalization of gene expression data. *Bioinformatics*, **18** (4), 555–565.

12 Shmulevich, I., Dougherty, R., Kim, S. and Zhang, W. (2002) Probabilistic

Boolean networks: a rule-based uncertainty model for gene regulatory networks. *Bioinformatics*, **18** (2), 261–274.

13 Friedman, N., Linial, M., Nachman, I. and Pe'er, D. (2000) Using Bayesian networks to analyze expression data. *Journal of Computational Biology*, **7** (3–4), 601–620.

14 Rabbat, M.G., Figueiredo, A.T. and Nowak, R.D. (2006) Network inference from co-occurrences. University of Wiscosin–Madison *Technical Report* ECE-06-2, April.

15 Zhu, D., Rabbat, M., Hero, A.O., Nowak, R. and Figerado, M. (2006) *De novo* reconstructing signaling pathways from multiple data sources in*New Research in Signaling Transduction*, Nova Publisher, Chapter 7.

16 Kauffman, S.A. (1993) *The Origins of Order: Self-Organization and Selection in Evolution*, Oxford University Press, New York.

17 Huang, S. (1999) Gene expression profiling, genetic networks, and cellular states: an integrating concept for tumorigenesis and drug discovery. *Journal of Molecular Medicine*, **77**, 469–480.

18 Schäfer, J. and Strimmer, K. (2005) An empirical Bayes approach to inferring large-scale gene association networks. *Bioinformatics*, **21**, 754–764.

19 Bickel, P.J. and Doksum, K.A. (2000) *Mathematical Statistics: Basic Ideas and Selected Topics*, 2nd edition,Prentice Hall, Upper Saddle River, NJ.

20 Hollander, A. and Wolfe, D. (1999) *Nonparametric Statistical Methods*, Wiley-Interscience, Hoboken, NJ.

21 Reiner, A., Yekutieli, D. and Benjamini, Y. (2003) Identifying differentially expressed genes using false discovery rate controlling procedures. *Bioinformatics*, **19**, 386–375.

22 Hero, A.O., Fleury, G., Mears, A. and Swaroop, A. (2004) Multicriteria gene screening for analysis of differential expression with DNA microarrays. *EURASIP Journal on Applied Signal Processing*, **1**, 43–52.

23 Benjamini, Y. and Hochberg, Y. (1995) Controlling the false discovery rate – a practical and powerful approach to multiple testing. *Journal of the Royal Statistical Society Series B – Methodological*, **57**, 289–300.

24 Benjamini, Y. and Yekutieli, D. (2004) False discovery rate adjusted multiple confidence intervals for selected parameters. *Journal of the American Statistical Association*, **100**, 71–80.

25 Zhu, D., Li, H. and Hero, A.O. (2007) Reconstructing condition-specific signal transduction hierarchy using Bayesian networks.Proceedings of RECOMB'07, San Francisco, CA.

26 Ma, H.W. and Zeng, A.P. (2003) The connectivity structure giant strong component and centrality of metabolic networks. *Bioinformatics*, **19**, 1423–1430.

27 Cormem, T.H., Leiserson, C.E. and Rivest, R.L. (1990) *Introduction to Algorithm*, MIT Press,Cambridge.

28 Eisen, M., Spellman, P., Brown, P.O. and Botstein, D. (1998) Cluster analysis and display of genome-wide expression patterns. *Proceedings of the National Academy of Sciences of the United States of America*, **95**, 14863–14868.

29 Hartigan, J.A. and Wong, M.A. (1979) A *k*-means clustering algorithm. *Applied Statistics*, **28**, 100–108.

30 Ideker, T., Thorsson, V., Siegel, A.F. and Hood, L.E. (2000) Testing for differentially-expressed genes by maximum-likelihood analysis of microarray data. *Journal of Computational Biology*, **7**, 805–817.

31 Wieczorke, R., Krampe, S., Weierstall, T., Freidel, K., Hollenberg, C.P. and Boles, E. (1999) Concurrent knock-out of at least 20 transporter genes is required to block uptake of hexoses in *Saccharomyces cerevisiae*. *FEBS Letters*, **464**, 123–128.

32 Akimoto, M., Cheng, H., Zhu, D., Brzezinski, J.A., Khanna, R., Filippova, E., Oh, C.T.E., Jing, Y., Linares, J.L.,

Zareparsi, S., Mears, A., Hero, A.O., Glaser, T. and Swaroop, A. (2006) Targeting of GFP to newborn rods by Nrl promoter and temporal expression profiling of flow-sorted photoreceptors. *Proceedings of the National Academy of Sciences of the United States of America*, **103** (10), 3890–3895.

33 Swaroop, A., Xu, J.Z., Pawar, H., Jackson, A., Skolnick, C. and Agarwal, N. (2002) A conserved retina-specific gene encodes a basic motif/leucine zipper domain. *Proceedings of the National Academy of Sciences of the United States of America*, **89**, 266–270.

34 Bolstad, B.M., Irizarry, R.A., Astrand, M. and Speed, T.P. (2003) A comparison of normalization methods for high density oligonucleotide array data based on bias and variance. *Bioinformatics*, **19**, 185–193.

35 Ashburner, M., Ball, C.A., Blake, J.A., Botstein, D., Butler, H., Cherry, J.M., Davis, A.P., Dolinski, K., Dwight, S.S., Eppig, J.T., Harris, M.A., Hill, D.P., Issel-Tarver, L., Kasarskis, A., Lewis, S., Matese, J.C., Richardson, J.E., Ringwald, M., Rubin, G.M. and Sherlock, G. (2000) Gene ontology: tool for the unification of biology. The Gene Ontology Consortium. *Nature Genetics*, **25**, 25–29.

36 DequÕant, M.L., Glynn, E., Gaudenz, K., Wahl, M., Chen, J., Mushegian, A. and PourquiÕ, O. (2006) A complex oscillating network of signaling genes underlies the mouse segmentation clock. *Science*, **314** (5805), 1595–1598.

37 Zhu, D. and Qin, Z.S. (2005) Structural comparison of metabolic networks in selected single cell organisms. *BMC Bioinformatics*, **6**, 8.

38 Morrison, J.L., Breitling, R., Higham, D.J. and Gilbert, D.R. (2005) GeneRank: using search engine technology for the analysis of microarray experiments. *BMC Bioinformatics*, **6**, 233.

3
Finding Verified Edges in Genetic/Gene Networks: Bilayer Verification for Network Recovery in the Presence of Hidden Confounders

Jason E. Aten

3.1
Introduction: Gene and Genetic Networks

Complex diseases such as metabolic syndrome, atherosclerosis, obesity, and diabetes in which environmental and clinical covariates contribute significantly along with genetic susceptibility factors pose special challenges to biology and genetics. First, modern quantitative trait loci (QTL) mapping strategies in humans, mice, and rats are limited in resolution simply due to the infrequency of crossover events. Thus, despite technical advances in Single Nucleotide Polymorphism (SNP) typing such as the Illumina [1] and ParAlelle [2] technologies that leave little doubt as to the exact crossover points at which recombination did occur, the fact remains that we often cannot fully resolve QTLs to within fewer than 100 candidate genes. This is a significant problem for the establishment of a functional genome, as well as discovering the specific roles of genes in complex diseases. Second, due to technical or ethical constraints, we are often unable to experimentally control clinical and environmental factors, and thus must rely on observational and longitudinal studies alone. Fisher's idea of randomized experiments and ANOVA data analysis cannot be ethically or practically realized in the context of complex disease.

Hence, although we have the list of genomic parts, for the vast majority of genes, we lack their complete functional role in the biological blueprint. For many issues of serious and pressing medical importance, we cannot undertake precise experimental studies. We therefore ask how much our analytical, algorithmic, and probabilistic modeling can realize from observational data and semiexperimental genomic technologies that have emerged in recent years. These technologies include mRNA microarrays, antibody protein arrays, and comparative full genome analysis.

We would like to utilize Bayesian network learning algorithms [3–8] to help decode the correlated genes in microarray data and to integrate genetic marker data with microarray measures. However, such learning algorithms, when applied to microarray data sets, face a significant missing data problem that greatly

Analysis of Microarray Data: A Network-Based Approach. Edited by F. Emmert-Streib and M. Dehmer

ISBN: 978-3-527-31822-3

confounds the recovery of accurate genetic-gene and gene–gene networks. This is because actual protein levels, the active entity of most genes, are typically unknown. These unknowns may actively confound the correct derivation of a gene–gene interaction networks based on microarray data.

In an important synthesis of classical and new approaches, Schadt and coworkers [9] brought the power of classical genetic quantitative trait loci analysis to bear on gene selection from modern microarray data, producing expression-quantitative trait loci (eQTL). Although successful at selecting genes for further experimental investigation, this work is limited to a comparison of a handful of three-node models and does not learn gene–gene interactions.

As we further examined the consequences of using genetic data to help inform the recovery of larger gene–gene interaction networks from microarray data, we encountered the issue of confounding such interactions by unknown variables, in particular protein levels. More generally, this analysis applies to hidden variables in any Bayesian network. In our intended application, the network describes the genetic marker–gene and gene–gene influence network within a tissue.

We describe here a new verification process by which confounding influence can, in recognizable cases, be ruled out when the observations flow from a faithful probability distribution. Such edges in the Bayesian network are verified edges, and their discovery is based on two interconnected layers of v-structures. Verified edges in a Bayesian network are associations that, given minimal distortion due to sampling variation and given a faithful generating process (see theory definitions and results), cannot be associations only due to hidden confounders.

Although common biological feedback mechanisms would initially suggest undirected or cyclic graphs, the restriction of learned networks to directed acyclic graphs (DAGs) is useful in that DAGs suggest priorities in terms of causal order and more readily suggest intervention points that may be useful in gene or siRNA therapy.

The goal of this work is to learn a model of the underlying genetic network sufficient enough to predict the best intervention point of an siRNA-based intervention. The criterion for judging best is the gene target that would specifically alter the clinical trait of interest, while minimally affecting the rest of the network.

What distinguishes this work uniquely is the ability to efficiently handle large-scale microarray-sized data sets, learn directional Bayesian networks even in the presence of confounders, and learn structure consistent with any DAG-consistent causal theory. We offer the ability to readily impose prior source/sink structure on the network and to integrate genetic marker and experimental and observational data into the learning algorithm.

This chapter is organized as follows. In Section 3.2, we review the essential terminology, definitions, and theory of learning Bayesian networks, and discuss influential prior work that illuminates our contribution. In Section 3.3, we derive a process of learning Bayesian networks by recursively finding two levels of v-structures in the data, a procedure we term bilayer verification. In Section 3.4, we discuss the details of the methods used in the application of the algorithms, and describe the microarray data that are analyzed. We report the results of simulation studies as to the false-positive rate for v-structure recovery in Section 3.5, and sketch the output of the algorithm when run on aortic lesion and microarray data. In conclusion, we offer a brief look at a further application in terms of functional role assignment for genes in the process of mediating the effect of HDL cholesterol on atherosclerosis.

3.2 Background and Prior Theory

3.2.1 Motivation

Models are needed that can integrate prior knowledge of molecular biology, microarray mRNA level measurements, genetic markers, DNA sequence information, and eventually proteomic measurements of multiple proteins in multiple circumstances.

Often, it is the discovery of a particular interaction that is more important than the relative strength of that interaction. If we know the players, we can more readily know where to look to understand their roles.

Bayesian networks marry graphs with probability theory. Graphs have been adopted for modeling because they have several distinct advantages.

1. Graphs allow clear expression of prior knowledge and assumptions, in particular, knowledge of conditional independence.
2. Graphs convey how to efficiently encode joint probability dependencies.
3. Graphs give explicit models for inferring from observations, and suggest efficient ways of doing so.
4. They provide natural means of integrating data from disparate sources.
5. Directed graphs provide a language for modeling prior causal relationships and describing the discovery of new causal relations. They have been used to put discussions of causality on a mathematical basis [3].

Although common biological feedback mechanisms might initially suggest undirected or cyclic graphs, the restriction of learned networks to directed acyclic graphs is useful in that DAGs suggest priorities, in terms of causal order, and more readily intervention points that may be useful in gene or siRNA therapy.

3.2.2 Bayesian Networks Theory

Bayesian networks [3,8,10] were invented originally with the intention of encoding data dependencies. They were conceived of as a general way of efficiently encoding a large-dimensional multivariate joint probability distribution. More recently, they have been used to specify causal models and subsequently put discussions of counterfactuals in causality on a mathematical basis [16].

Definition 1 (Bayesian network): A Bayesian network is a pair (G, θ) consisting of a directed acyclic graph G in which each node is a random variable, together with a set of parameters θ that encode for each node the conditional probability distribution of that node given that its parents are in the DAG.

The directed edges leading into a node X indicate the necessary and sufficient inputs to determine the probability distribution of Xs. Hence, the missing edges in the DAG typically encode the maximum information about the joint distribution: information about conditional independencies.

Information about independence in Bayesian networks is determined from the graph using two principles: the Markov property and d-separation (the d prefix is for directed; in contrast to the simpler property of separation in an undirected graph).

Definition 2 (the Markov property): Conditional upon knowing the state of its parents, a given node is independent of its nondescendants in the graph.

Equivalently, the Markov property is formulated as

$$P(V) = \prod_{\upsilon \in V} P(\upsilon|\text{pa}(\upsilon))$$

where V is all variables (all nodes) in the graph and $\text{pa}(\upsilon)$ is the set of parents for a given node υ. If $\text{pa}(\upsilon)$ is the empty set, then υ is termed a root or exogenous node. Nodes with parents are also called endogenous nodes ("of internal origin," meaning causes are given within the graph) in the structural equation modeling literature.

In addition to the Markov property, the d-separation criterion allows one to read more complex, induced conditional independence statements from a graph. It is also the basis for the learning of Bayesian networks. Suppose we specify three disjoint sets of nodes X, Y, and Z, where Z might be the empty set, but X and Y must have cardinality of at least one. If Z d-separates X from Y, then the graph claims that in the probability distribution that it represents, X is conditionally independent of Y given Z. This is written symbolically as $X \perp\!\!\!\perp Y|Z$, indicating $P(X, Y|Z) = P(X|Z)P(Y|Z)$. The opposite, conditional dependence, is noted by $X \not\perp\!\!\!\perp Y|Z$. This notation is attributed to Dawid [11], while Pearl [8] uses this notation and further adopts three-tuple relation notation from predicate calculus that is also commonly seen to denote conditional independence: $I(X, Z, Y)$, meaning that X is independent of Y given Z.

The definition of d-separation is a condition for all paths in the network whose legs can be taken with or against the arrows on the directed edges. While traversing each path, three types of connection patterns may be encountered when going from

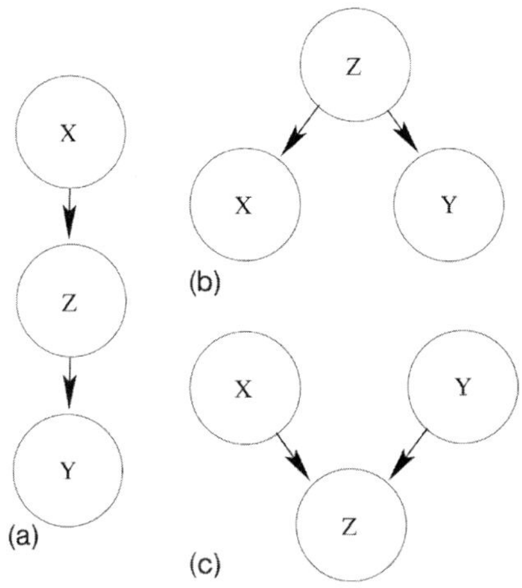

Figure 3.1 The three patterns of connection in Bayesian networks. (a) Chain or serial connection, (b) forked or diverging connection, and (c) v-structure or collider or converging connection.

a node in X to a node in Y. These are illustrated in Figure 3.1, and are called serial, diverging, and converging connections. They are also known as chain, fork, and collider connections.

Definition 3 (d-separation): In a DAG, two disjoint sets of variables X and Y are d-separated by a third set D, itself disjoint from X and Y, if and only if along every path from an X node to a Y node we find that each node Z encountered satisfies one of the following three criteria:

- Z is at a converging connection, and neither Z nor any of its descendants (along directed paths) are in D;
- Z is at a diverging connection, and $Z \in D$;
- Z is at a serial connection, and $Z \in D$.

For example, considering Figure 3.1, if $Z \in D$, then in both the serial (a) and diverging (b) connections we would have d-separation between X and Y. However, at the collider (c) we only have d-separation if $Z \notin D$ and this requirement of not being in D would apply to any of Zs descendants as well.

If the two parent nodes of a collider are not directly connected by an arrow, this structure is termed an unshielded collider or v-structure.

3.2.2.1 d-Separation at Colliders

To illustrate and provide intuition for the dependence-inducing effect of conditioning on a causal child, consider a recessive lethal gene, with lethal allele "a" and dominant allele "A." Suppose that mother and father are both heterozygous Aa, and that we can actually observe which allele was contributed from the mother and which from the father.

		Dad gives	
		A	a
Mom gives	A	AA	Aa
	a	Aa	aa

When we do not condition upon survival of the offspring, the "A" and "a" alleles are passed along in independent 50/50 proportions from both parents.

However, conditioning upon survival induces dependence in what were previously independent alleles. For if we know the offspring survived and that one parent gave "a," then there is no doubt that the other parent gave "A."

		Dad gave	
		A	a
Mom gave	A	33%	33%
	a	33%	0%

Hence, conditioning on the child induced dependence between the independent parents of that child. The d-separation rule corresponds exactly to this situation: we do not want to condition on any causal children or children's descendants if we wish to d-separate the parents.

3.2.2.2 Placing Genetic Tests Within the Bayesian Network Framework

The transmission disequilibrium test [12] is a commonly used genetic test that illustrates the power of the Bayesian network representation for analysis and creation of new inference procedures. Consider Figure 3.2 that illustrates the test as originally utilized for diabetes candidate gene testing by Spielman *et al.* [12].

The analyst specifies a Bayesian network with causal relationships hypothesized as drawn in Figure 3.2a. If the starred edge (labeled *) does not exist, then the d-separation properties of the graph indicate that it is sufficient to condition the parental genotypes to eliminate confounding by ethnic stratification. By confirming conditional independence between the Gm locus genotype and clinical diabetes, the authors of the test were utilizing the d-separation properties of the graph to infer that Figure 3.2b, which lacks the *-edge, was the true graph. The paradigm of testing edges in a learned or partially specified graph gives us a powerful methodology for discovering, discarding, and verifying candidate edges in a given or recovered network.

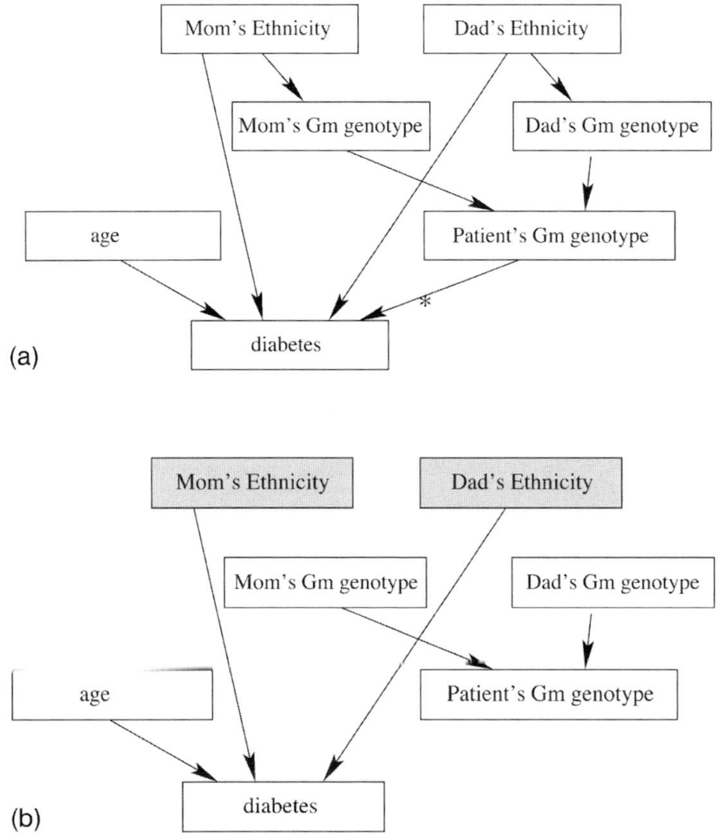

Figure 3.2 The transmission/disequilibrium test checks for d-separation (conditional independence) after conditioning on parent genotypes. (a) Testing *-edge and (b) d-separation implies that *-edge does not exist.

Definition 4 (faithful probability distribution): A probability distribution is faithful if each d-separation and d-dependence observed in the underlying causal graph is mirrored in the observed joint probability distribution.

A fundamental assumption in the structure learning algorithms, such as the PC and local causal discovery (LCD) algorithms, is that the observed data follow from a faithful probability distribution. The interested reader is referred to [3,8]. Other names for faithfulness [7] are stability [3] and DAG isomorphism ([8], p. 128). The primary difficulty being addressed by the faithfulness assumption is that the particular probability distribution being observed may have independencies that are due to an unusual coincidence of parameters rather than being a result of the structure of the edges in the graph. Peculiar parameter induced independencies will fool learning algorithms, but are expected to be uncommon.

The last essential component of the theory of Bayesian networks for interpreting the causal discovery algorithms is Verma and Pearl's [13] theorem on the equivalence of models under observation alone. Using observational data alone, Theorem 1.2.8 (from [3], p. 19) tells us that the v-structures allow us to distinguish directionality in the undirected Markov Random Field or skeleton graph.

Theorem 1 (observational equivalence [13]) Two DAGs are observationally equivalent if and only if they have the same skeletons and the same sets of v-structures, that is, two converging arrows whose tails are not connected by an arrow.

3.2.3 Learning Network Structure from Observed Conditional Independencies

To provide context for our algorithms, we survey two influential causal discovery algorithms from which we draw ideas. The PC algorithm and the LCD algorithm can handle very different sized data sets, given the time complexity of their execution. The PC algorithm will never finish on 100 variables, whereas the LCD will return a little structural information about many billions of variables, without necessarily tying any two microobservations together. Our algorithms, recursive v-structure location (RVL) and recursive v-structure verification (RVV) fall in the middle, in that they will capture more of the structure of the network than is learned by LCD, albeit less than that learned by PC, and yet will operate efficiently on large data mining problems such as those posed by microarray data sources.

3.2.4 Prior Work: The PC Algorithm

The PC algorithm of Spirtes *et al.* [7,14] learns, to the extent possible, the global Bayesian network structure from an observed data set by testing for conditional independence between various sets of nodes.

Given the results of these tests, a network pattern is constructed so that the Markov property holds and d-separation claims the resulting graph mirroring those conditional independencies found in the data. There are two phases to the PC algorithm. In the first phase, an undirected graph is learned. This is known as the skeleton of the Bayesian network. In the second phase, arrowheads are added to some of the edges where they can be inferred. The output graph may not be fully oriented and is called a pattern. When the pattern contains undirected edges, these indicate that the data are consistent with models in which either orientation is possible [14].

3.2.4.1 PC Algorithm

1. Begin with the complete undirected graph on V. That is, create a fully connected skeleton model in which each node is connected to every other node by an undirected edge.

2. Track the cardinality of the conditioning sets in k. Initialize $k \leftarrow 0$. For each pair of nodes X and Y, instantiate a set DSep(X, Y) and initialize it to the empty set $\{\phi\}$.

3. Chisel the skeleton: For every adjacent single nodes X and Y, remove the arc between them if for all subsets Z of size k containing nodes adjacent to X but not containing Y, we find that $X \perp\!\!\!\perp Y \mid Z$. When $k=0$ this is simply a marginal independence test. Add the nodes in Z to DSep(X, Y).

4. If any arcs were deleted, increment K and repeat the previous step. This results in X and Y being connected iff $X \perp\!\!\!\perp Y \mid \{$ every subset of $V\{X, Y\}\}$.

5. Orientation phase: Collider induction. For each triple of 3 individual nodes $X \rightarrow Y \leftarrow Z$ such that X and Y are connected, Y and Z are connected, and yet X and Z are not connected, orient the chain as $X \rightarrow Y \leftarrow Z$ iff $Y \notin$ DSep(X, Z).

6. Orientation phase: Consequences of collider induction. Iterate through all undirected arcs $Y - Z$ in the graph and:
   (a) If $X \rightarrow Y$, $Y - Z$, and X is not adjacent to Z, then orient $Y \rightarrow Z$.
   (b) If we have $Y \rightarrow Z$ and there is already a directed path from Y to Z, then orient $Y \rightarrow Z$. Otherwise a directed cycle would be introduced, violating the DAG property. Depth-first search along the directed edges will determine if a violating path would be created.
   (c) Continue at step 6a cycling through all undirected $Y \rightarrow Z$ arcs until one pass fails to direct any arcs.

The major drawback of the PC algorithm is that its execution time is exponential in the number of variables observed. Two observations will allow us to modify the PC algorithm for speed and applicability to genetic data. First, we may be solely interested in determining the impact of various sources of genetic variation on one specific sink variable. Second, despite a large number of correlated variables, we may specifically be interested only in those that we can determine fall into the pattern of direct causal influences of our sink variable. Furthermore, among those direct causes, those that appear to be under genetic control are the most interesting. We therefore proceed to develop an algorithm based on these observations that significantly speeds up the causal analysis, at the expense of missing some parts of the complete network in return for focusing on determining most relevant candidate genes for further experimental investigation. Our formulation extends a similar local structure discovery algorithm by Cooper, which we briefly review in the following section.

3.2.5
Prior Work: The Local Causal Discovery Algorithm

Cooper [5] proposed a polynomial time constraint based local causal discovery algorithm for discovering relationships in large data sets, based on his observation that *information about statistical independence and dependence relationships among a*

Table 3.1 Analysis of the depicted graphs with respect to two specific independence properties.

Graph	$X1 \perp\!\!\!\perp X4 \mid X2$	$X1 \perp\!\!\!\perp X4$
A1	—	—
A2	—	Yes
B	Yes	—
C	Yes	—
D	—	Yes
E	—	—
F	Yes	—
G	—	—
H	—	—
I	—	—
J	—	—

set of variables can be used to constrain (sometimes significantly) the possible causal relationships among a subset of the variables [5,15]. As Silverstein and coworkers write.

A simple example of such a constraint is that if attributes A and B are independent, then it is clear that there is no causal relationship between them. It has been shown that, under some reasonable set of assumptions about the data (to be discussed later), a whole array of valid constraints can be derived on the causal relationships between the variables.

The LCD algorithm proceeds as follows. This formulation is given by Silverstein and coworkers [15]; however, it corrects an error in their conclusion (x and y were interchanged at line 7 in their Table 3.1).

3.2.5.1 LCD Algorithm

```
1. Input: A set V of variables. w is a given variable know to have
   no causes among the other variables. A data set D. Tests for
   dependence, independence, and conditional dependence. (D,
   I, and CI respectively).
2. Output: A list of causal relationships supported by the data
3. For all variables x ≠ w
4. If D(x, y)
5. For all variables y ∉ { x, w}
6. If D(x, y) and D(y, w) and x ⊥⊥ w | y
7. output 'y causes x'
```

The structure being tested for at line 6 is for the chain or serial connection. Directionality of information flow in chains can be distinguished when one can make the assumption that there exists w, a variable with no prior causes in the network. Lacking such a variable w, the other means of inducing directionality is by locating the unshielded collider or v-structure. This forms the basis of the directional arrow assignment in all causal discovery algorithms without prior knowledge of exogenous (no parents in the graph) variables. Both chain location (termed CCC causality by Silverstein) and v-structure location (termed CCU causality by Silverstein) [15] can be used in genetic network analysis, given that we can assume that the DNA changes inherited since conception preceded the transcription of their changing genes. Although maternal environment effects may be correlated with the mothers DNA in ways that are not present in the child's DNA, we neglect such situations, not because they are not expected, but simply because we lack data to make such distinctions at this point.

3.3 New Theory

3.3.1 Novel Algorithm: The RVL Algorithm for Learning DAGs Efficiently

Here we describe an algorithm that recursively learns directional arrows without first learning the complete undirected skeleton. We outline an algorithm for building the network from the bottom up, starting with the desired outcome or response variable as the sink node in the network.

Other prior work in which a sink node is chosen and structural equation models are learned includes the FBD and FTC algorithms [6]. The RVL algorithm below uses a more stringent link addition criteria than FBD and FTC do, and learns two parents at once, rather than the single parent addition at a time. Both FBD and FTC are susceptible to incorrectly inferring the directionality of the links, even using the strength of the multiple regression coefficients R^2, as FTC does to try and select more direct links. However, the intuition behind the FTC algorithm is useful. FTC suggests using the strength of R^2 as an indicator of how direct an influence one variable is having on another: "variables with high R^2 are less likely to have latent influences, so they are preferred as dependent variables" [6]. Although not specifically reflected in the algorithm below, it would be a simple variant to prioritize successive sink nodes by drawing them from a priority queue Q in terms of the maximal R^2 available.

3.3.1.1 Algorithm: Recursive v-Structure Location

1. Choose a sink node Z. In our case a downstream clinical trait such as aortic lesions will be used to demonstrate.
2. Compute all pairwise marginal dependencies and note the degree of each node. That is, note how many pairwise marginal dependencies involve each node (variable). For instance count the correlations of Z with any other variable that has a correlation coefficient p-value less than some threshold of significance.

3. Considering the set U of neighbors of Z, pare down the set U by eliminating indirect relationships as follows. For distinct nodes $u_i \in U$ and $u_j \in U$ if $u_i \perp\!\!\!\perp Z \mid u_j$, then u_i is at best an indirect cause of Z, and so eliminate ui from U for the purposes of v-structure detection.

4. Now detect upstream causes by applying the unshielded collider (v-structure) test. For each distinct pair (X, Y), where both X and Y are drawn from the chiseled down set U, test for both $X \perp\!\!\!\perp Y \mid Z$ and for $X \perp\!\!\!\perp Y$ marginally. If both tests hold true, then add both X and Y as parents of Z. If either X or Y has not yet been considered as a sink, add it to a queue Q of sinks to check. Repeat step 4 until all distinct pairs (X, Y) have been checked for v-structure.

5. Recursively select a new sink node Z from the queue Q and start again at step 3.

The algorithm terminates when all sinks have been considered in step 5. Alternatively, it is easy to add additional stopping criteria that halt the algorithm after a particular depth of network has been learned.

RVL does not detect all causes, but we do proceed upstream as long as we can locate natural controls in the form of v-structures. Moreover, this algorithm is applicable to much larger data sets in which an outcome of interest, such as a clinical degree of disease, is the effect to which we wish to find direct causes. However, our algorithm is applicable to much larger sets, and may be compared with the LCD algorithm and adaptations [5,15].

3.3.2
Novel Theory: Bilayer Verification and the RVV Algorithm for Verifying Graphs in the Presence of Unobserved Confounders

We now formulate a theorem that will allow us to both verify direct links and detect hidden confounding influences in the learned graph. Its name derives from the fact that by examining two layers of learned network at once, we can verify the edges and check for confounding at the earlier learned edge. Prior work and inspiration includes Pearl's definitions of Potential and Genuine Cause ([3], pp. 54–56), Cooper's analysis [5] of the "+++" pattern, and the FTC/FBD algorithms [6]. Once understood, this theorem suggests a natural extension to the RVL algorithm if two layers of v-structure have been learned.

Theorem 2 (Bilayer verification). Let five observed random variables $X1$, $X2$, $X3$, $X4$, and $X5$ be drawn from a faithful probability distribution, so that independence between variables is entirely due to the underlying influence structure and not a rare cancellation of parameter values and observed frequencies. Suppose that the following pattern of dependence, characteristic of two interlocked v-structures (collider patterns), is observed.

First v-structure:

$$X1 \not\perp\!\!\!\perp X2, \tag{3.1}$$

$$X1 \not\perp\!\!\!\perp X3, \tag{3.2}$$

$$X2 \perp\!\!\!\perp X3, \tag{3.3}$$

$$X2 \not\perp\!\!\!\perp X3 \mid X1. \tag{3.4}$$

Second v-structure:

$$X2 \not\perp\!\!\!\perp X4, \tag{3.5}$$

$$X2 \not\perp\!\!\!\perp X5, \tag{3.6}$$

$$X4 \perp\!\!\!\perp X5, \tag{3.7}$$

$$X4 \not\perp\!\!\!\perp X5 \mid X2. \tag{3.8}$$

In the absence of confounding, such dependencies would be the evidence for Figure 3.3 graph C to be the underlying causal structure [3,7]. However, even in the presence of confounding, the **-edge from $X2$ to $X1$ can be either verified, ambiguous (possibly confounded), or dismissed as surely confounded according to the following rules for edge deletion and verification:

1. If ($X1 \not\perp\!\!\!\perp X4|X2$) and ($X4 \perp\!\!\!\perp X1$), then delete the **-edge $X2 \rightarrow X1$ as it was only due to confounding.
2. If ($X1 \perp\!\!\!\perp X4|X2$) and ($X4 \not\perp\!\!\!\perp X1$), then mark **-edge $X2 \rightarrow X1$ as verified (not confounded).

Since $X2 \not\perp\!\!\!\perp X1$, there exists a marginally d-connected path between $X2$ and $X1$. Similarly, $X3 \not\perp\!\!\!\perp X1$ shows that there exists a marginally d-connected path between $X3$ and $X1$. Because $X2 \perp\!\!\!\perp X3$ and $X2 \not\perp\!\!\!\perp X3|X1$ are together characteristic only of a collider at $X1$, we know that all paths from $X2$ to $X3$ that account for the conditional dependence $X2 \not\perp\!\!\!\perp X3|X1$ must transit $X1$ via two distinct directional edges both heading into $X1$ from either side. Hence, any final edge from $X2$ to $X1$ such as ** in the figures must be oriented with the arrowhead into $X1$.

Having used $X3$ and the properties of d-separation at a v-structure to establish the directionality of the final **-edge in any paths from $X2$ to $X1$, it now suffices to analyze just the left side of the v-structure in terms of confounding. The right-side paths (from $X3$ to $X1$) may have additional confounders, but they will not alter the conclusion about the directionality of the **-edge. An identical analysis holds for the second layer of v-structure utilized in Equations 3.5 to 3.8, and hence confounding involving $X5$ and $X2$ is also safely ignored.

Now consider Table 3.1 in conjunction with Figures 3.3 to 3.5, which exhaust the possibilities for confounding of the three variables $X1$, $X2$, and $X4$. Circles represent latent, unobserved confounders. Squares represent observed variables.

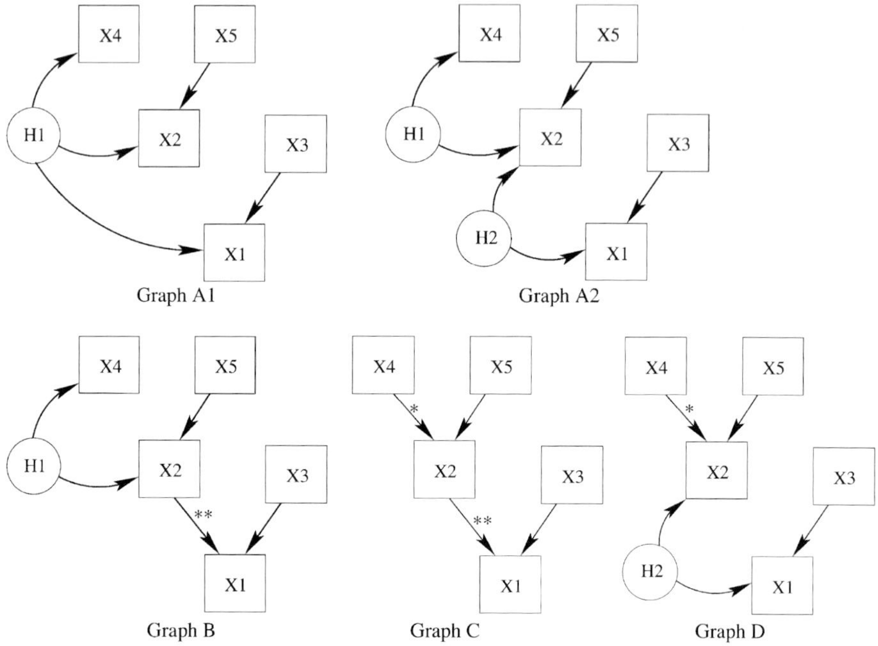

Figure 3.3 Illustrating bilayer verification, graphs A1, A2, B–D.

The table is marked yes if the independence statement holds, and "—" if it does not.

Graphs B, C, and F share the $X1 \perp\!\!\!\perp X4 \mid X2$ and $X1 \not\perp\!\!\!\perp X4$ property, and in each of these three graphs the **-edge $X4 \rightarrow X1$ is direct and unconfounded. Hence, if this pattern of independence is observed, we can confirm or mark the **-edge as verified (not confounded). Graphs A2 and D share the $X1 \not\perp\!\!\!\perp X4 \mid X2$ and $X1 \perp\!\!\!\perp X4$ pattern, and in both cases there is no true direct cause from $X2 \rightarrow X1$, instead just confounding exists. Graphs I and J share the same properties in the table as the other indeterminant cases A1, E, G, and H. Moreover, these properties hold for all permutations of the presence and absence of the *- and **-edges of graphs I and J; this conclusion follows from a simple enumeration (not shown) and checking of the four possibilities for the *- and **-edge states (present/absent).

More broadly, note that $X1 \not\perp\!\!\!\perp X4$ reveals that there is some marginally d-connected path between $X1$ and $X4$, and $X1 \perp\!\!\!\perp X4 \mid X1$ reveals that $X2$ intercepts (is present on) all such paths. More specifically, on all such paths $X2$ is not the location of a collider, but rather is a chain or fork connection. A fork guarantees that the path departs $X2$ and arrives directionally at $X1$, and any chain must be oriented to flow from $X2$ toward $X1$ by the directionality argument for **-edge that began the proof. Hence, we know that a directed path from $X2$ to $X1$ exists and can be denoted by **-edge.

Additionally, in the case when the grandparent ($X4$ above) is a genetic marker and hence is reasonably modeled as exogenous (does not have any input arrows from

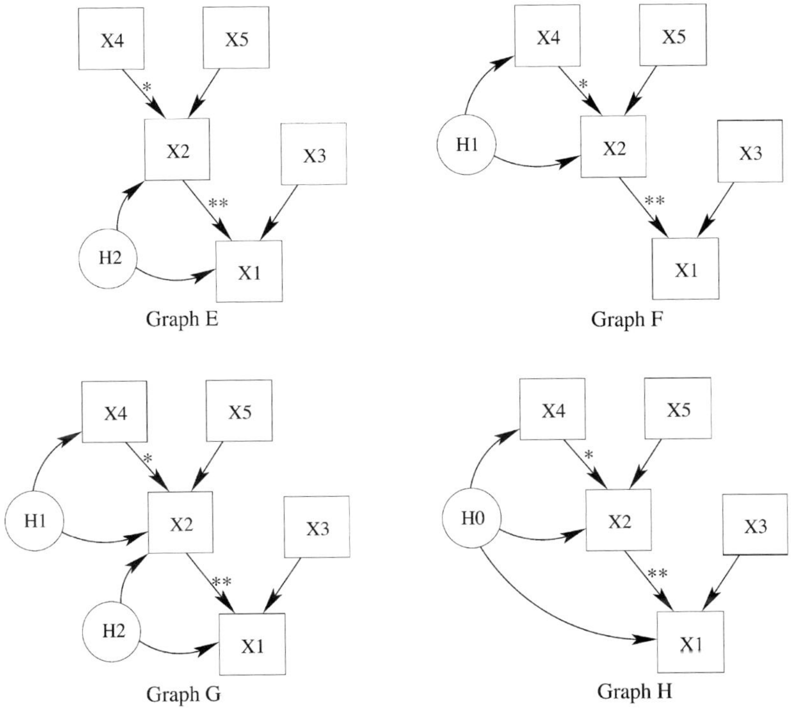

Figure 3.4 Illustrating bilayer verification, graphs E–H.

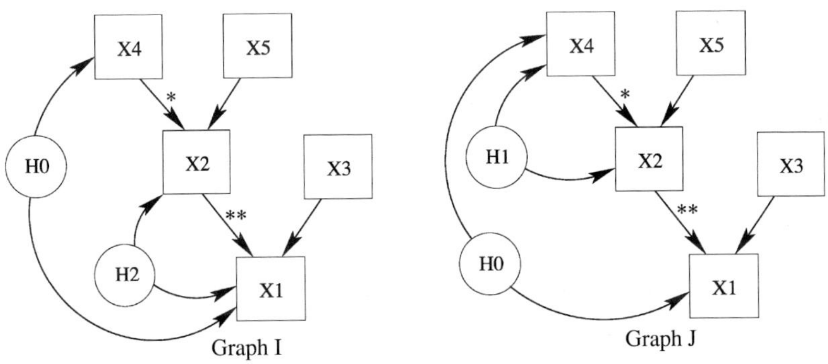

Figure 3.5 Illustrating bilayer verification, graphs I and J.

observed or confounding unobserved variables), we note that only (referring again to Figures 3.3 to 3.5 and Table 3.1 graphs C, D, and E apply, and yet the exact same conclusion is reached. Indeed, since this case is a subset of the above analysis, the conclusions of the theorem continue to hold for any parentless grandparent variables. However, we gain from the additional knowledge of no directed arrows into $X4$. Graphs C, D, and E now correspond in an invertible manner to unique patterns of independence. If we encounter the pattern associated with graphs C or E, we can immediately conclude that **-edge is verifiably present when $X1 \not\perp\!\!\!\perp X4$, even though there is also confounding present in graph E. This follows from examining graphs C, D, and E, which exhaust the possibilities when learning second layer relationships in which $X4$ has no inputs. Notice that once we conclude that $X1 \not\perp\!\!\!\perp X4$, we can immediately rule out graph D. In the remaining graphs (C and E) we are assured of the presence of the **-edge.

This reasoning results in the following corollary to the bilayer verification.

Corollay 1: Corollay parentless grandparent verifies parent–child relationship. If a variable M is known to have no parents (as with a genetic marker), then observing $M \perp\!\!\!\perp X2$ and $M \perp\!\!\!\perp X1|X2$ at a stage in learning recursive v-structures at which we have learned $X2 \rightarrow X1$ means that we can immediately mark $X2 \rightarrow X1$ as a verified edge, because the **-edge must be present even if the $X2 \rightarrow X1$ relationship is also confounded (as in graph E).

As in the LCD algorithm, the $M \perp\!\!\!\perp X1|X2$ condition must also hold here, in addition to $M \perp\!\!\!\perp X2$, so that M itself is not confounding the relationship between X_1 and X_2[1].

The same relationship analysis that holds for $X4 \rightarrow X2 \rightarrow X1$ holds by symmetry for the other grandparent learned by RVL, such as the pictured ($X5 \rightarrow X2 \rightarrow X1$) relationship. As a result of incorporating these rules into RVL, as each possible grandparent relationship is learned, we term the Recursive V-structure Verification (RVV) Algorithm.

RVL has been implemented in C++ and a sample network learned is pictured in Figure 3.6. The pruning rules from the bilayer verification theorem are implemented in RVV, which also adds in root (no parent) connections (e.g., QTL information) in the manner of the LCD algorithm.

Bringing together RVL and the bilayer theorem yields a new means of learning network structure that suggests when edges can be verified. We term the resulting learning algorithm RVV for recursive v-structures with verification.

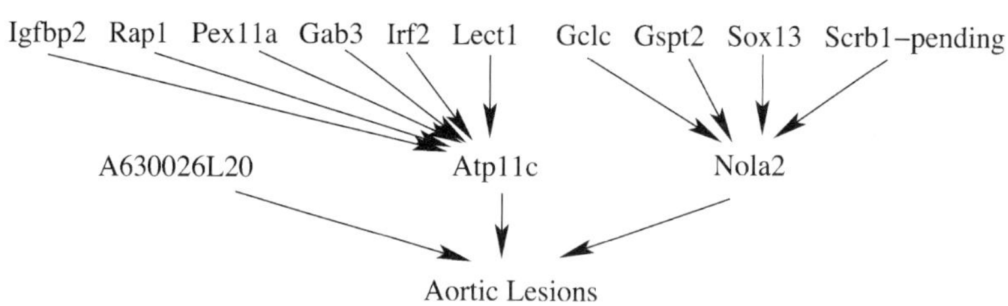

Figure 3.6 Truncated results of the RVV algorithm run on aortic lesions in the BxH-Apoe data, male mice, and liver mRNA profile.

[1] I thank Peter Langfelder for discussion that illustrated this point.

3.3.2.1 Algorithm: Recursive v-Structures with Verification

1. Choose a sink node Z.

2. Compute all pairwise marginal dependencies.

3. Considering the set U of neighbors of Z, pare down the set U by eliminating indirect relationships as follows. For distinct nodes $u_i \in U$ and $u_j \in U$, if $u_i \perp\!\!\!\perp Z \mid u_j$, then u_i is at best an indirect cause of Z, and so eliminate u_i from U for the purposes of v-structure detection. If m survives this test and u_i is a genetic marker (or a known exogenous node) then add the marker immediately as a parent of Z.

4. Now detect upstream causes by applying the unshielded collider (v-structure) test. For each distinct pair (X, Y), where both X and Y are drawn from the chiseled down set U, test for both $X \not\perp\!\!\!\perp Y \mid Z$ and for $X \perp\!\!\!\perp Y$ marginally. If both tests hold true, then add both X and Y as parents of Z. If either X or Y has not yet been considered as a sink, add it to a queue Q of sinks to check.

5. Verification: once there are two or more layers in the network, verification becomes possible. If $X \rightarrow Z^1 \rightarrow Z^0$, then test for two conditional independencies: if $X \perp\!\!\!\perp Z^0 \mid Z^1$ and $X \perp\!\!\!\perp Z^0$ then mark $Z^1 \rightarrow Z^0$ as confirmed. If $X \not\perp\!\!\!\perp Z^0 \mid Z^1$ and $X \perp\!\!\!\perp Z^0$ then delete edge $Z^1 \rightarrow Z^0$ as it was due to confounding.

6. Root connections: if exogenous (root) variables such as genetic markers M or predicted genotypes at specified loci are in the graph, for each Y learned as a parent of Z, check for $Y \not\perp\!\!\!\perp M$ (QTL existence) and $M \perp\!\!\!\perp Z \mid Y$. If both these conditions hold, then add M as a parent of Y.

7. Repeat step 4 until all distinct pairs (X, Y) have been checked for v-structure.

8. Recursively select a new sink node Z from the queue Q and start again at step 3.

The root connections test in step 6 examines whether $M \rightarrow Y \rightarrow Z$. This conditional independence test can be accomplished for the gene expression microarray data as follows, given that Y and Z are continuous gene expression values, while M, being a genotype in our context, takes on either two (in the case of SNP genetic marker data) or three (for microsatellite genetic marker data) discrete values. We simply regress Z on Y and compute the residuals $\{E(Z_i \mid Y_i)\}$ for $i = 1, \ldots, n$ after Y's influence on Z has been taken into account.

3.4 Methods

3.4.1 C3H/HeJ × C57BL/6J Microarray Data

The C3H × B6 data set used in the network recovery trials was kindly provided by collaborators and has been studied in various aspects in earlier works [9,16–21]. Additional background for the experimental methods that provided the data can be seen in the work on the network analysis of eQTL [22] and in the implication by such approaches of the Insig2 [23] and the 5-lipoxygenase genes in specific diseases [24].

The data set for these analyses consists of the F2 generation from an intercross of a C57BL/6J (APOE null) inbred strain with a C3H/HeJ (APOE null) inbred strain. C57BL/6J is susceptible to a variety of atherosclerosis, diabetes, and heart disease related traits to which C3H is resistant. Particularly on the APOE null background, the F2 mice are expected to show significant spectrum of atherosclerosis and metabolic syndrome response to being fed a high-fat Western diet until 5 months of age. A total of 334 mice were bred, genotyped with classical microsatellite and newer SNP markers at high density, and had a variety of clinical covariates measured, including fat mass determination via NMR, aorta sectioning and measurement of aortic lesions, insulin, glucose, free fatty-acid levels in the blood, and cholesterol fractions (unesterified, HDL, and LDL + VLDL) [16]. Significantly, the liver, brain, adipose tissue, and skeletal muscle tissue from each F2 were assayed via Rosetta Inpharmatics [25] spotted oligo microarray, returning some 23 000 mRNA levels for each of the four tissues. Significant differences in the gene expression profiles between the male and female F2s were observed [17].

Initial genotyping was carried out at 1392 loci across the 19 mouse autosomes and the X chromosome, with an average of 70 markers per chromosome. Additional ParAllele SNP genotyping was also done at 1822 SNPs [2].

3.4.2 Parameters and the v-Structure Test

Finite samples and noisy data leave open the possibility of false positives to any data-based learning algorithm. To evaluate the type I error rate or probability of a false-positive verification, we first describe in detail the mechanics of the procedure for locating v-structures. Specific parameters are used to control the actual implementation of the v-structure test process and subsequently to compute the probability of this test being fooled by a sampling variation. Here, we outline the relevant portion of the logic used in the test and describe the process and parameters in detail. In the subsequent results section, we will present simulation studies measuring the relative importance of these parameters in small sample sizes typically encountered.

There are several parameters that control the central v-structure test:

- η_1 = pvalue_for_correlation
- η_2 = pvalue_for_uncorrelated
- η_3 = rho_below_for_uncorrelated
- η_4 = induced_conditional_dependence_jump

The meaning of these parameters is as follows. Referring to Figure 3.1c, the value of η_1 gives the threshold *p*-value above which a parent *X* to child *Z* edge or parent *Y* to child *Z* is rejected as not significant. The value of η_2 gives the *p*-value threshold for the association test below which the variables *X* and *Y* are considered not independent (hence resulting in voiding the v-structure). The value of η_3 is a second requirement for concluding that the parents *X* and *Y* are independent; their correlation must fall below η_3 in absolute value. Lastly, the characteristic of conditional dependence is one of the uniquely defining attributes of an unshielded collider or v-structure, and the parameter η_4 gives the minimum jump in correlation that we will consider as indicated the conditional dependence between the two parents *X* and *Y*. A further minor parameter, min_number_cases_in_induced_-dependence, is simply a minimum number of cases (typically set to 20) to allow the test to proceed.

In pseudocode the first test for v-structure is implemented as follows, given three random variables *X*, *Y*, *Z* to be checked for v-structure as labeled in Figure 3.1c.

3.4.2.1 Mechanics of the v-Structure Test

```
Compute the Pearson correlation rho_xy and its p-value p_xy.
Compute rho_xz and p_xz.
Compute rho_yz and p_yz.
Compute rho_xy|z and p_xy|z.

if (p_xy ≥ pvalue_for_uncorrelated

   AND |rho_xy| ≤ rho_below_for_uncorrelated
   AND p_xz ≤ pvalue_for_correlation
   AND p_yz ≤ pvalue_for_correlation
   AND p_xy|z < pvalue_for_correlation
   AND |rho_xy|z| > induced_conditional_dependence_jump +|rho_xy|
   AND cases ≥ min_number_cases_in_induced_dependence)
then
   {
   1st layer v-structure detected, for bilayer verification,
   repeat same check for 2nd layer
   v-structure.
   If 2nd layer passes, then apply the bi-layer verification
   rules from the theorem. e.g.:
```

```
Let p1 = p-value for grandparent–grandchild | parent
test;
Let p2 = p-value for grandparent-grandchild test
if (p1 < pvalue_for_correlation AND p2 > pvalue_for_uncor-
related) then delete parent-child edge as confounded
else if (p1 > pvalue_for_uncorrelated AND p2 pvalue_for_
correlation) then conclude: parent-child is a verified
edge.
}
```

3.4.3
Special Handling for Genotypes

Genotypes in an F2 mouse intercross commonly constructed for QTL analysis [26] are discrete valued variables that do not have parents in the graph, as their status is fixed from conception. Hence, they are readily presumed to be upstream causal factors relative to all gene expression traits in the network. When the marker is at the beginning of a chain of marker M followed by two traits X and Y, as in $M \rightarrow X \rightarrow Y$, then we check for correlation with an additive, dominant, and recessive model between the marker M and the conditioned trait $Y|X$. Since both X and Y are continuous, conditioning is easily accomplished by computing the residuals $Y - E(Y|X)$ after a linear regression of Y on X. Since in the graph we are checking $M \rightarrow X \rightarrow Y$, if we remove the influence of the direct parent X on the variation of Y, then we should see no further relationship between M and the residuals $Y|X$.

3.5
Results and Further Application

3.5.1
Estimating α False-Positive Rates for the v-Structure Test

Given the small sample sizes involved in our study and in most microarray experiments, it seemed most useful to examine the behavior of the v-structure test with simulation of small sample sizes. Here, we report simulation results obtained when varying both sample size ($N = 25$ and $N = 50$ were tested) as well as each of the four η parameters described above.

In the following discussion we introduce the notation $\tilde{\not\perp\!\!\!\perp}$ and $\tilde{\not\perp\!\!\!\perp}$ to discuss the results of statistical tests on finite sample data for dependence and independence. Specifically, A $\tilde{\not\perp\!\!\!\perp}$ B means that A and B have been found to be statistically dependent, despite the fact that this may be due to type I (false positive) error in the statistical dependence test. Similarly, A $\tilde{\perp\!\!\!\perp}$ B means that A and B look independent to our tests and our sample, even though in reality they may (in the true underlying population distribution) exhibit dependence.

To facilitate the analysis of the test, we distinguish between three parts of the computation, namely, V_1, V_2, and V_3 below. By simulating under the Gaussian distribution to evaluate the false-positive rate of three distinct portions of the test, we combine analysis and simulation to estimate the false-positive rates for the bilayer verification process under the assumption that our starting variables $X1, \ldots, X5$ are identically and independently distributed Normal(0,1). Analytically, we can express the probability of type I error as follows. It may help to follow the notation by referring to the labeling of Figure 3.3, graph C.

$$
\begin{aligned}
&P(\text{false-positive verification}) = \\
&P(\text{verification}\,|X1, X2, X3, X4, X5\ \text{are i.}i.d.\ N(0,1)) = P(X1 \mathrel{\widetilde{\not\perp\!\!\!\perp}} X2, \\
&\qquad X1 \mathrel{\widetilde{\not\perp\!\!\!\perp}} X3, \\
&\qquad X2 \mathrel{\widetilde{\perp\!\!\!\perp}} X3, \\
&\qquad X2 \mathrel{\widetilde{\not\perp\!\!\!\perp}} X3|X1, \\
&\qquad X2 \mathrel{\widetilde{\not\perp\!\!\!\perp}} X4, \\
&\qquad X2 \mathrel{\widetilde{\not\perp\!\!\!\perp}} X5, \\
&\qquad X4 \mathrel{\widetilde{\perp\!\!\!\perp}} X5, \\
&\qquad X4 \mathrel{\widetilde{\not\perp\!\!\!\perp}} X5|X2, \\
&\qquad X4 \mathrel{\widetilde{\not\perp\!\!\!\perp}} X1, \\
&\qquad X4 \mathrel{\widetilde{\perp\!\!\!\perp}} X1|X2, \\
&\qquad |\ \text{given that}\ X1, \ldots, X5 \\
&\qquad \text{are i.}i.d.\ N(0,1)).
\end{aligned}
\tag{3.9}
$$

To manage notation and make plain the approximation that we will use, let us define V_1 as the shorthand for the proposition of the first four statements above, such that the first v-structure is found.

$$
\begin{aligned}
V_1\quad \text{denotes}\quad &(X1 \mathrel{\widetilde{\not\perp\!\!\!\perp}} X2, \\
&X1 \mathrel{\widetilde{\not\perp\!\!\!\perp}} X3, \\
&X2 \mathrel{\widetilde{\perp\!\!\!\perp}} X3, \\
&X2 \mathrel{\widetilde{\not\perp\!\!\!\perp}} 3|X1)
\end{aligned}
$$

Likewise, defining V_2 as an indicator for the proposition of the second v-structure

$$
\begin{aligned}
V_2\quad \text{denotes}\quad &(X2 \mathrel{\widetilde{\not\perp\!\!\!\perp}} X4, \\
&X2 \mathrel{\widetilde{\not\perp\!\!\!\perp}} X5, \\
&X4 \mathrel{\widetilde{\perp\!\!\!\perp}} X5, \\
&X4 \mathrel{\widetilde{\not\perp\!\!\!\perp}} 5|X2)
\end{aligned}
$$

Next, we define V_3 as an indicator for the proposition that we apply the verification rules.

$$
\begin{aligned}
V_3\quad \text{denotes}\quad &(X1 \mathrel{\widetilde{\not\perp\!\!\!\perp}} X4, \\
&X1 \mathrel{\widetilde{\perp\!\!\!\perp}} X4|X2)
\end{aligned}
$$

Lastly, let I_0 be an indicator for the proposition that $X1, \ldots, X5$ are i.i.d. $N(0,1)$.

Then, we can write Equation 3.10 in a simplified notation to illustrate our method.

$$\begin{aligned} P(V_1, V_2, V_3|I_0) &= P(V_1|I_0)\cdot P(V_2|V_1, I_0)\cdot P(V_3|V_1, V_2, I_0) \\ &= P(V_1|I_0)\cdot P(V_2|I_0)\cdot P(V_3|V_2, I_0) \\ &\approx P(V_1|I_0)\cdot P(V_2|I_0)\cdot P(V_3|(X4 \tilde{\not\perp\!\!\!\perp} X2), (X2 \tilde{\not\perp\!\!\!\perp} X1), I_0) \\ &= \{P(V_1|I_0)\}^2\cdot P(V_3|(X4 \tilde{\not\perp\!\!\!\perp} X2), (X2 \tilde{\not\perp\!\!\!\perp} X1), I_0). \end{aligned} \tag{3.10}$$

The third line of Equation 3.10 is a conservative approximation for generating p-values in the sense that the actual probability of false-v-structures V_1 and V_2 is smaller than the probabilities of $(X4 \tilde{\not\perp\!\!\!\perp} X2)$ and $(X2 \tilde{\not\perp\!\!\!\perp} X1)$ that we substitute. The substituted terms capture the most salient feature of the relationship between the two v-structures that is subsequently tested during the application of the bilayer verification rules, and the probabilities of these two terms holding under I_0 can be addressed in reasonable time by simulation, whereas simulations looking for both V_1 and V_2 to hold are, as the results below demonstrate, such rare events that their occurrence cannot easily be addressed by simulation. Moreover, we may actually be quite interested in the type I error rate in the presence of partially acknowledged dependence between the variables, in the case if which this approximation is most appropriate.

The fourth line of Equation 3.10 follows as the two terms $P(V_1|I_0)$ and $P(V_2|I_0)$ are identical under I_0. The values under various parameter choices for the V_1 portion of the test are given in Tables 3.2 to 3.7 for small sample sizes. We also obtained results by simulation for the last term, $\mathrm{P}(\mathrm{V}_3|(\mathrm{X}_4 \tilde{\not\perp\!\!\!\perp} \mathrm{X}_2)), (\mathrm{X}_2 \tilde{\not\perp\!\!\!\perp} \mathrm{X}_1), \mathrm{I}_0)$, and these are rendered in Tables 3.8 and 3.9.

In detail, after appropriate transformation, our microarray data can be reasonably treated as Gaussian [9]. To determine $P(V_1|I_0)$, we therefore simulate three independent and identically distributed Normal(0,1) variables and check for v-structure under varying parameter values. The raw counts reported in Tables 3.2 to 3.7 are the number of independent simulations of size N to locate 10 false-positive v-structures. Six

Table 3.2 Simulation results addressing v-structure false-positive rates in small sample sizes.

N	η_1	η_2	η_3	η_4	Raw	Var	α (10/raw)	Rounded α	Clustered
25	0.05	0.1	0.2	0.1	218 862	4.15E+08	4.57E−05	5.00E−05	a
25	0.05	0.2	0.2	0.1	216 535	3.54E+08	4.62E−05	5.00E−05	a
25	0.05	0.1	0.3	0.1	146 164	8.26E+07	6.84E−05	5.00E−05	a
25	0.05	0.2	0.3	0.1	153 541	2.48E+08	6.51E−05	5.00E−05	a
25	0.05	0.1	0.2	0.2	191 842	3.04E+08	5.21E−05	5.00E−05	a
25	0.05	0.2	0.2	0.2	289 193	6.85E+08	3.46E−05	5.00E−05	a
25	0.05	0.1	0.3	0.2	169 346	1.18E+08	5.91E−05	5.00E−05	a
25	0.05	0.2	0.3	0.2	130 867	6.27E+07	7.64E−05	1.00E−04	a

Here $N=25$ and $\eta=0.05$. Raw = count of number of independent simulations of size N to reach 10 false positives; var = variance of the count of simulations between each false positive; σ = observed frequency of type I error or falsely predicted v-structure.

Table 3.3 Sample size of $N = 25$, increasing the stringency by taking $\eta = 0.01$.

N	η_1	η_2	η_3	η_4	Raw	Var	α (10/raw)	Rounded α	Clustered
25	0.01	0.1	0.2	0.1	11 286 400	9.12E+11	8.86E−07	1.00E−06	b
25	0.01	0.2	0.2	0.1	22 599 200	1.30E+13	4.42E−07	5.00E−07	c
25	0.01	0.1	0.3	0.1	15 264 300	1.02E+12	6.55E−07	5.00E−07	c
25	0.01	0.2	0.3	0.1	12 657 100	1.33E+12	7.90E−07	1.00E−06	b
25	0.01	0.1	0.2	0.2	12 356 600	1.04E+12	8.09E−07	1.00E−06	b
25	0.01	0.2	0.2	0.2	19 434 800	2.17E+12	5.15E−07	5.00E−07	c
25	0.01	0.1	0.3	0.2	13 468 000	2.26E+12	7.43E−07	1.00E−06	b
25	0.01	0.2	0.3	0.2	9 328 720	1.69E+11	1.07E−06	1.00E−06	b

Table 3.4 Sample size of $N = 50$, $\eta_1 = 0.05$.

N	η_1	η_2	η_3	η_4	Raw	Var	α (10/raw)	Rounded α	Clustered
50	0.05	0.1	0.2	0.1	171 428	1.41E+08	5.83E−05	5.00E−05	a
50	0.05	0.2	0.2	0.1	181 414	2.77E+08	5.51E−05	5.00E−05	a
50	0.05	0.1	0.3	0.1	101 932	5.57E+07	9.81E−05	1.00E−04	d
50	0.05	0.2	0.3	0.1	211 450	6.82E+08	4.73E−05	5.00E−05	a
50	0.05	0.1	0.2	0.2	3 761 080	7.41E+10	2.66E−06	5.00E−06	e
50	0.05	0.2	0.2	0.2	4 380 960	1.03E+11	2.28E−06	5.00E−06	e
50	0.05	0.1	0.3	0.2	3 148 920	4.48E+10	3.18E−06	5.00E−06	e
50	0.05	0.2	0.3	0.2	3 663 490	4.82E+10	2.73E−06	5.00E−06	e

Table 3.5 Sample size of $N = 50$, $\eta_1 = 0.01$.

N	η_1	η_2		η_3	η_4	Raw	Var	α (10/raw)	Rounded α	Clustered
50	0.01	0.1	0.2	0.1		21 629 600	3.23E+12	4.62E−07	5.00E−07	f
50	0.01	0.2	0.2	0.1	1	40 282 500	3.32E+12	2.48E−07	1.00E−07	f
50	0.01	0.1	0.3	0.1		10 390 000	8.95E+11	9.62E−07	1.00E−06	f
50	0.01	0.2	0.3	0.1		23 242 900	3.33E+12	4.30E−07	5.00E−07	f
50	0.01	0.1	0.2	0.1		24 225 400	7.45E+12	4.13E−07	5.00E−07	f
50	0.01	0.2	0.2	0.2		33 825 200	1.03E+13	2.96E−07	5.00E−07	f
50	0.01	0.1	0.3	0.2		25 620 500	3.71E+12	3.90E−07	5.00E−07	f
50	0.01	0.2	0.3	0.2		34 328 300	9.71E+12	2.91E−07	5.00E−07	f

significant figures are reported. The resulting estimates of the false-positive rate α for single v-structure location $P(V_1 | I_0)$ under various values of the control parameters are given. A rounding to nearest half decade is given as is a subjective indication of whether the results cluster near each other. As might be expected, the values of η_1 and N dominate the determination of the false-positive rate. We can see that although

Table 3.6 Sample size of $N = 100$, $\eta_1 = 0.01$, $\eta_4 = 0.1$.

N	η_1	η_2	η_3	η_4	Raw	Var	α (10/raw)	Rounded α	Clustered
100	0.05	0.1	0.2	0.1	9 120 470	5.69E+11	1.10E−06	1.00E−06	b
100	0.05	0.2	0.2	0.1	5 905 130	4.08E+11	1.69E−06	1.00E−06	b
100	0.05	0.1	0.3	0.1	6 215 150	5.57E+11	1.61E−06	1.00E−06	b
100	0.05	0.2	0.3	0.1	4 355 320	1.15E+11	2.30E−06	1.00E−06	b

Table 3.7 Sample size of $N = 100$, $\eta_1 = 0.01$, $\eta_4 = 0.1$.

N	η_1	η_2	η_3	η_4	Raw	Var	α (10/raw)	Rounded α	Clustered
100	0.01	0.1	0.2	0.1	88 889 800	7.26E+13	1.12E−07	1.00E−07	g
100	0.01	0.2	0.2	0.1	836 504 000	3.3.55E+15	1.20E−08	1.00E−08	h
100	0.01	0.1	0.3	0.1	110 320 000	3.3.54E+13	9.06E−08	1.00E−07	g
100	0.01	0.2	0.3	0.1	1 319 030 000	8.8.52E+15	7.58E−09	1.00E−08	h

Table 3.8 Sample size of $N = 25$.

N	η_1	η_2	η_3	α_{inline}
25	0.05	0.2	0.3	3.38E−02
25	0.05	0.2	0.2	4.85E−03
25	0.05	0.1	0.3	6.10E−02
25	0.05	0.1	0.2	8.59E−03
25	0.01	0.2	0.3	9.68E−03
25	0.01	0.2	0.2	1.78E−03
25	0.01	0.1	0.3	1.71E−02
25	0.01	0.1	0.2	1.65E−03

Simulation under Normal(0,1) distribution until 20 conditional false positives were obtained. The field α_{inline} gives the probability of falsely concluding that $\gamma \perp\!\!\!\perp \chi|\pi$ and also that $\gamma \tilde{\perp\!\!\!\perp} \chi$, while conditioning upon the fact that we have already falsely concluded that $\gamma \not\perp\!\!\!\perp |\pi$ and that $\pi \not\perp\!\!\!\perp \gamma$.

η_2, η_3, and η_4 appear to influence α to a small extent, their settings might be more reasonably chosen to maximize power, as their effect on α appears fairly minor. Note that the roles of X, Y, and Z as labeled in Figure 3.1c are fixed in advance of the simulation and test results reported in Tables 3.2 to 3.7, so that there is no multiple testing correction needed when X, Y, and Z are allowed to permute roles in the v-structure.

Joining the results from Tables 3.2 to 3.5 for $P(V1\,|I_0)$ together with the results from Tables 3.8 and 3.9 for $P(V_3|(X4 \tilde{\not\perp\!\!\!\perp} X2), (X2 \tilde{\not\perp\!\!\!\perp} X1), I_0)$ yields the composite final simulation results as presented in Tables 3.10 to 3.13 for sample sizes $N = 25$ and $N = 50$. The $N = 100$ results as presented in Tables 3.8 and 3.9 could not be extended

Table 3.9 Sample size of $N = 50$.

N	η_1	η_2	η_3	α_{inline}
50	0.05	0.2	0.3	5.28E−03
50	0.05	0.2	0.2	8.55E−03
50	0.05	0.1	0.3	3.57E−02
50	0.05	0.1	0.2	1.19E−02
50	0.01	0.2	0.3	4.32E−04
50	0.01	0.2	0.2	3.12E−04
50	0.01	0.1	0.3	4.32E−03
50	0.01	0.1	0.2	8.85E−04

Simulation under Normal(0,1) distribution until 20 conditional false positives were obtained. The field α_{inline} gives the probability of falsely concluding that $\gamma \perp\!\!\!\perp \chi|\pi$ and also that $\gamma \not\perp\!\!\!\perp \chi$, while conditioning upon the fact that we have already falsely concluded that $\gamma \not\perp\!\!\!\perp |\pi$ and that $\pi \not\perp\!\!\!\perp \gamma$.

Table 3.10 Sample size of $N = 25$ and $\eta_1 = 0.05$.

N	η_1	η_2	η_3	η_4	P(verification | I_0)
25	0.05	0.1	0.2	0.1	1.79E−11
25	0.05	0.2	0.2	0.1	1.04E−11
25	0.05	0.1	0.3	0.1	2.85E−10
25	0.05	0.2	0.3	0.1	1.44E−10
25	0.05	0.1	0.2	0.2	2.33E−11
25	0.05	0.2	0.2	0.2	5.80E−12
25	0.05	0.1	0.3	0.2	2.13E−10
25	0.05	0.2	0.3	0.2	1.98E−10

Composite false–positive probabilities for bilayer verification following Equation 3.10.

Table 3.11 Sample size of $N = 25$ and $\eta_1 = 0.01$.

N	η_1	η_2	η_3	η_4	P(verification | I_0)
25	0.01	0.1	0.2	0.1	1.30E−15
25	0.01	0.2	0.2	0.1	3.49E−16
25	0.01	0.1	0.3	0.1	7.33E−15
25	0.01	0.2	0.3	0.1	6.04E−15
25	0.01	0.1	0.2	0.2	1.08E−15
25	0.01	0.2	0.2	0.2	4.72E−16
25	0.01	0.1	0.3	0.2	9.42E−15
25	0.01	0.2	0.3	0.2	1.11E−14

Composite false–positive probabilities for bilayer verification following Equation 3.10.

Table 3.12 Sample size of $N = 50$ and $\eta_1 = 0.05$.

N	η_1	η_2	η_3	η_4	$P(\text{verification} \mid I_0)$
50	0.05	0.1	0.2	0.1	4.03E−11
50	0.05	0.2	0.2	0.1	2.60E−11
50	0.05	0.1	0.3	0.1	3.44E−10
50	0.05	0.2	0.3	0.1	1.18E−11
50	0.05	0.1	0.2	0.2	8.38E−14
50	0.05	0.2	0.2	0.2	4.46E−14
50	0.05	0.1	0.3	0.2	3.60E−13
50	0.05	0.2	0.3	0.2	3.93E−14

Composite false–positive probabilities for bilayer verification following Equation 3.10.

Table 3.13 Sample size of $N = 50$ and $\eta_1 = 0.05$.

N	η_1	η_2	η_3	η_4	$P(\text{verification} \mid I_0)$
50	0.01	0.1	0.2	0.1	1.89E−16
50	0.01	0.2	0.2	0.1	1.92E−17
50	0.01	0.1	0.3	0.1	4.00E−15
50	0.01	0.2	0.3	0.1	7.99E−17
50	0.01	0.1	0.2	0.2	1.51E−16
50	0.01	0.2	0.2	0.2	2.72E−17
50	0.01	0.1	0.3	0.2	6.59E−16
50	0.01	0.2	0.3	0.2	3.66E−17

Composite false–positive probabilities for bilayer verification following Equation 3.10.

more fully as the simulated events are so rare as to be prohibitively expensive in terms of compute time. Nonetheless, the final $P(\text{verification}|I_0)$ results do demonstrate the effects of sample size and various parameter adjustments in terms of taking the estimated false-positive rate from 1E–10 to 1E–17.

To summarize these results, we draw two primary conclusions. First, we have established small sample α rate tables (in particular Tables 9 to 12 for reference and incorporation into computer programs which utilize v-structure based DAG recovery and wish to report false positive rates for their algorithms and tests. Second, by taking the η_1 p-value threshold as the primary parameter of interest and conservatively locating the maximum p-value reported over all other parameter choices in each simulation table, we draw the following conclusion. At small sample sizes in the neighborhood of $N = 25$ and $N = 50$, for a critical parameter of $\eta_1 = 0.05$, the false positive rate for verification is conservatively less that 1×10^{-10}, independent of the choices for η_2, η_3, and η_4 that were studied. Under these same conditions, for $\eta_1 = 0.01$, the probability of false positive verification is less than 1×10^{-15}. These findings offer guidance to the utilization of v-structure based causality analysis, and may find particular utility when accounting for multiple comparisons.

3.5.2
Learning an Aortic Lesion Network

The RVV algorithm was run over the C3H × B6 liver microarray data given the starting point of the aortic lesion scores and parameters ($\eta_1 = 0.05$, $\eta_2 = 0.3$, $\eta_3 = 0.2$, and $\eta_4 = 0.1$). In the resulting network for the male mice, shown in Figure 3.6, three nodes stood as direct inputs to the aortic lesion nodes. These three direct parents had many more parents than could be pictured in the truncated output shown. The algorithm was halted after the two layers were learned. The data show many very highly correlated genes along with the primary parents, A630026L20, Atp11c, and Nola2. As a caveat, we suspect that the algorithm may be vulnerable to an order effect as to which genes are assayed first when a high degree of correlation among many hundreds of correlated genes is found. This will be investigated in future work.

3.5.3
Further Utilizing Networks: Assigning Functional Roles to Genes

Given a learned network of clinical covariates or a partial prespecification of a causal Bayesian network based on prior knowledge of the relationships between clinical traits, it then becomes possible to assign functional roles to genes by assaying which gene, when interposed in the network, best satisfies the causal constraints of d-separation.

Figure 3.7 is a model of the processes involved in metabolic syndrome. Consider the subgraph (Figure 3.8) from the protective HDL cholesterol via an unknown gene A to atherosclerosis. Which gene in a given tissue known to be of import, say liver,

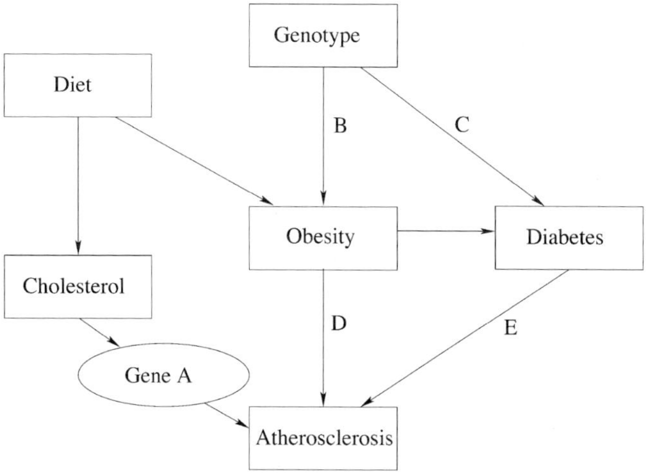

Figure 3.7 Relationships between clinical covariates, genes, and genotypes for metabolic syndrome.

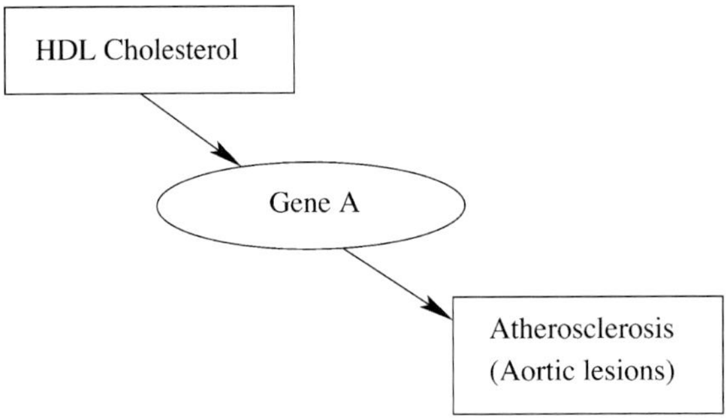

Figure 3.8 The subgraph for assigning gene functional roles.

best d-separates cholesterol from aortic lesions in the observed data? Such a gene might be a candidate for investigating the protective effects of HDL cholesterol on atherosclerosis.

Results for an analysis of the subgraph appear in Table 3.14, where insulin-like growth factor 1 (Igf1) is located as the gene that most screens between HDL cholesterol and aortic lesions. All genes in livers for the male mice were ranked (approximately 23 000 genes); only the top ranked genes are shown. We note a very encouraging sharp numeric drop in partial correlation when going from Igf1 to the closest runner up, Serpinc1. HDL cholesterol levels and aortic lesion counts were initially correlated at −0.18 (Pearson correlation p-value: 0.019 from 158 mice). We find in the males that liver Igf1 levels renders HDL and aortic lesions conditionally independent to the

Table 3.14 Functional gene assignment given a specified clinical covariate network.

Gene	Partial correlation between HDL and aortic lesions given gene
Igf1	−0.0494
Serpinc1	−0.0813
Cyp2d9	−0.0936
Arsa	−0.0966
Mbl2	−0.0980
Eef1a1	−0.0985
Ahcy	−0.1020
CRAD-L	−0.1030
MMT00027866	−0.1088
2310014L17Rik	−0.1099
2810460C24Rik	−0.1099
Ugt1a1	−0.1106
5430417P09Rik	−0.1130
Omd	−0.1132
Ftl1	−0.1134
Treh	−0.1139

Ranked best candidates.

degree of −0.049 partial correlation, the smallest of any gene. The negative correlation is expected, as HDL cholesterol is known to have a protective effect with regard to heart disease.

Although derived from accurate experimental data, these results await secondary experimental confirmation. The principal value of these results is to suggest that the methodology here may be of value. By first inferring or describing a known clinical trait network, we in effect create roles within a known drama; we can then cast genes into those roles, investigating the microarray data at hand to see which genes might best fit which given roles. To do so, we utilize the predicted conditional independence (d-separation) properties from the learned or specified graph.

3.5.4 Future Work

Future work will extend this analysis further to (1) identify the best candidates for roles B–E in the metabolic syndrome network; (2) incorporate additional prior knowledge of the sterol regulatory element-binding protein network [27]; (3) test for genes intervening between an RVV-learned structure of the clinical covariates alone; and (4) integrate the genetic marker data as discrete exogenous variables (known parentless nodes) in the network.

Acknowledgments

J.E.A. acknowledges support from grant HG02536-04 from the National Human Genome Research Institute (NHGRI) of the National Institutes of Health and grant DGE9987641 from the National Science Foundation.

List of Abbreviations

SNP	Single Nucleotide Polymorphism, a single base change in the genomic DNA sequence of an organism, typically shared by more than 1% of individuals in a population
APOE or *Apoe*	the gene Apolipoprotein E
PC algorithm	the name of a causal discovery algorithm by Peter Spirtes and Clark Glymour
LCD	the Local Causal Discovery algorithm by G. F. Cooper
FBD and FTC	the names for two automatic Structural Equation Model building algorithms by Cohen *et al.*
CCC and CCU	the names for two causal discovery rules articulated by Silverstein *et al.*
DAG	directed acyclic graph
SEM	Structural Equation Model

References

1 Gunderson, K.L., Steemers, F.J., Ren, H., Ng, P., Zhou, L., Tsan, C., Chang, W., Bullis, D., Musmacker, J., King, C., Lebruska, L.L., Barker, D., Oliphant, A., Kuhn, K.M. and Shen, R. (2006) Whole-genome genotyping. *Methods in Enzymology*, **410**, 359–376.

2 Wang, Y., Moorhead, M., Karlin-Neumann, G., Falkowski, M., Chen, C., Siddiqui, F., Davis, R.W., Willis, T.D. and Faham, M. (2005) Allele quantification using molecular inversion probes (MIP). *Nucleic Acids Research*, **33**, 183.

3 Pearl, J. (2000) *Causality: Models, Reasoning, and Inference*, Cambridge University Press, Cambridge, UK.

4 Neapolitan, R.E. (2004) *Learning Bayesian Networks*, Pearson Prentice Hall, Upper Saddle River, NJ.

5 Cooper, G.F. (1997) A simple constraint-based algorithm for efficiently mining observational databases for causal relationships. *Data Mining and Knowledge Discovery*, **1**, 203–224.

6 Cohen, P.R., Gregory, D.E., Ballesteros, L.A. and Amant, R.S. (1995) Two algorithms for inducing structural equation models from data. Preliminary Papers of the Fifth International Workshop on Artificial Intelligence and Statistics 1995, 129–139.

7 Spirtes, P., Glymour, C. and Scheines, R. (2000) *Causation Prediction, and Search*, 2nd edn,The MIT Press, Cambridge, MA.

8 Pearl, J. (1988) *Probabilistic Reasoning in Intelligent Systems*, revised 2nd edn, Morgan Kaufmann Publishers, Inc, San Francisco, CA.

9 Schadt, E.E., Lamb, J., Yang, X., Zhu, J., Edwards, S., GuhaThakurta, D., Sieberts, S.K., Monks, S., Reitman, M., Zhang, C., Lum, P.Y., Leonardson, A., Thieringer, R., Metzger, J.M., Yang, L., Castle, J., Zhu, H., Kash, S.F., Drake, T.A. and Sachs, A. and Lusis, A.J. (2005) An integrative genomics approach to infer causal associations between gene expression and disease. *Nature Genetics*, **37**, 710–717.

10 Jensen, F.V. (2001) *Bayesian Networks and Decision Graphs*, Springer-Verlag, New York, NY.

11 Dawid, A.P. (1979) Conditional independence in statistical theory. *Journal of the Royal Statistical Society, Series B*, **41**, 1–31.

12 Spielman, R.S., McGinnis, R.E. and Ewens, W.J. (1993) Transmission test for linkage disequilibrium: the insulin gene region and insulin-dependent diabetes mellitus (IDDM). *American Journal of Human Genetics*, **52**, 506–516.

13 Verma, T. and Pearl, J. (1990) Equivalence and synthesis of causal models. Proceedings of the 4th Workshop on Uncertainty in Artificial Intelligence July 1990, 220–227. Reprinted in P. Bonissone, M. Henrion, L.N. Kanal and J.F. Lemmer (Eds.), *Uncertainty in Artificial Intelligence*, **6**, Elsevier, Amsterdam, pp. 255–268.

14 Korb, K.B. and Nicholson, A.E. (2004) *Bayesian Artificial Intelligence*, Chapman & Hall/CRC, Boca Raton, FL.

15 Silverstein, C., Brin, S., Motwani, R. and Ullman, J. (2000) Scalable techniques for mining causal structures. *Data Mining and Knowledge Discovery*, **4**, 163–192.

16 Wang, S., Yehya, N., Schadt, E.E., Drake, T.A. and Lusis, A.J. (2006) Genetic and genomic analysis of fat mass trait with complex inheritance reveals marked sex specificity. *PLoS Genetics*, **2**, e15.

17 Yang, X., Schadt, E., Wang, S., Wang, H., Arnold, A.P., Ingram-Drake, L., Drake, T.A. and Lusis, A.J. (2006) Tissue-specific expression and regulation of sexually dimorphic genes in mice. *Genome Research*, **16**, 995–1004.

18 Drake, T.A., Schadt, E.E., Davis, R.C. and Lusis, A.J. (2005) Integrating genetic and gene expression data to study the

metabolic syndrome and diabetes in mice. *American Journal of Therapeutics*, **12**, 503–511.

19 Su, Z., Li, Y., James, J.C., McDuffie, M., Matsumoto, A.H., Helm, G.A., Weber, J.L., Lusis, A.J. and and Shi, W. (2006) Quantitative Trait Locus Analysis of Atherosclerosis in an Intercross Between C57BL/6 and C3H Mice Carrying the Mutant Apolipoprotein E Gene. *Genetics*, **172**, 1799–1807.

20 Drake, T.A., Schadt, E.E. and Lusis, A.J. (2006) Integrating genetic and gene expression data: application to cardiovascular and metabolic traits in mice. *Mammalian Genome*, **17**, 466–479.

21 Doss, S., Schadt, E.E., Drake, T.A. and Lusis, A.J. (2005) Cis-acting expression quantitative trait loci in mice. *Genome Research*, **15**, 681–691.

22 Ghazalpour, A., Doss, S., Sheth, S.S., Ingram-Drake, L.A., Schadt, E.E., Lusis, A.J. and Drake, T.A. (2005) Genomic analysis of metabolic pathway gene expression in mice. *Genome Biology*, **6**, R59.

23 Cervino, A.C., Edwards, S., Zhu, J., Laurie, C., Tokiwa, G., Lum, P.Y., Wang, S., Castellini, L.W., Lusis, A.J., Carlson, S., Sachs, A.B. and Schadt, E.E. (2005) Integrating QTL and high-density SNP analyses in mice to identify Insig2 as a susceptibility gene for plasma cholesterol levels. *Genomics*, **86**, 505–517.

24 Mehrabian, M., Allayee, H., Stockton, J., Lum, P.Y., Drake, T.A., Castellani, L.W., Suh, M., Armour, C., Edwards, S., Lamb, J., Lusis, A.J. and Schadt, E.E. (2005) Integrating genotypic and expression data in a segregating mouse population to identify 5-lipoxygenase as a susceptibility gene for obesity and bone traits. *Nature Genetics*, **37**, 1224–1233.

25 He, Y.D., Dai, H., Schadt, E.E., Cavet, G., Edwards, S.W., Stephaniants, S.B., Duenwald, S., Kleinhanz, R., Jones, A.R., Shoemaker, D.D. and Stoughton, R.B. (2003) Microarray standard data set and figures of merit for comparing data processing methods and experimental designs. *Bioinformatics*, **19**, 956–965.

26 Schadt, E.E., Monks, S.A., Drake, T.A., Lusis, A.J., Che, N., Colinayo, V., Ruff, T.G., Milligan, S.B., Lamb, J.R., Cavet, G., Linsley, P.S., Mao, M., Stoughton, R.B. and Friend, S.H. (2003) Genetics of gene expression surveyed in maize mouse and man. *Nature*, **422**, 297–302.

27 Horton, J.D., Goldstein, J.L. and Brown, M.S. (2002) SREBPs: activators of the complete program of cholesterol and fatty acid synthesis in the liver. *The Journal of Clinical Investigation*, **109**, 1125–1131.

4
Computational Inference of Biological Causal Networks – Analysis of Therapeutic Compound Effects

Jing Yu, Gabriel Helmlinger, Muriel Saulnier, and Anna Georgieva

4.1
Introduction

The central goal of biology is to reveal the causal chains of events that control complex biological functions. A critical component of medicine research is to understand the biological pathways involved in the disease progression and mechanism of action of a therapeutic compound across multiple levels of biological organization, from molecular to tissue level and all the way to clinical outcome. To address this goal, functional network inference algorithms are being developed and applied to various types of data in order to predict causal functional pathways in biological systems.

In the field of computational biology, a variety of inference algorithm approaches have recently been developed to model and predict causal molecular pathways from gene expression data, and to capitalize on the use of high-throughput genomics data and recover fragments of gene regulatory networks or biological pathways. These include a Boolean model [1–3], a linear model [4], a differential equation model [5,6], a Bayesian network (BN) model [7,8], and others [9,10].

BN inference algorithms have shown particular promise in this area [11–14], due to the following abilities that are more suitable for solving biological problems: (1) BNs can be generalized and applied to other types of data [11,13,15] and are easy to add levels of biological organization; (2) they are capable of capturing both linear and nonlinear combinatorial relationships; (3) they can robustly handle noise and stochastic processes due to probabilistic nature [16]; (4) they have a built-in mechanism to guard against overfitting [13,17]; and (5) they produce a graph-based model of conditional dependence and independence relationships, which can easily be used to interpret the causal interactions among variables. Static Bayesian network (SBN) inference algorithms have been used to analyze time-independent gene expression data sets [3,6–11], while dynamic Bayesian network (DBN) inference algorithms have recently been applied to time-series data collected in lower organisms such as yeast, *Saccharomyces cerevisiae*, or *E. coli* [18–23] to study regulation of the cell cycle. Recently, BNs have also been used to reverse engineer protein signaling pathways [24–26].

Analysis of Microarray Data: A Network-Based Approach. Edited by F. Emmert-Streib and M. Dehmer

ISBN: 978-3-527-31822-3

BN inference algorithms have the potential to help us understand the causal relationships in medical research for therapeutic compound effects, especially addressing the causal effects over time by using time-series data, which we present as a sample application here. Chronic inflammatory diseases such as rheumatoid arthritis, inflammatory bowel disease, systemic lupus erythematosus, multiple sclerosis, and type 1 diabetes are debilitating and are becoming increasingly common in our aging society. Therefore, it is imperative to better understand the molecular basis of the inflammatory processes underlying a given disease. In particular, a better characterization of the interaction between the main players in these processes would yield more effective compound treatments. To study the anti-inflammatory mechanism of a discovery compound at Novartis Pharmaceuticals Research, time-series gene expression data were collected from joint tissues with adjuvant-induced arthritis (AIA), a rat model for rheumatoid arthritis.

Here, we apply our DBN inference algorithm [27,28], developed and tested in a simulation framework, to a time-series gene expression data set collected from AIA rats. The goal of this computational study was to recover compound-modulated pathway fragments with regard to the anti-inflammatory effects of the compound. First, we introduce some basic theory of Bayesian networks; in Section 4.3, we describe the experimental design, data preprocessing, and DBN analysis; we then present both the computational results and biological findings followed by discussion and conclusions.

4.2 Basic Theory of Bayesian Networks

BNs can generally be described as probabilistic models that depict causal relationships between variables [29–31]. A static BN [7,17,32] is a directed acyclic graph (DAG) that encodes a joint probability distribution over χ, where $\chi = \{X_1, \ldots, X_n\}$ is a set of discrete random variables X_i ($i \in \{1, \ldots, n\}$). A BN for χ is specified as a pair $\langle G, \Theta \rangle$. The variable G represents a directed acyclic graph whose nodes, also known as vertices, correspond to the random variables $X_1, \ldots, X_n$ and whose directed links (also known as edges or arcs) from X_i to X_j indicate a statistical conditional dependence of X_j on X_i. All variables that have a directed link to X_i are known as its parents $\text{Pa}(X_i)$; all variables to which X_i has a directed link are known as its children; and recursively all variables receiving links from such targeted variables (children) and their targets are known as its descendants. Each variable X_i is independent of its nondescendants given its parents in G. The variable Θ represents a set of parameters that collectively quantify the probability distributions associated with the variables in the graph. Each parameter of Θ is specified by $\theta_{x_i|\text{pa}(X_i)} = P(X_i = x_i|\text{Pa}(X_i) = \text{pa}(X_i))$: namely, the probability of X_i taking on the value x_i given its parents $\text{Pa}(X_i)$ have the values in a particular instantiation of the parents, $\text{pa}(X_i)$, for all x_i and $\text{pa}(X_i)$. Thus, a BN specifies a unique joint probability distribution over χ given by

$$P(X_1,\ldots,X_n)=\prod_{i=1}^{n}P(X_i|\mathrm{Pa}(X_i)). \tag{4.1}$$

However, there are two limitations of the static BNs. First, the graph must be acyclic, because it cannot have cycles in the factoring of joint probability distribution. However, reciprocal interactions are common among biological systems. Second, the unique joint probability distribution of a BN can sometimes have several different equivalent factorings that differ only in the directions of some links. These graphs form a Markov equivalence class and are indistinguishable probabilistically [7], thus leading to uncertainty about the direction of interactions between variables.

For overcoming these two limitations, a DBN framework can be built with time-series data [33]. A DBN extends a static BN framework by including the dimension of time. A first-order Markov assumption is used, meaning that variables at one time slice are considered affected only by those in the immediately previous time slice. Such a DBN is a graphical representation of a joint probability distribution over χ', where $\chi'=\{X_{1(t)},\ldots,X_{n(t)},X_{1(t+\Delta t)},\ldots,X_{n(t+\Delta t)})\}$ is a set of discrete random variables X_i measured at both time t and time $t+\Delta t$. Just like a static DBN, a DBN is specified as a pair $\langle G,\Theta\rangle$. The graph G is restricted in a DBN in the sense that links are allowed only to go forward in time, that is, from a variable $X_i(t)$ to $X_j(t+\Delta t)$. Additionally, it is implied that all variables have directed links from themselves at time t to themselves at time $t+\Delta t$, that is, all $X_{i(t)}$ link to $X_{i(t+\Delta t)}$, which assumes dependence between adjacent time points. The collection of parameters Θ consists of $\theta_{X_i(t+\Delta t)|pa(X_{i(t+\Delta t)})}$, as above, for all $X_{i(t+\Delta t)}$ in χ'. The DBN representation for an underlying causal network with cycle is shown in Figure 4.1. So, in the DBN framework, there is no problem of cycle or uncertainty about direction of interaction.

The problem of discovering a static or dynamic BN from a collection of observed data can be stated as follows. Given a data set $D=\{Y_1,\ldots,Y_n\}$ of observed instances of χ, find the most probable graph G for explaining the data contained in D. One common approach to this problem is to introduce a scoring metric that evaluates how probable it is for each graph G to explain the data in D. In the presence of such a

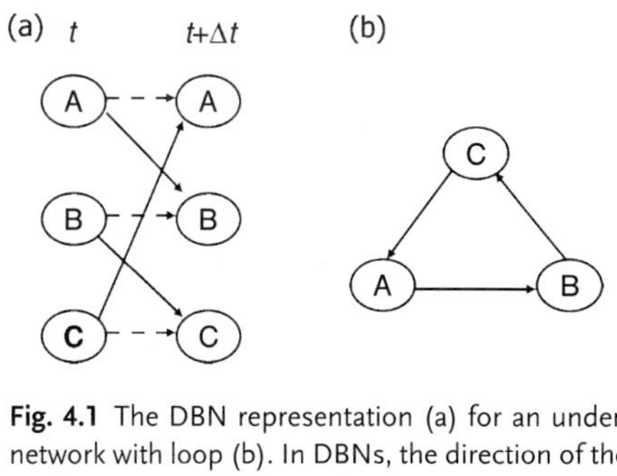

Fig. 4.1 The DBN representation (a) for an underlying causal network with loop (b). In DBNs, the direction of the link can only go from time t to time $t+\Delta t$.

scoring metric, the problem of discovering a BN reduces to the problem of searching for a graph that yields the highest score (top graph) or a collection of the graphs with highest scores. Since the number of possible graphs is not a polynomial function of the number of nodes, this is an NP-complete problem [34], which is computationally intractable. For solving this problem, heuristic search methods have been developed.

4.2.1 Bayesian Scoring Metrics

To score how well a graph explains the data, a Bayesian scoring metric is used, which can be generally described (according to Heckerman [17]) as follows:

$$\text{Score}(G:D) = \log P(G|D) = \log P(D|G) + \log P(G) - \log P(D). \qquad (4.2)$$

This states that the score of the graph G given data D is the log of the probability of G given D. This log probability can be further decomposed by Bayes rule into logs of the probability of D given G and prior probabilities of G and D. The marginal probability of the data $P(D)$ is equal for all G with a given D, and thus is not necessary to compare with different G. Besides, if we do not have any expectation to prefer some networks over others, the prior over graphs $P(G)$ is uniform for all graph G. Thus, only the probability of D given G needs to be calculated as follows:

$$P(D|G) = \int P(D|G,\Theta)P(\Theta|G)\,\mathrm{d}\Theta. \qquad (4.3)$$

This is the marginal likelihood, which averages the probability of the data over all possible parameter assignments to G.

There are mainly two types of scoring metrics that employ different assumptions: the Bayesian Dirichlet equivalent (BDe) and the Bayesian Information Criterion (BIC) scoring metrics. Both scoring metrics involve the generation of a conditional probability table (CPT) for each node. The tables store the probabilities estimated from all combinations of parent–child values extracted from the discretized data $\theta_{ijk} = P(x_i = k|\text{Pa}(x_i) = j)$ [17]. The probabilities in the tables are used to calculate a score for each node in the graph. Scores for all nodes are then summed to generate the score for the entire graph. Both scoring metrics have a penalty for overcomplexity to guard against overfitting of data. The BDe score captures the full Bayesian posterior probability $P(G|D)$. With this metric, the prior over graphs needs to be specified (usually use a uniform prior) and the prior over parameters is Dirichlet, a distribution over multinomial distributions describing the conditional probability of each variable in the network. We define N_{ijk} as the number of cases in D in which $X_i = x_i^k$ and $\text{Pa}(x_i) = pa_i^j$, $k \in [1, q_i]$, where r_i is the number of values x_i of X_i and q_i is the number of parent instantiations $\text{pa}(X_i)$ of the parents of X_i. There is a similar definition for pseudocounts α_{ijk}, which is related to the equivalent sample size α and prior distribution by $\alpha_{ijk} = \alpha P(x_i' = k|\text{Pa}(x_i)' = j)$. With $N_{ij} = \sum_k N_{ijk}$,

$\alpha_{ij} = \sum_k \alpha_{ijk}$, the calculation for BDe score is as follows:

$$\mathrm{BDe}(G:D) = \prod_{i=1}^{n}\prod_{j=1}^{q_i}\frac{\Gamma(\alpha_{ij})}{\Gamma(\alpha_{ij}+N_{ij})}\prod_{k=1}^{r_i}\frac{\Gamma(\alpha_{ijk}+N_{ijk})}{\Gamma(\alpha_{ijk})}. \tag{4.4}$$

The BIC score is an asymptotic approximation to the full posterior probability $P(G|D)$, based on a penalized maximum likelihood estimation of the term $P(G|D)$. The calculation for BIC score is defined as

$$\mathrm{BIC}(G:D) = \sum_i\sum_j\sum_k N_{ijk}\log\widehat{\theta}_{ijk} - \frac{\log N}{2}\#G, \tag{4.5}$$

where $\widehat{\theta}_{ijk} = N_{ijk}/N_{ij}$, N is the data size, and $\#G = \sum_{i=1}^{n} q_i(r_i-1)$ specifies the complexity of the graph. BIC is a good approximation to the full posterior (BDe) score with large amounts of data and is faster to compute; however, with a small amount of data, it is known to overpenalize complexity [17].

4.2.2 Heuristic Search Methods

Identifying the highest scoring BN using scoring metrics is NP-complete [34]. Thus, heuristic search methods are employed to search the highest scoring graph among a set of graphs. There are mainly two types of heuristic search methods: greedy with random restarts and simulated annealing [17,35,36]. For each step of the search, C denotes the set of all possible changes to a graph and $\Delta(c)$ denotes the score change of a graph taking the structure change of $c \in C$, where c stands for every possible change, either addition or deletion or reversal of one link.

Greedy search with random restarts initializes itself by choosing a random graph, and then chooses a step of change by one of the two methods. One is called random greedy, which chooses a random change to a graph that increases the score. The second method is called global greedy, where the change in score $\Delta(c)$ associated with every possible change $c \in C$ is evaluated, and then finally the change for which $\Delta(c)$ is maximized is chosen, provided the maximal $\Delta(c)$ is positive. This process keeps going until no score improvement can be made. To escape a possible local maximum, the algorithm then restarts from different random graphs until a prespecified number of iterations are reached and then outputs the top graph with the highest score found so far. The greedy search method cannot theoretically guarantee convergence to the global optimum, but it usually works empirically well and has an advantage in running time.

Simulated annealing also initializes itself by choosing a random graph but is additionally given an initial temperature T_0, a search parameter. A possible change $c \in C$ is selected at random and the probability $p = \exp(\Delta(c)/T_0)$ is evaluated. If $p > 1$ (which occurs whenever $\Delta(c)$ is positive), then the change c is made; otherwise, the change c is only made with probability p. The procedure begins at a very high temperature so that almost every eligible change in the graph can be made. As the search progresses, the temperature gradually decreases until a very low temperature is reached where almost only the change that can increase the score of the graph will be made. The search then performs similar to the local searches of the random greedy method. Simulated annealing can theoretically guarantee convergence to the global optimum, with sufficient amount of running time [37].

4.2.3 Inference Score

Discrete BNs typically use combinatorial interaction models making it difficult to determine the sign (+/−) and relative magnitude of interactions between variables. Although continuous BNs [39] typically use additive interaction models to make it easy to deduce the sign and relative magnitude of interactions between variables, interactions between transcriptional regulations are known often to be combinatorial rather than additive. Yu *et al.* [27] developed an influence score as part of discrete BN framework in an attempt to predict the sign (+/−) and relative magnitude of regulatory influences. This score is computed from the parameter estimates of the conditional probability values $\theta_{ijk} = P(x_i = k|\mathrm{Pa}(x_i) = j)$ for the top network. Although the values θ_{ijk} are posterior mean estimates of the conditional probabilities, the actual influence score is not; it represents a summarization of these conditional probability estimates into a single number to approximate the sign and magnitude of the interactions between a child variable and each of its parents. The influence score ranges from −1 to 1. The more positive the influence score, the stronger the activation; the more negative the influence score, the stronger the repression. When the influence score is near 0, it is difficult to infer the type of regulation.

4.3 Methods

4.3.1 Experimental Design

Treatment was administered on a daily basis with compound doses of 9 mg/kg/day for 14 days following AIA induction in rats. Joints were collected at six different time points: days 1, 2, 3, 4, 5, and 8. Day 1 samples were taken 6 hours after the first treatment; the remaining samples were taken 3 hours after the daily dosing. Each time point included 10 animals, 5 treated with the compound and 5 untreated controls. Samples were profiled on RAE230A Affymetrix gene chips, each containing 15 923 probes.

4.3.2 Tissue Contamination

Many of the samples included RNA from adjacent tissues such as muscle or skin, not expected to be significantly affected by the disease or the treatment. Those

Tab. 4.1 Distribution of final data points for inference analysis.

Sample point	Replicates
0, 1d	5
0, 2d	3
0, 3d	3
0, 4d	5
0, 5d	5
0, 8d	3
cmpd, 1d	5
cmpd, 2d	4
cmpd, 3d	3
cmpd, 4d	4
cmpd, 5d	5
cmpd, 8d	1

"0, 1d" refers to samples from control animals and were collected on the first day; "cmpd, 8d" refers to samples from treated animals and were collected on the eighth day. cmpd = compound.

samples were considered as contaminated and were excluded from the analysis. This resulted in a different number of replicates (from 1 to 5) for each treatment group at the various time points, as shown in Table 4.1. A total of 46 samples remained for further analysis.

4.3.3
Gene List Prefiltering

Gene expression intensities were normalized (log-transformed and centered) using the MAS5 algorithm [38]. ANOVA was applied to MAS5-normalized gene intensities. Interaction terms were included in the two-way ANOVA model on time and treatment, and a P-value cutoff of 0.05 was used. Each gene was required to have a median expression level of ≥ 50 in at least one group, a minimum fold change of 1.5, and a flag of P in $\geq 75\%$ of the samples in at least one group in order to be selected. Then the genes capturing the most variance across different conditions were put in the gene list. In addition, principle component analysis (PCA) and partial least square projection to latent structures (PLS) were applied to find and exclude from the final list those genes most likely derived from skin or muscle tissues (considered as contamination). After prefiltering, a total of 573 genes passed the statistical tests and were chosen to form the data set for inference analysis.

4.3.4
Outlier Removal

After prefiltering, the removal of contaminated samples, and upon plotting the data, outlier data points were observed. Outliers were removed using the following criterion: if, for a given gene, the highest expression value among all the measurements was at least twice the second highest expression value, then the highest value was replaced by the second highest value.

4.3.5
Further Screening of the Gene List

Due to the paucity of data (thousands of variables but tens of observations), DBN (and, for that matter, other learning algorithms) would have difficulties in achieving a high level of accuracy. Since the purpose of this study was to recover significant and biologically meaningful interactions, the number of genes needed to be further reduced. Yet another difficulty was encountered when we tried to apply a DBN inference algorithm directly to this time-series data set. DBN assumes dependence between the same variable in adjacent time points. However, in this particular study, each data point was collected from a different animal, since one rat needed to be sacrificed per one collected sample. A significant variance among animals or probe sets was observed, even when such expression data were supposed to represent "biological replicates" at a given time point. The variance could result from both intrinsic biological variability and measurement error. A high variance would potentially violate the assumption about dependence between adjacent time points in DBN, which in turn would lead to poor recovery results. The variance among animals is observed as noise in the data. In a previous simulation study, we observed that the DBN algorithm could handle a certain noise level (about 20%) [28]. We evaluated the noise level of this compound-related data set as follows. For the replicates (either in treatment group or in control group) of gene i measured at time point j, we defined $\eta_{ij} = \text{stdev}(\{x_{ij}\})/\text{average}(\{x_{ij}\})$, where $i \in \{1, \ldots, 573\}$ for each gene and $j \in \{1, \ldots, 12\}$ for each treatment group (2 of them) at each time point (6 of them). Here, stdev refers to standard deviation and $\{x_{ij}\}$ denotes the set of gene expression values in the evaluated group (varying in size from 1 to 5). Thus, the total noise level η is the average of η_{ij}. For this data set,

$$\eta = \frac{\sum_{i=1}^{573} \sum_{j=1}^{12} \eta}{573 \times 12} = 28.9\%. \tag{4.6}$$

This value was considered too high to directly apply to our algorithm. Such a noise level was likely the result of variance not directly attributable to the compound itself. Therefore, only genes relatively insensitive to such unknown effects were sought for use in the final DBN analysis, since their expression levels were expected to be relatively consistent among the replicates for each given time point. In addition, emphasis was placed on genes whose expression patterns were modulated by compound treatment. These observations led to the following criterion for gene selection to be used in the DBN analysis. Two variables, "compound" and "time," were added to the prefiltered data from Section 4.2.4. We then used the BDe scoring function [27] to determine how well the data could explain the local structure of the dependences between variables. If the BDe score for a gene was increased when having "compound" and/or "time point" as its parents (as compared to the gene having no parent), then the two variables were

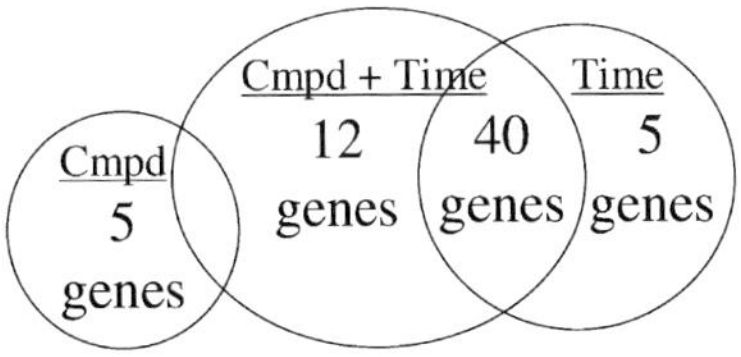

Fig. 4.2 Further screening of the gene list: gene selection based on predictions using the "compound" and/or "time point" variables.

considered as good predictors for this particular gene. This means that genes with such expression data were relatively consistent across the measured replicates at a given time point (since they could be well predicted by "time point"), and also that they were possibly involved in the pathways induced by the compound (since they could be well predicted by "compound").

Based on this analysis, three groups of genes were selected (Figure 4.2). (1) Genes that could be well predicted by both "compound" and "time point" variables (BDe score increased by a ratio of more than 1.1). A total of 52 genes, considered to be relatively insensitive to other unmeasured (unknown) effects and to be involved in the pathways affected by the compound, belonged to this category. (2) Genes that could be well predicted by the "time point" variable alone (score increased by a ratio of more than 1.1). This group consisted of 45 genes, with 40 of them overlapping with the first set. The five nonoverlapping genes were considered to be relatively insensitive to other unmeasured effects and not to have a strong response to the compound; however, they exhibited some expression changes over time. These genes could conceivably lie in more downstream portions of the compound-affected pathways, in such a way that their expression patterns did not follow an immediate, but rather a delayed, response following compound administration. (3) Genes that could be well predicted by the "compound" variable alone (score increased by a ratio of more than 1.2). This group consisted of five genes; no overlap with the other two categories was found. These genes were affected by compound treatment; however, their expression values remained relatively constant across all time points. Such genes might presumably be involved in more upstream portions of the pathways affected by the compound. With all three categories merged, a set of 62 "significant" genes was obtained. The noise level η for this data set was 21.55%, sufficiently close to the threshold derived in [28] considered suitable for DBN analysis. A pairwise t-test comparing the noise level ($\eta_j = \sum_1^{62} \eta_{ij}$) from this data set with the noise level ($\eta_j = \sum_1^{573} \eta_{ij}$) from the original data set without the screening step (where $j \in \{1, \ldots, 12\}$ enumerates each treatment group at each time point) indicated that the noise level was significantly decreased ($P = 0.00002$).

4.3.6
Data Pair-Up for Using DBN

After final screening of the gene list, the expression values of the selected 62 genes were expected to display small variances across animals from the same group.

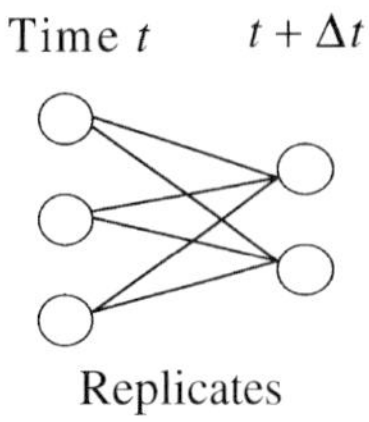

Fig. 4.3 Pairing up of all possible data points from adjacent time points, demonstrated by a case where there are three replicates at time t and two replicates at time $t + \Delta t$.

However, since each data point originated from a different animal, there was no prescribed way to pair up data from adjacent time points in the DBN analysis. Therefore, a list of all possible pair-ups between the measurements from adjacent time points was assembled, to make full use of the data and without bias. With this data pair-up method, 46 data points were mapped into 132 pair-ups of data for adjacent time points, as demonstrated in Figure 4.3.

4.3.7 Applying DBN

We finally applied DBN analysis to a data set that included 62 genes with 132 pair-ups in order to infer causal connections among the genes and to elucidate the pathways affected by the compound.

Our DBN inference algorithm [27] was written in C++ and was designed to search for a highest scoring network that best describes the probabilistic relationships between the variables in the data set. In this study, every node in the DBN network represented a gene with measured expression levels. Every directed link between two nodes represented a conditional statistical dependence of a child node on the parent node in this study, showing gene regulation relationships across time. Given the data set reduction, described previously, we used a first-order Markov DBN, where every variable at a given time point was influenced by itself and its parents at the immediate previous time point. An influence score [27], which allowed to assign a putative strength (between 0 and 1) and sign (+ excitatory/− inhibitory) to each dependence relationship, was calculated for each link in the final highest scoring network..

We used the following DBN configurations: three-category discretization and BDe scoring metrics with equivalent sample size of 1. Given the amount of data available in this set (132 pair-ups), we restricted the number of possible parents that could be recovered for each variable to 2, in order to reduce the number of possible false positive connections [28]. With this restriction, an exhaustive search was used to find the optimal graph with the DBN algorithm, which allows loops. To remove possible false positives, links with influence score of less than 0.001 were not considered in the final network [27]. For the particular problem addressed here, all inferred links connected to the compound were assigned a directionality originating from the compound.

4.4 Results

4.4.1 Computational Results

The overall network recovered via the application of the DBN algorithm is shown in Figure 4.4 as a compound-induced pathway structure (both topology and nature of interactions). For brevity, genes were named with numbers ranging from 1 to 62. One striking feature of the recovered network was the common observation of some specific genes displaying numerous fan-outs (children), such as the gene named *mercaptopyruvate sulfurtransferase* (node 25 in Figure 4.5). This may be a common phenomenon in molecular networks with a limited number of control nodes, for example, if the expressed gene relates to a key enzyme in metabolic pathways or a key transcription factor.

4.4.2 Biological Findings

The DBN algorithm linked the "compound" node directly to three children nodes. The three corresponding probe sets (nodes 58–60 in Figure 4.4) refer to the same expressed gene, Potential Target (the real name is masked for confidentiality reasons), which is a well-known cell adhesion molecule implicated in several inflammatory diseases. This finding obtained via an unsupervised learning scheme is promising in terms of consistency versus biological knowledge found already in the literature. Cascaded gene-to-gene relationships were further discovered by DBN. As mentioned earlier, only a small number of genes appeared to regulate many other genes in this recovered network. These genes were Potential Target (node 58; direct target of the compound in Figure 4.4), *urokinase-type plasminogen activator* (node 52; three tiers down from the compound), *chondroitin sulfate proteoglycan 2* (node 46; four tiers down from the compound), *mercaptopyruvate sulfurtransferase* (node 25; five tiers down from the compound), *fibroblast growth factor receptor 2* (node 57; six tiers down from the compound), and *heterogeneous nuclear ribonucleoprotein A1* (node 19; seven tiers down from the compound). We believe these genes play important roles within the compound-modulated network. All genes that could be well predicted solely by the "compound" variable (genes found in the third category of the gene list in Section 4.2.5) were located in the upstream portion of the compound-modulated pathway, either as a direct "target" or as a grandchild of the compound itself (rectangular nodes in Figure 4.4). All genes that could be well predicted solely by the "time point" variable (genes found in the second category of the gene list in Section 4.2.5) were in the downstream portion of the compound-modulated pathway (triangular nodes in Figure 4.4).

The inferred network information was in agreement with known biology. Figure 4.5 shows a fragment of the inferred network as extracted from Figure 4.4, with actual

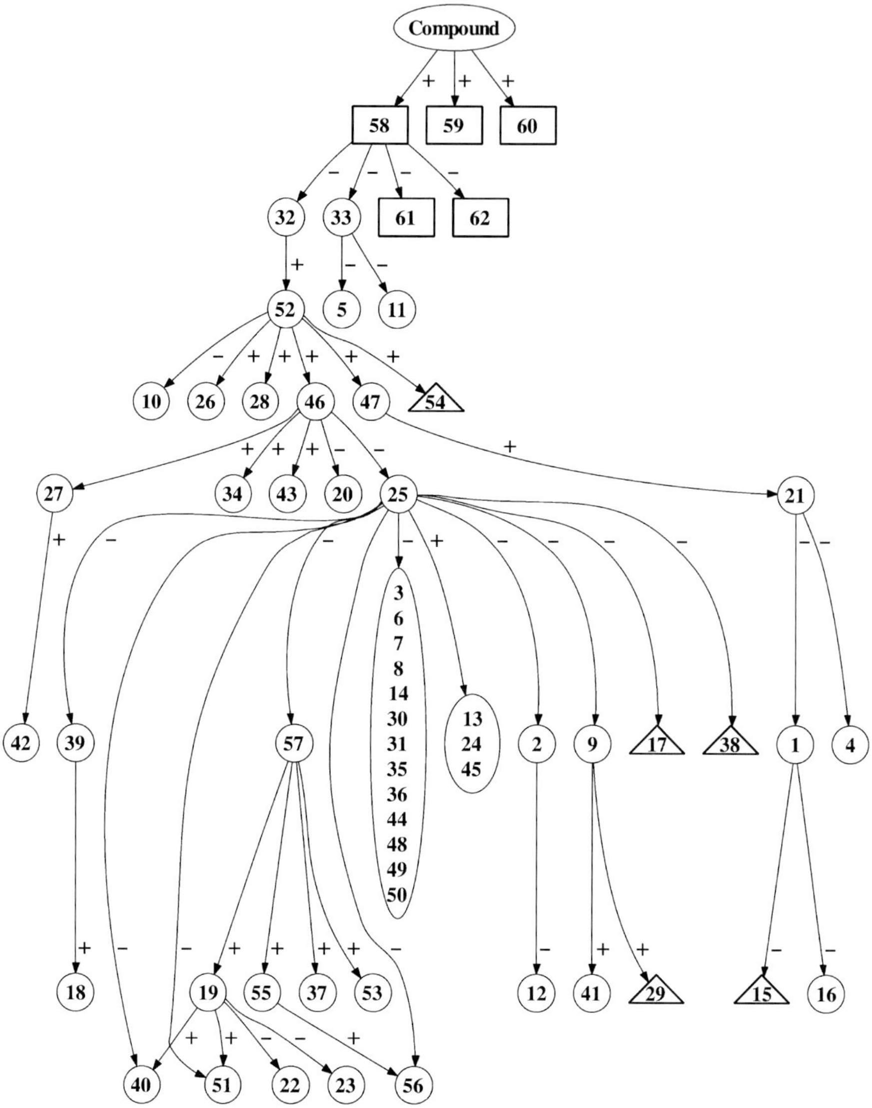

Fig. 4.4 Recovered network topology from the DBN algorithm. Actual gene names have been replaced by numbers for compactness of representation. Signs next to arrows represent influence scores for the links and indicate the nature of the relationship (excitatory or inhibitory). Rectangular nodes represent expressed genes that could be well predicted solely by the "compound" variable (genes described in the third category of the gene list in Section 4.2.5). Triangular nodes represent expressed genes that could be well predicted solely by the "time point" variable (genes described in the second category of the gene list in Section 4.2.5). Circle nodes represent expressed genes that could be well predicted by both "compound" and "time point" variables (genes described in the first category of the gene list in Section 4.2.5). Some nodes up/downregulated by node 25 are combined into one node to save space on the graph.

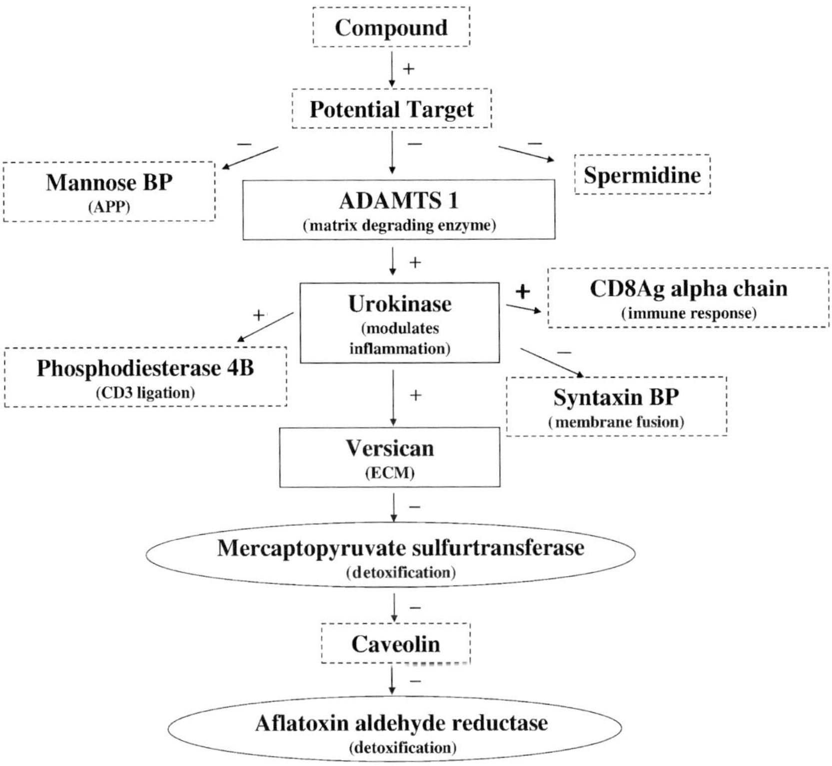

Fig. 4.5 A network fragment recovered from the Novartis data set using the DBN algorithm. Expressed genes are listed within the boxed nodes, with gene functions being described in parentheses. Solid-rectangular-boxed genes were found to be "downregulated" by the compound, in agreement with the current biological knowledge. Oblong-boxed genes were found to be "upregulated" by the compound, also in agreement with the current biological knowledge. Other variables are shown in dash-rectangular boxes.

gene names. *ADAMTS 1*, *urokinase*, and *versican* (solid-rectangular-boxed genes in Figure 4.5) were all found to be "downregulated" directly or indirectly by the "compound" node. These findings were consistent with the expected mode of action of the compound at the functional, pathophysiological level, such as inflammation modulation and repair of cartilage and bone extracellular (ECM) matrices (unpublished results). Further downstream, the inferred "upregulations" of *mercaptopyruvate sulfurtransferase* and *aflatoxin aldehyde reductase* (oblong-boxed genes) were also considered significant, in light of their involvement in detoxification and repair processes (unpublished results). The recovered network from DBN further suggested compound-induced molecular events, such as cytoskeleton reorganization (including stress fiber modulation) and cell signaling events. Some of the expressed genes that were recovered in the upstream portion of the pathways (e.g., ECM-related transcripts) were considered as closely related to the target(s) of the compound. Some of the entities recovered in the more downstream portion of the pathways (e.g., immune genes, whose names are not shown in Figure 4.5) were considered as a more indirect result from the compound modulatory actions.

4.5 Discussion and Conclusions

The time-series data from gene expression profiling studies in vertebrates *in vivo* come at a high cost and may be of a variable quality. However, such data are information rich and potentially hold a signature of the system dynamics. The present time-series data set was obtained from an *in vivo* study performed under high quality standards. This data set was analyzed via a DBN inference algorithm, in order to study time-dependent compound-induced effects in a rat disease model. This represents a first effort to apply a DBN algorithm, improved and tested in a simulation framework [27,28], to an experimentally measured gene expression data set collected from a vertebrate species.

Since thousands of genes are usually measured simultaneously in microarray experiments, and given the significant noise levels found in such *in vivo* settings, the prefiltering step yielding a final gene list for the inference algorithm is of high importance. The use of two-way ANOVA followed by PCA and PLS to control for tissue contamination reduced the list to hundreds of genes. Since the DBN framework assumes dependence between adjacent time points, strict requirements about the data needed to be considered for proper analysis. In this gene expression study, each data point represented a collection from a different animal. Previous simulation study [28] demonstrated that the particular DBN algorithm used in this study could handle a certain noise level (about 20%). However, *in vivo* data may present higher noise levels. A new screening method was therefore required and developed, to further select genes with lower variance and more likely to be involved in the pathways modulated by the compound of interest. This additional step yielded a reduced gene list that was more consistent with the assumptions made within the DBN framework.

A data pair-up method was developed here to handle a particular feature of *in vivo* gene expression data, namely, time-series data obtained from different animals. This method empirically increased the amount of data available for use in a DBN algorithm, yielding a better performance of the algorithm and as previously demonstrated with simulated data [28].

The influence score supplies detailed information about the connections predicted by DBNs in terms of the sign and strength of interactions. This is an important feature that helps in the proper interpretation of the inferred results. In a simulation study, it was shown that the sign of the influence score had a correctness of 100% [27]. In addition to correctly predicting up- or downregulation, the absolute value of the influence score also suggests how strong and how probable the inferred links are with respect to the underlying biology.

For the lack of complete knowledge of the biology itself, we could not rigorously evaluate the biological plausibility of the entire recovered network or quantitatively demonstrate the accuracy of the computational results. However, proprietary results from the biologists associated with the project show that some of the known biological connections were reproduced correctly with the algorithm. In particular, the inferred direct "target" of the compound, Potential Target, as predicted by the DBN algorithms, was expressed widely in a variety of mesenchymal cells (inflammatory cells, cartilage), as well as epithelial cells (synovium). Other expressed genes, predicted by the algorithms to be part of the compound-induced cascade of events, are also known to play important roles in anti-inflammatory and detoxifying functions. This points to promising applications of DBN to expression data sets in order to partially recover underlying pathways involved in, for example, anti-inflammatory as well as its side effects. This can help understand the mechanism of action of the drug as well as guide further development for alleviating the side effects. In addition, the inferred links also suggested novel biological predictions (hypotheses), which will guide further discovery via confirmatory experiments. Studies are in progress to experimentally confirm the validity of the direct compound target suggested by the results of this model-based analysis. In general, we believe that the Bayesian network inference algorithms, when applied to proper data sets, will be significant tools in deciphering causal pathways involved in complex biological functions.

Acknowledgment

We would like to thank Dr Dale Bodian for the statistical prefiltering analysis.

References

1 Akutsu, T. and Miyano, S. (2000) Inferring qualitative relations in genetic networks and metabolic pathways. *Bioinformatics*, **16**, 727–734.

2 Arkin, A., Shen, P. and Ross, J. (1997) A test case of correlation metric construction of a reaction pathway from measurements. *Science*, **277**, 1275–1279.

3 Liang, S., Fuhrman, S. and Somogyi, R. (1998) REVEAL a general reverse engineering algorithm for inference of genetic network architectures. *Pacific Symposium on Biocomputing*, **3**, 18–29.

4 D'haeseleer, P., Wen, X., Furham, S. and Somogyi, R. (1999) Linear modeling of mRNA expression levels during CNS development and injury. *Pacific Symposium on Biocomputing*, **4**, 41–52.

5 di Bernardo, D., Thompson, M.J., Gardner, T.S., Chobot, S.E., Eastwood, E.L., Wojtovich, A.P., Elliott, S.J., Schaus, S.E. and Collins, J.J. (2005) Chemogenomic profiling on a genome-wide scale using reverse-engineered gene networks. *Nature Biotechnology*, **23** (3), 377–383.

6 Gardner, T.S., di Bernardo, D., Lorenz, D. and Collins, J.J. (2003) Inferring genetic networks and identifying compound mode of action via expression profiling. *Science*, **301**, 102–105.

7 Friedman, N., Linial, M., Nachman, I. and Pe'er, D. (2000) Using Bayesian networks to analyze expression data. *Journal of Computational Biology*, **7**, 601–620.

8 Hartemink, A.J., Gifford, D., Jaakkola, T. and Young, R. (2001) Using graphical models and genomic expression data to statistically validate models of genetic regulatory networks. *Pacific Symposium on Biocomputing,* **6**, 422–433.

9 Weaver, D.C., Workman, C.T. and Stormo, G.D. (1999) Modeling regulatory networks with weight matrices. *Pacific Symposium on Biocomputing,* **4**, 112–123.

10 Xu, H., Wu, P., Wu, C., Tidwell, C. and Wang, Y. (2002) A smooth response surface algorithm for constructing a gene regulatory network. *Physiological Genomics,* **11**, 11–20.

11 Hartemink, A.J., Giford, D., Jakkoola, T. and Young, R. (2002) Combining location and expression data for principled discovery of genetic regulatory network models. *Pacific Symposium on Biocomputing,* **7**, 437–449.

12 Husmeier, D. (2003) Sensitivity and specificity of inferring genetic regulatory interactions from microarray experiments with dynamic Bayesian networks. *Bioinformatics,* **19**, 2271–2282.

13 Smith, V.A., Jarvis, E.D. and Hartemink, A.J. (2002) Evaluating functional network inference using simulations of complex biological systems. *Bioinformatics,* **18**, S216–S224.

14 Smith, V.A., Jarvis, E.D. and Hartemink, A.J. (2003) Influence of network topology and data collection on functional network influence. *Pacific Symposium on Biocomputing,* **8**, 164–175.

15 Jarvis, E.D., Smith, V.A., Wada, K., Rivas, M.V., McElroy, M., Smulders, T.V., Carninci, P., Hayashizaki, Y., Dietrich, F., Wu, X., McCnnell, P., Yu, J., Wang, P.P., Hartemink, A.J. and Lin, S. (2002) A framework for integrating the songbird brain. *Journal of Comparative Physiology A – Zoology,* **188**, 961–980.

16 Neapolitan, R.E. (2003) *Learning Bayesian Networks,* Prentice Hall, Harlow.

17 Heckerman, D. (1996) A tutorial on learning with Bayesian networks. *Technical Report* MSR-TR-95–06, Microsoft Research.

18 Kim, S.Y., Imoto, S. and Miyano, S. (2003) Inferring gene networks from time series microarray data using dynamic Bayesian networks. *Briefings in Bioinformatics,* **4** (3), 228–235.

19 Kim, S.Y., Imoto, S. and Miyano, S. (2004) Dynamic Bayesian network and non-parametric regression for nonlinear modeling of gene networks from time series gene expression data. *Biosystems,* **75** (1–3), 57–65.

20 Ong, I., Glasner, J. and Page, D. (2002) Modeling regulatory pathways in *E. coli* from time series expression profiles. *Bioinformatics,* **18** (Suppl. 1), 241–248.

21 Perrin, B.E., Ralaivola, L., Mazurie, A., Bottani, S., Mallet, J. and D'Alchebuc, F. (2003) Related articles, links gene networks inference using dynamic Bayesian networks. *Bioinformatics,* **19** (Suppl. 2), II138–II148.

22 Wang, S.C. (2004) Reconstructing genetic networks from time ordered gene expression data using Bayesian method with global search algorithm. *Journal of Bioinformatics and Computational Biology,* **2** (3), 441–458.

23 Zou, M., Conzen, S.D. (2004) A new dynamic Bayesian network (DBN) approach for identifying gene regulatory networks from time course microarray data. *Bioinformatics,* **21** (1), 71–79.

24 Li, Z. and Chan, C. (2004) Inferring pathways and networks with a Bayesian framework. *FASEB Journal,* **18** (6), 746–748.

25 Sachs, K., Gifford, D., Jaakkola, T., Sorger, P. and Lauffenburger, D.A. (2002) Bayesian network approach to cell signaling pathway modelling. *Science's Signal Transduction Knowledge Environment (STKE),* **148**, PE38.

26 Woolf, P.J., Prudhomme, W., Daheron, L., Daley, G.Q. and Lauffenburger, D.A. (2005) Bayesian analysis of signaling networks governing embryonic stem cell fate decision. *Bioinformatics*, **21**, (6), 741–753.

27 Yu, J., Smith, V.A., Wang, P.P., Hartemink, A.J. and Jarvis, E.D. (2004) Advances to Bayesian network inference for generating causal networks from observational biological data. *Bioinformatics*, **20**, 3594–3603.

28 Yu, J. (2005) Developing Bayesian network inference algorithms to predict causal functional pathways in biological systems. Ph.D. Dissertation.

29 Heckerman, D. (1999) *Learning in Graphical Models*, MIT Press, Cambridge, MA.

30 Jensen, F.V. (1998) *An Introduction to Bayesian Networks*, University College London Press, London.

31 Pearl, J. (2000) *Causality: Models, Reasoning, and Inference*, Cambridge University Press, Cambridge.

32 Cooper, G.F. and Herskovits, E. (1992) A Bayesian method for the induction of probabilistic networks from data. *Machine Learning*, **9** (4), 309–347.

33 Murphy, K. and Mian, S. (1999) Modeling gene expression data using dynamic Bayesian networks. *Technical Report*, University of California, Berkeley.

34 Chickering, D.M. (1996) Learning Bayesian networks is NP-complete, in *Learning from Data: Artificial Intelligence and Statistics*, **112**, (eds V.D. Fisher and H.J. Lenz), Springer-Verlag, pp. 121–130.

35 Buntine, W. (1991) Theory refinement on Bayesian networks.Proceedings of the Seventh Annual Conference on Uncertainty Artificial Intelligence, pp. 52–60.

36 Spirtes, P. and Meek, C. (1995) Learning Bayesian networks with discrete variables from data. KDD, pp. 294–299.

37 Dejori, M. and Stetter, M. (2003) Bayesian inference of genetic networks from gene expression data: convergence and reliability. Proceedings of the 2003 International Conference on Artificial Intelligence, pp. 323–327.

38 Hubbell, E., Liu, W.M. and Mei, R. (2002) Robust estimators for expression analysis. *Bioinformatics*, **18**, 1585–1592.

39 Imoto, S., Goto, T. and Miyano, S. (2002) Estimation of genetic networks and functional structures between genes using Bayesian networks and nonparametric regression. *Pacific Symposium on Biocomputing*, **7**, 175–186.

5
Reverse Engineering Gene Regulatory Networks with Various Machine Learning Methods

Marco Grzegorczyk, Dirk Husmeier, and Adriano V. Werhli

5.1
Introduction

During the last decade, the development of high-throughput postgenomic biotechnologies has resulted in the production of exponentially expanding sets of data, such as microarray gene expression profiles and yeast two-hybrid interactions. This has prompted substantial interest in reverse engineering molecular biological regulatory networks and signaling pathways in the cell from these novel types of data. One of the first seminal papers promoting this approach aimed at learning gene regulatory networks in *Saccharomyces cerevisiae* from microarray gene expression profiles [1]. Since then, several authors have applied various machine learning methods to reconstruct regulatory networks from different types of postgenomic data [2–7]. These methods are of enormous importance in the comparatively recent field of genetical genomics [8], where QTL marker analysis is first applied to identify putative sets of regulatory genes, from which then a more refined regulatory network is to be reverse engineered.

Given the diversity of proposed reverse engineering methods, it is important for the systems biology community to obtain a better understanding of their relative strengths and weaknesses. One of the first major evaluation studies was carried out in [9]. The authors simulated a complex biological system at different levels of organization, involving behavior, neural anatomy, and gene expression of songbirds. They then tried to infer the structure of the known true genetic network from the simulated gene expression data with Bayesian networks. A related study [10] evaluated the accuracy of reverse engineering gene regulatory networks with Bayesian networks from data simulated from realistic molecular biological pathways, where the latter were modeled with a system of coupled differential equations. This network was also used in an earlier study by Zak *et al.* [11], who investigated the inference accuracy of deterministic linear and log-linear models. Recently, a more comprehensive evaluation study was carried out in [12].

To complement these studies and to particularly evaluate the relative strengths and weaknesses of different machine learning approaches, we have carried out a

Analysis of Microarray Data: A Network-Based Approach. Edited by F. Emmert-Streib and M. Dehmer

ISBN: 978-3-527-31822-3

comprehensive cross-method comparison of three principled reverse engineering paradigms:

1. relevance networks (RNs), where pairwise association scores between potentially interacting nodes are inferred in a reductionist sense, independent of the remaining network;
2. graphical Gaussian models (GGMs), where a constraint-based approach to inference is applied in a holistic sense to infer direct associations between nodes, represented in an undirected graph; and
3. Bayesian networks (BNs), where a score-based approach to inference is applied to infer a directed graph, which under certain regularity conditions might indicate potentially causal interactions.

Our study is based on the RAF protein signaling pathway reported in [13], which describes the interaction of 11 phosphorylated proteins and phospholipids in human immune system cells; this is to ensure that the underlying network topology is biologically realistic and relevant. We used two types of data for evaluation: the measured protein concentrations reported in [13] and synthetically generated data. The latter were obtained from a multivariate Gaussian distribution, as well as a modified steady-state approximation to a differential equation description of chemical kinetics, as proposed in [14,15]. Of particular interest in our study is a comparison between passive observations and active interventions and a quantification of the improvement in network reconstruction accuracy obtained from the latter.

Our chapter is organized as follows. We start with a concise yet self-contained introduction to three machine learning approaches that have been applied to the reverse engineering of gene regulatory networks (Section 5.2). This is followed by a description of the gold-standard regulatory network, on which our comparative evaluation was based (Section 5.3), a discussion of the evaluation criteria (Section 5.4), and an overview of the benchmark data – both synthetic data and that from cytometry experiments (Section 5.5). We describe our simulation studies in Section 5.6, followed by a presentation of the results in Section 5.7, a discussion of our findings in Section 5.8, and a concluding summary in Section 5.9.

5.2 Methods

The objective of this section is to provide an overview of various machine learning methods that have recently been applied to the reconstruction of gene regulatory networks. A gene regulatory network is a graphical representation of interactions between genes. Thus, the relations between genes are usually represented as if the expressions of genes would directly affect the expression levels of other genes. It is not explicitly mentioned that the interactions between genes are actually mediated by proteins, metabolites, and/or other protein–metabolite complexes. That is, in gene networks, the "genes" must be interpreted as abstractions and may also include proteins encoded by these genes. As most genes code for proteins, an interaction between two genes in a gene network may indicate that the encoded proteins interact with each other, or it may be due to the fact that one gene codes for a protein that interacts with a transcription factor regulating the other gene. More precisely, each gene contains the information (code) for the synthesis of a particular protein, and the synthesis of the proteins is regulated by various control mechanisms. Various proteins serve as transcription factors, which can bind to regulatory sites of other genes. In this way, the synthesis of other specific proteins encoded by these genes is regulated. This results in complex molecular pathways in which the individual components interact. In these pathways, genes usually do not interact directly with each other. Rather, their activities (expressions) determine indirectly, that is, via the synthesis of proteins, the expressions of other genes.

From this point of view, gene networks must be interpreted as a simplification of the real molecular biological regulatory mechanisms within cells. But on the contrary, gene networks are capable of representing the indirect interactions between genes, that is, the final effect of the activity of one gene on the activities of other genes, whereby the exact detailed molecular biological mechanisms are omitted. Often only gene expression data are available, for example, from microarray experiments. Consequently, extracting gene networks is the first but important step toward uncovering the complete biochemical regulatory mechanisms in cells.

From a mathematical point of view, the following situation is given. There are different biological variables, X_i ($i = 1, \ldots, n$), for example, each measuring the expression of a certain gene in a cell, and a sample of size m of these variables is available ($x_{1j}, \ldots, x_{nj}$) ($j = 1, \ldots, m$). That is, x_{ij} is the value (e.g., expression) of the ith domain variable (e.g., gene or protein) in the jth sample. The goal is to find all dependencies between these n variables. Different graphical models for discovering these dependencies have been proposed in the literature. Although they are based on different statistical aspects, most of them lead to the same kind of result, that is, a mathematical graph representing all dependencies between the variables. Such a graph consists of a set of nodes, and each node corresponds to a variable. The edges between the nodes of the graph correspond to interactions or probabilistic dependencies between the variables. In other words, the variables $X_1, \ldots, X_n$ are in one-to-one correspondence with the nodes in a graph, and the set of edges determines the relationships between them. Hence, the terms "variables" and "nodes" can be used interchangeably. Furthermore, the terms "graph" and "network" will be used interchangeably. The exact meaning of the edges depends on the employed graphical model, and can be related to pairwise correlations, direct stochastic dependence relations conditional on the whole system, or causal interactions. We will discuss these aspects in more detail in the following subsections.

5.2.1 Relevance Networks

The method of *relevance networks*, proposed in [16], is exclusively based on pairwise association scores and therefore represents a very simple machine learning approach to reverse engineering regulatory networks. An association score is computed for all pairs of variables X_i and X_j ($i, j \in \{1, \ldots, n\}$) from the data. For continuous data, the Pearson correlation coefficient can be used:

$$\text{corr}(x,y) = \frac{(1/m)\sum_{i=1}^{m}(x_i-\bar{x})(y_i-\bar{y})}{\sqrt{(1/m)\sum_{i=1}^{m}(x_i-\bar{x})^2}\sqrt{(1/m)\sum_{i=1}^{m}(y_i-\bar{y})^2}}, \tag{5.1}$$

where $x=(x_1, \ldots, x_n)$ and $y=(y_1, \ldots, y_m)$ are the m-dimensional observations of two different variables with empirical means $\bar{x}$ and $\bar{y}$. Interpreting the variables as the nodes of a network, the pairwise association scores are compared with a predefined threshold value, and the nodes whose pairwise association scores exceed this threshold are linked by an undirected edge.

In a relevance network, the interactions are not inferred within the context of the whole system, that is, there is no distinction between direct and indirect interactions. Not rarely does a high correlation coefficient between two nodes indicate only a pseudo-association, for example, if both nodes depend on a common regulator. Hence, a high correlation coefficient between two nodes does not necessarily indicate a direct association, and with regard to the graphical representation of the network, only the direct interactions are of interest.

In statistical terminology, a relevance network based on the Pearson correlation is referred to as a *covariance graph*. The threshold value can be estimated by a randomization test so as to keep the number of false positive edges below an a priori specified tolerance level. For further details see [16]. Alternatively, instead of the Pearson correlation, the mutual information can be used to compute the pairwise association scores in relevance networks. Mutual information scores can be computed for discrete variables only, so that continuous data have to be discretized; this incurs a certain information loss. But an advantage is that this score can deal with nonlinear associations. See [16] for further details.

5.2.2 Gaussian Graphical Models

A more promising machine learning method is given by GGMs. These models are based on the assumption that the data are distributed according to a multivariate Gaussian distribution $N(\mu, \Sigma)$. But to avoid the shortcomings of relevance networks based on Pearson correlation coefficients (see Section 5.2), partial correlations are considered in Gaussian graphical models. That is, the strength of a direct association between two nodes X_i and X_j is measured by the partial correlation coefficient $\pi_{i,j}$, which describes the correlation between these two nodes that is conditional on all the other nodes in the system. From the theory of normal distributions, it is known that the partial correlation coefficients $\pi_{i,j}$ can be computed from the inverse $\Omega=\Sigma^{-1}$ of the covariance matrix Σ via the following relation [17]:

$$\pi_{i,j} = \frac{-\omega_{i,j}}{\sqrt{\omega_{i,i}\omega_{j,j}}}, \tag{5.2}$$

where $\omega_{i,j}$ are the elements of the matrix Ω.

Hence, to reconstruct a Gaussian graphical model from a given data set D, the following procedure can be employed. From the data D, the empirical covariance matrix is estimated and inverted to obtain Ω, and subsequently the entries $\pi_{i,j}$ of the partial correlation matrix Π can be computed using (5.2). As small elements $\pi_{i,j}$ in Π correspond to weak partial correlations, the corresponding nodes become unconnected. Conversely, high entries in the matrix correspond to strong partial correlations, which indicate direct associations between the corresponding nodes.

The disadvantage of this procedure is that the empirical covariance matrix can only be inverted if the number of observations exceeds the number of nodes in the network, that is, if the matrix is nonsingular. Even for nonsingular matrices, problems occur when the eigenvalue spectrum extends over too many orders of magnitude, that is, if the matrix is ill conditioned. Note that in applications to gene expression data, the number of variables usually exceeds the number of observations, rendering the empirical covariance matrix singular and, hence, noninvertible.

To learn a Gaussian graphical model from such data, Schäfer and Strimmer [7] have proposed the application of a shrinkage covariance estimator. The shrinkage estimator $\widehat{\Sigma}$ is guaranteed to be nonsingular so that it can be inverted to obtain a new estimator $\widehat{\Omega} = \left(\widehat{\Sigma}\right)^{-1}$ for the matrix Ω. The key idea is as follows. It is known that the (unconstrained) maximum likelihood estimator $\widehat{\Sigma}_{\mathrm{ML}}$ for the covariance matrix Σ has a high variance if the number of nodes exceeds the number of observations ($n > m$). On the contrary, there are several potentially constrained estimators that have a certain bias but a much lower variance. The shrinkage approach combines the maximum likelihood estimator with one of these constrained estimators $\widehat{\Sigma}_{\mathrm{C}}$ in a weighted average:

$$\widehat{\sum} = (1-\lambda)\widehat{\sum}_{\mathrm{ML}} + \lambda\widehat{\sum}_{\mathrm{C}}, \tag{5.3}$$

where $\lambda \in [0,1]$ denotes the shrinkage intensity. The authors show that the Ledoit Wolf lemma [18] can be applied to compute the shrinkage intensity λ that minimizes the expected mean square error (MSE) in the sense of classical statistical decision theory, that is, with respect to the distribution over the data. For practical applications, the authors approximate the optimal intensity A and replace the intractable expectation values by their corresponding sample averages.

In their paper [7], the authors focus on the constrained estimator $\widehat{\Sigma}_{\mathrm{C}}$ by assuming that the variables (nodes) are pairwise uncorrelated ($\Sigma_{i,k} = 0$ for $i \neq k$) but may have unequal variances ($\Sigma_{i,i} \neq \Sigma_{k,k}$ for $i \neq k$). The shrinkage covariance estimator $\widehat{\Sigma} = \left(\widehat{\Sigma}\right)_{\mathrm{i,j}}$ is then given by

$$\left(\widehat{\sum}\right)_{i,j} = \left\{ \begin{array}{ll} s_{ii}^2, & i = j, \\ s_{ij}^2 \min\{1, \max\{0, 1-\widehat{\lambda *}\}\}, & i \neq j, \end{array} \right\}, \tag{5.4}$$

where the shrinkage intensity $\widehat{\lambda^*}$ for an unbiased estimator is given by

$$\widehat{\lambda^*} = \frac{\sum_{i=1}^{n-1}\sum_{j=i+1}^{n}\widehat{\mathrm{Var}\left(r_{ij}^2\right)}}{\sum_{i=1}^{n-1}\sum_{j=i+1}^{n}\left(r_{ij}^2\right)^2}. \tag{5.5}$$

In Equation 5.4, s_{ij}^2 is the empirical covariance between variables X_i and X_j:

$$S_{ij}^2 = \frac{1}{m-1}\sum_{k=1}^{m}(x_{ik}-\overline{x_i})(x_{jk}-\overline{x_j}), \tag{5.6}$$

and r_{ij}^2 is the corresponding empirical correlation:

$$r_{ij}^2 = \frac{s_{ij}^2}{\sqrt{s_{ii}^2 s_{jj}^2}}. \tag{5.7}$$

The variances of the correlations in Equation 5.5 can be estimated as follows:

$$\widehat{\mathrm{Var}(r_{ij}^2)} = \frac{m}{(m-1)^3}\sum_{k=1}^{m}(w_{kij}-\overline{w_{ij}})^2, \tag{5.8}$$

with

$$w_{kij} = \left(\sqrt{s_{ii}^2 s_{jj}^2}\right)^{-1}(x_{ki}-\overline{x_i})(x_{kj}-\overline{x_j}) \quad i,j\in\{1,\ldots,n\}$$
$$\text{and } k\in\{1,\ldots,m\} \text{ and } \overline{w_{ij}} = \frac{1}{m}\sum_{k=1}^{m} w_{kij}$$

In these equations, x_{ik} ($k=1, \ldots, m$) is the kth observation of the ith domain variable X_i ($i=1, \ldots, n$). Computing the inverse of $\widehat{\Sigma}$ and applying Equation 5.2 leads to the shrinkage estimator $\widehat{\Pi}$ of the partial correlation coefficient matrix Π.

5.2.3 Bayesian Networks

5.2.3.1 Introduction to Bayesian Networks

BNs are interpretable and flexible models for representing probabilistic relationships between multiple interacting variables. At a qualitative level, the graph of a BN describes the relationships between the variables in the form of conditional independence relations. At a quantitative level, (local) relationships between variables are described by (conditional) probability distributions. Formally, a BN is defined by a graph $\mathcal{G}$, a family of (conditional) probability distributions F, and their parameters q, which together specify a joint distribution over all domain variables.

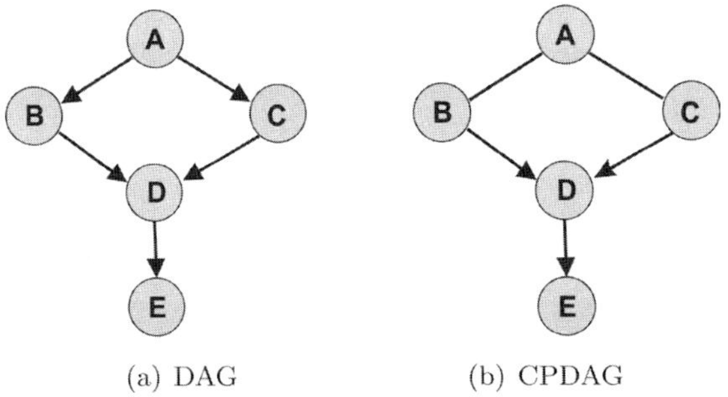

Figure 5.1 Bayesian network and CPDAG. (a) Example of a Bayesian network (DAG) with five nodes and five directed edges. (b) The CPDAG representation of the DAG presented in panel (a). Two edges of the DAG in panel (a) are reversible and become undirected in the CPDAG.

The graph $\mathcal{G}$ of a Bayesian network consists of a set of n nodes $X_1, \ldots, X_n$ and a set of directed edges. As before, each node represents a variable (for which we use the same symbol X_i), and the directed edges indicate dependence relations between the variables. If there is a directed edge pointing from node X to node Y, then X is called a parent (node) of Y, and Y is called a child (node) of X. If a node Z can be reached by following a path of directed edges, starting at node X, Z is called a descendant of X, while X is called an ancestor of Z. The set of all parents $\mathrm{pa}(X|\mathcal{G})$ of a node X is defined as the parent set of X in $\mathcal{G}$. In panel (a) of Figure 5.1, where an example of a simple Bayesian network over five variables is given, the nodes B and C are parents of node D, node E is a child of node D, and node E is a descendant of all four other nodes. That is, A, B, C, and D are ancestors of node E. The parent sets are given by $\mathrm{pa}(\mathrm{A}|\mathcal{G}) = \{\}$, $\mathrm{pa}(\mathrm{B}|\mathcal{G}) = \mathrm{pa}(\mathrm{C}|\mathcal{G}) = \{\mathrm{A}\}$, $\mathrm{pa}(\mathrm{D}|\mathcal{G}) = \{\mathrm{B}, \mathrm{C}\}$, and $\mathrm{pa}(\mathrm{E}|\mathcal{G}) = \{\mathrm{D}\}$.

BNs are based on directed acyclic graphs (DAGs), that is, graphs with directed edges in which no node can be its own descendant. Graphically, this means that there are no cycles of directed edges (loops) in DAGs. It is due to this acyclicity that the joint probability distribution in BNs can be factorized into a product of simpler local distributions [19]:

$$P(X_1, \ldots, X_n|\mathcal{G}) = \prod_{i=1}^{n} P(X_i|\mathrm{pa}(X_i|\mathcal{G})). \tag{5.9}$$

The factors in the product in Equation 5.9 are referred to as local probability distributions. For the DAG represented in panel (a) of Figure 5.1 the application of Equation 5.9 yields the following factorization of the joint probability distribution:

$$P(A, B, C, D, E|\mathcal{G}) = P(A)P(B|A)P(C|A)P(D|B, C)P(E|D). \tag{5.10}$$

Owing to this factorization, the joint probability distribution becomes a product of simpler conditional probability distributions.

More than one DAG can imply exactly the same set of (in)dependencies. For example, the following two DAGs $(\mathcal{G}_1): X_1 \rightarrow X_2$ and $(\mathcal{G}_2): X_1 \leftarrow X_2$ over the domain X_1, X_2 imply that the variables X_1 and X_2 are not stochastically independent. This leads to identical probability distributions for both DAGs:

$$P(X_1, X_2|\mathcal{G}_1) = P(X_1)P(X_2|X_1) = P(X_1, X_2) = P(X_2)P(X_1|X_2) = P(X_1, X_2|\mathcal{G}_2).$$

This implies that the graphs $\mathcal{G}_1$ and $\mathcal{G}_2$ only show alternative possibilities of describing the same set of conditional independence relations. Consequently, the independence assumptions of a Bayesian network cannot be uniquely represented by DAGs. If two DAGs over the same domain assert the same set of independence assumptions among the variables, those graphs are said to be *equivalent*. This relation of graph equivalence imposes a set of *equivalence classes* over DAGs. The directed acyclic graphs within an equivalence class have the same underlying undirected graph, but may disagree on the direction of some of the edges.

Verma and Pearl [20] prove that two directed acyclic graphs are equivalent if and only if they have the same *skeleton* and the same set of *v-structures*. The skeleton of a DAG is defined as the undirected graph resulting from ignoring all edge directions. And a v-structure denotes a configuration of two directed edges converging on the same node without an edge between the parents (see [21]). For example, the DAG in panel (a) of Figure 5.1 has only one v-structure, symbolically: $B \rightarrow D \leftarrow C$. Chickering [21] shows that equivalence classes of DAGs can be uniquely represented using *completed partially directed acyclic graphs* (CPDAGs). A CPDAG contains the same skeleton as the original DAG, but possesses both directed (compelled) and undirected (reversible) edges. Every compelled edge $X_i \rightarrow X_j$ in a CPDAG denotes that all DAGs of this class contain this edge, whereas every reversible edge $X_i - X_j$ in the CPDAG representation denotes that some DAGs contain the directed edge $X_i \rightarrow X_j$, while others contain the oppositely orientated edge $X_i \leftarrow X_j$. A directed edge of the DAG is compelled in the CPDAG if and only if its reversal changes the set of v-structures, otherwise it is reversible. An algorithm that takes a DAG as an input and outputs the corresponding CPDAG representation can be found in [22]. As an example, in panel (a) of Figure 5.1 the edges $B \rightarrow D$ and $C \rightarrow D$ are compelled as their reversal would delete the v-structure $B \rightarrow D \leftarrow C$. The edge $D \rightarrow E$ is also compelled, as its reversal would lead to two new v-structures, symbolically: $B \rightarrow D \leftarrow E$ and $C \rightarrow D \leftarrow E$. The CPDAG representation of the DAG shown in panel (a) of Figure 5.1 can be found in panel (b).

5.2.3.2 **Learning Causal Relationships**

Although Bayesian networks are based on directed (acyclic) graphs (DAGs), it is important to note that not all directed edges in a Bayesian network can be interpreted causally. Like a Bayesian network, a causal network is mathematically represented by a DAG. However, the edges in a causal network have a stricter interpretation: the parents of a variable are its immediate causes. In the presentation of a causal network, it is meaningful to make the causal Markov assumption: given the values of a variable's immediate causes, it is independent of its earlier causes. Under this assumption, a causal network can be interpreted as a Bayesian network in that it satisfies the corresponding Markov independencies. However, the reverse does not hold. The DAG on which the Bayesian network model is based just asserts a set of independence assumptions among the domain variables. More precisely, for each DAG, we have the condition that given a domain variable X and parent nodes pa(X), X is independent of all its other ancestors. However, the same set of independence assumptions can often be asserted by different (equivalent) DAGs having the same skeleton but edges with opposite orientations, as discussed above. Consequently, not every edge can indicate a causal relationship. The only way to interpret an edge causally is if we have no hidden variables and if all DAGs that are equivalent to each other (i.e., assert the same set of independence assumptions) agree on an edge direction, that is, if the respective edge is directed in the corresponding CPDAG representation. In Section 5.2.3.6, we will discuss ways to increase the number of directed edges in equivalence classes by active interventions; in this way, the number of putative causal interactions can be increased. However, a critical assumption made in this approach is the absence of any latent or hidden variables. If this assumption is violated, the observation that two variables depend on each other probabilistically can be explained by the existence of an unobserved common cause. Since we are usually unable to rule out the existence of latent factors, we interpret the existence of directed edges in CPDAGs as *putative* causal interactions, which ultimately require an experimental validation. For a more detailed treatment of this subject, see [23,24].

5.2.3.3 Bayesian Network Scoring Metrics

A stochastic model specifies the distributional form F and the parameters q of the local probability distributions $P(X_i|pa(X_i|\mathcal{G}))$ $(i = 1, \ldots, n)$. That is, it asserts a distribution to each domain node X_i in dependence of its parent nodes $pa(X_i|\mathcal{G})$, and the parent sets are implied by the DAG $\mathcal{G}$. Those local probability distributions together specify the joint probability distribution of all domain variables $P(X_i, \ldots, X_n|\mathcal{G})$ when the factorization rule (see Equation 5.9) is applied. Consequently, given a data set, stochastic models can be used to score DAGs with respect to their posterior probabilities $P(\mathcal{G}|D, F, q)$. Neglecting F and q, the posterior probability of a directed acyclic graph $\mathcal{G}$ given the data D can be represented as follows:

$$P(\mathcal{G}|D) = \frac{P(\mathcal{G}, D)}{P(D)} = \frac{P(D|\mathcal{G})P(\mathcal{G})}{\sum_{\mathcal{G}' \in \Omega} P(D|\mathcal{G}')P(\mathcal{G}')} = \frac{P(D|\mathcal{G})P(\mathcal{G})}{Z_c}, \tag{5.11}$$

where $P(\mathcal{G})$ $(\mathcal{G} \in \Omega)$ is the prior probability over the space Ω of all possible DAGs for the domain $X_1, \ldots, X_n$. $P(D|\mathcal{G})$ is the marginal likelihood, that is, the probability of the data D given the graph $\mathcal{G}$. As the number of possible DAGs in Ω grows exponentially with the number of domain nodes n, the denominator on the right-hand side of Equation 5.11, which is a sum over the whole DAG space Ω, is not tractable for high n, say $n > 6$. As it does not depend on $\mathcal{G}$, it is a normalization factor that can be denoted by Z_c.

The marginal likelihood of the data is defined as the integral of the likelihood $f(D|q,\mathcal{G})$ and the parameter prior distribution $f(q|\mathcal{G})$ over all possible parameter values q:

$$P(D|\mathcal{G}) = \int f(D,q|\mathcal{G})dq = \int f(D|q,\mathcal{G})f(q|\mathcal{G})dq, \tag{5.12}$$

whereby the parameter vector $q = q(F,\mathcal{G})$ and its dimension depend on the distributional form F as well as on the graph $\mathcal{G}$, which specifies the dependencies among the domain variables. The marginal likelihood as an integral over the whole parameter space protects against data overfitting, as it includes a penalty for model complexity. That is, it balances the ability of the Bayesian network model to explain the data with the ability to do so economically. A commonly used graph prior $P(\mathcal{G})$ ($\mathcal{G} \in \Omega$) is given by

$$P(\mathcal{G}) = \frac{1}{Z}\prod_{i=1}^{n}\binom{n-1}{|\mathrm{pa}(X_i\mathcal{G})|}^{-1}, \tag{5.13}$$

where Z is a normalization constant and $|\mathrm{pa}(X_i|\mathcal{G})|$ is the cardinality of the parent set $\mathrm{pa}(X_i|\mathcal{G})$ of X_i in $\mathcal{G}$. We note that the graph prior $P(\mathcal{G})$ consists of n local factors. Under certain regularity conditions, discussed, for instance, in [25,26], a closed-form solution can be derived for the marginal likelihood $P(D|\mathcal{G})$, so that the posterior probability $P(D|\mathcal{G})$ (see Equation 5.11) has a modular form:

$$P(\mathcal{G}|D) = \frac{1}{Z_c}P(D|\mathcal{G})P(\mathcal{G}) = \frac{1}{Z_c}\prod_{i=1}^{n}\psi[X_i|\mathrm{pa}(X_i|\mathcal{G}),D]. \tag{5.14}$$

Here, Z_c is a normalization factor and $\psi[X_i|\mathrm{pa}(X_i|\mathcal{G}),D]$ are local scores that result from a factorization of both the graph prior $P(\mathcal{G})$ and the likelihood $P(D|\mathcal{G})$. The likelihood is factorized with respect to Equation 5.9 so that there is one factor for each variable X_i and the corresponding parent set $\mathrm{pa}(X_i|\mathcal{G})$ is implied by the DAG $\mathcal{G}$. The functional forms of the factors depend on the stochastic model that is employed. Two widely applied stochastic models are 1. the linear Gaussian model with a normal-Wishart distribution as the conjugate prior (BGe model) and 2. the multinomial distribution with a Dirichlet prior (BDe model).

Details about the BDe model can be found in [26]. We discuss the BGe model in the following subsection.

5.2.3.4 The Gaussian BGe Scoring Metric

The continuous Gaussian model for Bayesian networks assumes Gaussian distributions for the variables, and the resulting scores are usually referred to as the BGe (Bayesian metric for Gaussian networks having score equivalence) score. Each variable X_i is interpreted as a Gaussian distributed random variable, whose mean value $E[X_i]$ depends on the values of its parent variables. That is, if a DAG $\mathcal{G}$ is given in which node X_i has u parent nodes $X_{i_1}, \ldots, X_{i_u}$, the distribution of X_i is given by

$$X_i \sim N\left(\mu_i + \sum_{j=1}^{n} b_{ij}(x_j - \mu_j), \sigma_i^2\right), \tag{5.15}$$

where μ_i is the unconditional mean of X_i, σ_i^2 is the conditional variance of X_i given the realizations $X_1 = x_1, \ldots, X_n = x_n$, and the coefficients b_{ij} reflect the strengths of the dependencies between X_i and the other variables. If $j \notin \{i_1, \ldots, i_u\}$ then $b_{ij} = 0$ so that the realization x_i of X_i is interpreted in dependence of the realizations $x_{i_1}, \ldots, x_{i_u}$ of the parent variables $X_{i_1}, \ldots, X_{i_u}$ only. Hence, each coefficient $b_{ij} \neq 0$ represents an edge in the DAG $\mathcal{G}$ that points from node X_j to node X_i.

The coefficients B_{ij} and the conditional variances σ_i^2 can be used to compute the *precision matrix* W of the joint multivariate Gaussian distribution with the following recursive formula [25]:

- Set $W(1) = 1$ and define $\vec{b}_i$ as the following column vector of length $i-1$: $\vec{b}_i = (b_{1,i}, \ldots, b_{i-1,i})^T \; (i = 1, \ldots, n)$.
- For $i = 1, \ldots, n-1$ compute $W(i+1)$ from $W(i)$, σ_i^2, and $\vec{b}_{i+1}$ as follows:

$$W(i+1) = \begin{pmatrix} W(i) + \left(\vec{b}_{i+1}\vec{b}_{i+1}^T\sigma_{i+1}^2\right) & -\vec{b}_{i+1}\sigma_{i+1}^2 \\ -\vec{b}_{i+1}^T\sigma_{i+1}^2 & \sigma_{i+1}^2 \end{pmatrix},$$

$W(n)$ is the precision matrix W for the joint Gaussian distribution of the variables $X_1, \ldots, X_n$, and the covariance matrix Σ is the inverse of the precision matrix: $\Sigma = W^{-1}$. Defining the unconditional mean vector $\mu = (\mu_1, \ldots, \mu_n)^T$, the joint Gaussian distribution is given by $(X_1, \ldots, X_n) \sim N(\mu, \Sigma)$. For the unknown parameters, we assume the conjugate normal-Wishart prior distribution, that is, we assume that the prior distribution over the unknown parameter vector μ is a Gaussian distribution with mean μ_0 and precision matrix νW with $\nu > 0$, whereby the matrix W in turn is Wishart distributed with $\alpha > n+1$ degrees of freedom and precision matrix T_0. The matrix T_0, the vector μ_0, as well as the degrees of freedom α and the factor ν are unknown hyperparameters that have to be specified in advance and can be used to include some background knowledge about the domain. See [25] for further details.

For the conjugate normal-Wishart parameter prior, we can derive a closed-form solution for the marginal likelihood when the data set D is complete (without any missing values) and the following two conditions of parameter independence and parameter modularity are satisfied [25]:

- parameter independence $P(\sigma_1^2, \ldots, \sigma_n^2, \vec{b}_1, \ldots, \vec{b}_n|\mathcal{G}) = \prod_{i=1}^{n} P(\sigma_i^2, \vec{b}_i|\mathcal{G})$,
- parameter modularity $P(\sigma_i^2, \vec{b}_i|\mathcal{G}) = P(\sigma_i^2, \vec{b}_i|pa(X_i|\mathcal{G}))$.

Parameter independence means that the unknown parameters of the local probability distributions are stochastically independent of each other. This is automatically satisfied when the normal-Wishart prior is used. Parameter modularity is an additional assumption and means that the prior distributions of the parameters of the ith local probability distribution depends only on the parent variables $\text{pa}(X_i|\mathcal{G})$ and not on the whole graph $\mathcal{G}$. The latter assumption is needed to derive a closed-form solution (see Equation 5.19) for an *arbitrary* DAG $\mathcal{G}$ from the closed-form solution Equation 5.16 for a *complete* DAG $\mathcal{G}_C$. A complete DAG is a graph in which each coefficient b_{ij} in Equation 5.15 is unequal to zero for $i<j$, that is, where the maximal number of edges is given. For a complete DAG, the closed-form solution is given by

$$P(D|\mathcal{G}_C) = (2\pi)^{-\frac{nm}{2}}\left(\frac{\nu}{\nu+m}\right)^{\frac{n}{2}}\frac{c(n,\alpha)}{c(n,\alpha+m)}|T_0|^{\frac{\alpha}{2}}|T_m|^{-\frac{\alpha+m}{2}}, \tag{5.16}$$

where m is the number of independent observations in D and $c(.,.)$ and T_m are given by

$$c(n,\alpha) = \left(2^{\frac{\alpha n}{2}}\pi^{\frac{n(n-1)}{4}}\prod_{i=1}^{n}\Gamma\left(\frac{\alpha+1-i}{2}\right)\right)^{-1}, \tag{5.17}$$

$$T_m = T_0 + \sum_{j=1}^{m}(D_{.j}-\bar{D})(D_{.j}-\bar{D})^{\mathrm{T}} + \frac{\nu m}{\nu+m}(\mu_0-\bar{x})(\mu_0-\bar{x})^{\mathrm{T}}. \tag{5.18}$$

In Equation 5.18 $\bar{D}$ is the mean vector of the m observation vectors $D_{.j}$ in the data set D. Finally, under the parameter modularity assumption, the BGe score of any Gaussian Bayesian network $\mathcal{G}$ is given by

$$P(D|\mathcal{G})P(\mathcal{G}) = P(\mathcal{G})\prod_{i=1}^{n}\frac{P(D^{(X_i,\text{pa}(X_i|\mathcal{G}))}|\mathcal{G}_C)}{P(D^{(\text{pa}(X_i|\mathcal{G}))}|\mathcal{G}_C)} \tag{5.19}$$

where $D^{(\text{pa}(X_i|\mathcal{G}))}$ is the data set D restricted to the variables in $\text{pa}(X_i|\mathcal{G})$, and $D^{(X_i,\text{pa}(X_i|\mathcal{G}))}$ is the data set D restricted to the variables in $\text{pa}(X_i|\mathcal{G})\cup X_i, \mathcal{G}_c$, which represents a complete DAG over the variables to which the corresponding data set $D^{(\cdot)}$ is restricted. Therefore, Equation 5.16 can be used on reduced data sets to compute the BGe score of any DAG $\mathcal{G}$ over the domain $X_1, \ldots, X_n$. With regard to the Markov chain Monte Carlo (MCMC) sampling schemes, it should be noted that the likelihood $P(D|\mathcal{G})$ as well as the graph prior $P(\mathcal{G})$ decomposes into a product of n local factors: one for each network node. Therefore, it can be seen from Equations 5.13 and 5.19 that the local scores $\psi[X_i|\text{pa}(X_i|\mathcal{G}), D]$ in Equation 5.14 are given by

$$\psi[X_i|\text{pa}(X_i|\mathcal{G}), D] = \binom{n-1}{|\text{pa}(X_i|\mathcal{G})|}^{-1}\frac{P(D^{(X_i,\text{pa}(X_i|\mathcal{G}))}|\mathcal{G}_C)}{P(D^{(\text{pa}(X_i|\mathcal{G}))}|\mathcal{G}_C)} \tag{5.20}$$

when the BGe scoring metric is employed.

5.2.3.5 **Structure Learning Via MCMC Simulations**

Our objective is to learn the dependency structure among the variables from data D. One possible method of learning is to search for the DAG $\mathcal{G}^*$ that maximizes the posterior distribution: $P(\mathcal{G}^*|D) \geq P(\mathcal{G}|D)$ for all directed acyclic graphs $\mathcal{G}$. But gene expression data are usually sparse, that is, the amount of different experimental conditions is small relative of the complexity of the BN model. The consequence is that $P(\mathcal{G}|D)$ is a diffuse distribution, which is not adequately represented by a single high-scoring graph $\mathcal{G}^*$ at the mode. Consequently, it is more appropriate to sample graphs from the posterior distribution $P(\mathcal{G}|D)$ and base the inference on more than one DAG. Since direct sampling from $P(\mathcal{G}|D)$ is impossible due to the intractability of the normalization factor $1/Z_C$, MCMC simulations can be adopted to generate a sample from the posterior distribution. Two different MCMC sampling schemes defined in the space of DAGs (structure MCMC [27]) and node orders (order MCMC [28]) have been proposed in the literature, and will be briefly described next.

In general, an MCMC sampling scheme can be used to generate a sample $s_1, s_2, \ldots$ from a discrete target distribution $P^*(.)$ with state space $S(|S| < \infty)$. This is accomplished by constructing a Markov chain in the space S that converges in distribution to the desired posterior distribution $P^*(.)$. The MCMC simulation scheme consists of evaluating at each step an acceptance probability with which a new state can replace the current state.

More precisely, the general mechanism of a Markov chain M_t with state space S is given by

$$P(M_{t+1} = x) = \sum_{y \in S} T(x|y) P(M_t = y), \tag{5.21}$$

for all $x \in S$ and $t \in N$. Here, $T(x|y)$ is the transition kernel, which denotes the probability of a transition from state y to state x. In addition, an initial distribution $P(M_1 = z)$ $(z \in S)$ is defined. If M_t is irreducible and aperiodic, that is, if $T(x, x) > 0$ for all states x, and if for all $x, y \in S$ there exists an integer k so that $P(M_{t+k} = x|M_t = y) > 0$, then it is guaranteed that the distribution of M_t converges to a stationary distribution P_∞ with $P_\infty(x) = \Sigma_{y \in S} T(x|y)$. That is, for all $z \in S$ the following convergence holds: $P(M_t = z) \rightarrow P_\infty(z)$ for $t \rightarrow \infty$; see [29] for further details. The stationary distribution is unique, depends only on the transition kernel $T(.|.)$, and is independent of the initial distribution. The *equation of detailed balance*

$$\frac{T(x|y)}{T(y|x)} = \frac{P^*(x)}{P^*(y)}, \tag{5.22}$$

for all states x and y is a sufficient condition for the stationary distribution $P_\infty(.)$ to equal the desired posterior distribution $P^*(.)$, as it follows from Equation 5.22 $T(x|y)P^*(y) = T(y|x)P^*(x)$, so that summing over the state space S yields the following:

$$P^*(x) = \sum_{y \in S} T(y|x) P^*(x) = \sum_{y \in S} T(x|y) P^*(y). \tag{5.23}$$

This proves that $P^*(.)$ is the stationary distribution, that is, $P_\infty(.) = P^*(.)$. Equation 5.22 can easily be satisfied when decomposing the transitions at each "time-index" t into two parts. In the first step, a new state x for M_{t+1} is proposed with a proposal probability $Q(x|y)$ that depends on the current state y of M_t. Afterwards, in the second step, x is accepted with an acceptance probability $A(x|y)$ as the new state of the Markov chain at $t+1$. If x is not accepted, the state at $t+1$ is set to the current state y. This procedure is reiterated for all $t > 0$. The transition probabilities are then given by

$$T(x|y) = Q(x|y)A(x|y), \tag{5.24}$$

for all $x, y \in S$ with $x \neq y$ and

$$T(x|x) = \sum_{\substack{y \in S \\ y \neq x}} (1 - A(y|x)Q(y|x)). \tag{5.25}$$

The equation of detailed balance is satisfied if the acceptance probability is chosen as follows: $A(x|y) = \min\{1, R(x|y)\}$, where

$$R(x|y) = \frac{P^*(x)Q(y|x)}{P^*(y)Q(x|y)}. \tag{5.26}$$

As $R(x|y)$ is equal to $1/R(y|x)$, it follows that $R(x|y) > 1 \Leftrightarrow R(y|x) < 1$ and the following two equivalence relations hold:

- $A(x|y) = 1 \Longleftrightarrow A(y|x) = \dfrac{1}{R(x|y)}$,
- $A(x|y) = R(x|y) \Longleftrightarrow A(y|x) = 1$.

This implies for the ratio of transition probabilities:

$$\frac{T(x|y)}{T(y|x)} = \frac{Q(x|y)A(x|y)}{Q(y|x)A(y|x)} = \frac{Q(x|y)}{Q(y|x)} R(x|y) = \frac{Q(x|y)}{Q(y|x)} \frac{P^*(x)Q(y|x)}{P^*(y)Q(x|y)} = \frac{P^*(x)}{P^*(y)}, \tag{5.27}$$

so that the equation of detailed balance (see Equation 5.22) is fulfilled. The proposal probabilities $Q(.|.)$ depend on the design of the transitions in the state space S, which will be described for two particular MCMC sampling schemes below. It takes "some time" until the Markov chain converges to its stationary distribution; hence, the idea is to sample from the chain for "long enough" to ensure that it has reached its stationary distribution. Usually, a number of iterations (called the burn-in phase) are specified in advance, and samples taken during this phase are not used for the subsequent analyses. As there is no sufficient condition to guarantee that convergence has been reached, there is a need for some convergence diagnostics, such as *trace plot diagnostics*. See [30] for further details.

The structure MCMC approach [27] is a MCMC sampling scheme that generates a sample of DAGs $\mathcal{G}_1, \mathcal{G}_2, \mathcal{G}_3, \ldots$ from the posterior distribution $P^*(.) = P(.|D)$. The state space S is the set of all valid DAGs. After having initialized the chain, for example, by a DAG without any edges, the proposal probabilities $Q(\mathcal{G}|\mathcal{G}^*)$ are defined by

$$Q(G|G^*) = \left\{ \begin{array}{cc} \frac{1}{|\Pi(G^*)|}, & G \in \Pi(G^*) \\ 0, & G \notin \Pi(G^*) \end{array} \right\}, \tag{5.28}$$

where $\Pi(\mathcal{G}^*)$ is the *neighborhood* of $\mathcal{G}^*$, that is, the set of all DAGs that can be reached from the current graph state $\mathcal{G}^*$ by deletion, addition, or reversal of a single edge. $|\Pi(\mathcal{G}^*)|$ is the cardinality of this set. As the new graph $\mathcal{G}$ has to be acyclic, it has to be checked which edges can be added to $\mathcal{G}^*$ and which edges can be reversed in $\mathcal{G}^*$ without violating the acyclicity constraint. See Figure 5.2 for an example, and [31] for further details and useful algorithms for checking acyclicity. If the acceptance probabilities are set to $A(\mathcal{G}|\mathcal{G}^*) = \min\{1, R(\mathcal{G}|\mathcal{G}^*)\}$, where $R(.|.)$ is defined in 5.26, it follows from Equation 5.26 that the Markov chain converges to the posterior distribution $P(\mathcal{G}|D)$.

In most applications of BNs to the reverse engineering of gene regulatory networks, a limit f on the cardinality of parent node sets is imposed, for example, $f = 3$. This limit is referred to as maximal fan-in. Each node in a DAG can have at most f parent nodes. This restriction reduces the state space and so improves convergence of the Markov chain. Fan-in restrictions can be justified in the context of expression data, as many experimental results have shown that the expression of a gene is usually controlled by a comparatively small number of active regulator genes; on the contrary, regulator genes themselves seem to be nearly unrestricted in the number of genes they regulate. When a fan-in restriction is imposed, all DAGs that contain nodes with more than f parents have to be removed from the neighborhood $\Pi(\mathcal{G}^*)$.

The order MCMC approach introduced in [28] is an MCMC sampling scheme that generates a sample of node orders $O_1, O_2, O_3, \ldots$ from the posterior distribution $P^*(.) = P(.|D)$ over node orders. So, the state space S is the set of all $n!$ possible orders of the domain variables. Afterward, in a second step, a sample of DAGs $\mathcal{G}_1, \mathcal{G}_2, \mathcal{G}_3, \ldots$ can be obtained by sampling DAGs out of the sampled node orders.

Each order $O = (X_{\sigma(1)}, \ldots, X_{\sigma(n)})$ of the variables $X_1, \ldots, X_n$ is implied through a permutation σ. The meaning of such an order in the context of order MCMC is as follows: O represents the set of all those DAGs $\mathcal{G}$ for which $X_{\sigma(i)} \notin \mathrm{pa}(X_{\sigma(j)}|\mathcal{G})$ if $\sigma(j)$ precedes $\sigma(j)$ in the permutation vector $\sigma = (\sigma(1), \ldots, \sigma(n))$. Only if $X_{\sigma(j)}$ succeeds $X_{\sigma(i)}$ in σ, the relation $X_{\sigma(i)} \in \mathrm{pa}(X_{\sigma(j)}|\mathcal{G})$ is valid. That is, the jth variable $X_{\sigma(j)}$ in the order O is not allowed to have parents that are standing to the right of $X_{\sigma(j)}$ in O. The valid parent sets of $X_{\sigma(j)}$ are restricted to variables that are standing to the left. Consequently, node $X_{\sigma(1)}$ must have the empty parent set $\mathrm{pa}(X_{\sigma(1)}|\mathcal{G}) = \varnothing$; node $X_{\sigma(2)}$ can have either the empty parent set $\varnothing$ or the set $\{X_{\sigma(1)}\}$; node $X_{\sigma(3)}$ is allowed to have one of the following four parent sets, $\varnothing$, $\{X_{\sigma(1)}\}$, $\{X_{\sigma(2)}\}\{X_{\sigma(1)}, X_{\sigma(2)}\}$; and so on.

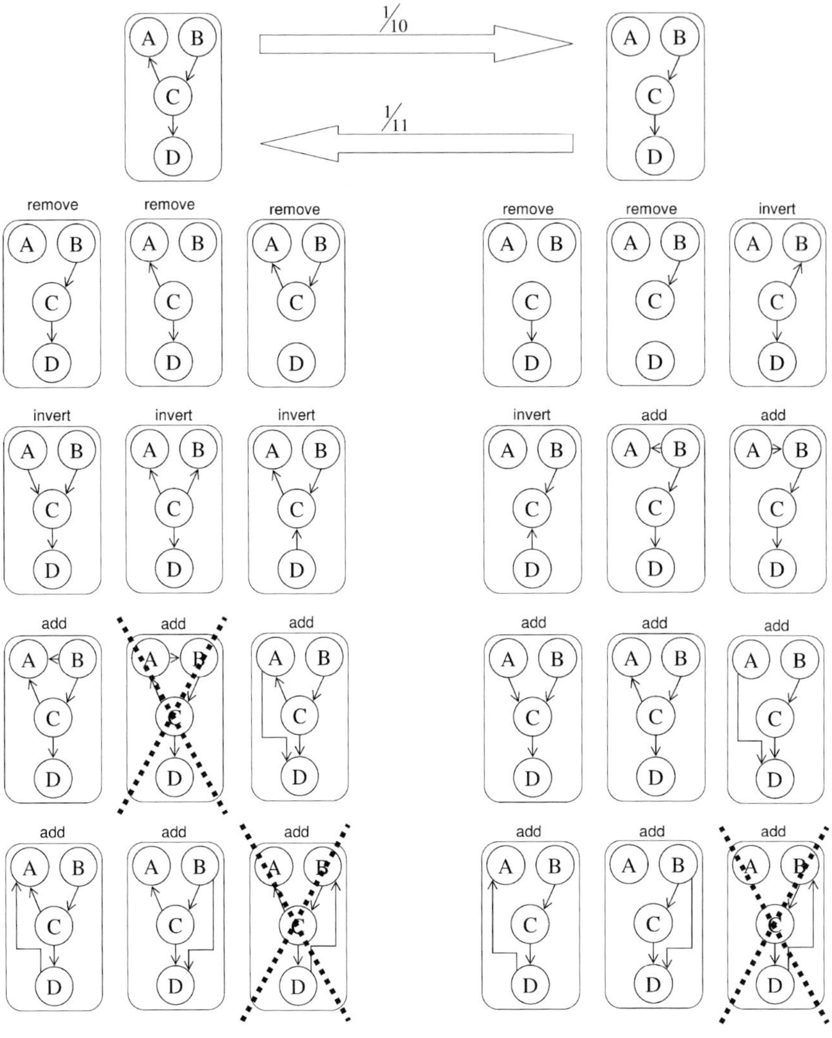

Figure 5.2 Structure MCMC: Proposal moves and Hastings factor. The figure illustrates the three elementary operations for proposing new graphs with structure MCMC: edge creation, edge deletion, and edge reversal. It presents an example of the neighborhoods of two graphs, that is, the set of graphs that can be reached with proposal operations from each of the selected graphs. Owing to the acyclicity constraint for DAGs, these neighborhoods are not necessarily of the same size. Consequently, the Hastings factor, that is, the ratio of the proposal probabilities of the forward and backward moves, may be different from 1. In the present example, the Hastings factor is 10/11.

The likelihood $P(D|O)$ of a given node order O can be computed efficiently, as the selection of the parent set for one node with respect to O does not lead to any additional restrictions for another. That is, for each node, its valid parent sets can be selected independently with respect to the order. As long as the restrictions implied through the order O are regarded for each node, it is guaranteed that no cycles will come into being. The likelihood $P(D|O)$ for the node order $O=(X_{\sigma(1)}, \ldots, X_{\sigma(n)})$ is given by

$$P(D|O) = \prod_{i=1}^{n} \sum_{P \in V_i(O)} \psi[X_i|\text{pa}(X_i|\mathcal{G}) = P, D], \tag{5.29}$$

where $V_i(O)$ denotes the system of all parent sets that are valid for domain variable X_i with respect to the given order O. If a fan-in restriction f is imposed, the system $V_i(O)$ is restricted to sets of cardinalities not higher than f.

The idea of order MCMC is to construct a Markov chain that converges to the posterior distribution over node orders $P(O|D)$. The equation of detailed balance (see Equation 5.22) states that the Markov chain converges to the posterior distribution if the ratio of transition probabilities satisfies

$$\frac{T(O|O^*)}{T(O^*|O)} = \frac{P(O|D)}{P(O^*|D)}, \tag{5.30}$$

where O is a node order that can be reached from the current node order O^*. As in Equation 5.24, the transition probability is decomposed into a proposal and an acceptance probability: $T(O_2|O_1) = Q(O_2|O_1)A(O_2|O_1)$. The acceptance probabilities depend on the proposal probabilities (see Equation 5.26), which in turn depend on the way the transitions are designed in the space of node orders. Friedman and Koller [28] recommend the use of a simple *flip-operator* that exchanges one node for another in the current order. This leads to the following proposal probabilities:

$$Q(O|O^*) = \left\{ \begin{array}{ll} \frac{2}{n(n-1)}, & O \in \Pi(O^*) \\ 0, & O \notin \Pi(O^*) \end{array} \right\}, \tag{5.31}$$

where $\prod(O^*)$ is the set of all node orders $O^\dagger = (X_{\sigma\dagger(1)}, \ldots, X_{\sigma\dagger(n)})$ that can be reached from O^* by flipping two nodes in O^* and leaving all other nodes in the ordering unchanged. The proposal probability $Q(.|Q^*)$ is a uniform distribution over all $[n(n-1)]/2$ possibilities of exchanging the positions of two nodes in the ordering. Convergence to the desired posterior distribution is guaranteed (for $t \to \infty$) if the acceptance probabilities are set to $A(O|O^*) = \min\{1, R(O|O^*)\}$, where $R(.|.)$ is defined in Equation 5.26. If a uniform prior over all $n!$ possible node orders is assumed, that is, $P(O) = 1/n!$ for every order O, the term $R(O|O^*)$ is given by (from Equation 5.26)

$$R(O|O^*) = \frac{P(O|D)\frac{2}{n(n-1)}}{P(O^*|D)\frac{2}{n(n-1)}} = \frac{\frac{P(D|O)P(O)}{P(D)}}{\frac{P(D|O^*)P(O^*)}{P(D)}} = \frac{P(D|O)}{P(D|O^*)}. \tag{5.32}$$

The likelihoods $P(O|D)$ and $P(O^*|D)$ can be computed using Equation 5.29. The orders $O = O(\sigma)$ and $O^* = O(\sigma^*)$ differ by the exchange of two nodes $X_{\sigma(j)}$ and $X_{\sigma(k)}$

only. Consequently, the factors for the nodes that precede $X_{\sigma(j)}$ or succeed $X_{\sigma(k)}$ in both orders do not change in Equation 5.29, that is, $V_i(\sigma) = V_i(\sigma^*)$ for $i<j$ as well as for $i>k$. This implies that the ratio of likelihoods reduces to

$$\frac{P(D|O)}{P(D|O^*)} = \prod_{i=j}^{k} \frac{\sum_{P \in V_i(O)} \psi[X_i|\text{pa}(X_i|G) = P, D]}{\sum_{P \in V_i(O^*)} \psi[X_i|\text{pa}(X_i|G) = P, D]}. \tag{5.33}$$

Order MCMC outputs a sample of node orders $O_1,\ldots,O_T$, which, if convergence of the Markov chain has been reached, is a sample from the posterior distribution over node orders $P(O|D)$. Afterward, for each sampled order O_i, a DAG $\mathcal{G}_i$ can be sampled from the posterior distribution $P(\mathcal{G}|O_i, D)$ over DAGs given the order O_i and the data D. Depending on a concrete node order, for each node its parent set can be sampled independently out of its valid parent sets in O_i:

$$P(\text{pa}(X_i|\mathcal{G}) = P_0|O_i, D) = \frac{I_{\{V_i(O_i)\}}(P_0)\psi[X_i|\text{pa}(X_i|\mathcal{G}) = P_0, D]}{\sum_{P \in V_i(O_i)} \psi[X_i|\text{pa}(X_i|\mathcal{G}) = P, D]}. \tag{5.34}$$

The indicator function $I_{\{V_i(O_i)\}}(P_0)$ is equal to 1 if the condition $P_0 \in V_i(O_i)$ is true, and zero otherwise. Subsequently, the complete DAG can be obtained straightforwardly: for each node and each of its parent nodes, there is an edge pointing from the parent node to the node itself. So, a DAG sample $\mathcal{G}_1, \ldots, \mathcal{G}_T$ is generated from the sampled node orders $O_1, \ldots, O_T$.

Friedman and Koller [28] show that order MCMC is superior to structure MCMC with regard to convergence and mixing of the resulting Markov chain. But there is a certain drawback of the order MCMC sampling scheme. Using order MCMC, the prior over DAGs, which can have a nonnegligible influence on the posterior distribution, especially for sparse data, cannot be defined explicitly. More precisely, the assumption that each ordering O has the same prior probability $P(O)$ leads to a change in the form of the originally specified prior over DAGs $P(\mathcal{G})$. DAGs that are consistent with more orders are more likely than DAGs consistent with fewer orders. For instance, the DAG without any edge can be sampled out of all $n!$ node orders, while a DAG of the type $X_{\sigma(1)} \rightarrow X_{\sigma(2)} \rightarrow \cdots \rightarrow X_{\sigma(n)}$ can only be sampled out of one single node order, namely, $O = (X_{\sigma(1)}, \ldots, X_{\sigma(n)})$.

5.2.3.6 Learning Bayesian Networks from Interventional Data

Although most of the available biological expression and pathway data are passively observed (so-called *observational data*), sometimes experimenters can actively intervene and externally set certain domain variables, using, for example, gene knockouts or overexpressions (*interventional data*). If these interventions are *ideal*, then the intervened variables are set deterministically using forces outside the studied domain; so their values no longer depend on the other domain variables. However, their assigned values can influence the values of other variables in the studied domain; consequently, the intervened data points are useful for discovering causal relationships (directed edges). Under fairly weak conditions, a combination of observational and ideal interventional data can be analyzed using Bayesian networks. These conditions are described in detail in [32] for the *BDe scoring metric* and in [33] for the *BGe scoring metric*. Two modifications are required. First, each factor of the posterior probability only depends on those data points where the variable itself was not intervened, that is, Equation 5.14 has to be replaced by the following expression:

$$P(\mathcal{G}|D) = \frac{1}{Z_C}\prod_{i=1}^{n}\psi[X_i|\mathrm{pa}(X_i|\mathcal{G}), D_{X_i \notin \mathcal{I}}], \tag{5.35}$$

where $\mathcal{I}$ is the set of interventions and $D_{X_i \notin \mathcal{I}}$ denotes the data set where data points are removed for those cases where node X_i has been intervened; see [32] and [33] for further details. Second, the definition of equivalence classes has to be generalized. While for pure observational data, two DAGs assert the same set of independence assumptions among the domain variables if and only if they have the same skeleton and the same set of v-structures, this definition of "equivalence" needs to be extended when considering a mixture of observational and ideal interventional measurements.

To see that, consider two domain variables A and B and two candidate DAG structures $\mathcal{G}_1 : \mathrm{A} \rightarrow \mathrm{B}$ and $\mathcal{G}_2 : \mathrm{A} \leftarrow \mathrm{B}$. Assume that node B is set deterministically to the fixed value b by an ideal intervention. This manipulation cannot influence the distribution of node A in DAG $\mathcal{G}_1 : \mathrm{A} \rightarrow \mathrm{B}$, as A does not depend on B in $\mathcal{G}_1$. That is, the probability $\mathrm{P}(\mathrm{A} = \mathrm{a}|\mathcal{G}_1, \mathrm{B} = \mathrm{b})$ reduces to $P(A = a)$. On the contrary, in DAG $\mathcal{G}_2 : \mathrm{A} \leftarrow \mathrm{B}$ the manipulation of node B does cause a change in the distribution of node A, as A depends on B. That is, $\mathrm{P}(\mathrm{A} = \mathrm{a}|\mathcal{G}_2, \mathrm{B} = \mathrm{b})$ is equal to $P(A = a|B = b)$. Thus, although both DAGs are equivalent with the same CPDAG representation A—B, the (in-)dependence relations differ for interventional data.

Tian and Pearl [34] show that two DAGs assert the same set of dependence/independence assumptions among the variables for a mixture of observational and ideal interventional data if and only if (i) they are *equivalent*, that is, they have the same skeleton and the same set of v-structures, and (ii) each variable that was manipulated in at least one observation possesses the same parent set. The resulting equivalence classes are referred to as *transition-sequence equivalence* or *TS-equivalence* classes. Two DAGs that assert the same set of independence assumptions among the variables for a mixture of observational and ideal interventional data are said to be *TS-equivalent*. More precisely, when the concept of *TS-equivalence* is used, then all edges that are connected to an intervened node become automatically directed in the CPDAG representation. This may lead to new v-structures, and further edges not entering or leaving an intervened node may become directed too.

As a simple way to automatically determine which edge directions of the DAG are enforced in the TS-equivalence class, we can add two dummy nodes to each intervened node, with directed edges pointing away from the dummy nodes, and then automatically determine the standard CPDAG structure for the resulting graph using standard algorithms [22]. The new directed dummy edges lead to new v-structures and thus enforce more edges of the original DAG to become directed.

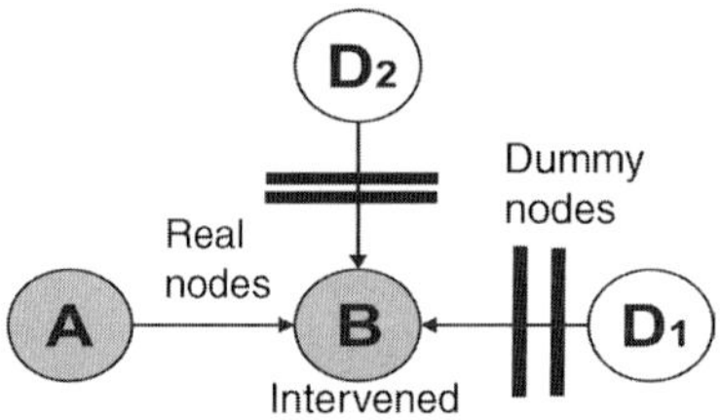

Figure 5.3 Effect of an intervention on the graph. The figure shows an illustration of how to convert a DAG into a TS-equivalence CPDAG when subjecting node B in the Bayesian network $A \rightarrow B$ to an intervention.

Finally, the deletion of all dummy nodes and dummy edges from the CPDAG gives the CPDAG representation of the original DAG in terms of *TS-equivalence*.

Figure 5.3 illustrates how to obtain the TS-equivalence CPDAG for the simple network "$A \rightarrow B$," in which node B was intervened. Owing to the addition of two dummy parents D_1 and D_2 for the intervened node B, the edge from A to B participates in two v-structures, $(A \rightarrow B \leftarrow D_1)$ and $(A \rightarrow B \leftarrow D_2)$, and so becomes directed. After removing the dummy components, the edge is left directed. If B had not been an intervened node, the edge would have been undirected in the CPDAG representation.

5.3
The RAF Signalling Pathway

The RAF signaling pathway, shown in Figure 5.4, is a biologically well-known regulatory network that describes the intracellular relationships among different molecules involved in *signal transduction*, that is, a cascade of cellular protein signaling events. Special enzymes (protein kinases) modify other target proteins (substrates) by adding phosphate groups to them (phosphorylation). This leads to a functional change of the targets so that further chemical reactions follow in the signaling cascade. As protein kinases are known to regulate the majority of cellular pathways and cell growth, deregulated kinase activity can lead to diseases such as cancer.

Sachs *et al.* [13] measured the expression levels of $n = 11$ phosphorylated proteins and phospholipids of the RAF signaling pathway in thousands of human immune system cells with cytometry experiments. In addition to about 1200 pure observational measurements, the $n = 11$ molecules in the cascade were also measured after nine different molecular cues. More precisely, the molecules were profiled 15 minutes after nine different stimulations of the network. For each of these molecular interventions, more than 600 measurements were made, whereby an effect on the molecules in the cascade could be observed for six of these perturbations. As it is known that these interventions predominantly influence only a single molecule in the cascade, they can be considered as ideal interventions. A brief summary of the

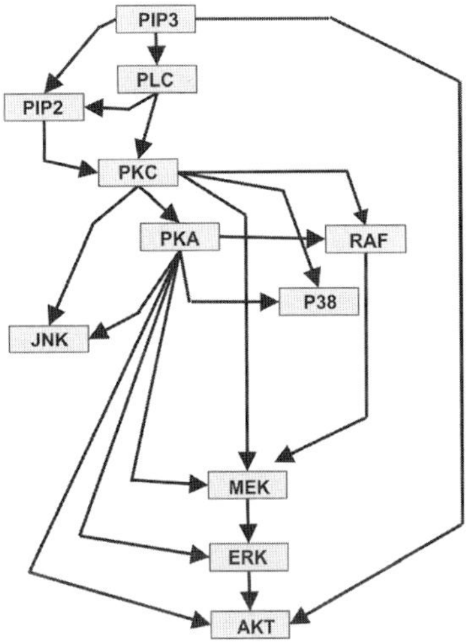

Figure 5.4 Raf signaling pathway. The graph shows the currently accepted regulatory network, taken from [13]. Nodes represent proteins, edges represent interactions, and arrows indicate the direction of signal transduction. In the interventional studies, the following nodes were targeted. Activations: PKA and PKC. Inhibitions: PIP2, AKT, PKC, and MEK.

effects of the six molecular interventions on the measured molecules activities can be found in Table 5.1. Three molecular interventions having no observable effect on the cascade were discarded from the analysis. More details on the probe preparations, the exact experimental conditions, as well as more information about the stimulatory agents can be found in [13].

It turned out that the interventional measurements could not be analyzed without preprocessing. Occasionally, there was a discrepancy between expected and observed concentrations for intervened nodes, for example, some inhibitions had not led to low concentrations, while some activations had not led to high concentrations.

Table 5.1 Interventions in the RAF network.

Reagent	Effect
AKT-inhibitor	Inhibits AKT
G06976	Inhibits PKC
Psitectorigenin	Inhibits PIP2
U0126	Inhibits MEK
Phorbol myristate acetate	Activates PKC
8-Bromo adenosine 3′, 5′-cyclic monophosphate	Activates PKA

The table shows the effects of the ideal interventions in the cytometry experiment on the RAF network.

This is because the measured concentrations do not reflect the true activities of the corresponding protein. It was therefore decided to replace in each interventional data set the values of the activated (inhibited) nodes by the maximal (minimal) concentration of that node measured under a general perturbation of the system. Afterward, quantile normalization was used to normalize each interventional data set. That is, for each of the 11 molecules, its m realizations were replaced by quantiles of the standard Gaussian distribution $N(0,1)$. More precisely, for each of the 11 molecules, the jth highest realization was replaced by the (j/m)-quantile of the Gaussian distribution, whereby the ranks of identical realizations were averaged.

5.4
Criteria for Evaluating Learning Performances

All the machine learning methods described in Section 5.2 can be used to generate new hypotheses about gene interactions, which either can be confirmed by the information in biological data bases (and/or traditional molecular biology experiments) or otherwise must be seen with caution. That is, although the algorithms presented in Section 5.2 may find relations that are supported by the data, there is no guarantee that these findings correspond to real biological interactions. It is therefore necessary to test and evaluate the learning performance of the various machine learning methods discussed in this chapter before applying them to real biological expression data.

The learning performance of a machine learning method can be evaluated by generating data from a known synthetic network and comparing the predicted network structure with this known structure. Alternatively, the network learned from real biological expression data can be compared with a biologically accepted "gold-standard network," which is a regulatory network that can be considered to be reliable, based on our current molecular biological knowledge. For our experimental cross-method evaluation described below, we chose the RAF signaling pathway, described in Section 5.3 and depicted in Figure 5.4; the network consists of 11 domain nodes and 20 directed edges.

Recall from Section 5.2 that RNs and GGMs extract undirected graphs, while Bayesian networks extract partially directed graphs. Therefore, to compare the performance of these methods, we apply two different criteria. The first approach, referred to as the *undirected graph evaluation* (UGE), discards the information about the edge directions altogether. To this end, the original true graph and the learnt graphs are replaced by their skeletons, where the skeleton is defined as the graph in which two nodes are connected by an undirected edge whenever they are connected by any type of edge. Note that the skeleton of an undirected network is the network itself. The second approach, referred to as the *directed graph evaluation* (DGE), compares the learnt graphs with the original true directed graph. In terms of DGE each predicted undirected edge is interpreted as the superposition of two directed edges, pointing in opposite directions. For both graph evaluation schemes and each method, the learning performance can be assessed by counting the number of true

Table 5.2 Comparison between UGE and DGE, example 1.

	$A \to B$	$A \leftarrow B$	$A - B$	$A\ B$
UGE				
TP	1	1	1	0
FP	0	0	0	0
TN	0	0	0	0
FN	0	0	0	1
DGE				
TP	1	0	1	0
FP	0	1	1	0
TN	1	0	0	1
FN	0	1	0	1

Comparison between the UGE and the DGE when the true connection between node A and node B is a directed edge pointing from node A to node B, symbolically $A \to B$. The top row indicates all four possible relationships between nodes A and B, which can potentially be predicted with a machine learning method. The table evaluates the prediction in terms of the UGE and DGE evaluation criteria by indicating if the prediction is counted as a (i) true positive (TP), (ii) false positive (FP), (iii) true negative (TN) and/or false negative (FN) edge.

positive (TP) and false positive (FP) edge findings of a specific graph prediction. The difference between the two graph evaluation schemes with regard to TP, true negative (TN), FP, and false negative (FN) counts is illustrated in Tables 5.2 and 5.3.

Each of the three reverse engineering methods compared in our study outputs scores associated with the edges in the network domain. These scores are of

Table 5.3 Comparison between UGE and DGE, example 2.

	$A \to B$	$A \leftarrow B$	$A - B$	$A\ B$
UGE				
TP	0	0	0	0
FP	0	0	0	0
TN	0	0	0	1
FN	1	1	1	0
DGE				
TP	0	0	0	0
FP	1	1	2	0
TN	1	1	0	2
FN	0	0	0	0

Comparison between the UGE and the directed graph evaluation when there is no connection between node A and node B in the true network, symbolically $A\ B$. The top row indicates all four possible relationships between A and B that can potentially be predicted with a machine learning method. The table evaluates the prediction in terms of the UGE and DGE evaluation criteria by indicating if the prediction is counted as a (i) true positive (TP), (ii) false positive (FP), (iii) true negative (TN) and/or false negative (FN) edge.

different nature: correlation coefficients for RNs, partial correlation coefficients for GGMs, and marginal posterior probabilities for BNs. However, all the scores define a ranking of the edges, and this ranking can be used to plot a receiver operator characteristics (ROC) curve, where the relative number of TP edges is plotted against the relative number of FP edges. The mathematical details are as follows. The true graph topology $\mathcal{G}_{true}$ can be interpreted as a set of binary indicator variables e_{ij} ($i, j \in \{1, \ldots, n\}$). In terms of DGE, e_{ij} ($i \neq j$) represents the directed edge pointing from node X_i to node X_j, and $e_{ij} = 1$ indicates that this directed edge is present in $\mathcal{G}_{true}$, while $e_{ij} = 0$ means that it is not. Correspondingly, in terms of UGE, e_{ij} ($i<j$) represents the undirected edge between X_i and X_j and $e_{ij} = 1$ indicates that this undirected edge is present in the skeleton of $\mathcal{G}_{true}$, while $e_{ij} = 0$ means that it is not. We note that the cardinality of the set of all possible directed edges e_{ij} ($i, j \in \{1, \ldots, n\}$, $i \neq j$) is $n(n-1)$, while the cardinality of the set of all possible undirected edges e_{ij} ($i, j \in \{1, \ldots, n\}$, $i<j$) is $n(n-1)/2$.

Each method outputs a confidence score $\psi(e_{ik})$ for the directed and undirected edges. By imposing a threshold θ on these confidence scores $\psi(e_{ik})$, we obtain a specific graph $\mathcal{G}$, where each edge is present if its score $\psi(e_{ik})$ is higher than θ, and it is absent otherwise. It depends on the evaluation criterion whether $\mathcal{G}$ is an undirected or a directed graph. As mentioned above, with regard to the DGE criterion, undirected edges are interpreted as the superposition of two directed edges having opposite directions.

For both DGE and UGE, let $\varepsilon(\theta) = \{e_{ik} | \psi(e_{ik}) > \theta\}$ denote the set of all edges whose confidence scores exceed a given threshold θ. Then, for the imposed threshold θ the number of TP, FP, TN, and FN edge findings can be counted, and the *true positive rate* TPR = TP/(TP + FN) and the *false positive rate* FPR = FP/(TN + FP) can be computed. The true positive rate TPR is also referred to as *sensitivity* and the false positive rate FPR is also referred to as *inverse specificity*. Rather than selecting an arbitrary value for the threshold θ, this procedure can be repeated for various values of θ and the ensuing TPR scores can be plotted against the corresponding FPR scores. This gives the ROC curves. Generally, such ROC curves show which rate of erroneously learnt edges (FPR) must be accepted to obtain a desired recovery rate of true positive edges (TPR).

Ideally, we would like to evaluate the methods on the basis of the whole ROC curves. Unfortunately, this approach would not allow us to concisely summarize the results obtained from applying several methods to many data sets. We therefore pursued two different approaches. The first approach is based on numerically integrating the ROC curve, for example, by the trapezoidal method, so as to obtain the area under the curve (AUC), with larger scores indicating an overall better performance. While this approach does not require us to commit ourselves to the adoption of any (arbitrary) decision criterion, it does not lead to a specific network prediction. It also ignores the fact that, in practice, one is particularly interested in the performance for low FP rates. Our second performance criterion, hence, is based on the selection of a threshold on the edge confidence scores, from which a specific graph prediction is obtained. The question, then, is how to define this threshold. Schäfer and Strimmer [6] discuss a method for converting the (partial) correlation

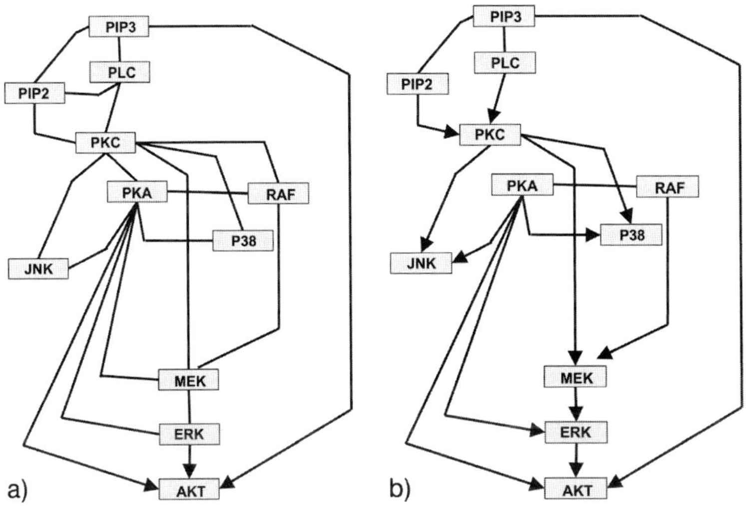

Figure 5.5 CPDAG representations of both the gold-standard and modified RAF pathway. The modified RAF pathway was obtained by deleting the following four directed edges from the (gold standard) RAF signaling pathway of Figure 5.4: (1) PKC → RAF, (2) PKA → MEK, (3) PLC → PIP2, and (4) PKC → PKA. It can be seen that the four edge deletions have led to a considerable increase in the number of directed edges in the CPDAG representation.

coefficients of RNs and GGMs into q-values (i.e., P-values corrected for multiple testing see [35]) and "posterior probabilities." However, these posterior probabilities are not equivalent to those defined for BNs. Imposing the same threshold on both leads to different rates of TPs and FPs and, hence, different operating points on the ROC curves that cannot be compared due to differing sensitivities and (inverse) specificities. We also found that controlling the false discovery rate at the typical value of $q = 0.05$ turned out to be too conservative; the number of predicted edges is very low, and sometimes zero. We therefore chose a threshold for each method under comparison such that it led to a fixed count of five FP edges. We then compared the numbers of TP counts associated with these thresholds. This procedure is guaranteed to compare the competing methods at the same operation point on the ROC curves, that is, it compares the number of true positive edges corresponding to the same number of false positive edges, namely five edges.

5.5 Data

We base the evaluation of the three reverse engineering methods (RNs, GGMs, and BNs) on the Raf signaling network, depicted in Figure 5.4. RAF is a critical signaling

Table 5.4 Comparison of machine learning methods.

Method	Model	Training scheme	Result type
RN	Relevance network	Pearson correlation	Undirected graph
GGM	Gaussian graphical model	Shrinkage-based estimator	Undirected graph
BN	Bayesian network	Order MCMC with BGe score	Partially directed graph

The table shows an overview of the three machine learning methods compared in our evaluation studies.

protein involved in regulating cellular proliferation in human immune system cells. The deregulation of the Raf pathway can lead to carcinogenesis, and the pathway has therefore been extensively studied in the literature (e.g. [13,36]). We use four types of data for our evaluation. First, we distinguish between passive observations and active interventions. Second, we use both real laboratory data and synthetic simulations. This combination of data is based on the following rationale. For simulated data, the true structure of the regulatory network is known; this allows us, in principle, to faithfully evaluate the prediction results. However, the model used for data generation is a simplification of real molecular biological processes, and this might lead to systematic deviations and a biased evaluation. The latter shortcoming is addressed by using real laboratory data. In this case, however, we ultimately do not know the true signaling network; the current gold standard might be disputed in light of future experimental findings. By combining both approaches, we are likely to obtain a more reliable picture of the performance of the competing methods. Below, we will briefly summarize the main features of the data. A more comprehensive description with further details can be found in [37,38].

Linear Gaussian Distribution. A simple synthetic way of generating data from the gold-standard network of Figure 5.4 is to sample them from a linear Gaussian distribution. The random variable X_i denoting the expression of node i is distributed according to

$$X_i \sim N\left(\sum_k \omega_{ik} x_k, \sigma^2\right), \tag{5.36}$$

where $N(.)$ denotes the normal distribution, the sum extends over all parents of node i, and x_k represents the value of node k. We set the standard deviation to $\sigma = 0.1$, sampled the interaction strength $|\omega_{ik}|$ from the uniform distribution over the interval [0.5, 2], and randomly varied the sign of ω_{ik}. For simulating (noisy) interventions, we replaced the conditional distribution in Equation 5.36 by the following unconditional distributions. For inhibitions, we sampled X_i from a zero-mean Gaussian distribution, $N(0, \sigma)$. For activations, we sampled X_i from the tails of the empirical distribution of X_i, beyond the 2.5 and the 97.5 percentiles.

Realistic Nonlinear Simulation. The following approach is a more realistic simulation of data typical of signals measured in molecular biology. The expression of a gene is controlled by the interaction of various transcription factors, which may have an inhibitory or activating influence. Ignoring time delays inherent in transcription and translation, these interactions can be compared to enzyme–substrate reactions in organic chemistry. From chemical kinetics, it is known that the concentrations of the molecules involved in these reactions can be described by a system of ordinary differential equations (ODEs) [39]. Assuming equilibrium and adopting a steady-state approximation, it is possible to derive a set of closed-form equations that describe the product concentrations as nonlinear (sigmoidal) functions of combinations of substrates. However, instead of solving the steady-state approximation to ODEs explicitly, as pursued in [12], we approximate the solution with a qualitatively equivalent combination of multiplications and sums of sigmoidal transfer functions. The resulting sigma–pi formalism has been implemented in the software package NetBuilder [14,15], which we have used for simulating the data from the gold-standard Raf networks; see [38] for further information. To model stochastic influences, we subjected all nodes to additive Gaussian noise and repeated the simulations for three different noise levels. Interventions were simulated by drawing values from a peaked Gaussian distribution ($\sigma^2 = 0.01$) around the maximum (activation) and minimum (inhibition) values of the domain.

Cytometry Data. Sachs *et al.* [13] have applied intracellular multicolor flow cytometry experiments to quantitatively measure protein expression levels. Data were collected after a series of stimulatory cues and inhibitory interventions targeting specific proteins in the Raf pathway. A summary is given in Table 5.1. In our study, we considered two types of training data on which the reconstruction of the regulatory network was evaluated: (i) observational data, in which no active interventions had been used; and (ii) interventional data, where the data were composed of seven subsets of equal size: one subset without interventions and one subset for each of the six interventions listed in Table 5.1. The data and the statistical preprocessing are described in Section 5.3. See also [13] for further and more detailed information. The original data (prior to preprocessing) are available from the Web site http://www.sciencemag.org/cgi/content/full/308/5721/519/DC1.

Data Set Size. Flow cytometry allows the simultaneous measurement of the protein expression levels in thousands of individual cells. Sachs *et al.* [13] have shown that for such a large data set, it is possible to reverse engineer a network that is very similar to the known gold-standard Raf signaling network. However, for many other types of current postgenomic data, such abundance of data is not available. We therefore sampled the data obtained in [13] down to 100 data points; this is a representative figure for the typical number of different experimental conditions in current microarray experiments. We averaged the results over five independent samples. We used the same sample size and the same number of replications for the synthetic data. For observational data, all nodes were unperturbed. Interventional data were obtained by perturbing each of the six target nodes listed in Table 5.1 in turn, taking 14 measurement for each type of intervention and including a further set of 16 unperturbed measurements.

5.6 Simulations

As opposed to GGMs, RNs and BNs do not require the assumption of a Gaussian distribution. However, deviations from the Gaussian incur an information loss as a consequence of data discretization (mutual information for RNs, and BDe score for BNs). Alternatively, when avoiding the discretization with the heteroskedastic regression approach in [40], the integral in Equation 5.14 becomes intractable and has to be approximated. It would obviously be interesting to evaluate the merits and shortcomings of these nonlinear approaches. However, the main objective of the present study is the comparison of three modeling paradigms:

1. pairwise association scores independent of all other nodes (RNs),
2. undirected graphical models with constraint-based inference (GGMs), and
3. directed graphical models with score-based inference (BNs).

To avoid the perturbing influence of additional decision factors, for example, related to data discretization and to enable a fair comparison with GGMs, we use the Gaussian assumption throughout. To minimize the deviation from this assumption, we subjected the data to a standardization, ensuring that all marginal distributions of individual nodes have mean 0 and variance 1.

Applying the Gaussian assumption to BNs, with the normal-Wishart distribution as a conjugate prior on the parameters, the integral in 5.14 has a closed-form solution, referred to as the BGe score; see Sections 5.2.3.3 and 5.2.3.4 for a derivation. As discussed in Section 5.2.3.4, the BGe score depends on various hyperparameters, which can be interpreted as pseudocounts from a prior network. To make the prior probability over parameters – $f(q|\mathcal{G})$ in Equation 5.12 – as uninformative as possible, we set the prior network to a completely unconnected graph with an equivalent sample size as small as possible subject to the constraint that the covariance matrix is nonsingular. For the prior over network structures – $P(\mathcal{G})$ in Equation 5.11 – we followed the approach in [28] and chose a distribution that is uniform over parent cardinalities subject to a fan-in restriction of 3; see Equation 5.13. We carried out MCMC over node orders, as described in Section 5.2.3.5. To test for convergence, each MCMC run was repeated from two independent initializations. Consistency in the marginal posterior probabilities of the edges was taken as an indication of sufficient convergence. We found that a burn-in period of 20 000 steps was usually sufficient and followed this up with a sampling period of 80 000 steps, keeping samples in intervals of 200 MCMC steps. For RNs, we computed the pairwise node associations with the Pearson correlation; see Equation 5.1. We computed the covariance matrix in GGMs with the shrinkage approach described in Section 5.2.2, choosing a diagonal matrix as the shrinkage target. Note that this target corresponds to the empty prior network; hence, the effect of shrinkage is equivalent to the selected prior for the computation of the BGe score in BNs. The practical computations were carried out with the software provided by [7]. The MCMC simulations were carried out with our own MATLAB programs, which are available on request.

5.7 Results

For a concise summary, we present our results visually in terms of scatter plots. A complete set of tables with statistical significance tests is available from [37,38] and the supplementary material can be found in [41].

Figure 5.6 compares the performance of BNs and GGMs on the synthetic Gaussian data and the protein concentrations from the cytometry experiment. The two panels on the left refer to the Gaussian data. Without interventions, BNs and GGMs achieve a similar performance in terms of both AUC and TP scores. Interventions lead to improved predictions with BNs. As a consequence of interventions, the number of correctly predicted undirected edges increases slightly from 15.8 to 18.5; this is not significant, though $P = 0.097$. However, the number of correctly predicted directed edges shows a significant increase from 4.9 to 18.4 ($P < 10^{-4}$). On the intervened data, BNs outperform GGMs, and this improvement is significant when the edge directions are taken into account (AUC. $P = 0.0002$; TP: $P = 0.0005$).

The two columns on the right of Figure 5.6 summarize the results obtained for the cytometry data. Without interventions, GGMs and BNs show a similar performance. As a consequence of interventions, the performance of BNs improves, but less substantially than for the Gaussian data. For instance, the number of correctly predicted directed edges increases from 3.3 to 6.9, which is just significant ($P = 0.013$). With interventions, BNs tend to outperform GGMs. This improvement is only significant for the DGE–TP score, though $P = 0.007$; while the UGE–AUC score for BNs is consistently better than for GGMs, its P-value of 0.055 is above the standard significance threshold.

To obtain a deeper understanding of the models' performance, we applied them to the nonlinear simulated data (NetBuilder) with different noise levels. The results are shown in Figure 5.7. When comparing the performance of BNs and GGMs on observational data, we observe the following trend. For low noise levels, GGMs slightly outperform BNs, although this difference is only significant for the DGE–TP score ($P = 0.008$); all other P-values are above 0.05. When increasing the noise level, the situation is reversed. BNs outperform GGMs, and the differences are significant for all scores except for DGE–TP (UGE–AUC: $p = 0.025$; DGE–AUC: $P = 0.029$, UGE–TP: $P = 0.016$; DGE–TP: $P = 0.067$). For large noise levels, GGMs and BNs show a similar performance, without a significant difference in any score. Interventions lead to an improvement in the performance of BNs when taking the edge direction into account. The improvement is significant in both scores, DGE–TP and DGE–AUC, for all noise levels, with $P < 0.002$. The improvement is most pronounced for the medium noise level, where the number of correctly predicted

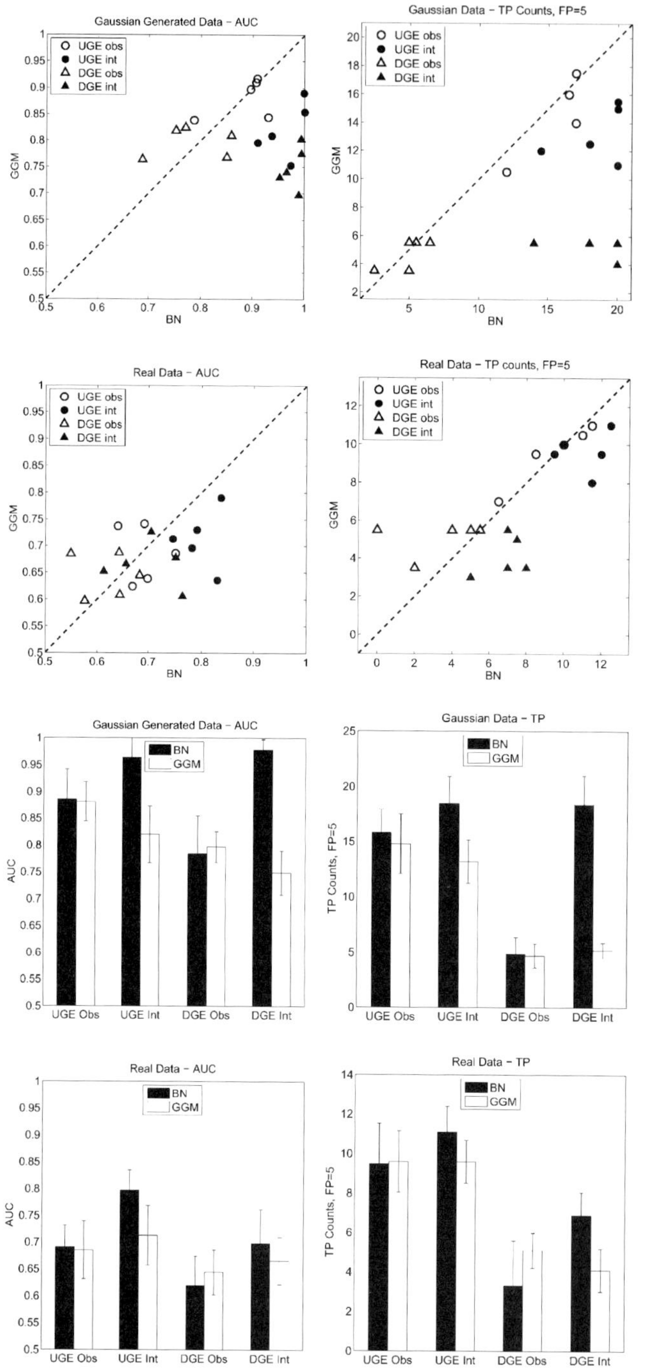

Figure 5.6 (legend see p. 131)

edges increases from 7.2 to 17.3 ($P \ll 10^{-4}$). A comparison between GGMs and BNs reveals that with interventions, BNs consistently outperform GGMs when taking the edge direction into account; all differences are significant with $P < 0.005$.

Figures 5.8 and 5.9 compare the performance of BNs and GGMs with RNs. On the Gaussian observational data, both GGMs and BNs consistently outperform RNs. However, there is no significant difference in the performance of the methods on the nonlinear simulated data (NetBuilder) and the cytometry protein concentrations when no interventions are used; in fact, the DGE–TP scores for BNs are actually worse than those obtained with RNs (see the next section for a discussion). With interventions, GGMs outperform RNs on the cytometry data (UGE: $P = 0.001$; DGE: $P = 0.001$), and they obtain higher TP counts than RNs do on the nonlinear simulated data ($P < 0.0002$ for both UGE and DGE). BNs consistently outperform RNs on all data sets with respect to all scoring schemes when interventions are used ($P < 0.001$).

5.8 Discussion

Dependence on the Noise Level. When varying the noise level on the nonlinear simulated data (Figure 5.7), we observe that when increasing the noise level, the performance with BNs first increases and then decreases. For instance, the average number of predicted true undirected edges increases from TP = 11 for $\sigma = 0.01$ to TP = 18 for $\sigma = 0.1$ and then decreases again to TP = 15.5 for $\sigma = 0.3$. To understand this behavior, consider a parent node that regulates several children, where the children do not have any direct interactions; see the top right panel of Figure 5.10. Without noise, the response of each child is a deterministic function of the parent. However, this implies a deterministic functional relationship between the children. Consequently, the true network cannot be distinguished from a network in which all children are connected by edges, and it is intrinsically impossible to learn the true network. The deterministic relationship between the children is

Figure 5.6 GGMs versus BNs on Gaussian and cytometry data. Top row: Scatter plots comparing the performance of GGMs (vertical axis) with BNs (horizontal axis). The diagonal line represents equal performance. Symbols above that line indicate that GGMs outperform BNs. Conversely, symbols below that line point to a better performance of BNs over GGMs. Each subfigure compares the results obtained from two different data types, using only passive observations (empty symbols) and including active interventions (filled symbols). Two different evaluation criteria have been applied, based on directed graphs (DGE, represented by triangles) and their undirected skeletons (UGE, represented by circles). Bottom row: Histograms showing the average AUC scores and TP counts for BNs (filled bars) and GGMs (empty bars). The codes under the histograms indicate the type of evaluation (UGE versus DGE) and whether observational (Obs) or interventional (Int) data have been used. Columns: The four columns refer to different data and scoring criteria. Left: Gaussian data, AUC score. Center left: Gaussian data, TP counts. Center right: Cytometry data, AUC score. Right: Cytometry data, TP counts. From [41], with permission from Oxford University Press.

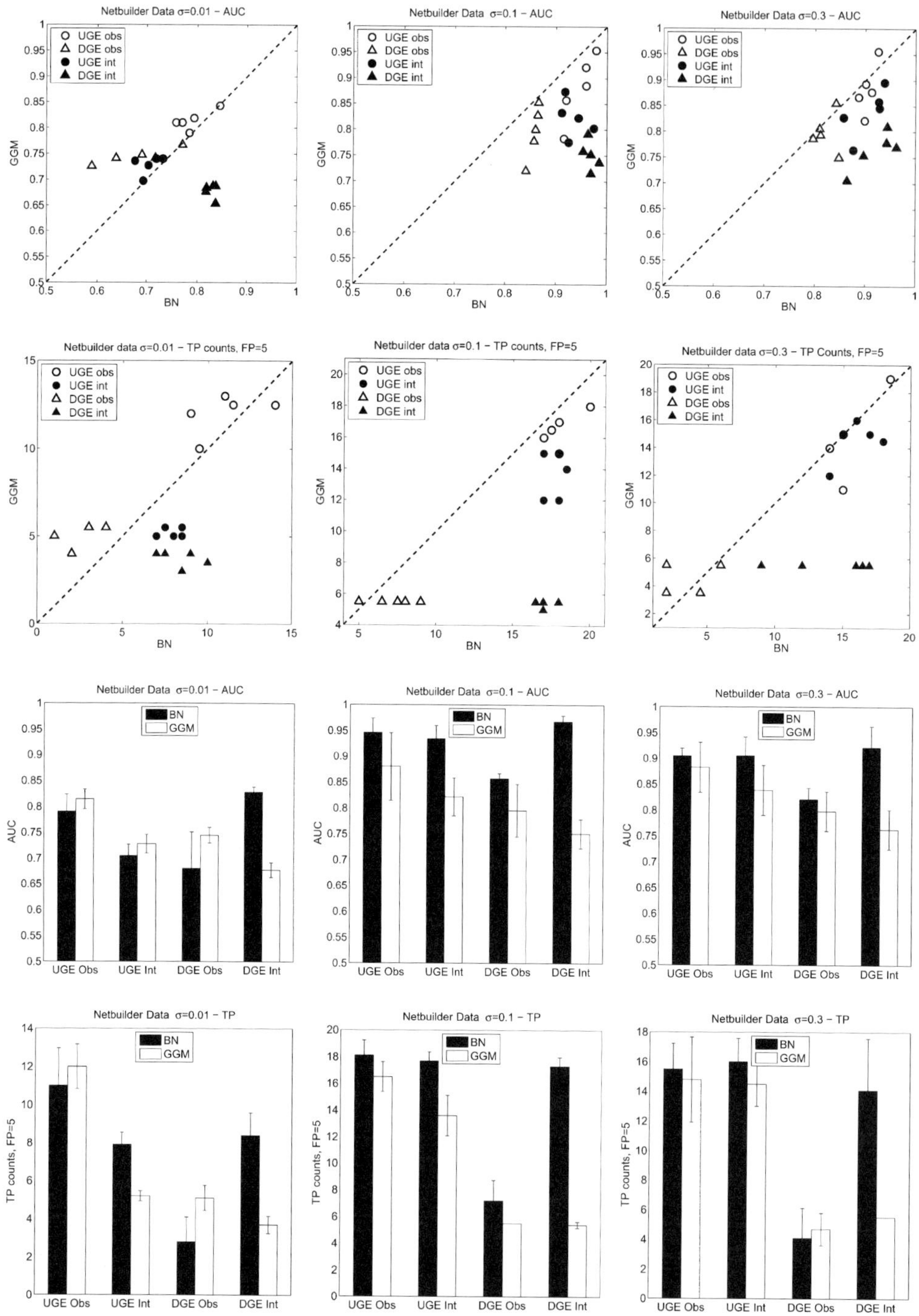

Figure 5.7 (legend see p. 133)

destroyed by the addition of noise, which renders, on average, the signal of a child more similar to that of its parent than that of a sibling. Consequently, some noise is useful and forms the basis for learning gene regulatory networks from data. However, when the noise level becomes so large that it hides the regular signal, successful learning will no longer be feasible. Hence, we would expect the accuracy of reconstructing regulatory networks to first increase and then decrease with increasing noise level. This trend was confirmed in an independent simulation study, shown in Figure 5.10.

GGMs versus RNs. To better understand the different performance of GGMs and RNs, we computed the average posterior probabilities of the true and false edges from the (partial) correlation coefficients according to the scheme described in [6]. The results are shown in Figure 5.11, which suggest that GGMs show a clearer separation of the true and false edges than RNs. This difference has not translated itself into an improved performance of GGMs over RNs in terms of AUC and TP scores for the unintervened non-Gaussian data. The reason is that although the separation between the scores is poorer for RNs than for GGMs, it has not affected the ranking of the edges. However, this finding suggests that inference with RNs is less stable than with GGMs. In fact, for interventions, RNs show a more substantial degradation in their performance than GGMs, which consistently outperform RNs on the intervened cytometry data ($P < 0.021$) and obtain significantly higher TP counts on the nonlinear simulated data ($P < 10^{-4}$).

Interventions for Low Noise Level. The left column of Figure 5.7 reveals a curious finding for the low-noise scenario: on interventions, the UGE score for BNs deteriorates. As discussed above, the ability to suppress spurious associations between unconnected nodes deteriorates for low noise levels. Interventions reduce the average noise level; so if the noise is already very low, this further reduction in the noise may lead to the prediction of spurious associations. The deterioration of the UGE (as opposed to the DGE) score can be explained by the fact that a spurious undirected edge is equivalent to two spurious directed edges (since there are twice as many directed as undirected edges in the graph) and that the UGE score does not benefit from any corrections of edge directions that result from the interventions.

Figure 5.7 GGMs versus BNs on data simulated with NetBuilder. This figure compares the performance of GGMs and BNs on the synthetic data generated with NetBuilder. The columns refer to different standard deviations of the additive Gaussian noise. Left column: $\sigma = 0.01$. Center column: $\sigma = 0.1$. Right column: $\sigma = 0.3$. The top panel (top two rows) shows scatter plots of GGM scores plotted against BN scores; a detailed explanation of the symbols is given in the caption of Figure 5.6 The bottom panel (bottom two rows) shows histograms with average AUC scores and TP counts for BNs (filled bars) and GGMs (empty bars); see the caption of Figure 5.6 for further explanations. In each panel, the two rows refer to different scoring criteria, discussed in Section 5.4. Top row: AUC score. Bottom row: TP count. From [41], with permission from Oxford University Press.

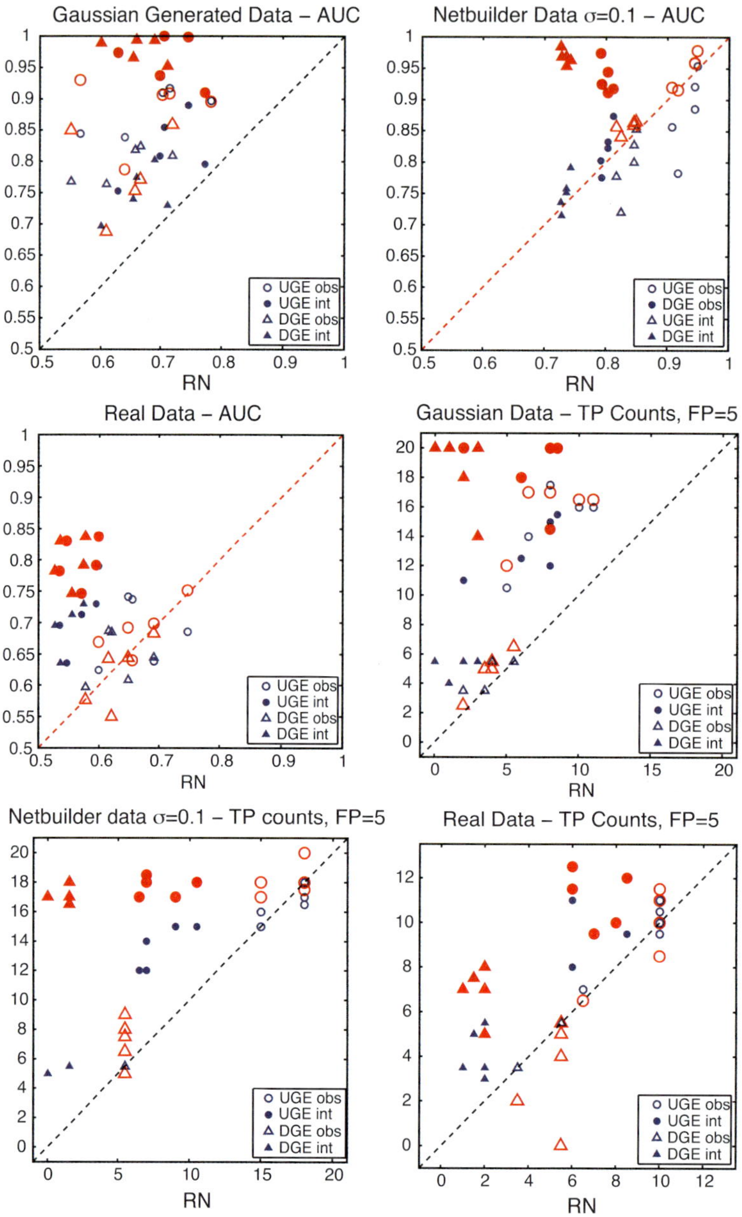

Figure 5.8 GGMs and BNs versus RNs. This figure compares the performance of GGMs and BNs (vertical axis) with RNs (horizontal axis). The columns refer to different data sets. Left column: Gaussian data. Center column: Data generated with NetBuilder, subject to additive Gaussian noise with $\sigma = 0.1$. Right column: Cytometry data. The two rows refer to different scoring criteria, discussed in Section 5.4. Top row: AUC score. Bottom row: TP count. The symbols of the six scatter plots are explained in the caption of Figure 5.6. The different symbol sizes refer to different comparisons. Large: BNs versus RNs. Small: GGMs versus RNs.

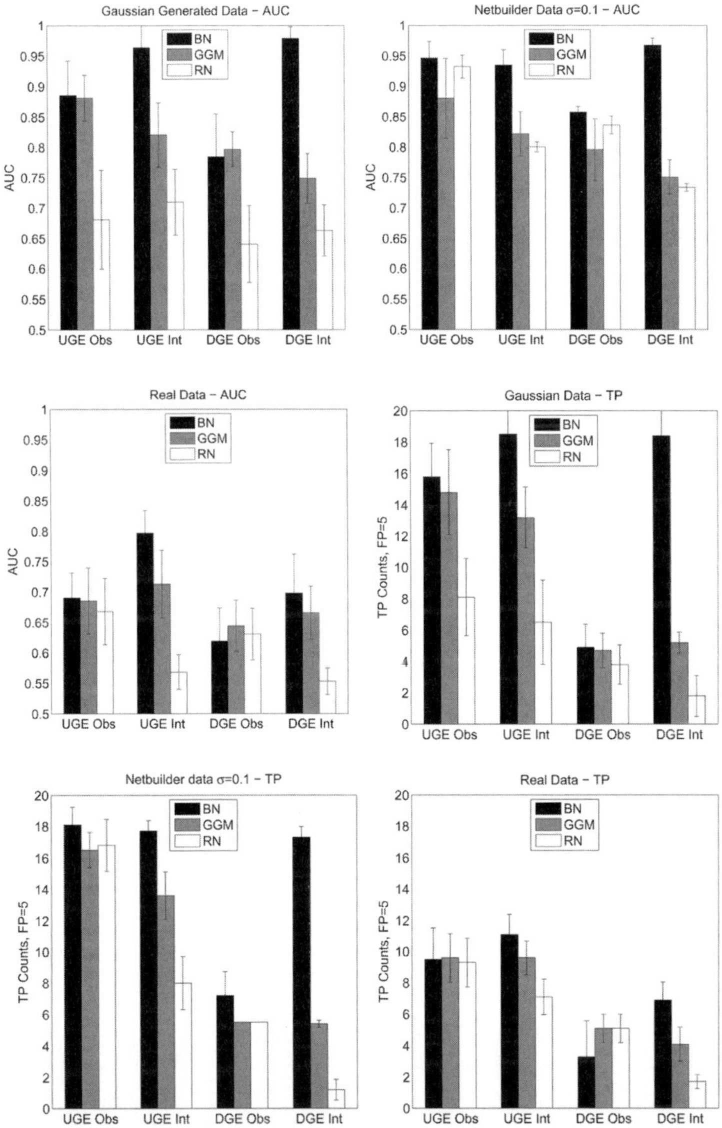

Figure 5.9 Cross-data comparison between BNs, GGMs, and RNs. The histograms show the average AUC scores and TP counts for BNs (black bars), GGMs (gray bars), and RNs (white bars). The codes under the histograms indicate the type of evaluation (UGE versus DGE) and whether observational (Obs) or interventional (Int) data have been used. The six panels refer to different data and scoring criteria. Top row: AUC scores. Bottom row: TP counts. Left column: Gaussian data. Center column: NetBuilder data. Right column: Cytometry data. From [41], with permission from Oxford University Press.

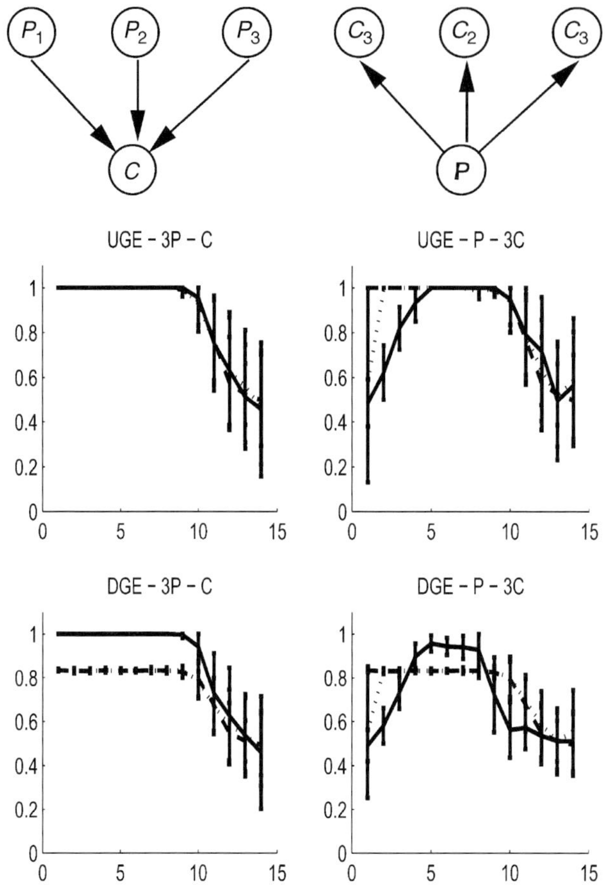

Figure 5.10 Effect of noise on learning network structures. The figure shows plots of the AUC means with error bars representing the AUC standard deviations for the three methods under comparison: Bayesian networks (solid line), Gaussian graphical models (dashed line), and relevance networks (dotted line). For each of the 14 different noise levels (σ), 25 Gaussian-distributed test data sets were generated using two simple graph topologies with 4 nodes each, shown at the top. In the first topology $G(P_i, C)$ (left panels), three parent nodes P_i have a common child node C with $C = 1/3$ $(P_1 + P_2 + P_3) + N(0, \sigma^2)$. In the second topology, $G(P, C_i)$ (right panels) one parent node P has three child nodes C_i with $C_i = 1/3 \cdot P + N(0, \sigma^2)$. For both networks and all three methods, the UGE (middle) and the DGE (bottom) figures of merit were computed. Bayesian networks outperform GGMs and relevance networks on the first topology, $G(P_i, C)$ (left column), in terms of the DGE score. Note that this topology contains two v-structures in which the edge directions are enforced. For the second network structure, $G(P, C_i)$ (right column), the reconstruction accuracy deteriorates for low and high noise levels, as discussed in Section 5.8. The deterioration at low noise levels is particularly pronounced for Bayesian networks.

(a)

	Gaussian			
	Obs		Int	
	DGE	UGE	DGE	UGE
BN	0.74	0.82	0.93	0.89
GGM	0.77	0.89	0.76	0.87
RN	0.49	0.49	0.53	0.54

(b)

	Netbuilder			
	Obs		Int	
	DGE	UGE	DGE	UGE
BN	0.78	0.86	0.88	0.78
GGM	0.74	0.84	0.72	0.82
RN	0.54	0.55	0.55	0.57

(c)

	Real			
	Obs		Int	
	DGE	UGE	DGE	UGE
BN	0.70	0.79	0.77	0.74
GGM	0.79	0.93	0.72	0.81
RN	0.77	0.88	0.57	0.60

Figure 5.11 Separation scores. The separation score is defined as $S = T/(T+F)$, where T is the average score of a true edge and F is the average score of a false edge. The perfect separation score of $S = 1$ is obtained when assigning a zero score to all false edges. Conversely, a method that cannot distinguish between true and false edges leads to an average separation score of $S = 0.5$. The abbreviations Obs and Int refer to observational and interventional data, respectively. The numbers are averages over all simulations carried out in the indicated category.

Dependence on the Network Topology. We investigated the influence of the network topology as follows. We removed four edges from the graph to create four v-structures and reran the whole analysis. The modified network structure is shown in Figure 5.5. Note that owing to the additional v-structures, a higher proportion of edge directions are resolved. We would therefore expect an increase in the performance of BNs relative to GGMs. The results are shown in Figures 5.12 and 5.13. The findings were, on the whole, similar to the results obtained for the original network. In the observational linear Gaussian data, the comparison of BNs versus GGMs showed a significant shift in favor of BNs, with $P < 0.05$ for all performance scores; this confirms our hypothesis. There was no significant difference between the performance scores of BNs and GGMs on the nonlinear data generated with NetBuilder, though.

Learning Directed Graphs from the Cytometry Data. BNs obtain poorer DGE scores on the unintervened cytometry data than RNs and GGMs, while there is no significant difference in the UGE scores. This suggests that while BNs learn the skeleton of the network as accurately as GGMs and RNs, some of the edge

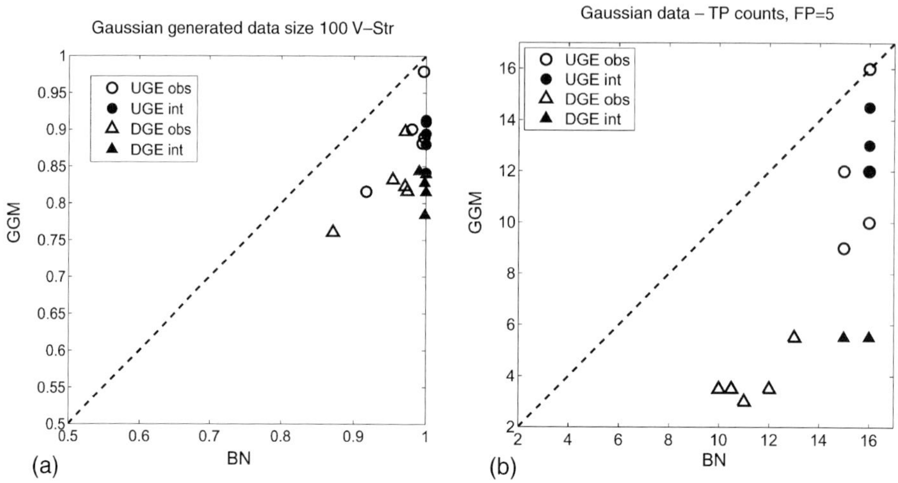

Figure 5.12 GGMs versus BNs on Gaussian v-structure data. Scatter plots comparing the performance of GGMs (vertical axis) with BNs (horizontal axis). The diagonal line represents equal performance. Symbols above the line indicate that GGMs outperform BNs. Conversely symbols below that line point to a better performance of BNs over GGMs. Each part compares the results obtained from two different data types, using only passive observations (empty symbols) and including active interventions (filled symbols). Two different evaluation criteria have been applied, based on directed graphs (DGE, represented by triangles) and their undirected skeletons (UGE, represented by circles). The two panels refer to two different scoring criteria. Left: AUC scores. Right: TP counts.

directions are systematically inverted. A possible explanation are errors in the gold-standard network. The recent literature provides evidence for a negative feedback loop between RAF and ERK via MEK. Active RAF phosphorylates and activates MEK, which, in turn, activates ERK. This corresponds to the directed regulatory path shown in Figure 5.4. However, through a negative feedback mechanism involving ERK, RAF is phosphorylated on inhibitory sites, generating an inactive, desensitized RAF. Details can be found in [36]. This feedback loop is not included in the gold-standard network reported by Sachs *et al.* [13], shown in Figure 5.4. Such yet unaccounted feedback loops could explain systematic deviations between the predicted and the gold-standard network, not only because the structure of a Bayesian network is constrained to be acyclic, but also because we ultimately do not have a reliable gold standard to assess the quality of the predictions. Negative feedback is also known to have a stabilizing effect with respect to interventions; this might explain why the improvement in the DGE scores for BNs is less pronounced than for the simulated data. This example points to a fundamental problem inherent in any evaluation based solely on real biological data and illustrates clearly the advantage of our combined evaluation based on both laboratory and simulated data.

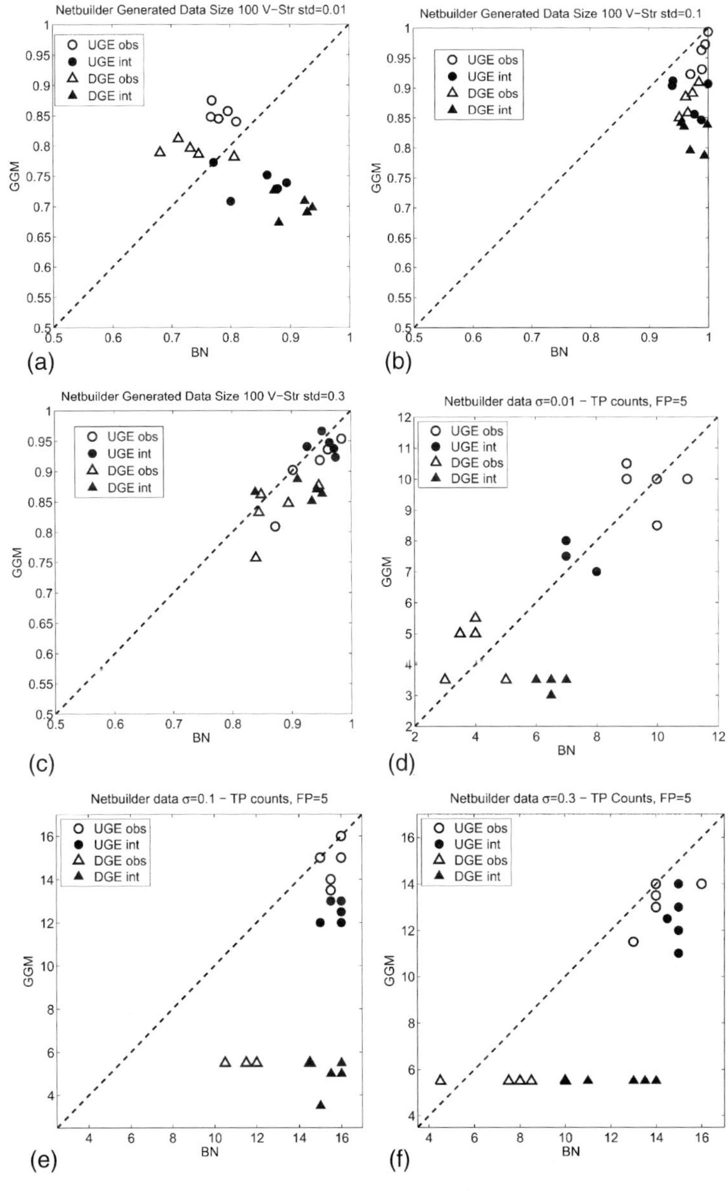

Figure 5.13 GGMs versus BNs on NetBuilder v-structure data. This figure compares the performance of GGMs and BNs on the synthetic data generated with NetBuilder, for the topology with some edges removed. The columns refer to different standard deviations of the additive Gaussian noise. Left column: $\sigma = 0.01$; center column: $\sigma = 0.1$; right column: $\sigma = 0.3$. The two rows refer to different scoring criteria. Top row: AUC score. Bottom row: TP counts. A detailed explanation of the symbols is given in the caption of Figure 5.12.

5.9 Conclusion

Our main findings can be summarized as follows. BNs and GGMs tend to outperform RNs, but the difference is less pronounced for the nonlinear simulated data (NetBuilder) and the measured protein concentrations (cytometry experiments) than for Gaussian data. Also, there is insufficient evidence for any significant difference between BNs and GGMs on observational data. These findings are different from those reported in [12], which seems to result from the improved inference algorithm for GGMs [7]. However, for interventional data, BNs outperform GGMs and RNs when taking the edge directions into account. This suggests that the higher computational costs of inference with BNs over GGMs and RNs are not justified for passive observations, but that active interventions in the form of gene knockouts and overexpressions are required to exploit the full potential of BNs.

Acknowledgments

Marco Grzegorczyk is supported by the British Biotechnology and Biological Sciences Research Council (BBSRC). Dirk Husmeier is supported by the Scottish Executive. Adriano Werhli is supported by Coordenacao de Aperfeicoamento de Pessoal de Nivel Superior (CAPES). We would like to thank Wolfgang Urfer for initiating our collaboration and for stimulating discussions over many years.

References

1 Friedman, N., Linial, M., Nachman, I. and Pe'er, D. (2000) Using Bayesian networks to analyze expression data. *Journal of Computational Biology*, **7**, 601–620.

2 Butte, A.S. and Kohane, I.S. (2003) Relevance networks: a first step toward finding genetic regulatory networks within microarray data in*The Analysis of Gene Expression Data*, (eds G. Parmigiani, E.S. Garett, R.A. Irizarry and S.L. Zeger), Springer, New York, pp. 428–446.

3 Hartemink, A.J. (2001) Principled computational methods for the validation and discovery of genetic regulatory networks. PhD thesis, MIT,http:// citeseer.ist.psu.edu/ hartemink01principled.html.

4 Imoto, S., Higuchi, T., Goto, T., Kuhara, S. and Miyano, S. (2003) Combining microarrays and biological knowledge for estimating gene networks via Bayesian networks. Proceedings IEEE Computer Society Bioinformatics Conference (CSB'03) August 11–14, 2003, Stanford, CA, pp. 104–113.

5 Nariai, N., Tamada, Y., Imoto, S. and Miyano, S. (2005) Estimating gene regulatory networks and protein–protein interactions of *Saccharomyces cerevisiae* from multiple genome-wide data. *Bioinformatics*, **21** (Suppl. 2), ii206–ii212.

6 Schäfer, J. and Strimmer, K. (2005) An empirical Bayes approach to inferring large-scale gene association networks. *Bioinformatics*, **21** (6), 754–764.

7 Schäfer, J. and Strimmer, K. (2005) A shrinkage approach to large-scale covariance matrix estimation and implications for functional genomics. *Statistical Applications in Genetics and Molecular Biology*, **4** (1), 32.

8 author>Bing, N. andHoeschele, I. (2005) Genetical genomics analysis of a yeast segregant population for transcription network inference. *Genetics*, **170**, 533–542.

9 Smith, V.A., Jarvis, E.D. and Hartemink, A.J. (2002) Evaluating functional network inference using simulations of complex biological systems. *Bioinformatics*, **18**, S216–S224. (ISMB02 special issue).

10 Husmeier, D. (2003) Sensitivity and specificity of inferring genetic regulatory interactions from microarray experiments with dynamic Bayesian networks. *Bioinformatics*, **19**, 2271–2282.

11 Zak, D.E., Doyle, F.J., Gonye, G.E. and Schwaber, J.S. (2001) Simulation studies for the identification of genetic networks from cDNA array and regulatory activity data.Proceedings of the Second International Conference on Systems Biology Pasadena, CA, pp.231–238.

12 Pournara, I.V. (2005) Reconstructing gene networks by passive and active Bayesian learning. PhD thesis, Birbeck College, University of London.

13 Sachs, K., Perez, O., Pe'er, D., Lauffenburger, D.A. and Nolan, G.P. (2005) Protein-signaling networks derived from multiparameter single-cell data. *Science*, **308**, 523–529.

14 Yuh, C.H., Bolouri, H. and Davidson, E.H. (1998) Genomic *cis*-regulatory logic: experimental and computational analysis of a sea urchin gene. *Science*, **279**, 1896–1902.

15 Yuh, C.H., Bolouri, H. and Davidson, E.H. (2001) *Cis*-regulatory logic in the endo16 gene: switching from a specification to a differentiation mode of control. *Development*, **128**, 617–629.

16 Butte, A.S. and Kohane, I.S. (2000) Mutual information relevance networks: functional genomic clustering using pairwise entropy measurements. *Pacific Symposium on Biocomputing*, **2000**, 418–429.

17 Edwards, D.M. (2000) *Introduction to Graphical Modelling*, Springer Verlag, New York.

18 Ledoit, O. and Wolf, M. (2004) A well conditioned estimator for large-dimensional covariance matrices. *Journal of Multivariate Analysis*, **88**, 365–411.

19 Jensen, F. (1996) *An Introduction to Bayesian Networks*, UCL Press, London, England.

20 Verma, T. and Pearl, J. (1990) *Equivalence and Synthesis of Causal Models*. Proceedings of the Sixth Conference on Uncertainty in Artificial Intelligence, vol. **6**, pp. 220–227.

21 Chickering, D.M. (1995) A transformational characterization of equivalent Bayesian network structures. *International Conference on Uncertainty in Artificial Intelligence (UAI)*, **11**, 87–98.

22 Chickering, D. (2002) Learning equivalence classes of Bayesian network structures. *Journal of Machine Learning Research*, **2**, 445–498.

23 Cooper, G. and Glymour, C. (1999) *Computation Causation, and Discovery*, MIT Press, Cambridge, MA.

24 Pearl, J. (2000) *Causality: Models, Reasoning and Intelligent Systems*, Cambridge University Press, London, UK.

25 Geiger, D. and Heckerman, D. (1994) *Learning Gaussian Networks*. Proceedings of the Tenth Conference on Uncertainty in Artificial Intelligence, July 29–31, 1994, Seattle, WA. Morgan Kaufmann, San Francisco, CA, pp. 235–243.

26 Heckerman, D. (1999) A tutorial on learning with Bayesian networks, in *Learning in Graphical Models, Adaptive Computation and Machine Learning*, (ed.M.I. Jordan), MIT Press, Cambridge, MA, pp. 301–354.

27 Madigan, D. and York, J. (1995) Bayesian graphical models for discrete data.

International Statistical Review, **63**, 215–232.

28 Friedman, N. and Koller, D. (2003) Being Bayesian about network structure. *Machine Learning*, **50**, 95–126.

29 Gilks, W., Richardson, S. and Spiegelhalter, D. (1996) *Markov Chain Monte Carlo in Practice*, Chapman & Hall, Suffolk.

30 Cowles, M.K. and Carlin, B.P. (1996) Markov chain Monte Carlo convergence diagnostics: a comparative review. *Journal of the American Statistical Association*, **91**, 883–904.

31 Giudici, P. and Castelo, R. (2003) Improving Markov chain Monte Carlo model search for data mining. *Machine Learning*, **50**, 127–158.

32 Yoo, C., Thorson, V. and Cooper, G. (2002) Discovery of causal relationships in a gene regulation pathway from a mixture of experimental and observational DNA microarray data. *Pacific Symposium on Biocomputing*, **7**, 498–509.

33 Pournara, I. and Wernisch, L. (2004) Reconstruction of gene networks using Bayesian learning and manipulation experiments. *Bioinformatics*, **20**, 2934–2942.

34 Tian, J. and Pearl, J. (2001) *Active Learning for Structure in Bayesian Networks*, vol. **17**, Seventeenth International Joint Conference on Artificial Intelligence, August 4–10, 2001, Seattle, WA, pp. 863–869.

35 Storey, J.D. and Tibshirani, R. (2003) Statistical significance for genomewide studies. *Proceedings of the National Academy of Sciences of the United States of America*, **100**, 9440–9445.

36 Dougherty, M.K., Muller, J., Ritt, D.A., Zhou, M., Zhou, X.Z., Copeland, T.D., Conrads, T.P., Veenstra, T.D., Lu, K.P. and Morrison, D.K. (2005) Regulation of raf-1 by direct feedback phosphorylation. *Molecular Cell*, **17**, 215–224.

37 Grzegorczyk, M. (2006) Comparative evaluation of different graphical models for the analysis of gene expression data. PhD thesis, University of Dortmund, https://eldorado.uni-dortmund.de/handle/2003/22855.

38 Werhli, A.V. (2007) Reconstruction of gene regulatory networks from postgenomic data. PhD thesis, Biomathematics & Statistics Scotland (BioSS) and University of Edinburgh.

39 Atkins, P.W. (1986) *Physical Chemistry*, 3rd edn,Oxford University Press, Oxford.

40 Imoto, S., Kim, S., Goto, T., Aburatani, S., Tashiro, K., Kuhara, S. and Miyano, S. (2003) Bayesian network and nonparametric heteroscedastic regression for nonlinear modeling of genetic network. *Journal of Bioinformatics and Computational Biology*, **1** (2), 231–252.

41 Werhli, A.V., Grzegorczyk, M. and Husmeier, D. (2006) Comparative evaluation of reverse engineering gene regulatory networks with relevance networks graphical Gaussian models and Bayesian networks. *Bioinformatics*, **22**, 2523–2531.

6
Statistical Methods for Inference of Genetic Networks and Regulatory Modules

Hongzhe Li

6.1
Introduction

The completion of the human genome project and the development of many high-throughput genomic technologies make it possible to systematically define the organization and function of gene, protein, and metabolite networks. Large-scale microarray gene expression data, promotor sequences data, and genome-wide chromatin immunoprecipitation (ChIP-chip) data provide the possibility of learning gene regulation and constructing the gene regulatory networks and pathways or cellular networks [1–3]. In a recent review, Bansal *et al.* [4] summarized the methods for inferring genetic networks into two broad classes: those based on the "physical interaction" approach that aims at identifying interactions among transcription factors and their target genes, and those based on the "influence interaction" approach that aims at relating the expression of a gene to the expression of the other genes in the cell, rather than relating it to the sequence motif found in its promotor. In this chapter, we review some recently developed statistical methods for several problems related to inferences of genetic network and regulatory modules, including both "physical interaction" networks using diverse data sets and "influence interaction" networks using gene expression data alone.

Early research on gene expression analysis has mainly focused on using clustering analysis to identify coregulated genes [5]. Recently, some efforts have been devoted to developing probabilistic models for modeling regulatory and cellular networks based on genome-wide high-throughout data, including both Bayesian network modeling [2,6] and Gaussian graphical modeling [7–10]. The goal of such probabilistic modeling is to investigate the patterns of association in order to generate biological insights plausibly related to underlying biological and regulatory pathways. It is important to note that the interaction between two genes in a gene network defined by such graphical models does not necessarily imply a physical interaction, but can refer to an indirect regulation via proteins, metabolites, and ncRNA that have been measured directly, and therefore its interpretation depends on the model formulations [4]. In this chapter, we will present some details on

Analysis of Microarray Data: A Network-Based Approach. Edited by F. Emmert-Streib and M. Dehmer

ISBN: 978-3-527-31822-3

Gaussian graphical models and methods for estimating the graphical structure in the high-dimensional settings.

It is now understood that much of a cell's activity is organized as a network of interacting modules that consist of genes coregulated by a set of regulators to respond to different conditions [6,11]. It is therefore important to identify such regulatory modules. Genome-wide expression profiles provide important information on cellular states and cell's activities and therefore provide information to infer regulatory modules. Segal *et al.* [6] present a probabilistic method for identifying regulatory modules from gene expression data using classification and regression tree (CART) methods. In this approach, a set of regulators including transcriptional factors and signaling proteins is first identified from literature. The model further assumes that both regulators and its targets must be regulated at the transcriptional levels, resulting in detectable changes in expression. These regulators can then be used as predictors for gene expression levels using CART. The genes that are regulated by the same set of regulators are then identified as regulatory modules. Bonneau *et al.* [12] recently proposed a similar framework for learning parsimonious regulatory networks and called the method "Inferelator." The method first clusters genes into groups and then essentially performs linear regression with the least absolute shrinkage and selection operator (Lasso) [13] to select the regulators that are related to the expression variations of the cluster of genes. In this chapter, we will present a sparse regression mixture modeling (SRMM) approach for identifying the gene regulatory modules. Since we expect only a small set of regulators that control the expression of a module, the regression model should therefore be sparse. Our approach is based on a combination of finite mixture modeling [14] and Lasso.

Another area of intensive research in recent years has been to integrate gene expression data with motif or ChIP-chip data in order to identify transcriptional networks [15–18]. The fundamental idea of these approaches is based on linear regression analysis with gene expression levels as responses and motifs or ChIP-chip data as predictors. Although these approaches work reasonably well in the discovery of regulatory motifs in lower organisms, they often fail to identify mammalian transcriptional factor binding sites [3]. Das *et al.* [3] proposed to correlate the binding strength of motifs with expression levels using the multivariate adaptive smoothing splines (MARS) of Friedman [19]. Ernst *et al.* [11] proposed an interesting approach based on the hidden Markov model for reconstructing dynamic regulatory networks using microarray time course (MTC) gene expression data and ChIP-chip data. We will present an approach based on varying coefficient models in order to identify the transcriptional factors that are involved in a given biological process.

In this chapter, we present the statistical formulations of the problems related to inference of genetic networks based on gene expression data, inference of regulatory modules based on both gene expression data and genome annotation including information on transcriptional factors and regulators, and inference of regulatory networks based on gene expression and sequence motif or ChIP-chip data. We review some statistical methods developed for these methods and focus on the approaches that we have developed. We illustrate these methods by presenting results from analysis of several real data sets. Finally, we present a brief discussion on future work in this important area.

6.2 Network Inference Based on Gaussian Graphical Models

Graphical models use graphs to represent dependencies between stochastic variables [20]. The graphical approach yields dependence models that are easily visualized and presented. One specific graphical model is the Gaussian graphical model, which assumes that the multivariate vector follows a multivariate normal distribution with a particular structure of the inverse of the covariance matrix, often called the precision or concentration matrix. For such Gaussian graphical models, it is usually assumed that the patterns of variation in expression for a given gene will be predicted by those of a small subset of other genes. This assumption leads to sparsity (i.e., many zeros) in the precision matrix of the multivariate distribution and reduces the problem to well-known neighborhood selection or covariance selection problems [21]. In such a concentration graph modeling framework, the key idea is to use partial correlation as a measure of independence of any two genes, rendering it straightforward to distinguish direct from indirect interactions. This is in contrast to the covariance graphical model where marginal correlations are used. It has been demonstrated in the literature that many biochemical and genetic networks are not fully connected [22–24], and many genetic interaction networks contain many genes with few interactions and a few genes with many interactions. Therefore, the genetic networks are intrinsically sparse and the corresponding precision matrix should be sparse.

In the literature, there are several approaches to covariance selection problems in the context of microarray data analysis. Schafer and Strimmer [7] proposed a naive approach to estimate the precision matrix by using a boosted G-inverse, then determining which off-diagonal elements are zero by a thresholding and false discovery procedure. The drawback of this approach is that the sparsity is not accounted for when estimating the precision matrix, so the procedure is expected to perform poorly. Meinshausen and Buhlmann [25] proposed a gene-by-gene approach by using Lasso [13] to find neighbors for each gene. Under a large set of assumptions, they showed that the neighbors can be consistently identified when the sample size goes to infinity, which is very rare for microarray gene expression data. Dobra *et al.* [9] proposed a Bayesian approach by converting the dependency networks into compositional networks using Cholesky decomposition. The graphs are then used to estimate the precision matrix. Since Cholesky decomposition of the precision matrix naturally imposes ordering restriction of the variables, the procedure is computationally quite intensive because it has to determine gene order in their model construction. Finally, Wille *et al.* [8] proposed to infer Gaussian graphs based on tri-graphs by considering all partial correlations conditioning on only one other variable. Strictly speaking, the resulting tri-graphs are not true Gaussian concentration graphs. In the following, we provide a brief introduction to Gaussian graphical models and review the threshold gradient descent (TGD) approach for identifying such graphs developed in Li and Gui [10].

6.2.1 Gaussian Graphical Models

We assume that the gene expression data observed are randomly sampled observational or experimental data from a multivariate normal probability model. Specifically, let X be a random normal p-dimensional vector and $X_1, \ldots, X_p$ denote the p elements, where p is the number of genes. Let $V = \{1, \ldots, p\}$ be the set of nodes (genes), and $X^{(k)}$ be the vector of gene expression levels for the kth sample. We assume that

$$X \sim N_p(0, \Sigma), \tag{6.1}$$

with positive definite variance–covariance matrix $\Sigma = \{\sigma_{ij}\}$ and precision matrix $\Omega = \Sigma^{-1} = \{\omega_{ij}\}$. This model can also be summarized as a graph model. Let $G = (V, E)$ be an undirected graph with vertex set $V = \{1, \ldots, p\}$ and edge set $E = \{e_{ij}\}$, where $e_{ij} = 1$ or 0 according to whether vertices i and j, $1 \le j \le p$, are adjacent in G or not. The Gaussian graphical model consists of all p-variate normal distributions $N_p(0, \Sigma)$, where Σ is unknown but where the precision matrix satisfies the following linear restrictions:

$$e_{ij} = 0 \Rightarrow \omega_{ij} = 0.$$

This model is also called a covariance selection model [21] or a Gaussian precision graph model.

Let $[-i]$ denote the set $\{1, 2, \ldots, i-1, i+1, \ldots, p\}$. In the Gaussian graphical model, it is well known that the partial regression coefficients of X_i on X_j in the normal linear regression $p(X_i|X_{[-i]})$ is $-\omega_{ij}/\omega_{ii}, j \in [-i]$ and the ijth partial correlation between the ith and the jth gene is $\rho_{ij} = -\omega_{ij}/\sqrt{\omega_{ij}\omega_{jj}}$. For a given gene g, we define the neighbor of this gene as

$$ne_g = \{j : \omega_{gj} \ne 0, j \in [-g]\},$$

which contains all the genes with a nonzero partial correlation with the gene g. From the multivariate normal distribution theory, we have the following conditional independence result:

$$X_g \perp X_{G \setminus (ne_g \cup g)} | X_{ne_g}.$$

6.2.2 Threshold Gradient Descent Regularization

We consider the estimation of the precision matrix Ω based on a sample of i.i.d. observations $X^{(k)} \in R^p$, $k \in N = \{1, \ldots, n\}$, where the set N can be interpreted as indexing the samples on which we observe the variables in V and $X^{(k)}$ is the kth observation. Li and Gui [10] developed a penalized procedure for estimating Ω using the idea of threshold gradient descent [26,27] to take into account the sparse nature of the precision matrix for genetic networks.

To utilize the sparse property of the precision matrix, we propose in this section to maximize the likelihood function based on model (6.1) subject to constraint by "sparse" precision matrix Ω. Let $\omega_{\mathrm{d}} \equiv \{\omega_{11}, \ldots, \omega_{pp}\}$ denote the vector of the diagonal elements of the matrix Ω and $\omega^{\mathrm{o}} \equiv \{\omega_{ij}\}_{i \neq j}$ denote the vector of $q = p(p-1)/2$ off-diagonal elements of the Ω matrix. The likelihood function can be written as

$$w(\omega^{\mathrm{d}}, \omega^{\mathrm{o}}) = \frac{n}{2} \log|\Omega| - \frac{1}{2} \sum_{k=1}^{n} X^{(k)'} \Omega X^{(k)}, \tag{6.2}$$

where $X^{(k)}$ is the kth observation. We assume that the variables are standardized. When $p < n$ the maximum likelihood estimate (MLE) of Ω is simply the inverse of the sample covariance matrix, and when $n < p$, the MLE of Ω is not unique.

To account for sparsity of the precision matrix Ω, Li and Gui [10] defined a loss function as the negative of the log-likelihood function (6.2),

$$l(\omega^{\mathrm{d}}, \omega^{\mathrm{o}}) = -w(\omega^{\mathrm{d}}, \omega^{\mathrm{o}}).$$

On the basis of Equation 6.2, the gradient of the loss function with respect to Ω is

$$\frac{\partial l}{\partial \Omega} = \frac{n}{2} \Omega^{-1} - \frac{1}{2} \sum_{k=1}^{n} X^{(k)} X^{(k)'}. \tag{6.3}$$

From this we can obtain the gradient of the loss function over the off-diagonal elements ω^{o}. Define $g(\omega^{\mathrm{o}}) = (g_1(\omega^{\mathrm{o}}), \ldots, g_q(\omega^{\mathrm{o}})) = -\nabla_{\omega^{\mathrm{o}}} l(\omega^{\mathrm{o}}, \omega^{\mathrm{d}})$ to be the negative gradient of l with respect to ω^{o}. To find an optimal path from all the paths from $\Omega = I$ to the MLE of Ω or to a precision matrix surface formed by $\Omega = S^{-}$ when $p > n$, we start from $\nu = 0$, $\omega^{\mathrm{o}} = (0, \ldots, 0)$ and $\omega^{\mathrm{d}} = (1, \ldots, 1)$ and update the elements ω^{o} by the following gradient descent step,

$$\hat{\omega}^{\mathrm{o}}(\nu + \Delta\nu) = \hat{\omega}^{\mathrm{o}}(\nu) + \Delta\nu h(\nu),$$

where $\hat{\omega}^{\mathrm{o}}(\nu)$ is the ω^{o} value corresponding to current ν, $\Delta\nu > 0$ is an infinitesimal increment, and $h(\nu)$ is the direction in the parameter space tangent to the path evaluated at $\hat{\omega}^{\mathrm{o}}(\nu)$. This tangent vector at each step represents a descent direction. To direct the path toward parameter points with diverse values, following Friedman and Popescu [26], we define $h(\nu)$ as

$$h(\nu) = \{f_j(\nu) \cdot g_j(\nu), j = 1, \ldots, q\},$$

where

$$f_j(\nu) = I[|g_j(\nu)| \geq \tau \cdot \max_{1 \leq k \leq q} |g_k(\nu)|],$$

where $I[\cdot]$ is an indicator function, and $0 \leq \tau \leq 1$ is a threshold parameter that regulates the diversity of the values of $f_j(\nu)$; larger values of τ lead to more diversity. $g(\nu)$ is the negative gradient evaluated at $\hat{\omega}^o(\nu)$ and current ω^d. Therefore, τ is the parameter that controls the degree of penalty and sparsity in the ω^o, with $\tau = 1$ giving the sparsest graphs. Instead of moving along the true gradient direction, the threshold gradient update moves along only those elements with large values of the gradient. After ω^o is updated, we update the diagonal elements of Ω, ω^d, by maximizing the log-likelihood function (6.2) with ω^o fixed at the current values, $\hat{\omega}^o$. This is done by using Newton–Raphson iterations.

In summary, for any threshold value $0 \leq \tau \leq 1$, the threshold gradient descent regularization algorithm for the sparse Gaussian graphical model involves the following six steps,

1. Set $\omega^o(0) = 0, \quad \omega^d(0) = 1, \quad \nu = 0.$
2. Calculate $g(\nu) = -\partial l/\partial \omega^o$ for the current ω^o and ω^d.
3. Calculate $f_j(\nu) = I[|g_j(\nu)| \geq \tau \cdot \max_{1 \leq k \leq q} |g_k(\nu)|]$ and $h(\nu)$.
4. Update $\omega^o(\nu + \Delta\nu) = \omega^o(\nu) + \Delta\nu \cdot h(\nu), \quad \nu = \nu + \Delta\nu.$
5. Update parameters ω^d by maximizing the log likelihood using Newton–Raphson iterations with ω^o fixed at $\omega^o(\nu + \Delta\nu)$.
6. Repeat steps 2–5.

For a given τ, it is easy to see that the likelihood function increases as the iterations increase, and different τ correspond to different paths for Ω from I to S^-. It should be emphasized that for a given τ, the threshold gradient iterations stop before it reaches S^-, and the number of gradient iterations at which to stop the algorithm can be determined by cross-validation (see Section 6.2.3). Li and Gui [10] particularly considered the algorithm with $\tau = 1$, which corresponds to the sparsest graph for a given TGD step, and called the proposed procedure the direct threshold gradient descent procedure. Such a procedure is expected to perform better for gene expression data, since most biological or genetic networks are expected to be very sparse [28].

6.2.3
Model Selection by Cross-Validation and Bootstrap

As the iterations continue, more and more nonzero elements are selected in the precision matrix and the corresponding undirected graphs grow larger. The final model should provide the best balance between coverage (correctly identified connections/total true connections) and false positives (incorrectly identified connections/total identified connections). Li and Gui [10] proposed to use K-fold cross-validation for choosing the number of TGD iterations, ν, where for each ν, the K-fold cross-validated log-likelihood criterion is defined as

$$\mathrm{CV}(\nu) = \frac{1}{K}\sum_{k=1}^{K}\left(-n_k \log|\Omega_{-k}| + \sum_{i\,\in\,V_k} X^{(i)}\Omega X^{(i)}\right),$$

where n_k is the size of the kth validation set V_k and Ω_{-k} is the TGD estimate of the precision matrix based on sample $V\backslash V_k$ evaluated at $\hat{\Omega}(\nu)$. Alternatively, we can use the Bayesian information criterion (BIC) criteria for selecting ν, where the degrees of freedom can be defined as the number of nonzero entries of the off-diagonal elements of the precision matrix. This is similar in spirit to Lasso in linear regression where the degrees of freedom is defined as the number of nonzero coefficients [29].

Since the number of the off-diagonal elements in the precision matrix is often quite large compared to the sample size, there is often considerable uncertainty in the edges chosen. As a final step in the procedure, we propose to use the bootstrap method to determine the statistical accuracy and the importance of each of the edges identified by the TGD procedure. In bootstrapping, B bootstrap data sets, $X^{*1}, \ldots,$ X^{*B}, are sampled with replacement from the original data set such that each bootstrap sample contains n observations. We then apply the TGD procedure to each bootstrap data set and examine which edges are in the final models. One can then choose only the edges with a high probability of being nonzero in the precision matrix over the bootstrap samples.

6.2.4
Simulation Results and Application to Real Data Set

Li and Gui [10] performed simulation studies to evaluate the proposed threshold gradient descent procedure and applied this to the analysis of isoprenoid metabolic pathways. Results indicate that by accounting for sparsity in estimating the precision matrix, one can obtain a better estimate of the precision matrix, and the TDG procedure can effectively identify the linked edges in the Gaussian graphical models.

Li and Gui [10] applied the TGD procedure to the analysis of the *Arabidopsis thaliana* isoprenoid pathway. The isoprenoid biosynthetic pathway provides intermediates of many natural products including sterols, chlorophylls, carotenoids, plastoquinone and abscisic acid, and so on. It is now known that plants contain two pathways for the synthesis of the structural precursors of isoprenoids: the mevalonate (MVA) pathway, located in the cytosol/ER, and the recently discovered methylerythritol 4-phosphate (MEP) pathway, located in the plastids. The pathway in plastids, which is mevalonate independent, occurs and is responsible for the subsequent biosynthesis of plastidiar terpenoids such as carotenoids and the side chains of chlorophyll and plastoquinone [8]. It is therefore important to understand the organization and regulation of this complex metabolic pathway, with the long-term goal of using the generated knowledge to undertake metabolic engineering strategies oriented to increase the production of isoprenoids with pharmaceutical and food applications, and also the design and development of new antibiotics.

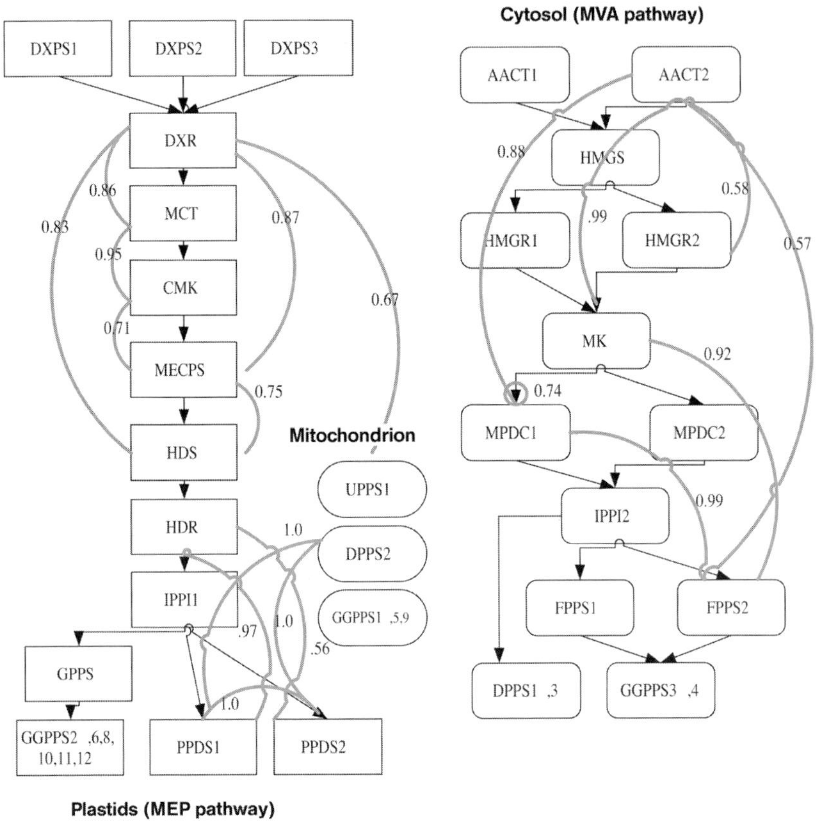

Fig. 6.1 Pathways identified by the TGD method for the 40 genes in the isoprenoid pathways, where the solid arrows are the true pathways and curved undirected lines are the estimated edges with bootstrap probability of greater than 0.5 for the TGD method. For this plot, the left panel represents a subgraph of the gene module in the MEP pathway and the right panel represents a subgraph of the gene module in the MVA pathway. The numbers on the estimated edges are the bootstrap probabilities.

To better understand the pathway and gain insights into the cross-link between the two pathways at the transcriptional level, Wille *et al.* [8] reported a data set including the gene expression patterns monitored under various experimental conditions using 118 GeneChip microarrays. For the construction of the genetic network, they focused on 40 genes, 16 of which were assigned to the cytosolic MVA pathway, 19 to the plastidal MEP pathway, and 5 genes encoding proteins located in the mitochondria. See the solid lines of Figure 6.1 for the MVA and the MEP pathways and the genes involved.

To demonstrate whether the proposed TGD method can identify the known isoprenoid pathways of these 40 genes based on the 118 gene expression measurements, Li and Gui [10] first estimated the precision matrix by the threshold gradient methods. Using 10-fold cross-validation, the TGD procedure resulted in 20 nonzero off-diagonal elements. We next used a bootstrap with the TGD procedure to estimate the confidence of the edges. With bootstrap probability of 0.50 or higher, we identified 19 pairs of genes that are connected with high confidence, of which 12 pairs have a bootstrap probability of 0.80 or higher. These 19 pairs are plotted on the true network in Figure 6.1. We find a module with strongly interconnected genes in each of the two pathways. For the MEP pathway, DXPS2, DXR, MCT, CMK, and MECPS are connected as the true pathway. Similarly, the genes in the MVA pathways, AACT2, HMGR2, MK, MPDC1, FPPS1, and FPP2, are closely connected. In addition, there are also several genes in the MEP pathway that are linked to proteins in the mitochondria.

6.3
Methods for Identifying Regulatory Modules

Since the networks inferred from gene expression data alone do not imply any physical interactions among genes, it is important to incorporate other data sources to infer regulatory networks. In addition, influence interactions may include physical interactions if the two interacting partners are a transcriptional factor and its target, suggesting that incorporating the transcriptional factors information can help in identifying the relevant regulatory networks and modules. Segal *et al.* [6] proposed to integrate gene expression data and knowledge of transcriptional factors to identify transcriptional modules using regression trees in a mixture-modeling framework. Lee *et al.* [30] further developed this idea for integrating microarray gene expression data, SNPs data, and regulatory information to identify the genetic variations that affect the regulatory modules. In this section, we reformulate the problem into SRMM and propose a Lasso-EM algorithm for identifying such regulatory modules.

6.3.1
The SRMM for Identifying Transcriptional Modules

Consider the microarray gene expression data setup of G genes over C experimental conditions. These G genes also include C transcriptional factors or signaling proteins. Let Y_{gc} be the log-expression level of the gth gene at the experimental condition c, for $g = 1, \ldots, G$ and $c = 1, \ldots, C$ and $Y = \{Y_{gc,}g = 1, \ldots, G, \quad c = 1, \ldots C\}$. Let X_{rc} be log-expression level of the rth regulator at the condition c, for $r = 1, \ldots, R$ and $c = 1,\ C, X_c = \{X_{1c,} \ldots, X_{Rc}\}$ be the expression level of the R regulators at the cth experiment condition, and $X = \{X_{rc}, r = 1 \ldots, R, \quad c = 1, \ldots, C\}$.

Assuming that there are K regulatory modules, let M_g be the module membership for the gene g. We assume the following model for the observed expression data for genes in the kth regulatory module:

$$
\begin{aligned}
Y_{gc}|\{M_g = k\} &= \sum_{r=1}^{R} X_{rc}\beta_{kr} + \epsilon_{gc}, \\
M_g &\sim \text{Multinomial}(\tau), \\
\text{with}\, \pi = (\pi_1, \ldots, \pi_K)^K, \pi_K &\leq 0, \text{and} \sum_{k=1}^{K} = 1,
\end{aligned}
\tag{6.4}
$$

where $\beta_k = (\beta_{k1}, \ldots, \beta_{kR})^T$ is the vector of module-specific parameters, ϵ_{gc} is the error term, which is assumed to follow $N(0, \sigma_k^2)$, and π_i is the prior probability that a gene belongs to the ith module. Since the expression level of a given gene is often regulated by a small set of regulators, for a given k, we should expect that many of the elements in vector β_k should be zero. We call model (6.4) the SRMM model. In this mixture model formulation, the unknown parameters include K, β_k and σ_k^2 for $k = 1, \ldots, K$, and π. Note that the SRMM model (6.4) can be easily extended to include interaction terms between the regulators,

$$Y_{gc}|\{M_g = k\} = \sum_{r=1}^{R} X_{rc}\beta_{kr} + \sum_{r,r'} X_{rc}X_{r'c}\beta_{krr'} + \epsilon_{gc},$$

where $\beta_{krr'}$ is used to model the interaction effect between the regulators r and r' on gene expression levels for genes in the kth module.

6.3.2
An EM Algorithm Based on Lasso

To deal with the problem of a large R and to account for the sparse nature of the parameters β_k in model (6.4), we propose to develop the following EM algorithm based on Lasso for estimating the model parameters. Let $M = \{M_{g1}, \ldots, M_{gK},\quad g = 1, \ldots G\}$ be the matrix of module-membership indicators for the G genes, where M_{gk} is 1 if the gth gene belongs to the kth module. The complete data log likelihood can be written as

$$l(\beta_k, \sigma_k^2 \pi; Y, M|X) = l_1(M; \pi) + l_2(Y|M, X; \beta_k, \sigma_k^2),$$

where

$$l_1(M; \pi) = \sum_{g=1}^{G}\sum_{k=1}^{K} M_{gk}\log(\pi_k),$$

$$l_2(Y|M, X; \beta_k, \sigma_k^2) = -\frac{1}{2}\sum_{g=1}^{G}\sum_{k=1}^{K}\sum_{c=1}^{C} M_{gk}\left\{\log(\sigma_k^2 + \frac{(Y_{gc} - X_c^T\beta_k)^2}{\sigma_k^2})\right\}. \tag{6.5}$$

It is easy to show that in the $(t+1)$th E-step, we have

$$\hat{M}_{gk} = E(M_{gk} = 1|X, Y) = \frac{\pi_k^{(t)} Pr(Y_g|M_{gk} = 1; \beta_k^{(t)}, \sigma_k^{(t)})}{\sum_{k=1}^{K} \pi_k^{(t)} Pr(Y_g|M_{gk} = 1; \beta_k^{(t)}, \sigma_k^{(t)})},$$

where $Pr(Y_g|M_{gk} = 1; \beta_k^{(t)}, \sigma_k^{(t)})$ is the normal density for Y_g if the gth gene belongs to the kth module. At the $(t+1)$th M-step, it is easy to check that the EM equation for updating the prior probability is given by

$$\hat{\pi}_k^{(t+1)} = \frac{\sum_{g=1}^{G} \hat{M}_{gk}}{G}.$$

However, the M-step for updating the parameter β_k needs to account for the sparsity of this parameter. From the expression of l_2 in Equation 6.5, we need to find the β_k that minimizes the following quantity:

$$\sum_{g=1}^{G}\hat{M}_{gk}\sum_{c=1}^{C}(Y_{gc}-X_c^{\mathrm{T}}\beta_k)^2=\sum_{g=1}^{G}\sum_{c=1}^{C}\left(\sqrt{\hat{M}_{gk}}Y_{gc}-\sqrt{\hat{M}_{gk}}X_c^{\mathrm{T}}\beta_k\right)^2,$$

subject to sparsity constraint of β_k,

$$|\beta_k|_1=\sum_{r=1}^{R}|\beta_{kr}|<s,$$

where s is a tuning parameter, assumed to be the same for all $k=1, \ldots, K$. This is equivalent to performing linear regression, with $\sqrt{\hat{M}_{gk}}Y_{gi}$ as responses and $\sqrt{\hat{M}_{gk}}X_{gi}^{\mathrm{T}}$ as regressors. We can then use Lasso to update β_k. However, since $G\times C$ is often very large, such an implementation can be very time-consuming. Alternatively, we can use a sparse version of the EM algorithm or the hard-EM algorithm to update β_k. Specifically, we first classify the gene g into the k_gth module, where $k_g=\mathrm{argmax}_k\hat{M}_{gk}$, for $g=1,\ldots,G$. We then estimate β_k using Lasso based on the data of the genes in the current kth module under the constraint that $|\beta_k|_1<s$. This can be efficiently implemented using the R Lars function [31].

Finally, after obtaining the update of β_k, we can update the error variance by

$$\hat{\sigma}_k^{2(t+1)}=\frac{\{\sum_{g=1}^{G}\hat{M}_{gk}(Y_g-X^{\mathrm{T}}\beta_k^{(t+1)})^T(Y_g-X^{\mathrm{T}}\beta_k^{(k+1)})\}}{\sum_{g=1}^{G}G\hat{M}_{gk}}.$$

We call this EM algorithm using Lasso, the Lasso-EM algorithm. After the convergence of the Lasso-EM algorithm, we can partition the genes into K regulatory modules. Specifically, we partition gene g into module k if $k=\mathrm{argmax}_{k'}\{Pr(M_g=k')\}$. In addition, we can obtain the regulation program of a module specifying the set of regulatory genes in the module that controls the module and the mRNA expression profile of the genes in the module as a function of the expression of the module's regulators. Specifically, for the kth module, the regulation program includes the regulator set $\{r:\hat{\beta}_{kr}\neq 0\}$, and the sign and the magnitude of $\hat{\beta}_{kr}$ determine how the regulation program controls the expression of the module. Note that our method allows some regulators to participate in the regulation programs of multiple modules and also a group of genes that is not regulated by any of the regulators.

6.3.3 Selection of the Number of Modules K and the Tuning Parameter s

To implement the proposed Lasso-EM algorithm, for each given number of the modules K, we need to determine the tuning parameter $s(K)$. We also need to determine the number of the regulatory modules K. As commonly used for mixture models, for a given number of regulatory modules k, we can choose the tuning parameter s by maximizing the BIC score, which is defined as

$$BIC(s(k)) = 1(s(k)) - p(s(k))\log(G \times C)$$

for the model with k cluster and tuning parameter s, where $p(s(k)) = \sum_{k=1}^{k} \sum_{r=1}^{R} I(\beta_{kr} \neq 0)$ is the total number of nonzero parameters in the model [29] and $G \times C$ is the number of observations, and the log likelihood for the model with k clusters and tuning parameter s can be written as

$$l(s(k)) = \log \sum_{g=1}^{G} \sum_{k=1}^{K} \pi_k (2\pi\sigma_k^2)^{-C/2} \exp\left\{ -\frac{\sum_{c=1}^{C} (Y_{gc} - X_c^{\mathrm{T}}\beta_k)^2}{\sigma_k^2} \right\}.$$

We then choose K as $K = \mathrm{argmax}_k l(s(k))$.

6.3.4
Application to Yeast Stress Data Set

To demonstrate the proposed Lasso-EM algorithm for identifying the transcriptional modules, we applied the proposed method to analyze the yeast data set reported in Segal *et al.* [6], consisting of 2355 genes, 466 candidate regulators (transcriptional factors and signaling proteins), and 173 arrays of the yeast stress data, measuring gene expression responses to various stress conditions. This set of 2355 genes has a significant change in gene expression at the measured stress conditions, excluding members of the generic environmental stress response cluster.

We started the Lasso-EM algorithm based on results from the hierarchical clustering. Allowing for at most five regulators for each module, the Lasso-EM algorithm identified 35 regulatory modules, each containing several significant GO processes, indicating that the same set of regulators may regulate different biological processes. As an example, Figure 6.2 shows a heatmap for the respiration/carbohydrate metabolism module, including four regulators, Hap4, GAC1, Reg2, and SNF.3 (YJL103C), whose effects on gene expression levels for genes in this module are given by the following linear model,

$$y = 0.12\mathrm{Hap4} + 0.015\mathrm{GAC1} + 0.047\mathrm{Reg2} + 0.06\mathrm{SNF3}.$$

This model specified the Hap4 transcriptional factor as an important activating regulator, consistent with the known role of Hap4 in the activation of respiration [6]. This model also suggests that the protein phosphatase type 1 regulatory subunit GAC1 and transcription factor Reg2 may also regulate the expression levels of this module. The results largely agree with the similar module identified by Segal *et al.* [6].

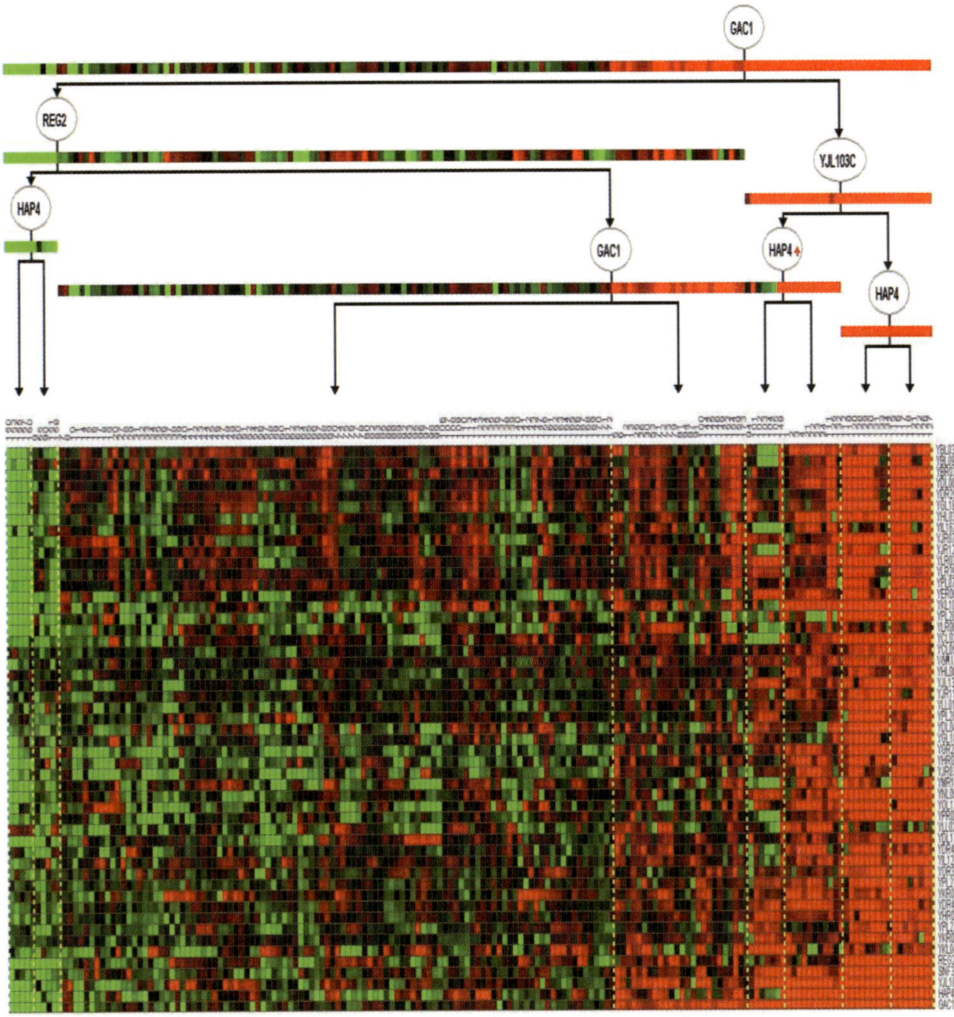

Fig. 6.2 The respiration and carbon module identified from the analysis of the yeast stress responses microarray gene expression data set using the proposed Lasso-EM algorithm. The plot was generated using the plotting tool provided in Segal *et al.* [6]. The heatmap is for gene expression, where genes are plotted as rows and arrays are columns. Arrays are arranged according to the regulation tree, where each node represents a regulator and the expression of the regulators is shown below their respective node.

6.4
Inference of Transcriptional Networks

Since many essential biological systems or processes are dynamic systems, it is important to study the gene expression patterns over time in a genomic scale to capture the dynamic behavior of gene expression. Research in analysis of such microarray time-course gene expression data has focused on two areas: clustering of MTC expression data [32,33]) and identifying genes that are temporally differentially expressed [34–37]. Although both problems are important and biologically relevant, they provide little information about our understanding of gene regulations. In this section, the methods based on the "physical interaction" approach that aim to identify interactions among transcriptional factors and their target genes through sequence motifs and ChIP-chip data found in promotor sequences are presented. In particular, Wang *et al.* [38] considered integrating microarray time-course gene expression data and motif or ChIP-chip data to identify the transcriptional factors that are involved in gene expression variations during a given biological process.

6.4.1
Functional Response Model with Time-Varying Coefficients for MTC Gene Expression Data

We consider a microarray time-course gene expression experiment. Let $Y_i(t)$ be the expression level of the ith gene at time t, for $i = 1, \ldots, n$. We assume the following regression model with functional response:

$$Y_i(t) = \mu(t) + \sum_{k=1}^{K} \beta_k(t) X_{ik} + \epsilon_i(t), \tag{6.6}$$

where $\mu(t)$ is the overall mean effect, $\beta_k(t)$ is the regulation effect associated with the kth transcriptional factor, X_{ik} is the matching score or the binding probability of the kth transcriptional factor on the promoter region of the ith gene, and $\epsilon_i(t)$ is a realization of a zero-mean stochastic process. Several different ways and data sources can be used to derive the matching score X_{ik}. One approach is to derive the score using the position-specific weight matrix (PSWM). Specifically, for each candidate TF k, let P_k be the positive specific weight matrix of length L, b with element $P_{kj}(b)$ being the probability of observing the base b at position j. Then, each L-mer l in the promoter sequence of the ith gene is assigned a score S_{ikl} as

$$S_{ikl} = \sum_{j=1}^{L} \log \frac{P_{kj}(b_{ilj})}{B(b_{ilj})},$$

where b_{ilj} is the nucleotide at position j on the lth sequence for gene i, and $B(b)$ is the probability of observing b in the background sequence. This score always assumes a value between 0 and 1. We then define $X_{ik} = \max_l S_{ikl}$, which is the maximum of the matching scores over all the L-mer in the promoter region of the ith gene. The maximum scores can then be converted into the binding probabilities using the method described in Chen *et al.* [39]. Alternatively, we can define the binding probability based on the ChIP-chip data [38,39].

6.4.2 Estimation Using B-Splines

We consider estimation of the nonparametric function in model (6.6) using the smoothing spline method by approximating $B_k(t)$ by using the natural cubic B-spline basis,

$$\beta_k(t) = \sum_{l=1}^{L+4} \beta_{kl} B_l(t), \tag{6.7}$$

where $B_l(t)$ is the natural cubic B-spline basis function, for $l = 1, \ldots, L+4$ where L is the number of interior knots. Replacing $B_k(t)$ by its B-spline approximation in Equation (6.7), model (6.6) can be approximated as

$$Y_i(t) = \mu(t) + \sum_{k=1}^{K} \left\{ \sum_{l=1}^{L+4} \beta_{kl} [B_l(t) X_{ik}] \right\} + \epsilon_i(t), \tag{6.8}$$

where we have K groups of parameters, with $\beta_k^* = \{\beta_{kl}, \ldots, \beta_{kL+4}\}$ being the parameters associated with the group k, and we want to select the groups with nonzero coefficients. This is the grouped variable selection problem considered in Yuan and Lin [40].

6.4.3 A Group SCAD Penalization Procedure

Wang *et al.* [38] proposed a general group SCAD (gSCAD) procedure for selecting the groups of variables in a linear regression setting. Selecting important variables in model (6.6) corresponds to the selection of groups of basis functions in model (6.8). Yuan and Lin [40] proposed several procedures for such group variable selection, including group LARS and group Lasso. Instead of using the L_1 penalty for group selection as in Yuan and Lin [40], we propose to use the SCAD penalty of Fan and Li [41]. Specifically, to select nonzero $B_k(t)$, we can minimize the following penalized loss function:

$$l(\beta) = \sum_{i=1}^{n} \sum_{j=1}^{T} [y_{ij} - \mu(t_j) - \sum_{k=1}^{K} \sum_{l=1}^{L+4} \beta_{kl} B_l(t_j) X_{ik}]^2 + nT \sum_{k=1}^{K} p\lambda(||\beta_k^*||_2), \tag{6.9}$$

where y_{ij} is the observed gene expression level for gene i at time t_j, $p_\lambda(\cdot)$ is the SCAD penalty with λ as a tuning parameter, which is defined as

$$p\lambda(|w|) = \begin{cases} \lambda |\omega| & \text{if} \quad |w| \leq \lambda \\ -\dfrac{(|\omega|^2 - 2a\lambda|\omega| + \lambda^2)}{2(a-1)} & \text{if} \quad \lambda < |w| < a\lambda \\ (a+1)\lambda^2{}_2 & \text{if} \quad |w| > a\lambda \end{cases} \tag{6.10}$$

and $||\beta_k^*||_2 = \sqrt{\sum_{l=1}^{L+4} \beta_{kl}^2}$. The penalty function (6.10) is a quadratic spline function with two knots at λ and $a\lambda$, where a is another tuning parameter. Fan and Li [41] showed that the Bayes risks are not sensitive to the choice of a and suggested using $a = 3.7$.

6.4.4
Numerical Algorithm, Properties, and Application

Wang *et al.* [38] proposed a local quadratic approximation algorithm, similar to that in Fan and Li [41], to perform the optimization problem of Equation 6.9 and developed a GCV procedure for selecting the tuning parameter λ in the SCAD penalty function (6.10) They also established the oracle property of the gSCAD estimates. Wang *et al.* [38] also performed simulation studies to evaluate the gSCAD procedure. Simulation results indicate that for a similar false-positive rate, the gSCAD procedure is more sensitive in identifying the relevant transcriptional factors than simple linear regression. In addition, the estimates of the transcriptional effects are less variable than those obtained from simple linear regression analysis.

We applied the proposed methods to the analysis of cell cycle MTC data corrected by Spellman *et al.* [42]. The cell cycle is one of life's most important processes, and the identification of cell cycle regulated genes has greatly facilitated the understanding of this important process. Spellman *et al.* [42] monitored genome-wide mRNA levels for 6178 yeast ORFs, simultaneously using several different methods of synchronization including an α-factor-mediated G_1 arrest, which covers approximately two cell cycle periods with measurements at 7-minutes intervals for 119 minutes with a total of 18 time points (http://genome-www.stanford.edu/cellcycle/data/rawdata/). Using data based on different synchronization experiments, Spellman *et al.* [42] identified a total of about 800 cell cycle regulated genes, some showing periodic expression patterns only in a specific experiment. Using a model-based approach [32], identified 297 cell cycle regulated genes based on the α-factor synchronization experiments. We applied the mixture model approach described in the previous section using the ChIP data of Lee *et al.* [43] to derive the binding probabilities X_{ik} for these 297 cell cycle regulated genes for a total of 96 transcriptional factors with at least one nonzero binding probability in the 297 genes.

We applied the gSCAD procedure with $L = 2$ and an additional L_2 penalty to identify the TFs that affect the expression changes over time for these 297 cell cycle regulated genes in the α-factor synchronization experiment. The gSCAD procedure identified a total of 71 TFs that are related to yeast cell cycle processes, including 19 of the 21 known and experimentally verified cell cycle related TFs. The estimated transcriptional effects of these 21 TFs are shown in Figure 6.3; except for the two TFs (CBF1 and GCN4) that were not selected by the gSCAD procedure and the TF LEU3, the other 18 TFs all showed time-dependent effects of these TFs on gene expression levels. In addition, the effects followed similar trends between the two cell cycle periods. It was not clear why CBF1 and GCN4 were not selected by the gSCAD. The minimum *p*-values of CBF1 and GCN4 over 18 times points from simple linear

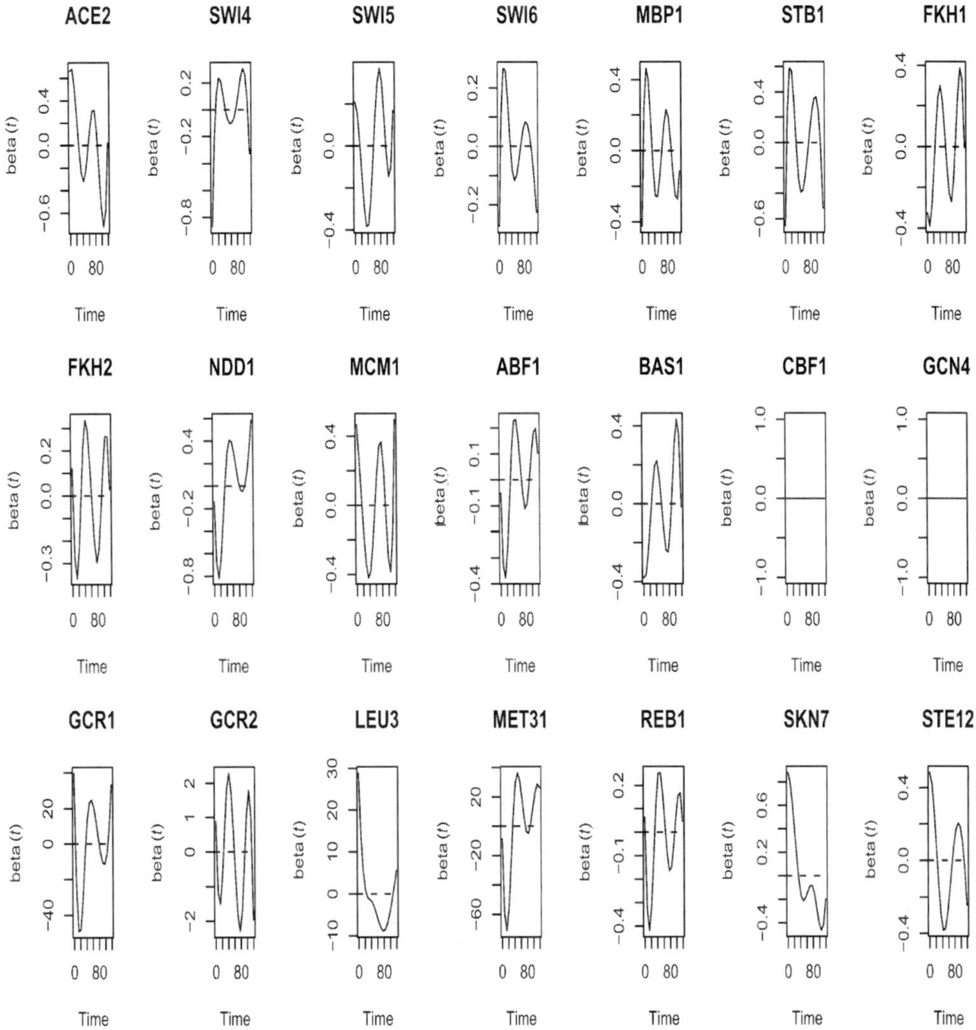

Fig. 6.3 Estimated time-dependent transcriptional effects for 21 known yeast transcriptional factors related to the cell cycle process using gSCAD. Note that CBF1 and GCN4 were not selected by gSCAD.

regressions are 0.06 and 0.14, respectively, also indicating that CBF1 and GCN4 were not related to expression variation over time. Overall, the model can explain 43% of the total variations of the gene expression levels.

To assess false identifications of the TFs that are related to a dynamic biological procedure, we randomly permuted the gene expression values across genes and time points and applied the gSCAD procedure again to the permuted data sets. We repeated this procedure 50 times. Among the 50 runs, 5 runs selected 4 TFs, 1 run selected 3 TFs, 16 runs selected 2 TFs, and the rest of the 28 runs did not select any of the TFs, indicating that our procedure indeed selects the relevant TFs with few false positives.

6.5 Discussion, Conclusions, and Future Research

In this chapter, we have reviewed several important problems and statistical methods related to the analysis of genetic networks and regulatory modules based on integrating microarray gene expression data and ChIP-chip data, including the problem of constructing genetic networks based on microarray gene expression data and Gaussian graphical models, the problem of identifying regulatory modules based on gene expression data and a predefined set of potential regulators, and the problem of identifying regulatory networks based on both microarray gene expression data and motif and ChIP-chip data. Our review mainly emphasizes the statistical formulation of the problems and our solutions to these problems. Applications to real data sets are also briefly discussed and presented. It should be emphasized that these algorithms only generate what can be loosely referred to as a "first approximation" to a gene regulatory network. The results of this method should not be interpreted as the definitive regulatory network but rather as a network that suggests (possibly indirect) regulatory interactions [12]. It should also be noted that this chapter emphasizes the use of gene expression data for inferences of genetic networks and regulatory modules; however, accurate protein-level measurements of TFs will invariably have a more direct influence on the mRNA levels of the genes they regulate.

The chapter only covers three problems in the inference of genetic networks and regulatory modules. As more and more large-scale genomic and epigenomic data are being generated, novel statistical and computational methods are required for many other problems related to the analysis of microarray gene expression data. We present two such areas that need further methodological development.

6.5.1 Incorporating Network Information into Analysis of Microarray Gene Expression Data

Although the focus of this chapter is on the inference of genetic pathways and networks, an equally important problem is to incorporate the pathway information derived from data or pathway databases into the analysis of clinical phenotype data to identify the pathways and networks that are related to various clinical phenotypes. There is a great need for methods that can link numerical microarray gene expression data measured on the networks to the phenotypes to obtain biologically interpretable results. As an example, Figure 6.4 shows the regulatory Apoptosis KEGG pathway [44], providing information on the regulatory relationship for genes on this pathway. We are now able to measure the gene expression levels for genes on this pathway for a sample of patients with different clinical phenotypes. We should

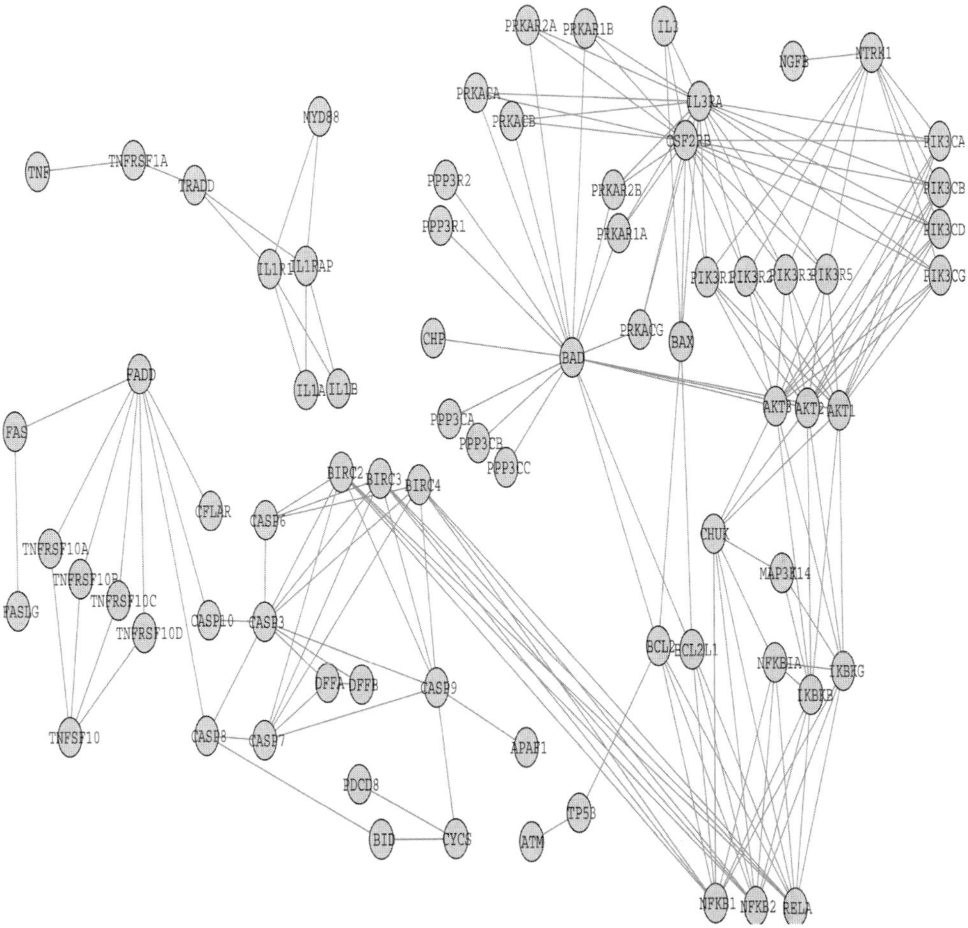

Fig. 6.4 KEGG Apoptosis regulatory pathway: nodes are genes, and edges represent the regulatory relationship between genes. Only genes with the corresponding probe-pairs found on the Affymetrix U 133A are plotted.

expect certain dependency of differential expression states for genes that are neighbors on this pathway. Incorporating such prior local dependency of gene expressions into the analysis of phenotype data can potentially gain power in identifying the relevant genes. Some preliminary work from our group indicates that such network-based analysis of gene expression data can greatly increase the sensitivity of identifying the relevant pathways or subnetworks [45,46]. For example, Wei and Li [46] developed a Markov random field (MRF) approach for identifying differentially expressed genes between two different experimental conditions, where a discrete MRF is used to model the local dependency of the differential expression states for genes on the network. By both simulations and analysis of several breast cancer gene expression studies, they demonstrated that such an MRF-based methods can identify more biologically interpretable genes and subnetworks.

Novel methods are also needed to formally incorporate network information into regression models. One solution to this problem is to perform a network-imposed smoothness penalty to obtain locally smoothed estimates of the regression parameters. Assume that Y follows a distribution in an exponential family with mean $\mu = E(Y)$ and variance $V = \mathrm{Var}(Y)$. The generalized linear model (GLM) [47] models the random component of Y through a link function g:

$$g(\mu) = \sum_{k=1}^{p} X_k \beta_k, \tag{6.11}$$

where X_k is the gene expression measurement of the kth genes and β_k is the regression coefficient corresponding to the kth gene, for a total of p genes on the genetic network $N = (G, E)$ with gene set G and edge set E. We further denote $\beta = \{\beta_1, \ldots, \beta_p\}$ as the vector of regression parameters of the model. Suppose that we have n i.i.d. observations (x_i, y_i), $i = 1, \ldots, n$ of a gene expression vector x_i and a response variable y_i. We can then define an estimate of β by minimizing the following regularized loss function,

$$\hat{\beta} = \operatorname{argmin}\left\{ l(\beta) + \lambda_1 \sum_{k=1}^{p} |\beta_k| + \lambda_2 \beta' L \beta \right\}, \tag{6.12}$$

where $l(\beta)$ is a loss function (e.g., negative of the log-likelihood function corresponding to the GLM (6.11), L is the Laplacian matrix as defined in Chung [48] and λ_1 and λ_2 are two tuning parameters. It is easy to verify that

$$\beta' L \beta = \sum_{i \sim j \in E} \left(\frac{\beta_i}{\sqrt{d_i}} - \frac{\beta_j}{\sqrt{d_j}} \right)^2, \tag{6.13}$$

where $i \sim j$ denotes that ith gene and jth gene are neighbors on the network N and d_i and d_j are the respective degrees. The scaling of the β parameters by their degrees is used to reflect the fact that the genes with more neighbors (e.g., the "hub" genes) tend to play a more important biological role and therefore should correspond to larger coefficients. In this regularized formulation (6.12), the first L_1 penalty leads to sparse solution and the second penalty $\beta' L \beta$ leads to a smoothness solution with respect to the network structure, that is, it enforces that the degree-scaled β estimates are similar for genes that are neighbors on the network. To account for both activation and inhibition effects, we can modify the smoothness penalty (6.13) as

$$\lambda_2 \left\{ \sum_{i \sim j \in E^+} \left(\frac{\beta_i}{\sqrt{d_i}} - \frac{\beta_j}{\sqrt{d_j}} \right)^2 + \sum_{i \sim j \in E^-} \left(\frac{\beta_i}{\sqrt{d_i}} - \frac{\beta_j}{\sqrt{d_j}} \right)^2 \right\},$$

where E^+ is the set of transcriptional activation edges ($\rightarrow$) and E^- is the set of transcriptional inhibition edges ($\dashv$). This modification is used to reflect the fact that β_i and β_j are expected to have the same sign if $(i \sim j) \in E^+$ and different signs if $(i \sim j) \in E^-$. Similar to the Elastic-Net penalty of Zou and Hastie [49], the optimization of Equation 6.12 can be efficiently solved by using the R Lars algorithm [31] and the tuning parameters λ_1 and λ_2 can be chosen using cross-validation.

6.5.2 Development of Statistical and Computational Methods for Integrating Gene Expression Data and Epigenomic Data

Another important area for future research is to develop rigorous statistical and computational methods for integrating microarray gene expression data with other types of genomic data for an even more detailed understanding of genetic networks, especially genetic regulatory networks. We have reviewed two such approaches: one uses the transcriptional factor annotation information for inferring regulatory modules, the other uses the sequence motif and ChIP-chip data on inference of regulatory models. However, other genome-wide data can be very useful. For example, we know that sequence polymorphisms affect gene expression by perturbing the complex networks of regulatory interactions. It is therefore important to simultaneously consider both single nucleotide data and the gene expression data to obtain both *cis-* and *trans*-effects on gene expression [50–52]. Standard methods attempt to associate each gene expression phenotype with genetic polymorphisms. Lee *et al.* recently developed an interesting method to understand the mechanisms by which genetic changes perturb gene regulation by combining SNP, gene expression, and transcriptional factors information.

Statistical and computational methodologies for genomic data analysis and integration are also needed for analyzing epigenomics data, with the aim to understand systems-level gene regulatory mechanisms. A multicellular organism contains only one genome, but different cell types contain different epigenomic patterns, including chromatin structure [53], histone modification [54], nucleosome positions [55], and DNA methylations [56]. These epigenomic markers are important for regulating protein–DNA binding activities and gene transcription. As more and more epigenomic data become available [54,56,57], it is important to develop novel statistical methods for analyzing such data together with gene expression data to estimate the direct regulatory effects of epigenomic factors.

6.5.3 Final Remarks

Elucidating genetic pathways and networks is one of the most important problems in modern biology research. Microarray gene expression data together with other high-throughput genomic, proteomic, and epigenomic data provide the opportunity to derive such networks and apply network knowledge to study other important biological systems such as disease initiation and progression. However, statistical and computational methods for analyzing these data are becoming even more important. We conclude this chapter by quoting Dr Collins's article published in Nature: Computational methods have become intrinsic to modern biological research, and their importance can only increase as large-scale methods for data generation become more prominent, as the amount and complexity of the data increases and as the questions being addressed become more sophisticated [58].

Acknowledgments

This research is supported by NIH grant ES009911 and a grant from the Pennsylvania Health Department. I thank my students Zhandong Liu and Caiyan Li for their work on SRMM; Zhi Wei for his work on the local MRF model; my postdoctoral fellows Dr Jiang Gui for his work on the TGD procedure and Dr Lifeng Wang for his work on gSCAD; and Edmund Weisberg, MS, for editorial help. I also thank Professor Dehmer and Dr Emmert-Streib for inviting me to contribute this chapter to the book.

References

1 Ideker, T., Thorsson, V., Ranish, J.A., Christmas, R., Buhler, J., Eng, J.K., Bumgarner, R., Goodlett, D.R., Aebersold, R. and Hood, L. (2001) Integrated genomic and proteomic analyses of a systematically perturbed metabolic network. *Science*, **292**, 929–934.

2 Friedman, N. (2004) Inferring cellular networks using probabilistic graphical models. *Science*, **30**, 799–805.

3 Das, D., Nahle, Z. and Zhang, M.Q. (2006) Adaptively inferring human transcriptional subnetworks. *Molecular Systems Biology*, **2**, 29.

4 Bansal, M., Belcastro, V., Ambesi-Impiombato, A. and di Bernardo, D. (2007) How to infer gene networks from expression profiles. *Molecular Systems Biology*, **3**, 78.

5 Tavazoie, S., Hughes, J.D., Campbell, M.J., Cho, R.J. and Church, G.M. (1999) Systematic determination of genetic network architecture. *Nature Genetics*, **22**, 281–285.

6 Segal, E., Shapira, M., Regev, A., Pe'er, D., Botstein, D., Koller, D. and Friedman, N. (2003) Module networks: identifying regulatory modules and their condition-specific regulators from gene expression data. *Nature Genetics*, **34** (2), 166–176.

7 Schafer, J. and Strimmer, K. (2005) An empirical Bayes approach to inferring large-scale gene association networks. *Bioinformatics*, **21**, 754–764.

8 Wille, A., Zimmermann, P., Vranova, E., Furholz, A., Laule, O., Bleuler, S., Hennig, L., Prelic, A., von Rohr, P., Thiele, L., Zitzler, E., Gruissem, W. and Bhlmann, P. (2004) Sparse graphical Gaussian modeling of the isoprenoid gene network in *Arabidopsis thaliana*. *Genome Biology*, **5**(11), R92,1–13.

9 Dobra, A., Jones, B., Hans, C., Nevis, J. and West, M. (2004) Sparse graphical models for exploring gene expression data. *Journal of Multivariate Analysis*, **90**, 196–212.

10 Li, H. and Gui, Li (2006) Gradient directed regularization for sparse

Gaussian concentration graphs with applications to inference of genetic networks. *Biostatistics*, **7**, 302–317.

11 Ernst, J., Vainas, O., Harbison, C.T., Simon, I. and Bar-Joseph, Z. (2007) Reconstructing dynamic regulatory maps. *Molecular Systems Biology*, **3**, 4.

12 Bonneau, R., Reiss, D., Shannon, P., Facciotti, M., Hood, L., Baliga, N. and Thorsson, V. (2006) The inferelator: an algorithm for learning parsimonious regulatory networks from systems-biology data sets *de novo*. *Genome Biology*, **7**, R36.

13 Tibshirani, R. (1996) Regression shrinkage and selection via the Lasso. *Journal of Royal Statistical Society, Series B*, **58**, 67–288.

14 McLachlan, G.J. and Basford, K.E. (1988) *Mixture Models: Inference and Applications to Clustering*, Marcel Dekker.

15 Bussemaker, H.J., Li, H. and Siggia, E.D. (2001) Regulatory element detection using correlation with expression. *Nature Genetics*, **27**, 167–171.

16 Keles, S. Van Der Laan, M. and Eisen, M.B. (2002) Identification of regulatory elements using a feature selection method. *Bioinformatics*, **18**, 1167–1175.

17 Gao, F. Foat, B.C. and Bussemaker, H.J. (2004) Defining transcriptional networks through integrative modeling of mRNA expression and transcription factor binding data. *BMC Bioinformatics*, **5**, 31.

18 Conlon, E.M., Liu, X.S., Lieb, J.D. and Liu, J.S. (2003) Integrating regulatory motif discovery and genome-wide expression analysis. *Proceedings of National Academy of Sciences*, **100**, 3339–3344.

19 Friedman, J. (2001) Multivariate adaptive regression splines. *Annals of Statistics*, **19**, 1–141.

20 Edwards, D. (2000) *Introduction to Graphical Modelling*, 2nd edn, Springer Verlag, New York.

21 Demspter, A.P. (1972) Covariance selection. *Biometrics*, **28**, 157–175.

22 Tegner, J., Yeung, M.K., Hasty, J. and Collins, J.J. (2003) Reverse engineering gene networks: integrating genetic perturbations with dynamical modeling. *Proceedings of National Academy of Sciences United States of America*, **100**, 5944–5949.

23 Jeong, H., Mason, S.P., Barabasi, A.L. and Oltvai, Z.N. (2001) Lethality and centrality in protein networks. *Nature*, **411**, 41–42.

24 Gardner, T.S., di Bernardo, D., Lorenz, D. and Collins, J. (2003) Inferring genetic networks and identifying compound mode of action via expression profiling. *Science*, **301**, 102–105.

25 Meinshausen, N. and Buhlmann, P. (2006) Consistent neighbourhood selection for high-dimensional graphs with the Lasso. *Annals of Statistics*, **34**, 1436–1462.

26 Friedman, J.H. and Popescu, B.E. (2004) Gradient Directed Regularization Technical report Stanford University.

27 Gui, J. and Li, J. (2005) Threshold gradient descent method for censored data regression with applications in pharmacogenomics. *Pacific Symposium on Biocomputing*, **10**, 272–283.

28 Barabasi, A.L. and Oltvai, Z.N. (2004) Network biology: understanding the cell's functional organization. *Nature Reviews Genetics*, **5**, 101–113.

29 Zou, H. Hastie, T. and Tibshirani, R. (2007) On the "Degrees of Freedom" of the Lasso. *Annals of Statistics*, in press.

30 Lee, S., Pe'er, D., Dudley, A.M., Church, G.M. and Koller, D. (2007) Identifying regulatory mechanisms using individual variation reveals key role for chromatin modification. *Proceedings of National Academy of Sciences United States of America*, **103**, 14062–14067.

31 Efron, B., Hastie, T., Johnstone, I. and Tibshirani, R. (2004) Least angle regression. *Annals of Statistics*, **32**, 407–499.

32 Luan, Y. and Li, H. (2004) Model-based methods for identifying periodically expressed genes based on time course microarray gene expression data. *Bioinformatics*, **20**, 332–339.

33 Ma, P., Castillo-Davis, C., Zhong, W. and Liu, J.S. (2006) A data-driven clustering method for time course gene expression data. *Nucleic Acids Research*, **34** (4), 1261–1269.

34 Hong, F. and Li, H. (2006) Functional hierarchical models for identifying genes with different time-course expression profiles. *Biometrics*, **62**, 534–544.

35 Yuan, M. and Kendziorski, C. (2006) Hidden Markov models for microarray time course data in multiple biological conditions. *Journal of American Statistical Association*, **101** (476), 1323–1340.

36 Tai, Y.C. and Speed, T.P. (2006) A multivariate empirical Bayes statistic for replicated microarray time course data. *Annals of Statistics*, **34**, 2387–2412.

37 Storey, J.D., Xiao, W., Leek, J.T., Dai, J.Y., Tompkins, R.G. and Davis, R.W. (2005) Significance analysis of time course microarray experiments. *Proceedings of National Academy of Sciences*, **102**, 12837–12842.

38 Wang, L. Chen, G. and Li, H. (2007) Group SCAD regression analysis for microarray time course gene expression data. *Bioinformatics*, **23**, 1486–1494.

39 Chen, G., Jensen, S. and Stockert, C. (2007) Clustering of genes into regulons using integrated modeling (COGRIM). *Genome Biology*, **8** (1), R4.

40 Yuan, M. and Lin, Y. (2006) Model selection and estimation in regression with grouped variables. *Journal of Royal Statistical Society Series B*, **68**, 49–67.

41 Fan, J. and Li, R. (2001) Variable selection via nonconcave penalized likelihood and its oracle properties. *Journal of American Statistical Association*, **96**, 1348–1360.

42 Spellman, P.T., Sherlock, G., Zhang, M.Q., Iyer, V.R., Anders, K., Eisen, M.B., Brown, P.O., Botstein, D. and Futcher, B. (1998) Comprehensive identification of cell cycle-regulated genes of the yeast *Saccharomyces cerevisiae* by microarray hybridization. *Molecular Biology of Cell*, **9**, 3273–3297.

43 Lee, T.I., Rinaldi, N.J., Robert, F., Odom, D.T., Bar-Joseph, Z., Gerber, G.K., Hannett, N.M., Harbison, C.T., Thompson, C.M. and Simon, I., *et al.* (2002) Transcriptional regulatory networks in *S. cerevisiae*. *Science*, **298**, 799–804.

44 Kanehisa, M. and Goto, S. (2002) KEGG: Kyoto encyclopedia of genes and genomes. *Nucleic Acids Research*, **28**, 27–30.

45 Wei, Z. and Li, H. (2007) Nonparametric pathway-based regression models for analysis of genomic data. *Biostatistics*, **8**, 265–284.

46 Wei, Z. and Li, H. (2007) A Markov random field model for network-based analysis of genomic data. *Bioinformatics*, **23**, 1537–1544.

47 McCullagh, P. and Nelder, J. (1989) *Generalized Linear Models*, 2nd edn, Chapman & Hall.

48 Chung, F. (1997) Spectral Graph Theory. CBMS Regional Conferences Series, Vol. 92, American Mathematical Society, Providence, RI.

49 Zou, H. and Hastie, T. (2005) Regularization and variable selection via the elastic net. *Journal of Royal Statistical Society Series B*, **67**, 301–320.

50 Brem, R.B., Storey, J.D., Whittle, J. and Kruglyak, L. (2005) Genetic interactions between polymorphisms that affect gene expression in yeast. *Nature*, **436**, 701–703.

51 Schadt, E.E., Lamb, J., Yang, X., Zhu, J., Edwards, S., Guhathakurta, D., Sieberts, S.K., Monks, S., Reitman, M., Zhang, C. Lum, P.Y., Leonardson, A., Thieringer, R., Metzqer, J.M., Yang, L., Castle, J., Zhu, H., Kash, S.F., Drake, T.A., Sachs, A. and Lusis, A.J. (2005) An integrative genomics approach to infer causal associations between gene expression and disease. *Nature Genetics*, **37**, 710–717.

52 Morley, M., Molony, C.M., Weber, T.M., Devlin, J.L., Ewens, K.G., Spielman, R.S. and Cheung, V.G. (2004) Genetic analysis of genome-wide variation in human gene expression. *Nature*, **430**, 743–747.

53 Steifeld, I. Shamir, R. and Kupiec, M. (2007) A genome-wide analysis in *Saccharomyces cerevisiae* demonstrates the influence of chromatin modifiers on transcription. *Nature Genetics*, **9** (3), 303–309.

54 Yuan, G.C., Liu, Y.J., Dion, M.F., Slack, M.D., Wu, L.F., Altschuler, S.J. and Rando, O.J. (2005) Genome-scale identification of nucleosome positions in *S. cerevisiae*. *Science*, **309** (5734), 626–630.

55 Segal, E., Fondufe-Mittendorf, Y., Chen, L., Thastrom, A., Field, A., Moore, I.K., Wang, J.Z. and Widom, J. (2006) A genomic code for nucleosome positioning. *Nature*, **442** (7104), 772–778.

56 Eckhardt, F., Lewin, J., Cortese, R., Rakyan, V.K., Attwood, J., Burger, M., Burton, J., Cox, T.V., Davies, R., Down, T.A., Haefliger, C., Horton, R., Howe, K., Jackson, D.K., Kunde, J., Koenig, C., Liddle, J., Niblett, D., Otto, T., Pettett, R., Seemann, S., Thompson, C., West, T., Rogers, J., Olek, A., Berlin, K. and Beck, S. (2006) DNA methylation profiling of human chromosomes 6, 20 and 22. *Nature Genetics*, **38**(12) 1378–1385.

57 Heintzman, N.D., Stuart, R.K., Hon, G., Fu, Y., Ching, C.W., Hawkins, R.D., Barrera, L.O., Van Calcar, S., Qu, C., Ching, K.A., Wang, W., Weng, Z., Green, R.D., Crawford, G.E. and Ren, B. (2007) Distinct and predictive chromatin signatures of transcriptional promoters and enhancers in the human genome. *Nature Genetics*, **39**, 311–318.

58 Collins, F.S., Green, E.D., Guttmacher, A.E. and Guyer, M.S. (2003) A vision for the future of genomics research. *Nature*, **422**, 835–847.

7
A Model of Genetic Networks with Delayed Stochastic Dynamics

Andre S. Ribeiro

7.1
Introduction

Since the discovery of the molecular structure of the DNA (deoxyribonucleic acid) and its significance for information transfer in living material by Watson and Crick [1], numerous studies have been conducted to understand better and better how this information is stored and used by cells.

One major step toward understanding this information storage system was the sequencing of the entire genome as a result of which genomes of many species, including the human genome, have now been sequenced. This effort, combined with other studies, provided not only genome sequences but also a better understanding of how genes function. A gene is a specific sequence of nucleotides encoded in the DNA. For a cell to extract the information contained in this code, a gene must be transcribed by an RNA polymerase (RNA*p*) into an RNA molecule [2]. From the RNA, ribosomes can produce proteins by following the code determined by the nucleotide sequence of the RNA, via a process called translation [2].

To initiate transcription, the RNA*p* has to first recognize and bind to the promoter region of the gene, which can have regulatory sites to which transcription factors and other coregulatory proteins can bind and can either activate or repress gene transcription. After the initial region, the gene has a coding sequence, which is the one transcribed by the RNA*p* into an RNA molecule. Transcription stops when the RNA*p* finds a termination sequence and unbinds from the DNA strand. Ribosomes read the RNA sequence produced and, for each codon, a corresponding amino acid is added to a polypeptide chain, which is then folded into a three-dimensional structure and becomes active and capable of performing a given task.

Transcription of DNA into RNA and translation of RNA into proteins are highly complex processes, involving a large number of molecules and reactions. In eukaryote cells, these two processes appear to be even more complex than in prokaryotes, since there are more levels of control of gene expression.

The DNA in eukaryotes is involved in a chromatin structure [2]. For a gene to be transcribed, a chromatin remodeling complex and histone acetylase must remodel

Analysis of Microarray Data: A Network-Based Approach. Edited by F. Emmert-Streib and M. Dehmer

ISBN: 978-3-527-31822-3

the chromatin structure, which is followed by many other steps, including the binding of an RNA*p* molecule that transcribes the gene. Also, the RNA molecule can undergo posttranscriptional modifications before it is translated. For example, alternative splicing mechanisms allow transcripts with different sequences to be obtained from the same gene and thus code for different proteins.

However, the current knowledge of almost the entire sequence of the genome and the mechanisms by which single genes are expressed does not provide a complete understanding of how genes function, since they are not independent units. A complicated web of interactions exists among genes. Proteins, RNA, or microRNA (miRNA) produced by a gene can act as transcription factors for another gene, regulating its expression level.

Not all components of this network are known and neither is the role of some known components. For example, the role of miRNA [3] in gene regulation is not yet understood. They consist of single-stranded RNA molecules of about 21–23 nucleotides in length and are thought to regulate the expression of other genes. However, they are not translated into proteins, which means that they are probably fed back directly into genes, controlling their activity.

The complex network of interactions between genes, via proteins, microRNAs, and other substances, is referred to as a gene regulatory network (GRN). As an example of how complex such networks can be, the human genome has between 30 000 and 45 000 genes whose activities are regulated by a network of their own products. Some gene expressions are known to be influenced by dozens or more factors, which include many other gene expression products. For this reason, the genome can be seen as a parallel processing nonlinear dynamical system.

One approach, used here, to understand gene regulatory networks consists in modeling and simulating the dynamics of these networks. The design of these network models aims to capture as many relevant features as possible, so that the simulation is as realistic as possible.

So far, only very small gene networks, involving two or three genes, usually artificially introduced in bacteria, have been experimentally observed in detail and modeled with a reasonable success (see [4–6]).

Due to many limitations such as computational ones and incomplete knowledge of some of the mechanisms involved, instead of attempting to model processes such as transcription and translation at a very detailed level, the goal is to model them correctly accounting for their effects on GRN dynamics. Therefore, and for additional reasons explained later on, the model of GRN presented here has to take into account, for example, the time transcription and translation processes take to complete but avoids chemical reactions involved in these processes. Also, the model has to include interactions between genes via their products of expression, modeled as correctly as possible from the systems' dynamics point of view. Although many details of regulating levels of the real GRNs are not included, as said, additional features and "layers of control" can be easily added following this modeling strategy.

Several approaches have been proposed to model GRNs. The first approach used Boolean networks [7] as models of GRNs. Afterward, models of coupled ordinary differential equations (ODEs) [8] and systems of chemical reactions [4], driven by the stochastic simulation algorithm (SSA) [9], were proposed to explain the increasing number of experimental observations of different aspects of GRN dynamics. The model presented here consists of an improved version of the last modeling strategy, but its dynamics is driven by the "delayed SSA" [10], to allow modeling multiple time-delayed reactions.

This chapter is organized as follows. First, it presents a description of recent experimental results, revealing the key elements necessary to account for when modeling reactions representing gene expression and interactions between genes. These experiments consist of measurements, at single molecules level, of single gene expression and of genes within small GRNs, and provide the best evidence for the nature of the dynamical behavior of GRNs. More important, they indicate which features cannot be ignored when modeling transcription and subsequent translation.

Following the experimental observations and their interpretation, the "delayed" stochastic simulation algorithm is described, which simulates the dynamics of GRNs capturing its main features: stochastic dynamics and the existence of events whose time duration for completion, once initiated, cannot be ignored.

Next, a model of single gene expression is described. In the last section, it is shown how this model is able to reproduce with high accuracy, and in all aspects, a recent experimental measurement of a single gene expression at the level of single molecules.

Provided this framework, a model of gene regulatory networks is proposed that can be used to model large-scale gene networks and simulate their dynamics. This model allows to use the "ensemble approach" [11], which consists in simulating the dynamics of many networks with similar features, so that one can extract, from the resulting time series, the general properties of the dynamical behavior of these networks.

Finally, besides the already mentioned simulation of the set of chemical reactions to reproduce the experiment by which measurements of single gene expression at the single molecules level were obtained, two examples of gene networks are presented. One is a toggle switch (TS), a small GRN that has been associated with cell differentiation process and has been used here to show the relevance of introducing time delays in the reactions of gene expression. The other GRN models the P53–Mdm2 feedback loop system of chemical reactions. This system of chemical reactions is associated with many important biochemical pathways in cells, including those controlling cell apoptosis and cell response to external stresses. A few examples of these systems dynamics, obtained with the modeling strategy described here, are shown.

7.2 Experimental Observations of Gene Expression

There are two key features in the gene expression process, which should be considered when creating a dynamical model of GRNs. Real genes expression measurements have shown that the underlying dynamics is subject to nonnegligible stochastic fluctuations. One important reason why the stochastic nature of the dynamics cannot be ignored is that genes exist in very small copy numbers in the genome (from one to a few), and some of them express at very low rates. In these conditions, approximations such as those using ODE models become less reliable.

Also, transcription and translation are multiple-step processes, involving a very large number of reactions. Each of these events takes a significant, nonnegligible, time to complete, each time it is initiated. This time duration has, therefore, to be accounted for when simulating GRN dynamics. This section describes the several experimental observations that support these statements.

7.2.1 The Stochastic Nature of Gene Expression

Stochastic fluctuations of gene expression were proven to have a significant role at the single cell level [12,13] by measuring the resulting concentrations of proteins.

For example, the stochastic nature of gene expression is the cause of probabilistic pathway of differentiation selection of a cell [12]. In this study, the authors studied the close relationship between differentiation pathway selection (i.e., if a cell of type A differentiates to either a cell type B or a cell type C) and the stochastic nature of gene expression. They showed that fluctuations in the rates of gene expression produce highly erratic time patterns of protein production in individual cells and a wide diversity in protein concentrations across cell populations.

The relation between fluctuations in concentrations at the molecular level and macroscopic phenotype selection was analyzed using the phage λ lysis-lysogeny decision circuit as a model. Experimentally, it is observed that an initially homogeneous cell population separates into distinct phenotypic subpopulations [12], thus showing the relevance of stochastic fluctuations in the dynamics of cells.

Importantly, they also showed that the regulatory proteins of this process exist in very low cellular concentrations and "compete" in the control of the pathway switch point, that is, the moment at which the cell becomes fully committed to differentiate in one of the two possible differentiation pathways. This fact unavoidably makes the choice of pathway stochastic, that is, it cannot be explained by deterministic processes.

A simplified version of the model, based on the original SSA to correctly account for the noisy nature of chemical reactions, presented here (with no time delays in gene expression reactions) was able to match the experimentally observed ratios of cells choosing each of the differentiation pathways and also properly model this system dynamics at the single cell level [12].

The same stochastic modeling strategy also mimicked correctly the production of proteins from an activated promoter in short bursts of variable numbers of proteins whose occurrence was separated by time intervals of random duration [4].

The conclusion, regarding modeling strategies of GRNs, is that the conventional deterministic kinetics, even with the addition of noise terms such as additive or multiplicative "Gaussian-like noise" cannot be used to predict the statistics of regulatory systems that produce probabilistic outcomes. However, it remains to be shown that ODE models with white or colored noise capture the true nature of chemical kinetics noise. Rather, a stochastic kinetics modeling strategy should therefore be used [12].

The relevance of having a model that correctly accounts for noise in the dynamics is enhanced by the discrete nature of the transcription factors and their binding sites, genes' promoter regions, which exist in the cell in very low copy numbers [14,15].

Recent studies reporting experimental results confirm the fact that noise cannot be neglected in GRN dynamics. Quantitative fluorescence measurements of gene expression products [16,17] also showed that GRN dynamics and cell differentiation depend highly on the noise. The results established that certain types of cellular differentiation are probabilistic and transient. They reported the experimental observation of cases where a cell was going back and forward from one "cell type" to another, without being subjected to external perturbations.

Other experiments [18–24] also reported similar observations of populations of genetically identical cells with phenotypically distinct individuals, exposed to similar environmental conditions, implying that GRNs are intrinsically stochastic.

Finally, recent numerical and analytical studies of the dynamics of models of toggle switches [25] compared the dynamics of these small gene networks using ODE models, with noise terms, and models based on the SSA. They showed that bistability (the spontaneous toggling between the two "stable" states) is attainable even without the need of cooperative binding (i.e., when cooperative binding exists, it is assumed that proteins have to form dimers before they can act as repressors of a gene) given careful parameters tuning. The bistability was shown to be the result of stochastic effects, even when having high protein concentrations. That is, bistability in these conditions is not observable in deterministic models. Importantly, this study stresses the necessity of using stochastic methods for modeling GRNs, even in the regime of high concentrations, as other modeling strategies were not able to mimic the results.

In summary, a few works – out of many that stress the necessity of using a simulation method that correctly accounts for the stochastic nature of gene expression when modeling GRNs – and their results were described. The described experimental and "simulation-based" results, and the conclusions drawn from them, clearly indicate that GRNs have a highly stochastic dynamics and that the SSA, or other similar algorithm based on it, should be used to simulate model of GRN dynamics.

7.2.2
Time Delays in Transcription and Translation

Gene expression has, aside its stochastic nature, another important dynamical feature. Among the many steps involved in the gene expression process, some are time consuming, such as transcripts creation and modification, mature mRNA transport to the cytoplasm (in the case of eukaryotes), mRNA translation to a protein molecule [26], and posttranslation protein modifications and folding. The time duration of these events creates time delays between different stages of the processes they are involved in. These delays often play important roles in biochemical dynamics and so must be incorporated in mathematical models that include a genetic regulatory component [27–29].

Through transcription elongation, the RNA*p* slides along the template strand and adds bases to the transcript, according to the DNA sequence that is read. This is a time-consuming process, whose duration depends on the gene length and the RNA*p* transcription speed. Also, it may vary between different events even for the same gene, because it depends on the rate by which the reactions involved occur, and these are stochastic events. Measurements of elongation times showed that the velocities of different transcription events followed a normal distribution [30].

Many reactions are involved in the transcription of a single gene. Modeling all these reactions as independent bimolecular reactions becomes prohibitive, even for a single gene, especially at the level of gene networks with more than a few genes. It was then proposed to reduce this multiple-step elongation process to a single delayed process [31], concentrating on the input and output. This way, multistep biochemical processes such as transcription and translation, each consisting a large number of elementary reactions, can be modeled as single-step delayed reactions, in which the initiating events are separated from the appearance of products by some interval of time. It was shown that delayed stochastic simulations closely approximate a detailed transcription model, except when two-body effects (e.g., collisions between RNA*ps* on a template strand) are important [10].

Since transcription and translation time durations vary from one event to the next, the time delay values should be randomly drawn from some delay distributions each time a reaction is chosen to occur. Unfortunately, the distribution of delays in gene transcription and translation is not yet known and, for that reason, usually, it is assumed that delays should be drawn from a normal distribution [5], whose mean value should be a function of the gene length and the RNA*p* average transcription velocity.

Recent experiments measured the elongation rates of single *Escherichia coli* RNA*p* molecules *in vitro*, by measuring the movement of single RNA*ps* over a number of DNA base pairs [30,32,33]. These experiments report that the elongation rates measured in a population of identical cells are heterogeneous. However, these results have been interpreted in two different ways. In one work, the heterogeneity was interpreted as being caused by cell population's heterogeneity and reactions' stochasticity, and not due to a variable RNA*p* transcription rate, which was considered uniform [32]. In similar works, the interpretation was that RNA*p* transcription rate varies from one transcription event to the next one [30,33]. At the single cell level modeling, it is necessary to know which of these two is the cause of the observations.

In the first stochastic models of gene expression, this process was assumed to be instantaneous [12]. Yet, experimental data support that transcription and translation take a considerable time to generate one RNA and, from it, a polypeptide, depending on the gene length (which varies significantly from gene to gene).

Although stochastic models of GRNs using only nondelayed reactions can explain experimental data regarding gene expression fluctuations (see [23]), these studies focused on steady-state dynamics, where "delayed" and "nondelayed" models have the same results, after an initial transient [6]. To accurately model more complex GRNs (e.g., when involving feedback mechanisms), it is necessary to model transcription and translation as time-delayed reactions, since once initiated they take a nonnegligible time duration.

Thus, in the model used here, the time duration of transcription and translation [5,34,35] is included to capture the features of transients [36,37]. For example, a study of the dynamics of a single TS found that protein production delays can cause a long transient, such that the two genes synchronize before the system reaches a "stable" state [38]. Also, it has been shown that time delay in eukaryotes plays a key role in the production of "delay-driven" oscillations of some protein concentrations [5,35].

Two parameters are important to define the dynamics of a delayed reaction. The reaction rate constant determines how often the reaction occurs according to the rules of the original SSA [9,39]. Introducing a time delay for products to appear, after the reaction occurrence, accounts for the time the reaction takes to complete, once it is initiated.

Delayed reactions affect the dynamics significantly, especially if some reactants involved in the reaction exist in very low quantities. As shown in Section 7.6, this proves to be a critical parameter in toggle switches dynamics.

Suppose a reaction represents a single gene expression, and there occurs another reaction by which the proteins decay. Also, assume that these reactions are nondelayed. If no other reactions are possible, and assuming k_t to be the protein production rate and k_d the decay rate, the expected number of proteins in the system is k_t/k_d, after an initial transient time to reach equilibrium.

In this case, modeling gene expression as an instantaneous event or a multiple time-delayed event will not change the protein level at which the system reaches equilibrium, if and only if k_t^{-1} is much larger than the time delay on the promoter release (represented by τ_1). The only difference between the "delayed" and the "nondelayed" cases is the time each takes to reach equilibrium. Notice, nevertheless, that such transient time to reach equilibrium might affect the system dynamics significantly in more complex systems of reactions [38]. Also, notice that if $\tau_1 \simeq k_t^{-1}$, the concentration of proteins in equilibrium is far smaller in the delayed case as the upper bound of protein production rate is significantly smaller.

However, delayed and nondelayed models' dynamics differ in more aspects. The next example is a case where the experimentally observed dynamics can only be correctly modeled using the "delayed" approach. Recently, the real-time production of single protein molecules under the control of a repressed lac promoter in individual *E. coli* cells was directly monitored through an epifluorescence microscope [40]. The gene chosen is known to, in the experiment conditions, transcribe at a very small rate since it is maintained, most of the time, strongly repressed (in this state the gene is completely unable to express). The proteins resulting from translation of the RNA, obtained from that gene transcription, travel to the cell membrane, where they then emit a light signal significantly above the natural fluorescent radiation level of the cell [40].

The protein concentrations of single cells were measured at the single molecule level. It was found that the proteins are produced in bursts, with the distribution of the bursts per cell cycle fitting a Poisson distribution. The number of proteins produced in several bursts follows a geometric distribution. The bursts also display particular temporal spreads [40]. The model of gene expression used here [6] that includes multiple time-delayed reactions reproduces the observed kinetics in more detail. In Section 7.6, the results of these model simulations are presented.

Next, the original SSA is described after which the delayed SSA is also presented.

7.3
The Delayed Stochastic Simulation Algorithm

7.3.1
Stochastic Simulation Algorithm

The stochastic simulation algorithm [9,39] is a Monte Carlo simulation of the chemical master equation. The algorithm is an exact procedure for numerically simulating the time evolution of a well-stirred reaction system. In the SSA, each chemical species' quantity is treated as an independent variable and each reaction is executed explicitly. The time evolution consists of a number of steps with each step being the execution of a specific reaction at a specific time. After a reaction is executed, the number of molecules of each of the affected species is updated according to the reaction formula and the algorithm advances to the next step. Each reaction and the time for the next reaction to occur are assumed to be independent of the preceding ones, and so the evolution of the system is a Markov process.

The algorithm is exact in the sense that each time one simulates a system of chemical reactions, under the conditions required by the SSA, the temporal trajectory obtained is exact, matching one of the system's possible trajectories in its state space.

The necessary condition for the SSA to be valid for any chemical system is that such a system is kept "well mixed" during the simulation, either by direct stirring or by requiring that nonreactive molecular collisions occur far more frequently than reactive molecular collisions [9]. For the meeting probability of two molecules to be spatially homogeneous, one must assume that each time a reaction occurs due to a collision between two potentially reacting molecules it will be followed by many collisions that do not result in reactions, so that the molecules once again become uniformly distributed in space, before the next reactive event occurs.

Each reaction rate constant, c_μ, is dependent on the reactive radii of the molecules involved in the reaction and their average relative velocities [9]. The velocities depend on the temperature of the system and the individual molecular masses. After setting the initial species populations X_i and reaction rate constants c_μ, the SSA calculates the propensity $a_\mu = a_\mu \cdot h_\mu$ for all possible reactions. The variable h_μ is a function of the reaction dependency on reactant concentrations, that is, the number of distinct molecular reactant combinations available at a given moment in time. After, the SSA generates two random numbers r_1 and r_1, used to calculate τ, the time interval for the next reaction to occur, and μ, that defines which reaction occurs. Finally, it increases the system time t by τ and adjusts the X_i quantities to account for the occurrence of reaction μ, assuming that the reaction occurred instantaneously. This process is repeated until no more reactions can occur, or during a user-defined time interval.

Assuming the system well stirred in a three-dimensional space of volume V, the propensity of a reaction between molecules of types 1 and 2 is the probability that a collision between two molecules, 1 and 2, will occur somewhere inside V, in the next infinitesimal time interval $(t,\ t+\mathrm{d}t)$. This quantity is given by

$$\frac{\mathrm{dV}_{\text{collision}}}{V} = V^{-1}\pi r_{12}^{2} v_{12} P_{12}^{\text{react}}\,\mathrm{d}t. \tag{7.1}$$

In Equation 7.1, $\mathrm{d}V_{\text{collision}}$ is the volume the two molecules must be in for the reaction to occur, V is the total volume, r_{12} is the relative radius of the reaction between 1 and 2, v_{12} is the relative velocity between the two molecules, and P_{12} is the probability that, if a collision occurs, a reaction occurs (this quantity is much smaller than 1, so that the "well-stirred" condition applies).

The probability that such collision will occur, between molecules of types 1 and 2, somewhere inside V in the next infinitesimal time interval $(t,\ t+\mathrm{d}t)$ is, given that X_1 and X_2 are the number of existing molecules of types 1 and 2, respectively:

$$\frac{\mathrm{dV}_{\text{collision}}}{V}(X_1, X_2) = X_1 X_2 V^{-1}\pi r_{12}^{2} v_{12} P_{12}^{\text{react}}\, dt. \tag{7.2}$$

As seen in the formulation of the algorithm, which follows, the probabilities for events to occur are then converted into the expected time one has to wait until they actually occur. That allows for computing the system state temporal evolution. The original formulations of the SSA to simulate the stochastic temporal evolution of a chemically reacting system are as follows [9]:

Step 0 (initialization). Input the desired values for the M reaction rate constants c_1, ..., c_M, and the N initial molecular population numbers $X_1, \ldots, X_N$. Set both the time variable t and the reaction counter n to zero. Initialize the unit interval uniform random number generator (URN).

Step 1. Calculate and store the M quantities, $a_1 = h_1 c_1, \ldots, a_M = h_M c_M$ for the current molecular population numbers, where h_μ is the number of distinct molecular reactant combinations available, given the system current state $(X_1, X_2, \ldots, X_N)$ $(\mu = 1, \ldots, M)$. Also, calculate and store as a_0 the sum of the M a_i values.

Step 2. Generate two random numbers r_1 and r_2 from a unitary uniform distribution and calculate τ according to $\tau = (1/a_0)\ln(1/r_1)$, and μ is an integer such that $\sum_{v=1}^{\mu-1} a_v < r_2 a_0 \leq \sum_{v=1}^{\mu} a_v$.

Step 3. Using the τ and μ values obtained in step 2, increase t by τ and adjust the molecular population levels to reflect the occurrence of the reaction chosen to occur. Then increase the reaction counter n by 1 and return to step 1.

As mentioned, the problem in using this algorithm to simulate the dynamics of gene expression is that transcription and translation involve a large number of bimolecular reactions, and thus it is a very expensive procedure computationally. For that reason, the "delayed SSA" was proposed [10].

7.3.2
The Delayed Stochastic Simulation Algorithm

It was proposed [31,36] that complex multiple-step biochemical processes or multiple-step biochemical functions could be modeled by single-step time-delayed reactions. In these reactions, products are not released in the system the moment the reaction occurs, as their appearance is delayed according to the average time necessary to complete their production.

Time delays should be included, not only for transcription and translation in gene networks [5,34,35], but also for modeling any transient between reaction and products appearance in the system [36,37]. For example, one should use time-delayed reactions to model diffusion through nonnegligible distances (assuming that the substance cannot react while diffusing in that region of the space) to capture the features of transients [41]. To simulate time-delayed reactions, one needs a "waiting list." Each product that takes time τ^* to be created after the reaction occurrence is placed, while the reaction occurs, in a waiting list until time τ^* elapses. At that moment, the product is released in the system and becomes available for possible reactions. Take, as an example, reaction 7.3:

$$\mathrm{A} + \mathrm{B} \rightarrow \mathrm{A} + \mathrm{C}(\tau^*). \tag{7.3}$$

When reaction 7.3 is selected to occur, the number of molecules of A is kept constant, a single molecule B is immediately removed from the system, and a molecule C appears in the system τ^* seconds after the reaction occurs. To do this, the delayed SSA [10], a generalization of the algorithm proposed in [36], proceeds as follows:

1. Set $t \leftarrow 0$, $t_{\mathrm{stop}} \leftarrow$ stop time, read initial number of molecules and reactions, and create empty waiting list L.
2. Do an SSA step for input events to get next reaction event R_1 and corresponding occurrence time t_1.
3. If $t_1 + t < t_{\min}$ (the least time in L), set $t \leftarrow t + t_1$. Update number of molecules by performing R_1, adding delayed products into L as necessary.
4. If $t_1 + t \geq t_{\min}$, set $t \leftarrow t_{\min}$. Update number of molecules by releasing the first element in L.
5. If $t < t_{\mathrm{stop}}$, go to step 2.

Notice that this algorithm is still exact because the distribution from which the time for the next reaction event to occur is generated is "memoryless" (since it is an exponential distribution). Therefore, no error is introduced. If one generates a time delay τ but the release of a substance from the wait list is scheduled to occur first, the system evolves up to the moment at which the substance is released, and then generates a new τ. When there are no reactions with time delays, the delayed SSA is identical to the original SSA.

This algorithm provides a methodology to model complex multiple-step processes as single-step delayed events, otherwise computationally too expensive to simulate. As shown next, the delayed SSA is able to model single gene expression correctly, as well as far more complex networks of interactions between genes, that is, gene regulatory networks dynamics.

7.4 Modeling Gene Expression as a Multiple Time-Delayed Stochastic Event

As previously stated, to model gene expression one has to account for the stochastic nature of this process and, to avoid modeling all reactions involved, the time the process, once initiated, takes to complete. Therefore, transcription and translation are modeled as "non-Markovian" events. The delayed SSA allows for multiple time delays in a single reaction and for time delays whose values are randomly drawn from a distribution, rather than just constant values, each time the reaction occurs.

Here, two models of transcription and translation are presented. In the first case, the two processes are modeled as a single-step multiple delayed reaction. The second model, more detailed, considers transcription and translation separately, each as a multiple time-delayed reaction.

In the first model, when an RNA*p* binds to a gene promoter region, if reacting, the output will be, besides the gene promoter region and the RNA*p*, an active protein.

Reaction 7.4 represents the production of a protein resulting from a gene expression, where $\mathrm{Pro}_i(t)$ is gene *i* promoter region. An RNA*p*, the RNA polymerase, must bind to this region of the gene for transcription to begin.

$$\mathrm{RNA}\,p(t) + \mathrm{Pro}_i(t) \xrightarrow{k_{i,t}} \mathrm{Pro}_i(t+\tau_{i,1}) + \mathrm{RNA}\,p(t+\tau_{i,2}) + n_i \times \mathrm{P}_i(t+\tau_{i,3}). \tag{7.4}$$

Substance P_i, in reaction 7.4, is the resulting protein created from the translation of the RNA created in the transcription. The probability rate constant for the RNA*p* to bind to the gene promoter region is, in this case, represented by k_i. The values τ are the times taken for each of the products of the reaction to become available in the system. This reaction has three time delays, in the release of promoter ($\tau_{i,1}$), the RNA*p* ($\tau_{i,2}$), and the protein $\tau_{i,3}$, such that $\tau_{i,1} < \tau_{i,2} < \tau_{i,3}$. The delay $\tau_{i,3}$ must, in this formulation, include the time transcription and translation take to complete.

The n_i constant is an integer, associated with the rate of translation of each RNA. The higher its value, the more times will a single RNA be translated. This constant can be drawn from a distribution of integers, each time the reaction is chosen to occur, if desired. The value of the integer n_i can also be different for each gene of a GRN.

To model transcription and translation as separate events, the following reactions are used (reactions 7.5 and 7.6), assuming one is modeling gene expression in prokaryotes [6]:

$$\mathrm{Pro}(t) + \mathrm{RNA}p(t) \xrightarrow{k_{\mathrm{transc}}} \mathrm{Pro}(t+\tau_1) + \mathrm{RBS}(t+\tau_1) + \mathrm{RNA}p(t+\tau_2) + \mathrm{R}(t+\tau_2), \tag{7.5}$$

$$\mathrm{Ribosome}(t) + \mathrm{RBS}(t) \xrightarrow{k_{\mathrm{transl}}} \mathrm{RBS}(t+\tau_3) + \mathrm{Ribosome}(t+\tau_4) + \mathrm{P}(t+\tau_5). \tag{7.6}$$

Reaction 7.5 models transcription. An RNA*p* molecule binds to a gene's promoter region and then transcribes the gene. The promoter remains unavailable for more reactions during τ_1 seconds after the reaction takes place, and the part of the RNA (the ribosome binding site, "RBS"), to which ribosomes can bind to and begin translation of the RNA, is also produced in that time duration. At the end of the transcription process, at τ_2, the RNA*p* and the complete RNA molecule (here represented by R) are released in the system.

When modeling cell gene expression in prokaryotes, one would use R as the substrate for the translation reaction 7.6, instead of the "RBS", since in this case the RNA must be transported out of the nucleus, in order to be translated. This event usually occurs after the RNA is fully formed. In prokaryotes, translation can begin as soon as the RBS is completed. In this case, R is not necessary to be included in reaction 7.5, and the RBS is used as subtract of the translation reaction.

In reaction 7.6, a ribosome binds to the RBS and translates it, producing a protein P in τ_5 seconds, while the RBS is available for more translation reactions at τ_3. The ribosome is released after τ_4 seconds, becoming available in the system to initiate more translation reactions.

Notice that an index *i* was included to identify the gene, since in the next section reactions 7.5 and 7.6 are generalized so that each different gene can be transcribed and translated and each gene has multiple operator sites, to generate models of gene regulatory networks.

7.5
A Gene Regulatory Network Model

In real cells, the GRN is involved in most cellular complex biological processes. Some of the products of gene expression consist of transcription factors and cofactors that directly affect the activity of downstream genes. Other products of gene expression regulate cellular chemistry, which via feedback chemical pathways regulate the expression of several genes.

Many chemical pathways in cells are not known in detail, and even if they were, it would not be possible to model and simulate all these processes. Reasonable reduction methods are needed to study the dynamics of GRNs.

To reduce the complexity of the GRN model, it is assumed here that gene expression products (the proteins) are fed back directly into GRN genes. That is, to model the regulation, by a gene expression product, of another gene expression rate or its protein quantity in the system, it is assumed here that the protein expressed by the first gene is either a transcription factor that binds to one operator site of the second gene (and changes its propensity of expressing) or the first gene proteins that, for example, degrade the other gene's proteins (interactions between proteins are called "indirect", while interactions between proteins and genes promoter regions are referred to as "direct"). Also, proteins can form homodimers or heterodimers (or higher order polymers) that can feed back into the network, acting as transcription factors, and so on. The set of interactions among the genes, via their products of expression, define the network topology.

In the model described here, genes are represented by their promoter regions. The promoter region includes the initiation sequence, to which the RNA*p* can bind and begin transcription, and the operator sites region, to which transcription factors can bind and change the genes' expression reaction propensity (including complete repression).

When a gene expression product is a transcription factor of another gene, it can act as an activator or an inhibitor of this second gene transcription reaction. Also, genes can have multiple operator sites (the regions of the gene to which transcription factors can bind), and the effect of multiple transcription factors can be, in general, combinatorial. For example, a certain protein can act as inhibitor if it is the only transcription factor bound to the gene, but invert its effect, if another specific transcription factor is also bound to the gene, in another operator site.

Since genes in a GRN can have many operator sites, the following notation is used: $\text{Pro}_{i,\{\vec{\text{op}}\}}$, such that i is the gene identification index; that is, each gene of the network is unique and is distinguished by the index i. The second index, $\{\vec{\text{op}}\}$, is a "vector" of all operator sites, and its value represents the state of each of the gene's operator sites. Such a state consists of what transcription factors bound to each operator site of the gene, if any, at any given moment.

For each possible combination of "input states" (promoter state) a regulating function is assigned that determines, depending on the input state, the gene expression capability. For example, imagine that gene 7 has two operator sites. Assume that proteins p_1 and p_2 can bind to operator site 1, and p_3 can bind to operator site 3, of gene 7. If, at any given moment in the simulation, p_1 is bound to site 1, and no protein is bound to site 2, the operator is in the state "$\text{Pro}_{i,(\text{p1},\ 0)}$". Another possible state, for example, would be "$\text{Pro}_{i,(\text{p2},\ \text{p3})}$". Given this set of possibilities, the gene network generator [42] must then assign a "state" for the promoter, given the combination of transcription factors bound, or not, to its operator sites. A state means whether the gene is repressed or activated, and when the gene is activated, what is the rate constant of transcription. For example, assume state $\text{Pro}_{i,(\text{p2},\ \text{p3})}$ implies that the gene is completely repressed, then no reaction allowing the RNA*p* to bind to that complex is created, when generating all possible reactions in the GRN. As another example, assume that in the state $\text{Pro}_{i,(\text{p1},\ 0)}$ the gene active; in this case a transcription reaction is allowed, with a rate constant different from the rate constant of the basic level of expression transcription reaction, if this last reaction exists. In general, when generating a GRN, a fraction of genes can be assigned to have basic level of expression (a promoter with no transcription factors bound to it can be transcribed by an RNA*p*), and other genes will not have that ability.

This procedure to "design" GRNs can be seen as a generalization of the procedure used to create random Boolean networks [7]. Here, it is possible to assign random Boolean functions, thereby creating a combinatorial logic. This level of complexity can be attained in the model proposed here in three ways: (i) either by allowing some reactions between the gene expression products, creating complex molecules, and then assigning such molecules to a randomly chosen gene promoter site, choosing randomly what effect that complex has on that gene expression level when connected to the promoter site, or (ii) by allowing a gene to have more than one operator site and randomly assigning the effects of all the possible binding combinations as activations or inhibitions, or (iii) by allowing different gene expression products to bind competitively to a single binding site, each with a different effect on the gene transcription rate.

It is known that different combinations of bound molecules to the various operator sites of a gene will have different effects on the gene transcription function. In general, the effects can be combinatorial although, usually, a set of transcription factors has a combined effect, which is (approximately) the sum of each transcription factor independent effect. Here, the general case is assumed; that is, the effect on a gene transcription rate of each distinct combination of transcription factors bound to the promoter can be randomly chosen.

These features let all Boolean functions and topologies of interactions have equivalents in this model. While doing a mapping from a Boolean network to the GRNs modeled here, an infinity of different GRNs can be created from a single Boolean network, since many parameters such as rate constants and time delays are not defined in Boolean networks.

As a gene can have as many operator sites as desired, without increasing significantly the system simulation time, the GRNs generated with this model are not restricted to topologies with low connectivity (the limits on increasing the order of polymers allowed in far more restrict, for example). For simplicity, in the present work, polymers' order is restricted to monomers and dimers, since the existence of higher order polymers is computationally expensive.

Finally, a fraction of genes can be assigned to have no direct inputs, and a fraction of genes that are not inputs to other genes.

In this model of GRNs, first proposed in [38], genes are treated as chemical species since in real GRNs they exist only in very small quantities (usually one to a few copies of the same gene). As gene promoter regions are also subject to time delays, combined with their very small quantity, stochastic effects must be taken into account using the delayed SSA [10] to simulate the dynamics.

GRNs are generated from the following reactions. For gene $i = 1, \ldots, N$, there is a reaction representing the basal transcription of promoter Pro_i by RNA polymerase (reaction 7.7).

It is know that eukaryotic gene regulatory proteins often assemble into complexes on DNA to enable transcription; thus, this model includes transcription reactions of $\text{Pro}_{i,\{\{\vec{\text{op}}\}}$, when the combined effect of the transcription factors bound to the operator sites results in an activated "state" (reaction 7.8) and translation of RNA by ribosomes (Rib) into proteins (p_i) (reaction 7.9). These are all time-delayed reactions. The time delays of each product are represented by τ variables and differ between different products of each reaction and between similar reactions for different genes (since these have different lengths).

The binding/unbinding of a transcription factor protein p_w from operator site j of a gene i is represented in reactions 7.10 and 7.12. If the formed complex is unable to be used for transcription, these two reactions represent repression/unrepression reactions, else these will be activation/inactivation reactions. Notice that reaction 7.10 is bidirectional and therefore represents two reactions: the binding of the repressor and its spontaneous unbinding. Reaction 7.12 represents the "unrepression" not due to spontaneous unbinding of the repressor from the gene operator site, but rather due to an external "reactivator" protein that removes the repressor from the operator site.

Decay of RNA, represented by its ribosome binding site, and proteins (p_i's) occur via the reactions 7.13. Finally, protein polymerization (here limited to dimers for simplicity) and the inverse reaction occur via the bidirectional reaction 7.14. Unless time delays, τ's, are explicitly represented in the products of the reactions (here this is represented using the notation "$X(\tau)$"), all events, reactants' depletion and products' appearance, occur instantaneously at the moment the reaction takes place, t:

$$\text{RNA}p + \text{Pro}_i \xrightarrow{\text{kt}_{i,\text{bas}}} \text{Pro}(\tau_i^1) + \text{RBS}_i(\tau_i^1) + \text{RNA}\,p(\tau_i^2), \tag{7.7}$$

$$\text{Pro}_{i,\{\vec{\text{op}}\}} + \text{RNA}\,p \xrightarrow{\text{kt}_{i,\text{op}_i}} \text{Pro}_{i,\{\vec{\text{op}}\}}(\tau_i^1) + \text{RBS}_i(\tau_i^1) + \text{RNA}\,p(\tau_i^2), \tag{7.8}$$

$$\text{Rib} + \text{RBS}_i \xrightarrow{\text{ktr}_{i,\text{RBS}}} \text{RBS}_i(\tau_i^3) + \text{Rib}(\tau_i^4) + p_i(\tau_i^5), \tag{7.9}$$

$$\text{Pro}_{i,(\ldots,j,\ldots)} + p_w \underset{k_{w,ij}}{\overset{k_{ij,w}}{\rightleftarrows}} \text{Pro}_{i,\ldots,p_w,\ldots}, \tag{7.10}$$

$$\text{Pro}_{i,(\ldots,p_z,\ldots)} \xrightarrow{\text{kd}_{iz}} \text{Pro}_{i,(\ldots,0,\ldots)}, \tag{7.11}$$

$$\text{Pro}_{i,(\ldots,p_z,\ldots)} + p_w \xrightarrow{k_{ijz,w}} \text{Pro}_{i,(\ldots,0,\ldots)} + p_z + p_w, \tag{7.12}$$

$$\text{RBS}_i \xrightarrow{k_{\text{dr},i}} \varnothing, \quad p_i \xrightarrow{k_{\text{dp},i}} \varnothing, \tag{7.13}$$

$$p_i + p_j \underset{k_{i,\text{undim}}}{\overset{k_{i,\text{dim}}}{\rightleftarrows}} p_{i,j}. \tag{7.14}$$

Complete RNA molecules are not explicitly represented since the appearance of RBSs is what determines when translation can start. This is done assuming that this

GRN is of a prokaryote. In eukaryotes, only the complete RNA would be inputs of translation reactions.

Each reaction notation and their effect on the GRN dynamics are now described in more detail. In reactions 7.9–7.9, τ's superscripts distinguish the delays between products and subscripts distinguish the delays of products of reactions associated with different genes. For example, if reaction 7.7 occurs for gene *i*, at time *t*, Pro_i and one RNA*p* are removed from the system and placed in a waiting list of events. At $t+\tau_i^1$, Pro_i and RBS and at $t+\tau_i^2$ RNA*p* are released (unchanged), becoming available for reactions.

The initiation frequency of transcription (reactions 7.7 and 7.8 is controlled by their respective rate constants and the time delay on the promoter, τ_i^1 [6]. The rate constant of the transcription reaction, for example, $\mathrm{kt}_{i,\mathrm{op}_i}$, determines how frequently the binding event of RNA*p* to Pro happens on average, if Pro is available (assuming abundance of RNA*p*), and the delay $\tau_{i,j}^1$ determines how much time it takes for Pro to be available for another reaction. If the RNA*p* exists in such large quantities that the number of available RNA*p*s is approximately invariant, then $\tau_{i,j}^2$ can be neglected [6]. In addition, $\tau_{i,j}^2$ can be omitted as well if it is much smaller than $(\mathrm{RNA}p\,\mathrm{kt}_{i,\mathrm{op}_i})^{-1}$. The promoter clearance time is usually not considered in previous modeling strategies (to do so here we can set τ^1 null). One difference in the dynamics, caused by the delay on the promoter, is the limitation on the number of RNA*p*s that can transcribe the gene simultaneously in agreement with observations [10]. Also, as shown in Section 7.6, this delay is necessary for a TS to toggle, given there is no cooperative binding or self-activation reaction. Translation reaction 7.9 dynamics is similar to reactions 7.7 and 7.8.

Reaction 7.10 represents two reactions, that is, the binding of a transcription factor to a promoter operator site and the backward reaction. The complex formed by the forward reaction can be a repressed or an activated promoter, if a transcription reaction is then created for that specific promoter–transcription factor complex.

Reaction 7.11 allows for the protein to decay when bound to the promoter. If absent, binding to the promoter would act as a "protection" against protein decay and that would affect the dynamics dramatically. This reaction also indirectly affects the repression "strength." Suppose a very high decay on the proteins is imposed. Since this causes proteins to have a very short lifetime, the effects of their repression are very weak due to the short duration they remain bound to the promoter, independent of the value assigned to the rate constants of the coupling–uncoupling reactions.

In this modeling strategy for GRNs, the ability to generate an ensemble of networks comes from the fact that all indexes (*i*, *j*, *z*, and *w*) can be randomly chosen integer numbers (from *1* to *N*) in the reactions responsible for the interactions between genes. If randomly generated, these index values will differ for each "network topology generation" [42]. Also, one can choose randomly which dimers are possible to be formed in the system of reactions. Thus, when the network of interactions between the genes is being created, since the number of operator sites of a gene and the indexes *j* and *w* are randomly chosen, a different wiring diagram of influences is generated at the beginning of each independent simulation. Inserting equations of reactions between the p_i's allows the creation of both homogeneous and heterogeneous dimers, trimers, and so on, which would act as activators and inhibitors. In general, polymers can be formed from any combination of p_i's. The random combination between the p_i's in polymers and their assignment as activators or inhibitors of genes and/or the existence of multiple operator sites in each gene also help define any desired topology for a GRN. Allowing the effect of these polymers on the genes' expression level to be randomly chosen is another way one can obtain any desired transfer functions. The flexibility with which one can generate these models of GRNs enables the use of the ensemble approach [11], as in random Boolean networks models, to characterize the general dynamical behavior of these systems.

These GRNs can be implemented using SGNSim [42]. A short description of how this simulator implements the models of GRNs just described follows.

SGNSim [42] consists of a GRN or chemical system generator and a dynamics simulator. It receives as input a file where reactions and elements' initial quantities are specified. This simulator output is a tab-delimited text file with the system time series and corresponding Fourier spectrum of all distinct chemical species involved in the system, given a preset sampling frequency. If networks are randomly generated, the set of reactions that composed them and the initial concentrations are also output, in a separate tab-delimited text file. The delays can be drawn from several distributions and the reaction rates from complex functions or from physical parameters. SGNSim can generate ensembles of GRNs, within a set of user-defined parameters, such as topology. It can also be used to model specific GRNs and other systems of chemical reactions.

SGNSim creates a GRN by generating a graph with the desired topology, imposing in-degree and out-degree distributions. Inputs are monomers or combined into multimers and set as direct or indirect. Then, each direct input is assigned to an operator site, while indirect inputs are given a target. Finally, a function is assigned to each gene, defining the gene's response to a combination of transcription factors (promoter state). The functions can be unate, or a transcription rate can be randomly assigned to each combination of promoter states, since the combined effect of transcription factors on transcription rates is not necessarily unate.

The information defining the set of reactions that the resulting GRN consists of is stored in an adjacency matrix. A site a_{ij} of the adjacency matrix contains the set of values (x_1, x_2, x_3, x_4). The value x_1 defines the type of interaction from j to i: 0 (no interaction), 1 (monomer), or 2 (dimer). In general, this value can grow to include higher order multimers. The value x_2 defines if it is a direct interaction (+1), that is, if p_j binds to Pro_i, or indirect (−1). The value x_3 is not null in the case of a dimer interaction and equal to the number identifying the protein that binds to p_j and forms the heterodimer. If higher order multimers are allowed, this field becomes a list of all participating monomers. Finally, x_4 identifies which operator site j of Pro_i the complex binds to and is 0 in the case of an indirect interaction.

Whether using the ensemble approach or a user-defined system of reactions, SGNSim can test the system in many ways. For instance, in GRNs, it can do "mock" perturbation experiments like gene deletion, overexpression, copy, and mutation. For example, to model a gene overexpression t seconds after the experiment started, a specific inducer is introduced at that time in the simulation [42]. Importantly, reaction rates can be some complex functions, such as Hill functions, that depend on concentrations at each moment. This allows, for example, for modeling active transport of proteins via a membrane [43].

One can include, if desired, more reactions in the GRNs, such as reactions representing alternative splicing – a process by which a pre-mRNA transcribed from a gene can lead to different mature mRNA molecules and thus different proteins. This can be achieved by introducing an extra reaction to the set of reactions forming the GRN where a precursor of the RNA is transformed into the final RNA that will afterward be transcribed: $\text{PreRBS}_i \xrightarrow{k_{i,\text{splice}}} \text{RBS}_{i,j}(\tau_i^{\text{splice}})$.

Given the modeling strategy of GRNs and the method of how to implement their structure and simulate the dynamics, a few applications are now presented as examples.

7.6 Applications

7.6.1 Modeling Single Gene Expression

Recently, the real-time production of single protein molecules under the control of a repressed lac promoter in individual *E. coli* cells was directly monitored through an epifluorescence microscope [40].

This important work [40] reports the construction of an *E. coli* strain SX4, in which a single copy of the chimeric gene tsr-venus was incorporated into the *E. coli* chromosome, replacing the native lacZ gene, while leaving intact the endogenous tsr gene of *E. coli*.

The introduced gene promoter is kept in a highly repressed state, and thus is unable to express most of the time. Since the endogenous tsr gene expresses in high quantities, the addition of the extra gene (because it is kept repressed most of the time) does not affect the normal cell behavior.

When the infrequent and spontaneous dissociation event of the repressor from the operator region of the gene occurs, transcription by RNA*p* begins. Usually this event generates a single mRNA molecule due to the short period the gene is available for transcribing. When the mRNA is produced, a few ribosome molecules bind to it, and proteins are produced. Meanwhile, the repressor rebinds to the operator. The proteins produced can be detected after the completion of their assembly process, which includes protein folding, incorporation onto the inner cell membrane, and maturation of the Venus fluorophore [40].

Observing the radiation emission events, it was found that the proteins are produced in bursts, with the distribution of the bursts per cell cycle fitting a Poisson distribution, and that the number of proteins produced per burst follows a geometric distribution [40]. Here we present a model to mimic this experiment, which was proposed in [6]. This important example is used to validate the gene expression model at the single molecules level.

$$\mathrm{Pro}(t) + \mathrm{RNA}p(t) \xrightarrow{k_{\mathrm{transc}}} \mathrm{Pro}(t+\tau_1) + \mathrm{RBS}(t+\tau_1) + \mathrm{RNA}p(t+\tau_2) + R(t+\tau_2), \tag{7.15}$$

$$\mathrm{Ribosome}(t) + \mathrm{RBS}(t) \xrightarrow{k_{\mathrm{transl}}} \mathrm{RBS}(t+\tau_3) + \mathrm{Ribosome}(t+\tau_4) + \mathrm{P}(t+\tau_5), \tag{7.16}$$

$$\mathrm{RBS}(t) \xrightarrow{\mathrm{RBS_{decay}}} \varnothing, \tag{7.17}$$

$$\mathrm{Pro}(t) + \mathrm{Rep}(t) \xrightarrow{k_{\mathrm{rep}}} \mathrm{Pro{\cdot}Rep}(t), \tag{7.18}$$

$$\mathrm{Pro{\cdot}Rep}(t) \xrightarrow{k_{\mathrm{unrep}}} \mathrm{Pro}(t) + \mathrm{Rep}(t). \tag{7.19}$$

Reactions 7.15 and 7.16 model, respectively, prokaryotic transcription and translation. The substance "R" represents a complete RNA molecule and does not intervene in any other reactions (it is included in the model to allow the exact counting of the number of transcription events occurring). Since a decay reaction for R is not defined, the number of R molecules in the system equals, at any given moment, the total number of transcription events that occurred.

The RBS (ribosome binding site of the RNA) is the part of the RNA to which the ribosomes bind to and initiate the RNA translation. In prokaryotes, which is the case here, this can occur as soon as the RBS is produced (τ_1 seconds after the transcription event occurs), even though transcription is not yet completed. The RBS is subject to decay via reaction 7.17, avoiding the possibility of creating an infinite number of proteins out of a single RNA molecule.

Reaction 7.18 models the repression event of the gene promoter region by a repressor (Rep). Reaction 7.19 models the inverse, that is, the unbinding of the repressor from the promoter region. Only when the promoter is free can the transcription begin, and since this reaction rate constant is very small, this occurs at very sparse intervals. (Thus, during the small time intervals the promoter is not being repressed, usually only a single transcription occurs.)

Since, once chosen by the delayed SSA to occur, these two reactions 7.18 and 7.19 are instantaneous, we can compute the expected fraction of time the gene will be free to express. This fraction of time is given by the expected time the repression reaction takes to occur, divided by the total expected time for a repression reaction followed by an unrepression reaction to occur. The time delay of the gene promoter region must be taken into account here. Given the rate constants of repression of reactions 7.18 and 7.19, k_{rep}, and unrepression, k_{unrep}, since there are 100 repressors "Rep" in the system and 1 promoter, and given the time delay of the promoter once a transcription reaction occurs, τ_1, if the state of the promoter at random moments during the simulation is measured, the fraction of times that the promoter is going to be observed as available for reactions is given by

$$1+\frac{1}{k_{\mathrm{unrep}}}+\tau\cdot k_{\mathrm{rep}}\cdot \mathrm{Rep}^{-1}.$$

Reactions 7.15 to 7.19, when simulated by the SSA, reproduce the observations from the experimental study described in [40], in which a single gene expression is monitored under the control of a repressed lac promoter in *E. coli* cells, demonstrating the ability of this modeling strategy to quantitatively model gene expression. The rate constants are set at [6] $k_{\mathrm{transc}}=0.01\,\mathrm{s}^{-1}$, $k_{\mathrm{transl}}=0.000\,42\,\mathrm{s}^{-1}$, $\mathrm{RBS}_{\mathrm{decay}}=0.01\,\mathrm{s}^{-1}$, $k_{\mathrm{rep}}=1\,\mathrm{s}^{-1}$, and $k_{\mathrm{unrep}}=0.1\,\mathrm{s}^{-1}$. Initially, it is set that RNAP $=40$, Pro $=1$, R $=0$, Ribosome $=100$, RBS $=0$, P $=0$, Pro·Rep $=0$, and Rep $=100$. Finally, the time delays are set at $\tau_1=40$ s, $\tau_2=90$ s, $\tau_3=2$ s, $\tau_4=58$ s, and $\tau_5=420\pm 140$ s, randomly generated from a normal distribution of mean value 420 and standard deviation of 140 (with cutoff at 0).

In Figure 7.1, the total number of proteins produced is plotted as a function of time. As seen, the proteins are produced in bursts by the gene under a repressed condition, as reported in [40]. Each time the repressor unbinds the promoter, a single RNA*p* can bind to the promoter, producing usually one or two RNA molecules, which are then translated into several proteins before decaying. Since proteins do not decay in this model, the total number of proteins equals the number of translation reactions. The bursts visibly occur at ~400, 6000, and 9000 s.

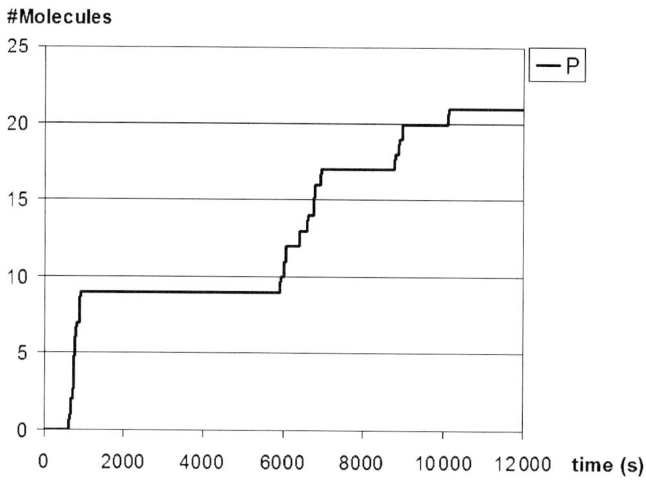

Fig. 7.1 Measurement from a single simulation of the time series of the number of proteins. The protein production occurs in bursts during the 200 min (about four cell cycles). Bursts are visible at ~400, 6000, and 9000 s. Since proteins do not decay in this model, the total number of proteins equals the number of translation reactions.

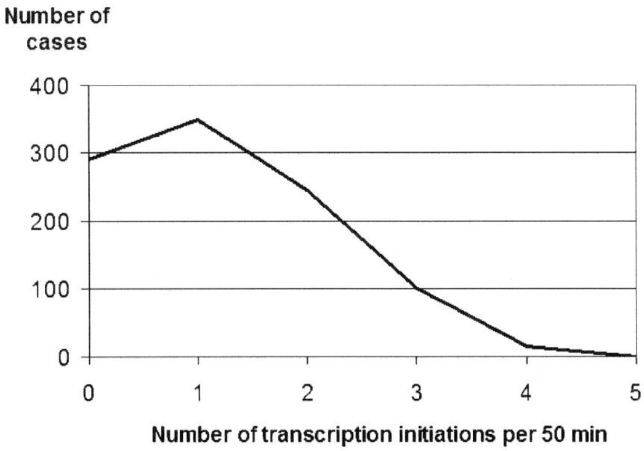

Fig. 7.2 Observed number of transcription initiations in 50 min, in each experiment. Data collected from 1000 independent simulations.

When running several independent simulations, one observes that the moment at which the bursts occur and the number of resulting proteins from each event vary significantly, due to the stochastic nature of the system's dynamics. In Figure 7.2, the distribution of the number of transcription initiations is plotted, over 1000 independent simulations. The resulting distribution of the burst sizes per cell cycle fits well a Poisson distribution.

In Figure 7.3, from the same 1000 simulations, it is shown that the number of translation reactions fits an exponential distribution, as reported in [40].

Notice that, using an ODE model of gene expression, since strong repression results only in diminishing a gene's expression rate, one would observe a proteins' time series approximately constant in time. The addition of noise terms, such as Gaussian-like noise would only make this level to vary around the mean

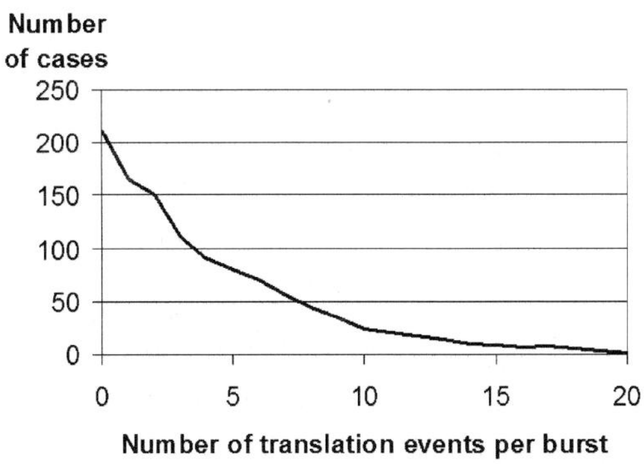

Fig. 7.3 Observed number of translation initiations per burst, in each experiment. Data collected from 1000 independent simulations.

concentration at equilibrium. This modeling strategy would not be able to reproduce the production "by bursts," since it is not appropriate to model systems with very few molecules and where single events, though sparse in time, are relevant.

The time delays also play an important role in the dynamics of this model, by limiting the number of RNA*p* molecules that can bind to a gene when it is not repressed. Similarly, it has been shown [43] that in the stochastic framework, it is not always equivalent (for the resulting proteins levels) to have a time delay, or, instead, have a smaller transcription rate constant.

In the next example, consisting of the simulation of the dynamics of a simple genetic circuit, the toggle switch, the importance of including time delays in transcription–translation reactions when simulating GRNs dynamics is clear.

7.6.2 Bistability of a Toggle Switch as a Result of Time Delays in Transcription

The genetic TS [21] consists of two genes, and each gene produces proteins that can repress the other gene expression. Namely, the protein transcribed by one gene binds to the other gene promoter region and inhibits transcription. Genetic TSs are among the most studied small GRNs, due to their simplicity and relevant role. They have been interpreted as decision circuits in cell differentiation and as cellular memory units [21].

It has long been hypothesized that cell differentiation is based on bistable genetic subcircuits that control many downstream genes [44]. In this process, a stem cell turns into a stable phenotype, in agreement with the hypothesis that ergodic sets of the GRN could correspond to cell types [45].

Recently, it was shown that a cell can use the TS to adopt different stable states [12,21,46]. Due to this property, models of differentiation pathways have been proposed using several TSs as decision subcircuits at each bifurcation [47]. Interestingly, using models that did not include time-delayed reactions, the findings in [47] pointed out the existence of multistep processes to explain the observations not accounted for by the model used. These multistep processes can be correctly modeled, from the system global dynamics point of view, by time-delayed reactions [31].

Recent studies of the TS dynamics [25] showed the TS to be bistable, that is, it can go from one "stable state" to the other, due to noise-related fluctuations even without assuming cooperative binding (i.e., the proteins do not combine into dimers) given careful parameters tuning. The necessary condition for bistability is the existence of self-activation reactions of the gene expression. These self-activation reactions consist in the binding of the protein produced by a gene, to its promoter, and enhancing its expression rate. Their results also stress the necessity of using stochastic methods to simulate GRNs, rather than using ODE models, even in the regime of high protein concentrations [25].

This example shows that a bistable TS can be obtained using our model of GRNs [38], without having to include cooperative binding or even self-activation reactions.

By modeling transcription and translation in a single step, through a multiple time-delayed reaction, the dynamics of a single TS is simulated. The effects of including time delays in the reaction responsible for transcription and translation are analyzed. The following set of reactions (reactions 7.20 to 7.26 is simulated:

$$\mathrm{RNA}p + \mathrm{Pro}_1 \xrightarrow{k_t} \mathrm{Pro}_1(\tau^1) + \mathrm{RNA}\,p(\tau^2) + \mathrm{p}_1(\tau^3), \tag{7.20}$$

$$\mathrm{RNA}p + \mathrm{Pro}_2 \xrightarrow{k_t} \mathrm{Pro}_2(\tau^1) + \mathrm{RNA}\,p(\tau^2) + \mathrm{p}_2(\tau^3), \tag{7.21}$$

$$\mathrm{Pro}_1 + \mathrm{p}_2 \underset{k_\mathrm{u}}{\overset{k_\mathrm{c}}{\rightleftarrows}} \mathrm{Pro}_1\mathrm{p}_2, \tag{7.22}$$

$$\mathrm{Pro}_2 + \mathrm{p}_1 \underset{k_\mathrm{u}}{\overset{k_\mathrm{c}}{\rightleftarrows}} \mathrm{Pro}_2\mathrm{p}_1, \tag{7.23}$$

$$\mathrm{Pro}_1\mathrm{p}_2 \xrightarrow{k_\mathrm{d}\,p} \mathrm{Pro}_1, \tag{7.24}$$

$$\mathrm{Pro}_2\mathrm{p}_1 \xrightarrow{k_\mathrm{d}\,p} \mathrm{Pro}_2, \tag{7.25}$$

$$\mathrm{p}_1, \mathrm{p}_2 \xrightarrow{k_\mathrm{d}} \varnothing. \tag{7.26}$$

Given reactions 7.20 to 7.26, the dynamics of a TS is simulated for four distinct cases: (A) setting all τ's to null, that is, in this case time delays are not included (Figure 7.4); (B) a nonnull time delay is included only on the protein production, namely, $\tau^3 = 100$ s, while $\tau^1 = \tau^2 = 0$ s (Figure 7.5); (C) delays are set at $\tau^1 = 2$ s, $\tau^2 = 20$ s, and $\tau^3 = 100$ s (Figure 7.6); (D) same settings as (B) but with a transcription rate constant ($k_t = 0.005\ \mathrm{s}^{-1}$), 100 times smaller (Figure 7.7).

If all time delays are set to null (case A), the system does not toggle as seen in the protein quantities time series of Figure 7.4. In this case, after a long transient, the system settles into one of the two stable states (one gene "on" and the other

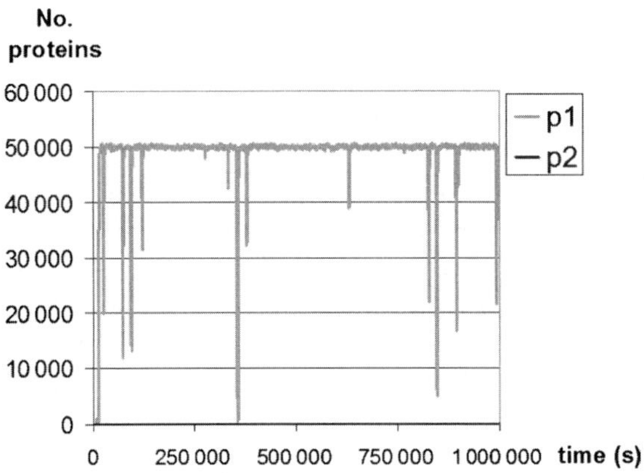

Fig. 7.4 Time series of a TS without delays or cooperative binding. After the transient ~14 000 s, the system remains stable. p_2 never grows beyond 20.

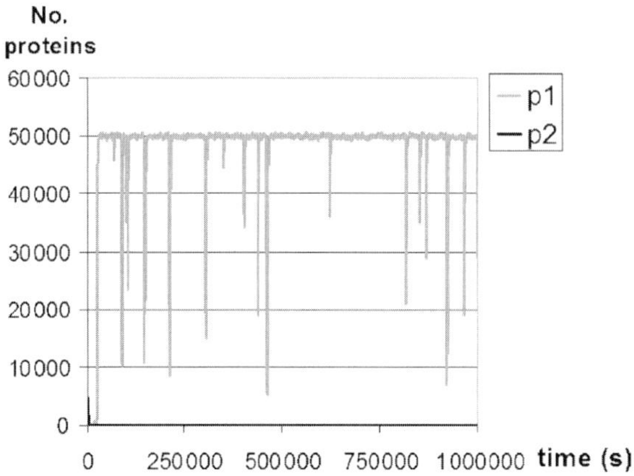

Fig. 7.5 Time series of a TS with 100 s delays on the protein release and no cooperative binding. As in case A, after a transient, the system remains stable and the number of proteins of the repressed gene (p_2) never goes beyond 20, except during the initial transient (unlike in case A), due to the delay on the protein release.

gene "off"), each equally probable. The choice depends solely on the initial stochastic fluctuations of protein concentrations. Once the choice is made, the system does not toggle. The average transient of 100 independent experiments is ~14 000 s with a standard deviation of ~7000 s. Decay and production reactions of proteins equilibrate at ~50 000 proteins, in agreement with an ODE model of similar reactions [6].

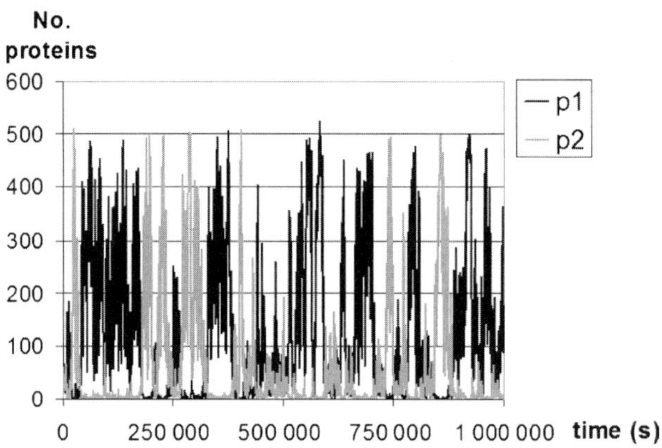

Fig. 7.6 Time series of a TS with multiple delayed transcription–translation reactions and no cooperative binding. Delays: $\tau^1 = 2$ s (promoter), $\tau^2 = 20$ s (RNA*p*), and $\tau^3 = 100$ s (proteins). The system toggles due to the delay on the promoter region release, which also causes no protein quantities increasing above 500.

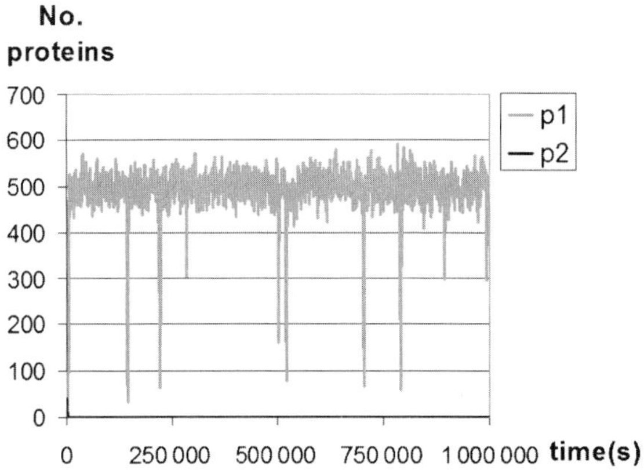

Fig. 7.7 Time series of a single TS with delays on the p's release only (set to 100 s), and $k_t = 0.005\ s^{-1}$. The system does not toggle since there is no delay on the promoter, even though proteins never increase above 500, due to the small rate of transcription–translation.

In Figure 7.5, the time series of both proteins, p_1 and p_2, is plotted, resulting from a simulation of model B. This TS (case B) differs from the previous one (case A) since we now introduced nonnull time delays in the protein production. Comparing this time series with the one in Figure 5.4, we see that the delays in protein production caused almost no changes in the dynamics. The protein level of the gene "on" appears to be more subject to fluctuations than in the previous case, but no generalization can be made in this respect by comparing only two time series. The system still reaches a single steady state, rather than having a toggling behavior. The average transient time to reach equilibrium and its standard deviation to attain the stable state increases, as expected, in comparison with case A, and is $\sim$19 000 s with a standard deviation of $\sim$12 500 s (averages attained from 100 independent simulations). Interestingly, due to the proteins initially being produced and still on the wait-list, there is a small burst of both genes' proteins at the beginning, that is, both genes are "on" at this stage. Once this first set of proteins is produced and released in the system, they start repressing the gene promoters, while most of them decay. This lasts on average 4000 s and is only possible due to the existence of delays on the protein release. After the transient, one of the genes becomes "on" and the protein level stabilizes at $\sim$50 000 proteins, while the other gene remains "off." That is, in this case, the delays on protein production only affect the initial transient and after that the steady-state solution is the same as if no delays existed.

In the case C, all delays assumed by the model are nonnull (reaction 7.20). As seen in Figure 7.6, the system dynamics changes drastically in comparison with the previous two cases. First, the maximum level that proteins reach is 500 (in comparison with 50 000), due to the delay on the promoter after each transcription reactions, which limits the number of RNA*p*s that can be transcribing the gene simultaneously. Since $\tau_1 = 2$ s, only one transcription can be initiated each 2 s.

The delay on the RNAp release diminishes the transcription reaction propensity (approximately by ~20%) since a fraction of the RNAps is not available, while occupied transcribing a gene and, thus, on the wait-list.

Importantly, the time series of p_1 and p_2 toggle (from p_1 being in larger quantity to p_1 and vice versa), after an average transient of ~4900 s for the first toggling to occur with a standard deviation of ~4050 s. The average number of toggles observed during Δt is 18.5 with a standard deviation of 3.75. Thus, the average toggling period is 50 000 s.

Although the time series looks "noisier" than when cooperative binding or self-activation reactions were implemented, the system clearly toggles.

The cause or causes for the toggling to occur is now investigated. It was observed here that there is no toggling if all delays are null or when there is a nonnull delay only on the protein release. Additionally, the delay on RNAp release is not causing the toggling. When setting the RNAp release delay to null, oscillations persisted (data not shown). Thus, the observed toggling in Figure 7.6 is either due to the delay on the promoter or is a consequence of having a far smaller maximum number of proteins of each gene, and, thus, stochastic fluctuations cause the system toggle. Notice that if it is the second case, then it is indirectly caused by the delay on the promoter, because this is the parameter responsible for the relatively small number of proteins in comparison to the previous cases A and B.

The results from case D show that the toggling is caused by the delay on the promoter and not by having a small maximum number of proteins of the gene "on." Case D consists in a TS with transcription/translation delays only on the protein release, as in case B, but with transcription rates 100 times smaller so that the level at which the protein is in highest quantity is the same as in case C. With that goal, k_t is now set at $0.01\ s^{-1}$. Such decrease, as seen in Figure 7.7, imposes that the protein levels do not go beyond 500 as in case C, but unlike this case, no toggling was ever observed.

The conclusion is that, given no cooperative binding or self-activation, the feature that allows toggling is the time delay on the promoter release. This simple case is a good example of how critical it is to include multiple time delays in transcription and translation reactions even when modeling small gene regulatory networks and also that tuning rate constants is not sufficient, in some cases, to obtain the same results as those obtained for a system with nonnull delays.

7.7
A Model of the P53–Mdm2 Feedback Loop Network

The tumor suppressor protein P53 has a fundamental role in cellular response to a variety of environmental stresses that can affect DNA structure and replication. Depending on the causes of stress, P53 can activate several genes responsible for regulating processes such as cell cycle arrest, DNA repair, and apoptosis [48]. Mutations in the gene that transcribes p53 RNA have been found in about 50% of human tumors [49].

Under normal conditions, P53 concentrations are kept low by an Mdm2 protein [50], a downregulator of P53. These two proteins form a negative feedback loop responsible for the oscillatory dynamics in their concentrations in cells exposed to radiation that induces DNA damage [51]. Under stress, P53 concentration can rapidly increase up to 16 times the basal concentration [52].

In this example, a model is presented that reproduces many features of the experimentally observed P53–Mdm2 dynamics, triggered by the appearance of DNA structural damages, which are externally induced by irradiating cells with γ or UV radiation.

The model used here is built from an ODE model proposed in [53]. The set of coupled ODEs is converted into a set of chemical reactions. The creation of P53 and Mdm2 proteins and RNA occurs via multiple time-delayed reactions.

The promoter regions of the two genes, from which p53 RNA and mdm2 RNA are transcribed, are explicitly represented by ProP53 and ProMdm2, respectively, to allow accounting for the time the promoter takes to be released after initiation of each transcription reaction. This has a relevant impact on the dynamics of the system [38], since it will cause the promoter to be disabled for a short period of time after a transcription reaction occurs, thus acting as a limiting factor on the transcription rate.

In this model, RNA*p*s are not explicitly represented. It is assumed here that they are never depleted and exist in a sufficient quantity so that the fluctuations of the total number of RNA*p*s do not affect any reaction propensity [39]. To produce an RNA, the RNA*p* must process the set of nucleotides of the gene. Thus, a time delay for the completion of the RNA production is also introduced [10].

Since the experimental measurements are made on eukaryote cells [53], translation of an RNA can only begin after the RNA has been completely transcribed, to account for the time it takes for the RNA to leave the nucleus so that it can be translated by ribosomes. Ribosomes, like the RNA*p*s, are also not explicitly represented. A more detailed model should include these two species.

Cells suffer simple and complex DNA double-strand breaks (DSBs) when irradiated by γ radiation. Simple breaks are those where both strands are broken at the same location, while complex ones are those where they break at different locations in the strands. Cells respond differently to simple and complex breaks in the DNA strand. However, this difference does not affect the set of reactions modeled here. Thus, the dynamical response of the P53–Mdm2 feedback loop network to the appearance of a single quantity ("DSB"), which is the sum of both types of DSBs, is tested. The amount of DSBs to be introduced is randomly generated from a Poisson distribution with a mean of 50 (35 for simples breaks and 15 for complex breaks) since the sum of two random variables following a Poisson distribution also follows a Poisson distribution where the mean is the sum of the mean values of the two initial distributions.

First, DSBs are added at given moments of the simulation, in quantities randomly generated following a Poisson distribution. A first-order reaction of DSBs decay (representing that they have been repaired) is added to the model, to mimic the dynamics of the DSB quantity in time.

The oscillations of P53 and Mdm2 concentrations are activated only when DSBs appear in the system; that is, they will not begin until DSBs are added, and die out some time after no more DSBs are present in the system. The model consists of the following reactions:

$$\mathrm{DSB} \xrightarrow{\mathrm{DSB_{decay}}} \varnothing, \tag{7.27}$$

$$\mathrm{ProP53} \xrightarrow{\mathrm{Sp53/5}} \mathrm{ProP53}(\tau_{\mathrm{ProP53}}) + \mathrm{pr53}(\tau_{\mathrm{pr53}}), \tag{7.28}$$

$$\mathrm{ProP53} +^{*} \mathrm{DSB} \xrightarrow{\mathrm{4Sp53}} \mathrm{ProP53}(\tau_{\mathrm{ProP53}}) + \mathrm{pr53}(\tau_{\mathrm{pr53}}), \tag{7.29}$$

$$\mathrm{pr53} \xrightarrow{\mathrm{Gp53}} \varnothing, \tag{7.30}$$

$$\mathrm{pr53} \xrightarrow{\mathrm{rp53}} \mathrm{pr53}(\tau^{\mathrm{RBS}}_{\mathrm{p53}}) + \mathrm{P53}(\tau_{\mathrm{P53}}), \tag{7.31}$$

$$\mathrm{P53} \xrightarrow{\mathrm{up53}} \varnothing, \tag{7.32}$$

$$\mathrm{P53} +^{*} \mathrm{MDM2} \xrightarrow{\mathrm{vp53}} \varnothing, \tag{7.33}$$

$$\mathrm{ProMdm2} \xrightarrow{\mathrm{Smdm2/10}} \mathrm{ProMdm2}(\tau^{1}_{\mathrm{ProMdm2}}) + \mathrm{mdm2}(\tau_{\mathrm{mdm2}}), \tag{7.34}$$

$$\mathrm{ProMdm2} +^{*} \mathrm{P53} \xrightarrow{\mathrm{Smdm2}} \mathrm{ProMdm2}(\tau^{2}_{\mathrm{ProMdm2}}) + \mathrm{mdm2}(\tau_{\mathrm{mdm2}}), \tag{7.35}$$

$$\mathrm{ProMdm2} +^{*} \mathrm{DSB} \xrightarrow{\mathrm{2Smdm2}} \mathrm{ProMdm2}(\tau^{3}_{\mathrm{ProMdm2}}) + \mathrm{mdm2}(\tau_{\mathrm{mdm2}}), \tag{7.36}$$

$$\mathrm{mdm2} \xrightarrow{\mathrm{Gmdm2}} \varnothing, \tag{7.37}$$

$$\mathrm{mdm2} \xrightarrow{\mathrm{rmdm2}} \mathrm{mdm2}(\tau^{\mathrm{RBS}}_{\mathrm{mdm2}}) + \mathrm{MDM2}(\tau_{\mathrm{MDM2}}), \tag{7.38}$$

$$\mathrm{MDM2} \xrightarrow{\mathrm{umdm2}} \varnothing, \tag{7.39}$$

$$^{*}\mathrm{P53} + \mathrm{MDM2} \xrightarrow{\mathrm{vp53}} \varnothing. \tag{7.40}$$

The rate constants are set at $\mathrm{DSB_{decay}} = 0.003\,\mathrm{s}^{-1}$, $\mathrm{Sp53} = 0.02\,\mathrm{s}^{-1}$, $\mathrm{Gp53} = 0.02\,\mathrm{s}^{-1}$, $\mathrm{Smdm2} = 0.045\,\mathrm{s}^{-1}$, $\mathrm{Gmdm2} = 0.02\,\mathrm{s}^{-1}$, $\mathrm{rp53} = 0.6\,\mathrm{s}^{-1}$, $\mathrm{up53} = 0.02\,\mathrm{s}^{-1}$, $\mathrm{vp53} = 9.2\,\mathrm{s}^{-1}$, $\mathrm{rmdm2} = 0.04\,\mathrm{s}^{-1}$, and $\mathrm{umdm2} = 0.14\,\mathrm{s}^{-1}$. The time delays are set at $\tau_{\mathrm{mdm2}} = 100\,\mathrm{s}$, $\tau_{\mathrm{MDM2}} = 10\,\mathrm{s}$, $\tau_{\mathrm{pr53}} = 100\,\mathrm{s}$, $\tau_{\mathrm{P53}} = 10\,\mathrm{s}$, $\tau_{\mathrm{ProP53}} = 1\,\mathrm{s}$, $\tau^{1}_{\mathrm{ProMdm2}} = 1\,\mathrm{s}$, $\tau^{2}_{\mathrm{ProMdm2}} = 0.01\,\mathrm{s}$, $\tau^{3}_{\mathrm{ProMdm2}} = 0.05\,\mathrm{s}$, $\tau^{\mathrm{RBS}}_{\mathrm{mdm2}} = 1\,\mathrm{s}$, and $\tau^{\mathrm{RBS}}_{\mathrm{p53}} = 0.1\,\mathrm{s}$. Also, the following initial quantities are imposed: P53 = 0, mdm2 = 0 (mdm2 RNA), MDM2 = 0, pr53 = 0 (p53 RNA), ProP53 = 1 (promoter region of the gene from which the p53 RNA is transcribed), and ProMdm2 = 1 (promoter region of the gene from which the mdm2 RNA is transcribed).

Additionally, there are transcription reactions that require the presence of activator molecules to model the effect of the presence of some substances known to be directly or indirectly responsible for activating the expression of one of the two proteins (P53 and Mdm2). For example, reaction 7.36 models the activation of Mdm2 transcription due to the presence of DSBs in the system, since it is known that when DSBs occur, signaling molecules detect them and will then begin a cascade of events that will eventually lead to a higher expression of P53 and Mdm2 [53,54]. Also, this reaction rate constant is set at a higher value than the rate constant of reaction 7.34, which models the basal level of transcription of mdm2 RNA (no activation required).

In Figure 7.8, the time series of P53, Mdm2, and DSBs of a single experiment are shown, where DSBs are added to the system at $t = 0$ s, that is, in the beginning of the simulation.

Qualitatively, the results are in agreement to those reported in [53]. When DSBs are introduced, P53 and Mdm2 oscillate between two and five times [54]. In this simulation, shown here as example, P53–Mdm2 oscillate twice four times as a response to the addition of DSBs. Also, the oscillations are damped and their ending is rather abrupt. Once they stop, only the addition of more DSBs can "restart" the oscillations of P53 and Mdm2.

It is interesting to note that the system responds quite diversely to each addition of DSBs (data not shown). The responses to identical additions of DSBs differ significantly in amplitude and number of oscillations. The similarity between any two responses shows their period of oscillations, also in agreement with experimental observations [53]. The number and intensity of oscillations also depend, on average, on the number of DSBs added. That is, adding more DSBs will originate a stronger response (larger amplitude oscillations), on average. Also, if the DSBs have a weaker decay, resulting in them remaining in the system for a longer time interval, the oscillations will also last longer.

When measuring the P53–Mdm2 feedback loop response of cells to irradiation, usually many cells are observed at the same time. Measurements at the single cell

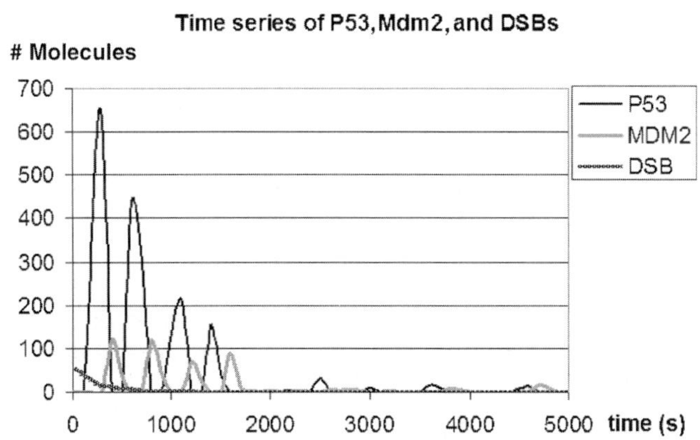

Fig. 7.8 One time series of the P53–Mdm2 feedback loop. Sampling period is 10 s. A single introduction of DSBs is made at $t = 0$ s.

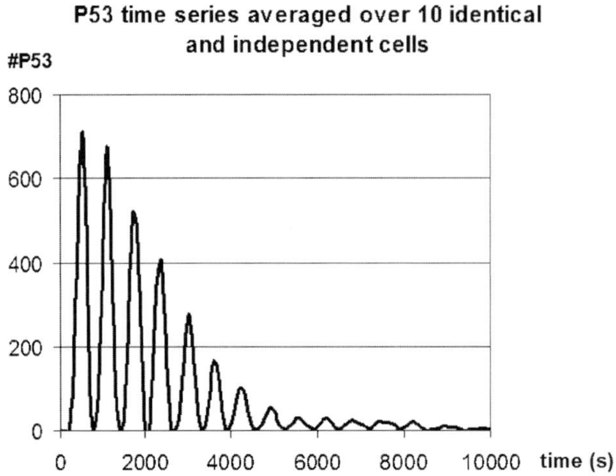

Fig. 7.9 Average quantity of P53 proteins of 10 independent cells subject to similar DSBs addition in the beginning of the simulation. Sampling period is 10 s.

level show that oscillations end abruptly, as the model here used mimics. However, observing a cell population's response, what we see is a clear damped oscillation that slowly dies out, rather than a "digital response" [53].

To mimic the observations at cell population level, several independent "cells" dynamics were simulated. Each cell contains the same set of reactions, 7.27 to 7.40, and is initially perturbed by the addition of DSBs. Thus, all cells are identical in terms of their reactions and initial quantities of substances, but each cell is subject to a distinct external perturbation (addition of DSBs).

In Figure 7.9, we have the average number of P53 proteins, observed over time in 10 independent cells.

The result is a damped oscillation, such that the amplitude decreases almost at constant rate, from one oscillation to the next one, in clear agreement with measurements from multiple cells [5], even though the P53–Mdm2 dynamics varies significantly from cell to cell, in the amplitude of the peaks and the number of oscillations.

Another very interesting observation was recently reported in [54]. When a cell divides, P53 and Mdm2 concentrations oscillate, and these oscillations continue to occur in the daughter cells at the same amplitude levels as in the mother cell. Also, the P53 and Mdm2 oscillations are highly synchronized between daughter cells of the same progenitor cell, followed by unsynchronization over time. Additionally, P53–Mdm2 oscillations in cells with different mother cells are less correlated than those from the same mother cell. The greater the separation in the lineage, the weaker the correlations.

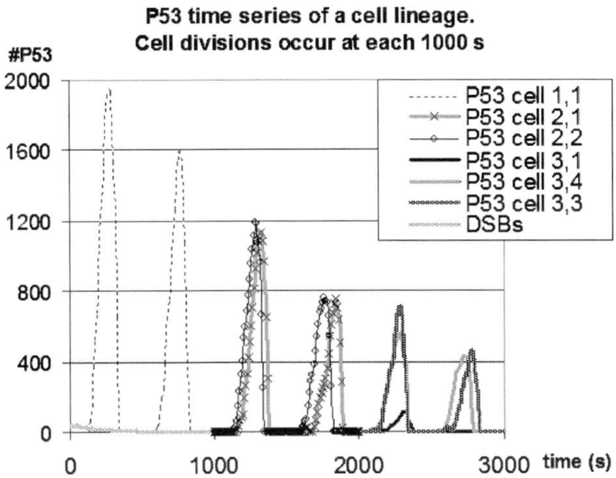

Fig. 7.10 Time series of P53 in each cell of the cell line (except cell (3,2) since its time series is very similar to cell (3,1)). At each 1000 s, cells divide and two daughter cells are created from each existing cell. Notice that cells (3,3) and (3,4) have similar dynamics (almost synchronized in both phase and amplitude) since they are daughters from the same mother cell (2,2), but are almost uncorrelated to cell (3,1) generated from a different mother cell (2,1). Sampling period is 10 s.

The results of a simulation to model this experiment are now presented. The simulation begins with a single "mother cell," with the P53–Mdm2 chemical network described. Next, its dynamics is simulated for a given time interval. At the end, two new cells are created (the daughter cells), which inherit the mother cell chemical reactions and concentrations at the moment the division occurs.

Only the mother cell is subject to an initial perturbation (by the addition of DSBs following a Poisson distribution at $t = 0$ s). Three generations of a cell lineage are simulated. The time series of the P53 protein of cells of the lineage are shown in Figure 7.10, using the following notation: a cell identified by (x, y) is cell number y of generation x.

From Figure 7.10, we observe that indeed the model mimics the observation that the oscillations in mother cells continue in the daughter cells as the experiments report [54]. Additionally, and also matching experimental measurements, as the cells get more farther apart in the cell line "tree," their P53 time-series dynamics differ more. The oscillations become more unrelated both in phase and in amplitude. In the extreme case, oscillations might have ceased in some lines, while persisting in other lines. As seen in Figure 7.10, another experimentally observed phenomenon, also present in the simulations, is that the oscillations can end abruptly at the single cell level [53]. The moment at which this abrupt ending occurs is entirely stochastic.

7.8
Summary, Conclusions, and Applications

A modeling strategy that is able to reproduce recent experimental observations of the dynamics of single genes and small gene networks was presented for gene regulatory networks.

Experimentally, it is established that gene expression is a stochastic process. Also, it is known that the stochastic nature of the reactions involved cannot be ignored, even when modeling GRNs where protein concentrations are always high during the simulation or when averaging over many simulations. Additionally, another set of experiments shows that to model reactions such as transcription and translation, one cannot ignore the time these reactions take to be concluded once initiated. Measurements of the rate at which RNA*p* molecules transcribe genes and of gene size provide reasonable estimations of the length of time delays one should introduce for these reactions, when modeling them as single-step multiple time-delayed reactions.

Simulations of models of these experiments validated the modeling strategy proposed here.

The set of examples of gene networks simulated here showed the flexibility of this modeling strategy and the richness of the resulting dynamics. The simulations are computationally feasible, and the results as realistic as currently possible. In fact, as shown [9], although the results of each simulation of a system of reaction differ due to the stochastic nature of the dynamics, each is exact; that is, it is a possible trajectory of the system in the state space, and not an approximation. The same holds when we use the delayed SSA, since generation of the time at which the next event occurs follows an exponential distribution.

A recent experiment [55] observed the dynamics of artificially constructed small gene networks with varying connectivity, between three genes and five promoters in *E. coli*. Interestingly, the behavior of such simple networks built out of a few, well-characterized components cannot always be inferred from the connectivity diagrams alone. This richness of the dynamics cannot be easily captured using other current modeling strategies. For example, each gene is unique in the sense that it can have unique rate constants, time delays, and it allows many different forms of interaction between genes. Even if identical, genes can behave differently due to stochastic fluctuations. Also, this modeling strategy allows for mimicking the key features of the random Boolean networks approach, such as having any desired topology and transfer functions. Finally, other intervening agents in GRN dynamics, such as protein networks or microRNA, can easily be incorporated in the model.

Notice that it is computationally more expensive to model and simulate a gene network dynamics using the strategy proposed here than, for example, using random Boolean networks as models of gene networks, with the same number of "genes." Simulations of gene networks of a few thousand genes, for a relatively longer experimental duration (a few minutes to hours, in system time), take a couple of hours in a Pentium 4. However, other models have not, so far, captured many features of gene networks dynamics that the methodology described here can.

Many applications are possible, provided by this framework. For example, it can be used to test, and further improve, existing algorithms of inference of structure and logic of gene networks. Also, we can use this framework to model specific gene networks, downstream chemical pathways, and then measure the effects caused by perturbations in these systems, such as the addition of gene expression inhibitors, or activators, for the purpose of, for example, driving the dynamics toward a desired goal. All these, and many other possible applications, have direct medical implications and could prove, in the near future, extremely valuable.

Acknowledgments

The author would like to thank Fred G. Biddle, Stuart A. Kauffman, Miguel Almeida, Jason Lloyd-Price, and Rui Zhu for their valuable contributions, discussions, and suggestions. The author also thanks iCORE, a funding agency of the Alberta government, Canada; the Foundation for Science and Technology of Portugal; the Institute for Biocomplexity and Informatics of the University of Calgary; and the Department of Physics and Astronomy of the University of Calgary.

References

1 Adelman, K., La Porta, A., Santangelo, T.J., Lis, J.T., Roberts, J.W. and Wang, M.D. (2002) Single molecule analysis of RNA polymerase elongation reveals uniform kinetic behavior. *Proceedings of the National Academy of Sciences of the United States of America*, **99**, 13538–13543.

2 Arkin, A., Ross, J. and McAdams, H.H. (1998) Stochastic kinetic analysis of developmental pathway bifurcation in phage λ-infected *Escherichia coli* cells. *Genetics*, **149**, 1633–1648.

3 Atkinson, M.R., Savageau, M.A., Myers, J.T. and Ninfa, A.J. (2003) Development of genetic circuitry exhibiting toggle switch or oscillatory behavior in *E. coli*. *Cell*, **113**, 597–607.

4 Vogelstein, B., Lane, D. and Levine, A.J. (2000) Surfing the p53 network. *Nature*, **408**, 307–310.

5 Lewis, J., Raff, M., Roberts, K., Alberts, B., Johnson, A. and Walters, P. (2004) *Molecular Biology of the Cell*, 4th edn, Garland Science, New York.

6 Bar-Or, R.L., Maya, R., Segel, L.A., Alon, U., Levine, A.J. and Oren, M. (2000) Generation of oscillations by the p53–mdm2 feedback loop: a theoretical and experimental study. *Proceedings of the National Academy of Sciences of the United States of America*, **97** (21), 11250–11255.

7 Becskei, A. and Serrano, L. (2000) Regulation of noise in the expression of a single gene. *Nature*, **405**, 590–593.

8 Bennett, W., Hussain, S., Vahakangas, K., Khan, M., Shields, P. and Harris, C. (1999) Molecular epidemiology of human cancer risk: gene–environment interactions and p53 mutation spectrum in human lung cancer. *The Journal of Pathology*, **187** (1), 8–18.

9 Blake, W.J., Kaern, M., Cantor, C.R. and Collins, J.J. (2003) Noise in eukaryotic gene expression. *Nature*, **422**, 633–637.

10 Bliss, R.D., Painter, P.R. and Marr, A.G. (1982) Role of feedback inhibition in stabilizing the classical operon. *Genome Informatics*, **97**, 177–193.

11 Bratsun, D., Volfson, D., Tsimring, L.S. and Hasty, J. (2005) Delay-induced stochastic oscillations in gene regulation. *Proceedings of the National Academy of Sciences of the United States of America*, **102**, 14593.

12 Buchholtz, F. and Schneider, F.W. (1987) Computer simulation of t3/t7 phage infection using lag times. *Biophysical Chemistry*, **26**, 171–179.

13 Busenberg, S.N. and Mahaffy, J.M. (1985) Interaction of spatial diffusion and delays in models of genetic control by repression. *Journal of Mathematical Biology*, **22**, 313–333.

14 Busenberg, S.N. and Mahaffy, J.M. (1988) The effects of dimension and size for a compartmental model of repression. *SIAM Journal on Applied Mathematics*, **48**, 882–903.

15 Chang, H.H., Oh, P.Y. and Ingber, D.E. and Huang, S. (2006) Multistable and multistep dynamics in neutrophil differentiation. *BMC Cell Biology*, **7** (11).

16 Davenport, R.J., White, G.J.L., Landick, R. and Bustamante, C. (2000) Single-molecule study of transcriptional pausing and arrest by *E. coli* RNA polymerase. *Science*, **287**, 2497–2500.

17 Elowitz, M.B. and Leibler, S. (2000) A synthetic oscillatory network of transcriptional regulators. *Nature*, **403**, 335–338.

18 Elowitz, M.B., Levine, A.J., Siggia, E.D. and Swain, P.S. (2002) Stochastic gene expression in a single cell. *Science*, **297**, 1183.

19 Gaffney, E.A. and Monk, N.A.M. (2006) Gene expression time delays and Turing pattern formation systems. *Bulletin of Mathematical Biology*, **68**, 99–130.

20 Gardner, T.S., Cantor, C.R. and Collins, J.J. (2000) Construction of a genetic toggle switch in *Escherichia coli*. *Nature*, **403**, 339–342.

21 Geva-Zatorsky, N., Rosenfeld, N., Itzkovitz, S., Milo, R., Sigal, A., Dekel, E., Yarnitzky, T., Liron, Y., Polak, P., Lahav, G. and Alon, U. (2006) Oscillations and variability in the p53 system. *Molecular Systems Biology*, **2**, 0033.

22 Gillespie, D.T. (1976) A general method for numerically simulating the stochastic time evolution of coupled chemical reactions. *Journal of Computational Physics*, **22**, 403–434.

23 Gillespie, D.T. (1977) Exact stochastic simulation of coupled chemical reactions. *Journal of Physical Chemistry*, **81**, 2340–2361.

24 Guet, C.C., Elowitz, M.B. and Hsing, W. and Leibler, S. (2002) Combinatorial synthesis of genetic networks. *Science*, **296**, 1466–1470.

25 Haupt, Y., Maya, R., Kazaz, A. and Oren, M. (1999) Mdm2 promotes the rapid degradation of p53. *Nature*, **387**, 296–299.

26 Oliner, J.D., Pietenpol, J.A., Thiagalingam, S., Gyuris, J., Kinzler, K.W. and Vogelstein, B. (1993) Oncoprotein mdm2 conceals the activation domain of tumour suppressor p53. *Nature*, **362** (6423), 857–860.

27 Kaern, M., Elston, T.C., Blake, W.J. and Collins, J.J. (2005) Stochasticity in gene expression: from theories to phenotypes. *Nature Reviews Genetics*, **6**, 451–464.

28 Kauffman, S.A. (1969) Metabolic stability and epigenesis in randomly constructed genetic nets. *Journal of Theoretical Biology*, **22**, 437–467.

29 Kauffman, S.A. (2004) A proposal for using the ensemble approach to understand genetic regulatory networks. *Journal of Theoretical Biology*, **230** (4), 581–590.

30 Ambros, V., Lee, R.C. and Feinbaum, R.L. (1993) The *C. elegans* heterochronic gene lin-4 encodes small RNAs with antisense complementarity to lin-14. *Cell*, **75**, 843–854.

31 Lewis, J. (2003) Autoinhibition with transcriptional delay: a simple mechanism for the zebrafish somitogenesis oscillator. *Current Biology*, **13**, 1398–1408.

32 Lipshtat, A., Loinger, A., Balaban, N.Q. and Biham, O. (2006) Genetic toggle switch without cooperative binding. *Physical Review Letters*, **96**, 188101.

33 Ma, L., Wagner, J., Rice, J.J., Hu, W., Levine, A.J. and Stolovitzky G.A. (2005) A plausible model for the digital response of p53 to DNA damage. *Proceedings of the National Academy of Sciences*, **102** (4), 14266–14271.

34 McAdams, H.H. and Arkin, A. (1997) Stochastic mechanisms in gene expression. *Proceedings of the National Academy of Sciences of the United States of America*, **94**, 814–819.

35 Mestl, T., Plahte, E. and Omholt, S.W. (1995) A mathematical framework for describing and analysing gene regulatory networks. *Journal of Theoretical Biology*, **176**, 291–300.

36 Monk, N.A.M. (2003) Oscillatory expression of Hes1, p53, and NFκB driven by transcriptional time delays *Current Biology*, **13**; 1409–1413.

37 Monod, J. and Jacob, F. (1961) Teleonomic mechanisms in cellular metabolism, growth, and differentiation. *Cold Spring Harbor Symposia on Quantitative Biology*, **26**, 389–401.

38 Ota, K., Yamada, T., Yamanishi, Y., Goto, S. and Kanehisa, M. (2003) Comprehensive analysis of delay in transcriptional regulation using expression profiles. *Genome Informatics*, **14**, 302–303.

39 Ozbudak, E.M., Thattai, M., Kurtser, I., Grossman, A.D. and van Oudenaarden, A. (2002) Regulation of noise in the expression of a single gene. *Nature Genetics*, **31**, 69–73.

40 Ptashne, M. (1992) *Genetic Switch: Phage λ and Higher Organisms*. Cell Press and Blackwell Scientific Publications, Cambridge, MA.

41 Raser, J.M. and O'Shea, E.K. (2004) Control of stochasticity in eukaryotic gene expression. *Science*, **304**, 1811–1814.

42 Raser, J.M. and O'Shea, E.K. (2005) Noise in gene expression: origins, consequences, and control. *Science*, **309**, 2010–2013.

43 Ribeiro, A.S. (2007) Effects of coupling strength and space on the dynamics of coupled toggle switches in stochastic gene networks with multiple-delayed reactions. *Physical Review E*, **75** (6), 061903.

44 Ribeiro, A.S. and Kauffman, S.A. (2007) Noisy attractors and ergodic sets in models of genetic regulatory networks. *Journal of Theoretical Biology*, **247** (4), 743–755.

45 Ribeiro, A.S. and Lloyd-Price, J. (2007) SGN Sim a stochastic genetic networks simulator. *Bioinformatics*, **23** (6), 777–779.

46 Ribeiro, A.S., Zhu, R. and Kauffman, S.A. (2006) A general modeling strategy for gene regulatory networks with stochastic dynamics (extended version). *Journal of Computational Biology*, **13**(9), 1630–1639.

47 Roussel, M.R. and Zhu, R. (2006) Validation of an algorithm for delay stochastic simulation of transcription and translation in prokaryotic gene expression. *Physical Biology*, **3**, 274–284.

48 Roussel, M.R. (1996) The use of delay differential equations in chemical kinetics. *Journal of Physical Chemistry*, **100**, 8323–8330.

49 Süel, G.M., Garcia-Ojalvo, J., Liberman, L.M. and Elowitz, M.B. (2006) An excitable gene regulatory circuit induces transient cellular differentiation. *Nature*, **440**, 545–550.

50 Süel, G.M., Kulkarni, R.P., Dworkin, J., Garcia-Ojalvo, J. and Elowitz, M.B. (2007) Tunability and noise dependence in differentiation dynamics. *Science*, **315**, 1717–1719.

51 Tolic-Nørrelykke, S.F., Engh, A.M., Landick, R. and Gelles, J. (2004) Diversity in the rates of transcript elongation by single RNA polymerase molecules. *Journal of Biological Chemistry*, **279**, 3292–3299.

52 Veflingstad, S.R., Plahte, E. and Monk, N.A.M. (2005) Effect of time delay on pattern formation: competition between homogenisation and patterning. *Physica D*, **207**, 254–271.

53 Crick, F. and Watson, J. (1953) Molecular structure of nucleic acids: a structure for deoxyribose nucleic acid. *Nature*, **171** (4356), 737–738.

54 Yu, J., Xiao, J., Ren, X., Lao, K. and Xie, S. (2006) Probing gene expression in live cells, one protein molecule at a time. *Science*, **311**, 1600–1603.

55 Zhu, R., Ribeiro, A.S., Salahub, D. and and Kauffman, S.A. (2007) Studying genetic regulatory networks at the molecular level: delayed reaction stochastic models. *Journal of Theoretical Biology*, **246** (4), 725–745.

8
Probabilistic Boolean Networks as Models for Gene Regulation

Yufei Huang and Edward R. Dougherty

8.1
Introduction

Understanding the role of genes in a biological system has been a major focus in life science research. Genes carry out their functions mainly in the form of *gene regulation* [1,2]. Gene regulation refers to the regulatory relations among genes that control the timing and the amount of gene expression. Messenger RNA (mRNA) and proteins are the two types of gene expression. When genes are expressed, the DNA that codes genes must be transcribed into mRNA (*transcription*) and the mRNA then is translated into proteins (*translation*), the building blocks of cells. Gene expression must be tightly regulated. Regulating gene expression is a vital process in the cell, which ultimately determines the cell structure and cell function. For instance, gene regulation is the basis of cellular differentiation, morphogenesis, apoptosis, and the versatility and adaptability found in an organism. Gene regulation is an active area of research in developmental biology, the biology of aging, and genetic diseases research; and is pivotal in cancer research. For instance, the transcriptional factor p53 is known to control cell death, or apoptosis, and activate cell suicide genes in cancer cells to prevent harmful mutations from being passed on. When p53 is defective or missing, cancers are more likely to occur. Despite the massive ongoing research effort, much is yet to be understood about gene regulation. For instance, even in an eukaryotic organism such as malaria parasite, 60% of the open reading frames are annotated to have "hypothetical" functions, let alone prokaryotic organisms. And here we must be extremely cautious about claiming a correspondence between a particular gene and its particular function. Genes interact in a highly complex nonlinear network, subject to feedback and to both internal and external cellular conditions. It is, in fact, the collective behavior of many genes that determines a cellular function, and this determination often depends on the cellular context.

The recent breakthroughs in the Human Genome Project and the subsequent mapping of genome sequences for around 180 organisms have, for the first time, spelled out the "dictionaries" of genes for human beings and many other organisms, thereby providing the necessary elements for studying system-level gene regulation.

Analysis of Microarray Data: A Network-Based Approach. Edited by F. Emmert-Streib and M. Dehmer

ISBN: 978-3-527-31822-3

At the same time, high-throughput biotechnologies, such as the mRNA expression microarray [3], mass spectrometry [4], the ChIP–Chip [5], and the protein microarray [6] produce transcriptomic and proteomic data that measure mRNA expressions, transcriptional factor binding information, protein abundance information, and protein–protein interactions at a system-wide level. The proliferation of data represents a shift in the landscape of life science research to the new era of computational systems biology, an era that underpins a system-level understanding of gene function through the extensive involvement of computational methods [7–11].

A primary issue in the computational study of system-level gene regulation is the mathematical modeling of regulatory events. Mathematical models are constructed to quantify the outcome of gene regulation and summarize the underlying biological mechanisms with mathematical functions. Even though mathematical modeling has long been a part of biological research, it comes to the fore in computational systems biology by focusing on system-level modeling. This calls for models that can describe intermolecular, and even intercellular relationships, in a multiple molecular–cellular system. These models are often referred to as *mathematical network models*, which make it possible to characterize causes of gene regulation, predict future activities of a cell system, and suggest intervention designs to alter system dynamics. For example, to reverse-engineer gene regulatory systems, a network model of gene regulation is first assumed, and then the features of the regulatory model, including regulatory connectivity and strength, are mathematically inferred based on experimental data. Future behavior of a regulatory system can be studied via predictions acquired from simulating the model network under specified conditions. Alternatively, we might characterize the long-run behavior of a system under different initial conditions or, in a related vein, apply control theory to design an optimal intervention strategy for the favorable altering of long-run behavior, say, away from metastasis in cancer cells.

In the process of mathematical (statistical) modeling of gene regulatory networks (GRNs), four aspects must be carefully scrutinized: accuracy, robustness, adaptivity, and a sense of objectivity. Accuracy dictates that models should be as faithful as possible to the underlying biology. This calls for employing increasingly complex models along with more advanced mathematical techniques. However, model complexity is limited by two factors: greater complexity requires more data and more finely quantized data. Moreover, it often leads to computational intractability, for instance, when applying dynamic programming to a network. Closely related to inference limitations on network complexity is robustness: a network should be robust with respect to design assumptions, which means that small changes in the assumptions under which the network is designed should not lead to catastrophic changes in model dynamics. In practice, the process of network inference is iterative: computational results will guide subsequent experiments, and the new experimental results will in turn be integrated into computations to improve the computational results. Consequently, models should be adaptive, that is, we must be able to effectively modify and extend their current structures to reflect the changes in knowledge. Finally, a sense of purpose is also important when choosing among different models. Differential equation models may be preferable for characterizing and simulating a gene regulatory process from the aspect of the biochemical reactions; in contrast, often coarse models such as Boolean networks defined on discrete states (the normal, the overexpressed, the underexpressed, etc.) are more appropriate for clinical purposes, because clinical objectives are naturally defined by a countable set of concepts, or data may naturally occur categorically.

According to De Jong [12], the existing gene regulation models can be classified hierarchically into different types based on their precision and complexities. At the bottom of the hierarchy are graph models, including most notably Bayesian networks, which use nodes to denote genes in a system and probabilistic distributions to represent gene interactions [13,14]. The graphical models are often intended to characterize the static behavior of the biological system of interest. One level up the hierarchy are the (probabilistic) Boolean networks [15,16]. Although Boolean networks admit only the discrete states of gene expression, they nevertheless model explicitly the (nonlinear) mechanism and the dynamic behavior of gene regulation, thus making a balanced trade-off between accuracy and complexity. At the top of the hierarchy are ordinary differential equations [17] and stochastic master equations [12,18], which provide a fine, continuous-time description of the dynamics and the stochastic outcome of biochemical reactions.

In this chapter, we provide an overview of probabilistic Boolean networks (PBNs) as a model for gene regulation. In Section 8.2, we introduce the basic concept of a PBN, its mathematical formulation, and the extensions to the context sensitive PBNs and PBNs with perturbation. We then focus in Section 8.3 on two Bayesian approaches for the construction of PBNs based on microarray data. In Section 8.4, we discuss in detail two optimal control problems using PBNs: finite-horizon control and infinite-horizon control. In each section, we provide examples of melanoma applications to illustrate the effectiveness of the discussed algorithms.

8.2 Modeling Gene Regulation with Probabilistic Boolean Networks

8.2.1 Preliminaries

The purpose of modeling system-level gene regulation is to map the biological activities of a gene regulatory system to a mathematical system that can capture (1) the regulatory interactions between genes of the target system and (2) the dynamics of gene regulation. Concepts of system-level modeling have been extensively studied in engineering and can be borrowed for the purpose of modeling gene regulatory systems. In engineering, a system consists of input signals, output signals, and a set of mathematical functions that relate the outputs to the inputs. The inputs are driving signals, and the outputs are responses of the system to the inputs. The system functions, describing the instantaneous relationship between the inputs and the outputs, accept the input signals as variables and calculate the system outputs. The block diagram of a perceived regulatory system

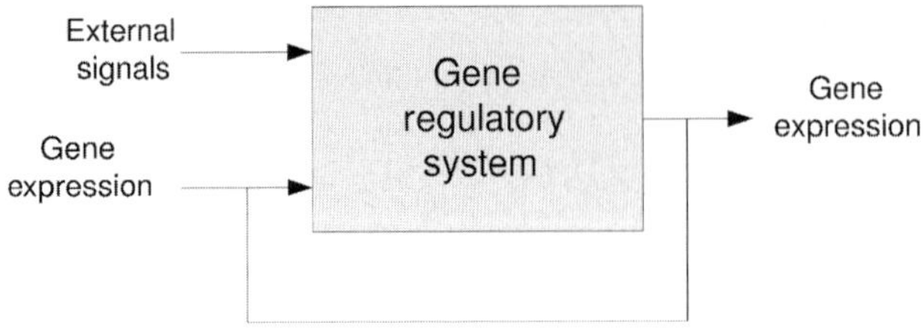

Figure 8.1 Block diagram of a gene regulatory system.

is shown in Figure 8.1. The outputs of a gene regulatory system are gene expression, for example, mRNA levels corresponding to those measured by microarray or other expression-based experiments. Since regulation is a mutual relationship among genes, feedback loops exit in a gene regulatory system. A feedback loop feeds outputs back to the inputs of the system. As a result, the inputs to the system include gene expressions fed back from the outputs. Additional inputs can arise from external chemical signals such as hormones that can intervene and alter short-term as well as long-term steady-state gene regulation. Such feedback loops, together with the signaling pathway, determine the dynamics of a regulatory system.

To define the system block for a gene regulatory network, a set of regulatory functions must be specified to model the input–output behavior. Given independent outputs, an illustration of the system block is shown in Figure 8.2. For a regulatory system consisting of G genes, there will be G regulatory functions and the gth function, F_g, admits the input expression levels of all or part of the G genes, along with the external signals as independent function variables, and calculates the output expression level of gene g. The most important issue in modeling is to define the form of the functions to reflect the gene regulatory system behavior as faithfully as possible, while at the same time taking into consideration modeling efficiency and robustness.

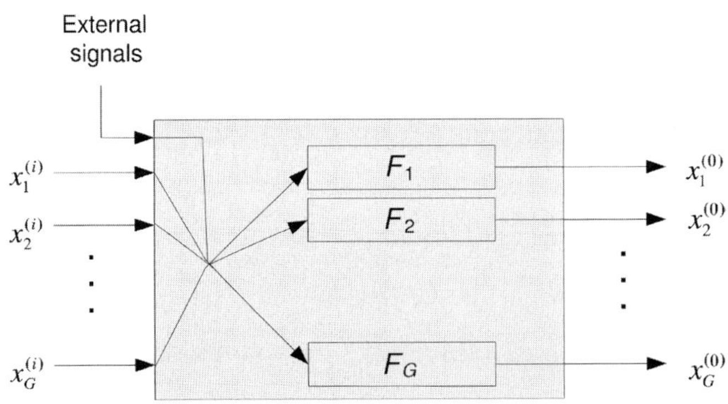

Figure 8.2 An illustration of system block in a gene regulatory system.

In a real biological system, gene regulation is continuous in time. In practice, it is possible to measure the system at discrete time instances, only. In electrical engineering, the process of converting a continuous-time system into a discrete time system is known as *sampling* [19]. Sampling frequency, that is, the frequency the discrete measurements are acquired, is a critical factor to consider when performing sampling. The discrete samples hopefully carry almost the same information as carried by the regular system; that is, we would like to have no, or minimal, information loss after sampling. The famous theory that guides the choice of the sampling frequency is known to the signal processing community as the Nyquist theorem: $f_s \geq 2f_{max}$, which says that the sampling frequency, f_s, must be greater or equal to twice f_{max}, the maximum fundamental frequency of the system. Despite its ubiquity in electrical engineering, the theorem is rarely consulted when sampling a biological system. Although this may be partly due to its anonymity to experimental biologists, the practical issue is that while economic constraints limit experimental procedures, the inability to obtain biological samples puts constraints on the sampling rate. These constraints notwithstanding, there has been research investigating the proper sampling frequency for time-series microarray experiments [20]. In this chapter, we are concerned with discrete-time regulatory systems, in which the system inputs and outputs are measured at discrete-time instances, and which do not take into account the effects of the sampling rate as they are based on practical limitations.

In addition to sampling, to model gene regulation with PBNs, gene expression measurements must also be quantized before they can be used for modeling. Through quantization, a real expression measurement is mapped into discrete levels. In a strict sense, a Boolean network admits only binary expression levels with 0 representing that the gene is not expressed and 1 that it is expressed; however, the extension to PBNs allows more than two quantization levels [15]. A particularly important case of a gene regulatory system arises in the case of ternary quantization, where the expression levels take on the values 1 (upregulated), -1 (downregulated), and 0 (invariant). We will use the ternary setting while considering the application of external control to melanoma. The terminology "probabilistic Boolean network" is applied to GRNs in general, under the supposition that the logical character of the networks is at issue. Quantization is an approximation progress that introduces loss of information; nevertheless, it can make the modeling more robust. The key to performing quantization is to determine a set of mapping rules that can minimize the loss of information. A few approaches have been proposed in the literature [21,22]. Without the loss of generality, we assume a binary quantization to the gene expression measurements and therefore the expression of a gene in a PBN will take on the values 0 and 1 only.

We conclude this subsection by introducing the concept of *system states*. A system state refers to a collection of system inputs or outputs measured at a sampling instant. In the context of a regulatory system, it corresponds to a collection of the input/output expression levels of all genes. Depending on the possible expression levels that a gene can visit, the system state space can be either finite or infinite. For a binary PBN in which each gene admits only two possible expression levels, the system state space is finite and contains a total of 2^G states. A particular state in the state space is a G-bit vector, often called a *gene activity profile* (GAP).

8.2.2
Probabilistic Boolean Networks

As discussed earlier, the key issue in modeling is to define the regulatory functions F_g (for any g) that can model the dynamics of gene regulatory relationships. PBNs were introduced in [15] to characterize the dynamics of large-scale gene regulation. Instead of modeling the kinetics of underlying chemical reactions, PBNs focus on the switch-like behavior of gene regulation and use logical (Boolean) functions as regulatory functions. The switch-like behavior of gene regulation is evident and is important when cells respond to external signals and when cells need to move from one state to another in a normal growth process. More generally, switch-like behavior is required in the discrete decision-making processes of the cell.

In defining a PBN, for the moment, we consider a feedback-only system and ignore external signals. In this case, the system input is the output fed back at the previous time instant. To put the concept on a firm mathematical ground, we define $\mathbf{x}(t) = [x_1(t), x_2(t), \ldots, x_G(t)]^T \in \{0, 1\}^G$ as the input binary state vector at time t and thus $\mathbf{x}(t+1) = [x_1(t+1), x_2(t+1), \ldots, x_G(t+1)]^T \in \{0, 1\}^G$ represents the output binary state vectors measured at time instant $t+1$ in response to $\mathbf{x}_2(t)$. There are totally $N_s = 2^G$ possible states in the system state space. For ease of composition, we will also denote a particular state by its decimal representation in the following. For example, the state 01 000 in a five-gene system will also be referred to as state eight. We further assume that, for each gene g, there are K_g Boolean functions and let f_{gk} represent the kth Boolean function of gene g, whose binary output is computed according to

$$b_{gk} = f_{gk}(\mathbf{x}(t)), \tag{8.1}$$

where $b_{gk} \in \{0, 1\}$. The original motivation for associating a collection of functions with each gene was to incorporate the uncertainty of system identification into the model. At time instant t, the PBN computes the output for each gene g according to the kth Boolean function with the probability c_{gk}, where $\sum_{k=1}^{K_g} c_{gk} = 1$. This is to say that, to calculate the output, a number between 1 and K is drawn according to the discrete distribution $\{c_{g1}, \ldots, c_{gK_g}\}$ with c_{gk} being the probability of selecting k. If k is selected, then f_{gk} will be used to compute the output. An illustration of the function F_g is shown in Figure 8.3. Since F_g is probabilistic, the input–output relationship must be defined by the conditional probability density function.

$$\begin{aligned} p(\mathbf{x}(t+1)|\mathbf{x}(t)) &= \prod_{g=1}^{G} p(x_g(t+1)|\mathbf{x}(t)) \\ &= \prod_{g=1}^{G}\sum_{k=1}^{K_g} p(x_g(t+1)|\, f_{gk} \text{ is chosen}, \mathbf{x}(t))\, p(f_{gk}) \\ &= \prod_{g=1}^{G}\sum_{k=1}^{K_g} \delta(x_g(t+1) - b_{gk}) c_{gk} \end{aligned} \tag{8.2}$$

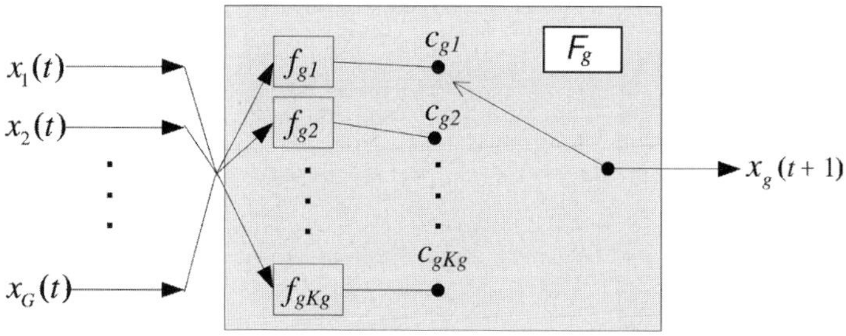

Figure 8.3 An illustration of the system function of gene g in a PBN.

where $\delta(\cdot)$ is the Kroneker Delta function and b_{gk} is the output of f_{gk} given the input $\mathbf{x}(t)$, which is defined by (8.1). In practice, the input–output Boolean relationship is more conveniently presented by a truth table.

We illustrate the preceding discussion by an example. Suppose we are given a PBN consisting of three genes x_1, x_2, and x_3. Gene 1 is associated with two Boolean functions f_{11} and f_{12} with the selection probability $c_{11}=0.5$ and $c_{12}=0.5$; gene 2 is associated with only one function f_{21}; gene 3 is again associated with two functions f_{31} and f_{32} with the respective selection probability $c_{31}=0.7$ and $c_{32}=0.3$. The input–output relationships of these functions are given by the truth table (Table 8.1). For instance, given the input state 110, the output expression state for gene 1 will be 1 no matter if f_{11} or f_{12} is chosen. However, for gene 3, the output can be 1 if f_{31} is chosen and 0 if f_{32} is chosen. Now, for the same input, the probability of producing 101 at the output can be calculated based on the truth table according to (8.2) as

$$p(100|110) = (c_{11}+c_{12})c_{21}c_{31} = (0.5+0.5)\times 1\times 0.7 = 0.7.$$

Now let us investigate the dynamic behavior of the feedback-only regulatory system modeled by PBN. In this system, the dynamics is due to the feedback of the outputs to the input, that is, the inputs to the system at time $t+1$ are actually the

Table 8.1 Truth table.

$x_1x_2x_3$	f_{11}	f_{12}	f_{21}	f_{13}	f_{23}
000	0	0	0	0	0
001	0	0	1	0	0
010	1	1	1	0	0
011	0	0	0	1	1
100	1	1	1	1	0
101	1	1	0	1	0
110	1	1	0	1	0
111	1	1	1	1	1
c_{gk}	0.5	0.5	1	0.7	0.3

output expression levels at time t. As a result, provided an initial input expression $\mathbf{x}(0)$, the system can evolve by itself according to F_g (for all g). Since the output at $t+1$ depends only on the output at t, this system exhibits the *Markov property* and its evolution can be defined by a discrete time-homogenous Markov chain with 8.2 as the transition probability. The dynamic behavior of the system is measured by the state distribution over a sequence of time instances. Now, let $\pi(t)$ represent an $N_S \times 1$ state distribution vector at time t, whose ith element specifies the probability of the chain visiting state i at time t. Also, as in the convention of the Markov chain theory, we define an $N_S \times N_S$ transition matrix $\mathbf{P}$, whose ijth element p_{ij} denotes the transition probability of the output state j given the input state i. Then the state distribution vector at time $t+1$ can be calculated by

$$\pi(t+1) = \mathbf{P}\pi(t). \tag{8.3}$$

In the context of network theory, states are viewed as nodes and transitions as arcs. The one-step transition among all possible states can be presented by the state transition diagram. The state transition diagram of the Markov chain corresponding to the PBN with truth table (Table 8.1) is shown in Figure 8.4. As can be seen from 8.3 and the state transition diagram, the dynamic behavior of the PBN depends on the structure of the transition matrix $\mathbf{P}$. We next discuss two different types of dynamic behavior of interest to gene regulation.

The first type is due to absorbing Markov chains. In an absorbing Markov chain, there are subsets of states, from which the transition probabilities of moving away are 0. Such a subset is also referred to as *attractor* cycle and states in attractor cycles

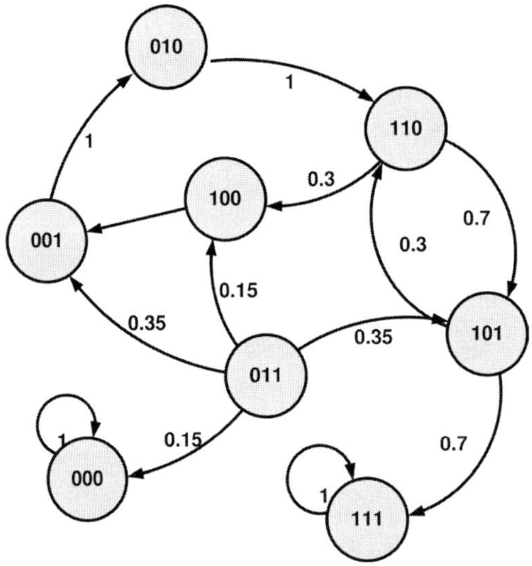

Figure 8.4 The state transition diagram of the Markov chain corresponding to the PBN with truth table (Table 8.1).

are called attractors, which play a key role in Boolean networks. Given a starting state, the network will enter an attractor cycle in a finite number of steps and cycle endlessly among the states in the cycle. The sets $\{0\,0\,0\}$ and $\{1\,1\,1\}$ in Figure 8.4 are absorbing sets (attractor cycles), and are attractors since both are singleton cycles. The set of states leading to an attractor cycle is called the *basin* of attraction. In contrast, nonattractor states are transient. They are visited not more than once on any network trajectory. Attractors characterize the long-run behavior of a Boolean network. Biologically, attractors have been conjectured to characterize phenotypes [23].

A second type of behavior corresponds to the time asymptotic behavior of an ergodic Markov chain. If the Markov chain defined by the transition matrix **P** is *irreducible* and *positive recurrent*, then the underlying gene regulatory system possesses a steady-state distribution, uniquely defining the probability distribution over the state space in the long run. In this case, if we simulate the gene regulation evolution from an arbitrary initial state according to (8.7) for a very long time, the gene regulatory system will visit states in the state space with a unique steady-state distribution. Now, let $\pi = [\pi_1, \pi_2, \ldots, \pi_{N_s}]^T$ be a vector describing the steady-state distribution, with π_i being the probability of the system visiting the ith state. Then, the steady-state distribution satisfies

$$\pi = \mathbf{P}\pi. \tag{8.4}$$

In a PBN corresponding to an ergodic Markov chain, all the states can be visited in finite time steps. As a result, no attractors exist for ergodic chains.

8.2.2.1 Context-Sensitive PBNs and PBNs with Random Perturbation

We now extend the feedback-only system to allow external signals. The external signals will affect the state transition and ultimately the dynamic behavior of the PBN. We discuss two scenarios. The first scenario, to allow an addition control on the selection of Boolean functions, is a generalization of the PBN discussed above. Specifically, at time t, an external binary signal s is applied with probability $p(s=1)=\lambda$. Particularly, if $s=1$, a *new* set of Boolean network functions is selected according to the selection probabilities for the state transition at time $t+1$; otherwise, the network functions remain the same at $t+1$. Note that, when $s=1$, the current network cannot be selected again for $t+1$. Now let $\mathbf{F}(t)=[F_1(t), \ldots, F_G(t)] \in F$, where F is the set of all possible regulatory-function vectors that G genes can adopt, and the size of F is N_f. If we assume that there are K_g functions associated with each gene g and that F is composed of the Cartesian product of the G regulatory sets for the genes in the network, then $N_f = \prod_{g=1}^{G} K_g$, but this need not be the case.

If $\mathbf{f}_k$ is the kth function vector in F, then, at $t+1$, the probability of selecting $\mathbf{f}_k$ given $s=1$ is

$$p(\mathbf{F}(t+1) = f_k | \mathbf{F}(t) = \mathbf{f}_l, s=1) = \frac{c_k}{\sum_{i=1, i\neq l} c_i}(1-\delta(\mathbf{f}_k - \mathbf{f}_l)). \tag{8.5}$$

This selection process implies Markov behavior in time on the network functions $\mathbf{F}(t)$ and the transition probability can be expressed as

$$p(\mathbf{F}(t+1) = f_k|\mathbf{F}(t) = \mathbf{f}_l) = \sum_{i\in\{0,1\}} p(\mathbf{F}(t+1) = \mathbf{f}_k|s = i, \mathbf{F}(t) = \mathbf{f}_l)\, p(s = i)$$
$$= \frac{c_k}{\sum_{i=1,i\neq l} c_i}(1-\delta(\mathbf{f}_k-\mathbf{f}_l))\lambda + \delta(\mathbf{f}_k-\mathbf{f}_l)(1-\lambda), \tag{8.6}$$

where $k, l \in= N_f = \{1, \ldots, N_f\}$. The second equality in (8.6) is arrived as a result that when $s = 0$ there is no new network generated at $t+1$, that is, $\mathbf{f}_k = \mathbf{f}_l$.

The original PBN corresponds to the case of $\lambda = 1$, and the original model is referred to as an *instantaneously random* PBN. In contrast, a PBN for which $\lambda < 1$ is referred to as a *context-sensitive* PBN. The context-sensitive PBN can model gene regulatory systems influenced by external signals. Given practical limitations on the size of a network model, there will almost always be external variables influencing genes in the model, and changes in these variables will appear as random changes in the function structure within the network. The external condition may remain unchanged for a period of time and the regulatory structure should remain unchanged accordingly. A PBN with smaller λ is used to model a system with less frequent changes in the regulatory structure. Therefore, for a stable system, a small λ should be assumed.

The dynamics of a context-sensitive PBN can still be understood by the corresponding Markov chain; however, due to the probabilistic external control in function selection, the transition distribution of the corresponding Markov chain changes with time and thus the Markov chain is no longer time homogeneous. Nonetheless, if we augment the state variable by including the system function $F_g(t)$ (for and g), then we will have a time-homogeneous Markov chain on the augmented state variable $[\mathbf{x}(t), \mathbf{F}(t)]$. The transition probability on the augmented state space is expressed as

$$\begin{aligned}
&p(\mathbf{x}(t+1), \mathbf{F}(t+1) = \mathbf{f}_k|\mathbf{x}(t), \mathbf{F}(t) = \mathbf{f}_l) \\
&\quad = p(\mathbf{x}(t+1)|\mathbf{x}(t), F(t) = \mathbf{f}_l)\, p(\mathbf{F}(t+1) = \mathbf{f}_k|\mathbf{x}(t), \mathbf{F}(t) = \mathbf{f}_l) \\
&\quad = \delta(\mathbf{f}_k(\mathbf{x}(t))-\mathbf{x}(t+1))\left(\frac{c_k}{\sum_{i=1,i\neq l} c_i}(1-\delta(\mathbf{f}_k-\mathbf{f}_l))\lambda + \delta(\mathbf{f}_k-\mathbf{f}_l)(1-\lambda)\right),
\end{aligned} \tag{8.7}$$

where the use of the delta function in the last equality is due to the fact that the transition probability is nonzero except that $\mathbf{x}(t)$ can move to $\mathbf{x}(t+1)$ according to function $\mathbf{F}(t+1) = \mathbf{f}_k$ in one step.

A second extension of the PBN definition results when a random perturbation is applied to the state of each gene in addition to the control signal s. In the Boolean context, the perturbation results in a flip of the gene state from 1 to 0, or vice versa.

This randomness applied to genes is biologically meaningful. It is known that genes may become either activated or inhibited due to external stimuli such as mutagens, heat stress, and so on. Mathematically, each gene is now assumed to be also controlled by an independent binary perturbation signal r_g, where $r_g = 1$ with probability q and $r_g = 0$ with $1 - q$. When $r_g = 1$, the state of gene g is flipped; otherwise, no perturbation is imposed. Perturbation affects the transition of gene states from t to $t = 1$. In particular, the transition will be determined by the network function **F** only when there is no perturbation, which has the probability $(1 - q)^G$. Otherwise, the new system state at $t + 1$ will be defined by perturbation and, if $r_g = 1$ for h genes, states of h genes will be flipped and states of the remaining $G - h$ genes will remain unchanged. Given $\mathbf{F}(t) = \mathbf{f}_l$, we can express the transition probability of system states as

$$
\begin{aligned}
&p(\mathbf{x}(t+1)|\mathbf{x}(t), \mathbf{F}(t) = \mathbf{f}_l) \\
&\quad = p(\mathbf{x}(t+1)|\mathbf{x}(t), \mathbf{F}(t) = \mathbf{f}_l, \text{no perturbation})\, p(\text{no perturbation}) \\
&\qquad + p(\mathbf{x}(t+1)|\mathbf{x}(t), \mathbf{F}(t) = \mathbf{f}_l, \text{no perturbation})\, p(\text{hgenes perturbed}) \\
&\quad = \delta(\mathbf{f}_k(\mathbf{x}(t)) - \mathbf{x}(t+1))(1-q)^G + (1-q)^{G-h} q^h (1-\delta(h))
\end{aligned} \tag{8.8}
$$

where h is the Hamming distance between $\mathbf{x}(t+1)$ and $\mathbf{x}(t)$. Then, the corresponding transitional probability on the augmented state $[\mathbf{x}(t), \mathbf{F}(t)]$ can be expressed as

$$
\begin{aligned}
&p(\mathbf{x}(t+1), \mathbf{F}(t+1) = \mathbf{f}_k|\mathbf{x}(t), \mathbf{F}(t) = \mathbf{f}_l) \\
&\quad = p(\mathbf{x}(t+1)|\mathbf{x}(t), \mathbf{F}(t) = \mathbf{f}_l)\, p(\mathbf{F}(t+1) = \mathbf{f}_k|\mathbf{x}(t), \mathbf{F}(t) = \mathbf{f}_l) \\
&\quad = (\delta(\mathbf{x}(t+1) - \mathbf{f}_k(\mathbf{x}(t)))(1-q)^G + (1-q)^{G-m} q^m (1-\delta(h-m))) \\
&\qquad \left(\frac{c_k}{\sum_{i=1, i\neq l} c_i} (1-\delta(\mathbf{f}_k - \mathbf{f}_l))\lambda + \delta(\mathbf{f}_k - \mathbf{f}_l)(1-\lambda) \right).
\end{aligned} \tag{8.9}
$$

In this section, we discuss the PBN modeling of feedback-only regulatory systems and feedback-with-external-input systems. In both cases, the system dynamics can be explored by applying Markov chain theory. We also defined two generalized systems: context-sensitive PBNs and PBNs with random perturbation. The two systems will be the focus for the optimal control of gene regulation, discussed in detail in Section 8.4. In the next section, we will discuss how a PBN is constructed based on expression data.

8.3 Reverse Engineering Regulatory Networks with PBN-Based Microarray Expression Data

An important task of modeling is to uncover regulatory networks based on gene-expression data. Although most of the regulation occurs at the protein level, owing to the abundance of expression microarray data and the close interaction between transcripts and proteins, it is still prudent and relevant to investigate the regulatory relationship at the transcriptional level. A transcriptional gene network can provide new leads for locating target networks subject to further wet lab study. The task is often referred to as *reverse engineering* for the fact that the microarray measurements are the outputs of the unknown regulatory system.

From a system viewpoint, the task of reverse engineering is a problem of system identification and is mathematically an inverse problem, with the main objective of determining the connectivity of the network; that is, to determine the genes that regulate target genes. In the context of PBNs, the connectivity of a target gene is defined by the possible input genes of the regulatory function that produce the expression of the target gene at the output. In graph theory, the input genes are the parents of target genes. To determine the connectivity, we must select the proper structure for regulatory functions and also determine the parameters of the functions. The overall problem falls into the realm of statistical parameter estimation and model selection. Solutions based on the *coefficient of determination* (CoD) [24–26], *best-fit extensions* [27,28], and Bayesian methods [29–31] have been reported. In this section, we focus on Bayesian solutions for determining the network connectivity and function parameters.

Before proceeding, some comments on data are in order. Most microarray data heretofore obtained are not the result of time-course experiments; rather, they are nondynamic and are assumed to come from the steady state of the network. This is typically the case for situations involving patients, where expression measurements are based on tissue samples. If a network is designed from steady-state data, then it does not reflect transient behavior, and the best we can hope for is to construct a network whose steady-state behavior reflects the steady-state behavior of the real biological system. In the event, we have time-course (dynamic) data; a constructed network can reflect transient behavior, depending on how the data have been gathered.

8.3.1 A Disjoint Bayesian Solution of Constructing Probabilistic Boolean Networks

In [29], a scheme was proposed for seeking solutions to the network connectivity, function structure and parameters, and the selection probabilities c_{gk} in separate steps. In the first step, the network connectivity is determined based on mutual information clustering. In the second step, a reversible-jump Markov-chain-Monte-Carlo (RJMCMC) sampling technique [32] is used to obtain the model order and the parameters for each f_{gk}. In the third step, the CoD is applied to determine the selection probability c_{gk}. We elaborate on each step.

The network connectivity is defined by the set of parent genes for each gene in the network. Given the fact that the target gene is regulated by the parent set, it is then reasonable to consider that the target gene and the parent set share information. Now, suppose that there is a total of r clusters with $\mathbf{X}_i$ representing the collection of gene expressions in the ith cluster, then a mutual information measure is defined as

$$E(s(r)) = \sum_{i \neq j} I(X_i, X_j), \tag{8.10}$$

where $s(r)$ represents an r-partition of the gene space and $I(X_i, X_j)$ is the pairwise mutual information between clusters i and j. To estimate $I(\cdot)$, a practical histogram-based approach is used in [29]. The optimal r-partition clustering is defined as the partition that minimizes the measure E, that is,

$$\hat{s}(r) = \arg \min_{s(r) \in S(r)} E(s(r)), \tag{8.11}$$

where $S(r)$ denotes the set of all possible r-partitions. The optimization in (8.11) is NP-hard, and a simulated annealing scheme is developed to approximate it (8.11) numerically. Then the overall optimal partition can be obtained by evaluating $s(r)$ for all possible r, that is,

$$\hat{s}(r^*) = \min_{2 \leq r \leq R} s(r),$$

where R is the upper limit on the number of clusters and r^* is the optimal partition number, which is obtained from

$$r^* = \arg \min_{2 \leq r \leq R} E(s(r)).$$

Based on PBN theory, even for the same target gene, the parent gene set could be different for different regulatory functions. Thus, a total of K_g sets of parent genes should be determined for each gene $g \in N_s = \{1, 2, \ldots, N_s\}$. To this end, instead of selecting only the single optimal partition, it is suggested to choose a superset of $L = m_1 + m_2 + 1$ different partitions $\{s(r^* - m_1), \ldots, s(r^* - m_2)\}$ for integers m_1 and m_2 and for $K_g \leq L$ for all g. Then, for gene g, K_g clusters are selected from the L partitions in the superset, which completely defines the network connectivity.

Once the parent gene set is defined for each regulatory function, the second step is to determine the functional structure and the associated parameters. In particular, a radial basis neural network model with $J_{\max}$ possible predictors is adopted for f_{gk}. As a result, there are totally $J_{\max} + 1$ possible models, and the Jth model can be expressed as

$$M_J \; : f_{gk}(\mathbf{x}^{(gk)}) = \sum_{j=1}^{J} a_{gkj} \phi(||\mathbf{x}^{(gk)} - \mu_{gkj})|| + b_g + \beta_{gk}^{\mathrm{T}} \mathbf{x}^{(gk)} + n, \quad 0 \leq J \leq J_{\max}, \tag{8.12}$$

where $\mathbf{x}^{(gk)}$ is a vector of parent genes for function f_{gk}, which are determined in the first step; $\phi(\cdot)$ is a radial basis function (RBF); $||\cdot||$ is a distance metric; a_{gkj}, μ_{gkj}, β_{gk} are the model parameters; and n is assumed to be white noise. If time-series data are available, then $\mathbf{x}^{(gk)}$ can be made a function of time, with $\mathbf{x}^{(gk)}(t)$ on the left-hand side and $\mathbf{x}^{(gk)}(t-1)\mathbf{x}^{(gk)}$ on the right-hand side.

Given a set of microarray expression data measured over a sequence of time points, or $\mathbf{X}$, the objective is to choose the model from the pool of $J+1$ candidate models that fit the data the best. From a Bayesian perspective, the best model is obtained to maximize the marginal likelihood function $p(\mathbf{X}|J)$:

$$J_{\text{opt}} = \arg\max_{J} p(\mathbf{X}|J). \tag{8.13}$$

Note that $p(\mathbf{X}|J)$ is obtained by marginalizing the model parameters from the full likelihood function. Since this is intractable due to the nonlinear radial basis function, a RJMCMC sampling is proposed to circumvent the difficulty. The RJMCMC sampling draws Monte Carlo samples from the joint model and the parameter space, which can be used to approximate the maximization in (8.13) as well as estimate the parameters of the selected model. This algorithm is run for each regulatory function of every gene in the network.

In the third and last step, the selection probability c_{gk} is estimated by the CoD, which measures the degree to which prediction of the expression level of a target gene is improved relatively to the best possible prediction in the absence of observations. For the kth function of gene g, the CoD is defined by

$$\theta_{gk} = \frac{\varepsilon_g - \varepsilon\left(x_g, f_{gk}^{(\text{opt})}\right)}{\varepsilon_g} \quad \forall k, \tag{8.14}$$

where x_g is the value of gene g, $\varepsilon_g = E[\|x_g - E[x_g]\|^2]$ is the mean squared error of the best (constant) estimate of x_g in the absence of any conditional variables, and $\varepsilon\left(x_g, f_{gk}^{(\text{opt})}\right)$ is an error measure of the optimal function $f_{gk}^{(\text{opt})}$, defined by

$$\varepsilon\left(x_g, f_{gk}^{(\text{opt})}\right) = E\left[\|b\left(\left(f_{gk}^{(\text{opt})}(\mathbf{x}^{(gk)}\right) - x_g\right)\|^2\right], \tag{8.15}$$

where $b(\cdot)$ is a $\{-1, 0, 1\}$-valued threshold function with $b(z) = -1$ if $z < -0.5$, $b(z) = 0$ if $|z| \leq 0.5$, and $b(z) = 1$ if $z > -0.5$, and $f_{gk}^{(\text{opt})}$ is the optimal function obtained in the second step. With the CoDs, an estimate about c_{gk} can be obtained as

$$c_{gk} = \frac{\theta_{gk}}{\sum_{k=1}^{K_g} \theta_{gk}} \quad \forall g. \tag{8.16}$$

It is easy to see that the functions according to the highest CoDs, or the fittest functions, will be selected more often in the evolution of the gene networks.

8.3.1.1 Experimental Results

In [29], the proposed scheme was tested on microarray data measuring responsiveness to three genotoxic stresses: ionizing radiation (IR), a chemical mutagen methane sulfonate (MMS), or ultraviolet (UV) [33]. Cell lines were chosen so that a sampling of both *p53* proficient and *p53* deficient cells would be assayed. In the end, 14 genes, each with 30 samples, were selected for the network inference and the continuous expression levels were also discretized to the ternary expression. The mutual information clustering produced a superset of parent gene sets of sizes from $k = 2$ to 7. Based on this superset, the RJMCMC sampling was performed to design the fittest functions. Further, the CoD of a predictor was estimated using cross-validation among the 30 samples: 20 samples were randomly chosen for training, the remaining 10 samples were used for test data. The results were compared with a previous study on the same data set obtained by using a full search restricted by a small number of predictor genes [34]. Good agreement was observed by the comparison, and results regarding four genes were discussed in detail. We mention just three examples. It is known that gene *p53* is influential, but not determinative, of the upregulation of genes *p21* and *MDM2*. Thus, some level of prediction of *p53* should be possible by a combination of these two genes. This expectation was met by the obtained inference results with *p21* and *MDM2* combining with two other genes to predict *p53*. The biological expectation is that *MDM2* is incompletely, but significantly, predicted by *p53*. The inference procedure found four predictor sets for *MDM2*, all including *p53*. Lastly, in [34], *BCL3* was shown to be strongly predicted by various predictor sets, with the strongest being {*RCH1*, *SSAT*, *p21*}. This is the best predictor set found for *BCL3* by the current method, with CoD 0.6256.

8.3.2 A Full Bayesian Solution

In the disjoint approach presented in the last section, errors committed in earlier steps creep into the later stages, and they cannot be corrected since the steps are disjoint. To overcome this problem, a full Bayesian solution was proposed in [31] to learn the connectivity of the network. As discussed before, the connectivity is assumed to be a joint set of the connectivity of each gene, which is determined by its regulators (parent set) defined on the basis of the inputs of the selected Boolean function. Let U_g represent the parent set of gene g. Under a Bayesian framework, $p(G|\mathbf{Y})$ is the a posteriori probability (APP) of the connectivity, that is, the probability of the network graph given the observations $\mathbf{Y}$. The optimal connectivity $\hat{G}$ is the one that maximizes $p(G|\mathbf{Y})$, namely,

$$\hat{G} = \arg\max_{G \in G} p(G|\mathbf{Y}), \tag{8.17}$$

where G represents the set of all possible graphs. The APP $p(G|\mathbf{Y})$ can be expressed according to the Bayes theorem as

$$p(G|\mathbf{Y}) \propto p(\mathbf{Y}|G)\, p(G), \tag{8.18}$$

where $p(G)$ is the prior distribution of G, which is assumed to be uniform over all possible connectivities, and $p(\mathbf{Y}|G)$ is the likelihood function, which is calculated according to

$$\begin{aligned} p(\mathbf{Y}|G) &= \sum_{\mathbf{F}} p(\mathbf{Y}|G,\mathbf{F})\, p(\mathbf{F}) \\ &= \prod_{g=1}^{G}\sum_{k=1}^{K_g} p(\mathbf{x}_g|\, f_{gk}, U_g)\, p(\, f_{gk}). \end{aligned} \tag{8.19}$$

Notice that to obtain $p(\mathbf{Y}|G)$ based on (8.19) requires summation over all possible regulation functions, which is computationally infeasible. Instead, an approximation to (8.19) is proposed as

$$p(\mathbf{Y}|G) \approx \prod_{g=1}^{G} p(\mathbf{x}_g|\,\hat{f}_g, U_g)\, p(\hat{f}_g), \tag{8.20}$$

where $\hat{f}_g$ is the optimal predictor (Boolean function) defined by (8.13) and obtained using the same approach as discussed in the previous section. Particularly, the radial basis neural network models defined in (8.12) are used and the RJMCMC sampling is applied to obtain the optimal predictor. Once the optimal predictor is obtained, the marginal likelihood $p(\mathbf{x}_g|\,\hat{f}_g, U_g)$ can be evaluated and so can the likelihood $p(\mathbf{Y}|G)$ up to a normalizing constant. However, the optimal connectivity $\hat{G}$ cannot be obtained exactly since the search space in performing (8.17) is still huge. To circumvent the difficulty, an MCMC sampling search strategy is developed. The algorithm starts with an initial graph, and in each iteration, a new network G' is proposed by a random single edge addition, deletion, or reversal to the network G obtained in the previous iteration. The proposed network G' is accepted with the probability

$$\tau = \min\left\{\frac{p(\mathbf{Y}|G')}{p(\mathbf{Y}|G)}, 1\right\}. \tag{8.21}$$

If G' is rejected, the current graph G will be kept as the graph in the current iteration. The iteration will proceed for $2T$ times, and the networks in the last T iterations will be used to construct the GRN. Specifically, it is assumed that K predictors will be used for each gene regulation. To determine the predictors for each gene, the K networks with the largest $p(\mathbf{Y}|G)$ are selected out of T kept networks from MCMC sampling. For each network, the predictors and the corresponding parent sets for each gene are given by optimal predictors discussed above. The probability of choosing the kth predictor, or c_{gk}, can be calculated as

$$c_{gk} = \frac{\varepsilon_k}{\sum_{j=1}^{K}\varepsilon_j}, \tag{8.22}$$

where $\varepsilon_k = \prod_{g=1}^{G} p(Y|\,\hat{f}_{gk})\, p\left(\hat{f}_{gk}\right)$, which is obtained in the RJMCMC algorithm.

8.3.2.1 Melanoma Application

The full Bayesian algorithm has been applied to reconstruct the gene regulatory network based on microarray gene-expression profiles. The microarray data used in this study resulted from the study of 31 malignant melanoma samples [35]. For the study, total messenger RNA was isolated directly from melanoma biopsies, and fluorescent cDNA from the message was prepared and hybridized to a microarray containing probes for 8150 cDNAs (representing 6971 unique genes). The 10 genes used here for the model were chosen from a set of 587 genes from the melanoma data set that have been subjected to an analysis of their ability to cross predict each other's state in a multivariate setting [36]: pirin, WNT5A, S100P, RET-1, MMP-3, PHO-C, MART-1, HADHB, synuclein, and STC2. The full Bayesian algorithm was applied to construct the ternary networks, and the top-scoring 25 networks were kept. Data state attractors and singleton attractors in these networks were examined. Five of the 25 networks have singleton attractors unique to themselves. The class of these five networks is denoted by Y and the other class of the rest 20 networks is represented by X. The steady-state distribution of the constructed network was investigated via simulation using all observed states as initial states. It is found that the steady-state distributions for the different initializations are essentially the same. The steady-state distribution is shown in Figure 8.5a. Owing to the concentration of mass in four states, the rest of the distribution appears negligible. Figure 8.5b shows the steady-state distribution with the masses of the dominating states truncated to 0.04. It reveals that there are other states with appreciable mass. The states with highest probabilities are labeled from (a) to (t). It is observed that, for the most part, the highly occurring states are the observed states. It is more striking to note that two singleton attractor states A and B carry over half (54%) of the steady-state mass. Almost another quarter of the steady-state mass is carried by another two states, C and D, each carrying 11.3% of the mass. Altogether, states A, B, C, and D carry almost 75% of the steady-state mass and they agree for six of the component genes, differing only among the genes RET-1, MMP-3, PHO-C, and STC2. In addition to the steady-state analysis, the connectivities of the constituent networks are also studied. The relationship between connectivity and attractors of the 25 networks is of particular interest for this study since it is the attractors of the constituent networks that comprise the attractors of the constructed networks. Comparison of connectivity relations between X and Y indicates a very strong congruency of connectivity between the two classes of networks having very different singleton attractors. The key observation is that optimizing connectivity leads to fairly consistent connectivity; however, it produces two classes of networks with significantly different attractors. This shows that focusing on connectivity does not necessarily preclude discovering a rich attractor structure. It is conjectured that this phenomenon is in agreement with real networks because, even in widely varying contexts, one would not expect a significant change in connectivity.

8.4 Optimal Control of Context-Sensitive PBN

8.4.1 Introduction to Network Intervention

Once a PBN characterizing gene regulation is constructed from the data, analysis can be carried out on the network to screen functionally prominent (marker) genes or subnetworks that determine specific cellular phenotypes (especially disease). A major

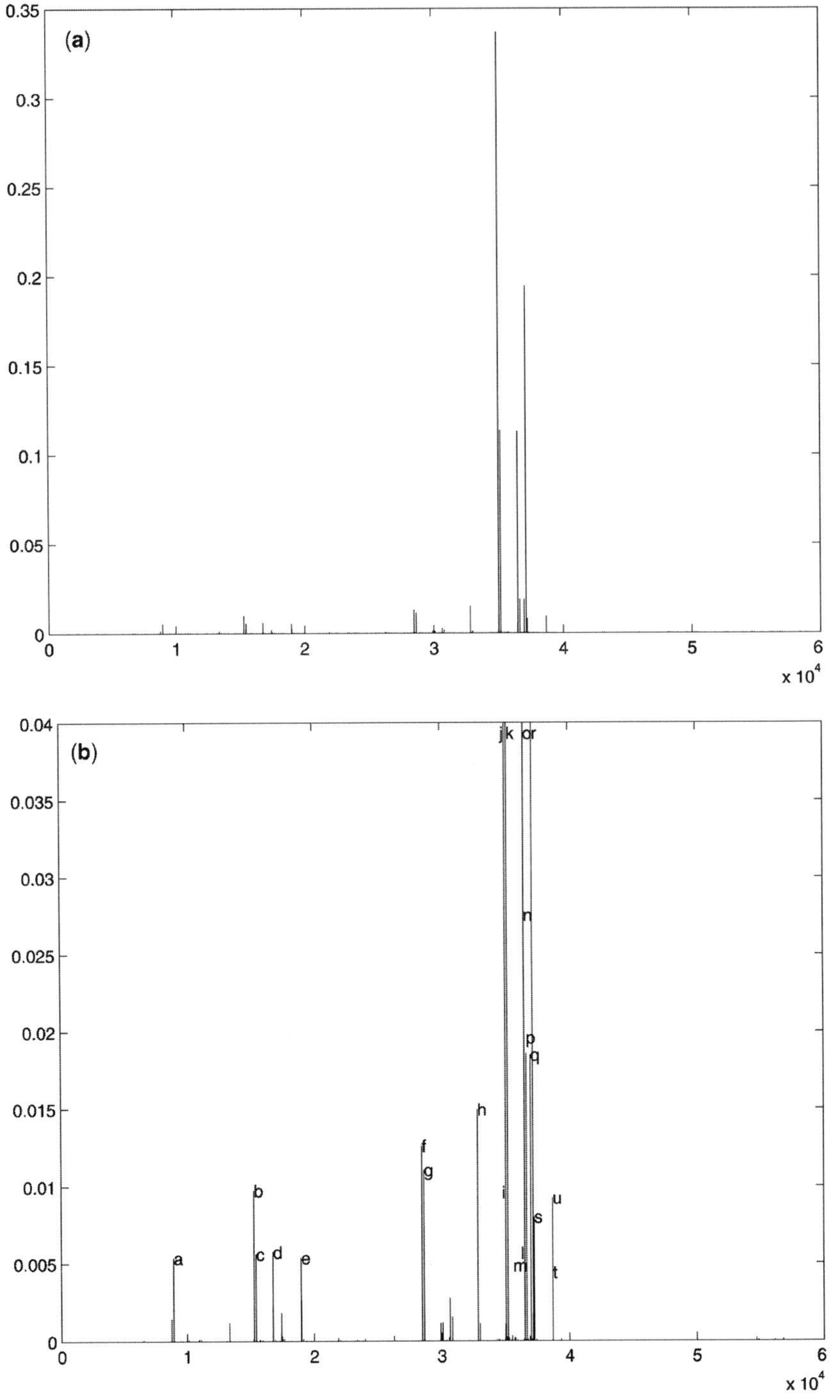

Figure 8.5 Estimated distribution after long run:
(a) steady-state distribution; (b) truncation at 0.04 [31].

goal is to develop intervention strategies to modify gene regulation in such way as to drive the dynamics of the network away from malignant phenotypes. For instance, tumorigenesis is associated with one type of steady-state behavior of the underlying gene network that is due to changes in network connectivity or functional relationships among the genes via mutations or rearrangements [37]. Even if reversing the cancerous phenotype is impossible, corrective therapeutic intervention can be applied to either cause the cancerous cells to self-destruct or prevent subsequent metastasis.

To date, intervention studies based on PBNs have used three different approaches: (i) resetting the state of the PBN through perturbation to a more desirable initial state and letting the network evolve from there [38,39]; (ii) changing the steady-state (long run) behavior of the network by minimally altering its rule-based structure [39,40]; and (iii) manipulating external (control) variables that alter the transition probabilities of the network and can, therefore, be used to desirably affect its dynamic evolution [39,41]. The control-theoretic approach is being extensively developed [41–45]. The optimal intervention algorithm has been modified to accommodate the case where the entire state vector or GAP is not available for measurement, and, whereas the original control-theoretic approach has been developed in the framework of instantaneously random PBNs, the intervention results have been extended to context-sensitive PBNs. We focus on the optimal external control of a context-sensitive PBN in both finite and infinite horizons.

8.4.2 Defining the Transition Probability of a Context-Sensitive PBN

We first derive the state transition probability of a context-sensitive PBN. The control strategies for modifying the dynamic behavior will be developed on the basis of transition probability. For convenience of composition, let $z(t) \in \{0, 1, \ldots, N_s - 1\}$ be the decimal correspondence of the G dimensional binary gene state $\mathbf{x}(t)$. In what follows, we will use $z(t)$ and $\mathbf{x}(t)$ interchangeably, whichever is more convenient. As discussed in Section 8.1, the general context-sensitive PBNs admit an internal perturbation with probability q and a control signal s at each time with probability λ. We have also pointed out that, to characterize the dynamics of a context-sensitive PBN, a time-homogeneous Markov chain can be constructed on the augmented state space of $[\mathbf{x}(t), \mathbf{F}(t)]$, whose transition is defined in (8.7). However, in practice, it is unlikely to know the specific network functions at each time that the gene expression profile is being generated. Realistically, we can measure the gene states only. Therefore, it is more practical to derive directly the transitions between the gene states $\mathbf{x}(t)$. Under the decimal representation, the transition distribution $p(z(t+1)|z(t))$ can be obtained as

$$\begin{aligned} p(z(t+1)|z(t)) &= p(\mathbf{x}(t+1)|\mathbf{x}(t)) \\ &= \sum_{\mathbf{f}_k} p(\mathbf{x}(t+1)|\mathbf{x}(t), \mathbf{F}(t) = \mathbf{f}_k)\, p(\mathbf{F}(t) = \mathbf{f}_k) \\ &= \sum_{\mathbf{f}_k} \delta(\mathbf{f}_l(\mathbf{x}(t)) - \mathbf{x}(t+1))(1-q)^G + (1-q)^{G-h} q^h (1-\delta(h)) c_k, \end{aligned} \tag{8.23}$$

where again h is the Hamming distance between $\mathbf{x}(t)$ and $\mathbf{x}(t+1)$. Note this transition distribution is time varying. Let $\pi(t)$ be the $N_s \times 1$ decimal state distribution vector of time t. The evolution of the state distribution can be calculated according to

$$\pi(t+1) = \mathbf{P}(t)\pi(t), \tag{8.24}$$

where $\mathbf{P}(t)$ is the $N_s \times N_s$ time-varying transitional matrix, whose i, jth element is $p(z(t+1)) = i|z(t) = j)$.

Now let $\mathbf{u}(t) \in \{0, 1\}^m$ be a vector of m external binary control inputs and $v(t)$ be the corresponding decimal representation of $\mathbf{u}(t)$, which is defined on a space $C = \{1, 2,\ldots, 2^m\}$. The external control inputs are employed to alter the gene states. When viewing the evolution of PBNs as a Markov chain, the external control inputs alter system evolution by changing the transitional distributions, or the transitional matrix. As a result, in the presence of external control inputs, the state distributions evolve as

$$\pi(t+1) = \mathbf{P}(v(t))\pi(t), \tag{8.25}$$

where the transitional matrix $\mathbf{P}(v(t))$ becomes a function of the control inputs. A Markov chain defined by (8.25) is referred to in the control literature as a controlled Markov chain or a Markov decision process [16]. Markov chains of this type occur in many real life applications, the most notable example being the control of queues. Given a controlled Markov chain, the control objective is to find a sequence of control inputs, referred to as a control strategy (policy), so that an appropriate cost function is minimized over the entire class of allowable control strategies. To arrive at a meaningful solution, the cost function must capture the costs and the benefits of using any control. The design of a good cost function depends on application and is likely to require considerable expertise.

8.4.3 External Intervention with Finite-Horizon Control

We first consider a finite-horizon problem, where the control is applied only over a finite number of steps, and the goal is to drive the state of the PBN to a desired state at the end of control. In the case of diseases such as cancer, treatment may be applied over a finite time horizon. For instance, in the case of radiation treatment, the patient may be treated with radiation over a fixed interval of time following which the treatment is suspended for some time to evaluate effects. After that, the treatment may be applied again but the important point to note is that the treatment window at each stage is finite. Suppose that the number of steps over which the control input is to be applied has been a priori determined to be M and we are interested in controlling the behavior of the PBN over the treatment period $t = 0, 1, 2, \ldots, M-1$ by applying control input $v(t)$ over the period. The net result of the control actions $v(0), v(1), \ldots, v(M-1)$ is that the state of the PBN will transition according to (8.25). To this end, we define a *per-stage* cost $C_t(z(t), v(t))$ as being the cost of applying the control input $v(t)$ when the state is $z(t)$. Note that even if the network starts from a deterministic initial state $z(0)$, the subsequent states will be random because of the stochastic nature of the state evolution. Consequently, the cost must be defined using expectation. The expected cost of control over the entire treatment horizon is defined as

$$E\left[\sum_{t=1}^{M-1} C_t(z(t), v(t))|z(0)\right], \tag{8.26}$$

where $z(0)$ is the initial state of the PBN, and the expectation $E[\,]$ is with respect to the $z(t)$s, whose distributions are $\pi(t)$s and are calculated according to (8.25).

Since the ultimate goal is to move the state of the PBN to some desired states in M steps, in addition to the expected cost (8.26) defined for the network transient period, we also need to assign higher costs to undesired terminal states to penalize the move of the Markov chain to undesired states at the terminal time M. Now, assume that all controls $v(t)$ $\forall t$ are set to zero. Then, at time M, we divide the states into different categories depending on how desirable or undesirable they are and assign higher terminal costs to the undesirable states. For instance, a state associated with rapid cell proliferation leading to cancer should be assigned a high terminal penalty, whereas a state associated with normal behavior should be assigned a low terminal penalty. This corresponds to a terminal cost, $C_M(z(M))$, on the state space of the PBN. Just like the transient cost, the terminal cost must be determined in consultation with clinicians [44]. For the purpose of this section, we assume that the assignment of terminal penalties has been carried out, and we have at our disposal a terminal cost $C_M(z(M))$ that is a function of the terminal state. Note that, even with the transient control inputs, the particular state that the PBN attains at step M is still a random variable and so is the terminal cost $C_M(z(M))$. Thus, we must consider the expected terminal cost. Combining the expected transient and the terminal costs, we arrive at the complete cost function

$$J(\mathbf{v}) = E\left[\sum_{t=1}^{M-1} C_t(z(t), v(t)) + C_M(z(M))|z(0)\right], \tag{8.27}$$

where $\mathbf{v}^\top = [v(0), \ldots, v(M-1)]$ is the vector of the control sequence. To proceed further, let us assume that at time t the control input $v(t)$ is a function of the current state $z(t)$, namely,

$$v(t) = \mu_t(z(t)). \tag{8.28}$$

Then, the cost function $J(\mathbf{v})$ can be considered as a cost function of the μ_ts. The optimal control problem can now be stated as: Given an initial state $z(0)$ and the network transition (8.26), find a control law $\mu = \{\mu_0, \mu_1, \ldots, \mu_{M-1}\}$ that minimizes the cost function (8.27). Mathematically, this can be expressed as

$$\mu^* = \arg \min_{\mu \in C^M} J(\mu) \tag{8.29}$$

subject to the constraint

$$p(z(t+1) = i | z(t) = j) = p_{ij}(v(t)), \tag{8.30}$$

where $p_{ij}(v(t))$ is the ijth entry of the matrix $\mathbf{P}(v(t))$.

Optimal control problems of this type can be solved using the technique of dynamic programming. This technique, pioneered by Bellman in the 1960s, is based on the so-called Principle of Optimality that can be explained as follows. Consider an optimization problem where we are interested in optimizing a performance index over a finite number, M, of steps. At each step, a decision is made and the objective is to devise a strategy or sequence of M decisions that is optimal in the sense that the cumulative performance index over all the M steps is optimized. In general, such an optimal strategy may not exist. However, when such an optimal strategy does exist, the principle of optimality asserts: if one searches for an optimal strategy over a subset of the original number of steps, this new optimal strategy will be given by the overall optimal strategy, restricted to the steps being considered. Although intuitively obvious, the principle of optimality can have far reaching consequences. In [43], a useful proposition is derived from the principle of optimality.

Proposition 1 Consider a dynamic programming algorithm that proceeds backward from time step $M-1$ to time step 0:

$$J_M(z(M)) = C_M(z(M)) \tag{8.31}$$

$$\begin{aligned} J_t(z(t)) &= \min_{v(t) \in C} E[C_t(z(t), v(t)) + J_{t+1}(z(t+1)) | z(t)] \\ &= \min_{v(t) \in C} \left\{ C_t(z(t), v(t)) + \sum_{j=1}^{2^G} p_{z(t),j}(v(t)) J_{t+1}(j) \right\} \end{aligned} \tag{8.32}$$

for $t = 0, 1, M-1$. If $v^*(t) = \mu_t^*(z(t))$ minimizes the right-hand side of (8.32) for each $z(t)$ and t, then the control law $\mu^* = \{\mu_0^*, \mu_1^*, \ldots, \mu_{M-1}^*\}$ is optimal. Also, let $J^*(\mathbf{v}^*)$ be the optimal value, the smallest value, of the cost functional in (8.27) with $\mathbf{v}^*$ being the optimal control strategy. Then

$$J^*(\mathbf{v}^*) = J_0(z(0)). \tag{8.33}$$

This proposition provides the algorithm for finding the optimal control strategy for a finite horizon.

8.4.3.1 **Melanoma Application**

In this section, we describe an application of the optimal finite-horizon control algorithm to a context-sensitive PBN derived from gene expression data collected in the same study of metastatic melanoma as in Section 8.1. In this study, the abundance of mRNA for the gene WNT5A was found to be highly discriminating among cells with properties typically associated with high versus low metastatic competence. These findings were validated and expanded in a second study, in which experimentally increasing the levels of the Wnt5a protein secreted by a melanoma cell line via genetic engineering methods directly altered the metastatic competence of that cell as measured by the standard *in vitro* assays for metastasis [46]. Furthermore, it was found that an intervention blocking the WNT5A protein from activating its receptor, the use of an antibody that binds the Wnt5a protein, could substantially reduce WNT5A's ability to induce a metastatic phenotype. This suggests a control strategy that reduces the WNT5A gene's action in affecting biological regulation, since the data suggest that the disruption of this influence could reduce the chance of a melanoma metastasizing, a desirable outcome.

A seven-gene network was considered in [43] with genes WNT5A, pirin, S100P, RET1, MART1, HADHB,, and STC2. To obtain the PBN, the full Bayesian approach presented in Section 8.2 was employed to construct four highly probable Boolean networks that are used as the constituent Boolean networks in the PBN. The four

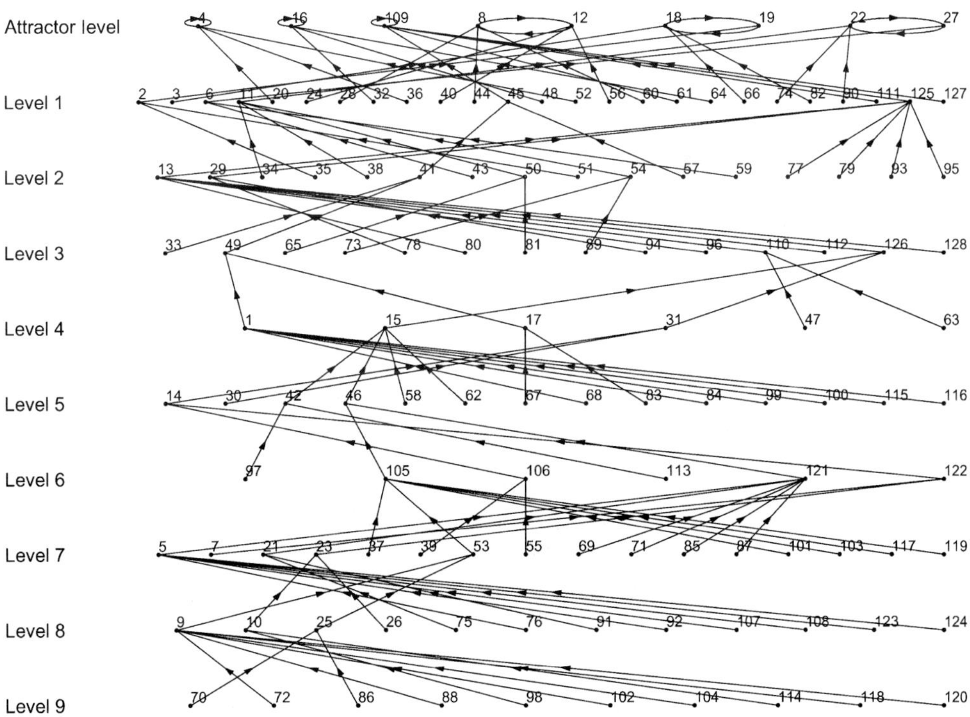

Figure 8.6 Network 1 [43].

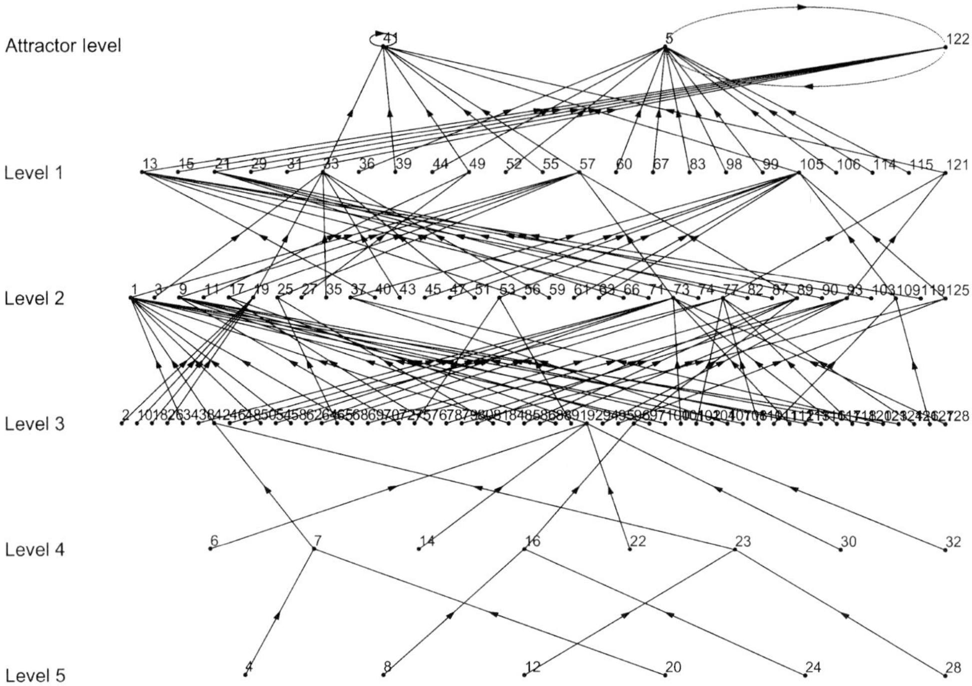

Figure 8.7 Network 2 [43].

generated Boolean networks are shown in Figures 8.6 to 8.9, where the states are labeled from 1 to $2^7 = 128$. Each constituent network is assumed to be derived from steady-state gene-expression data, and the attractor states and the level sets are shown in the figures. Observe that in each of these networks the state enters an attractor cycle in a small number of steps (at the most nine), which is consistent with what is expected in real networks. The control strategy of this section has been applied to the designed PBN with pirin chosen as the control gene and $p = q = 0.01$. Figure 8.10 shows the expected cost for a finite-horizon problem of length 5 originating from each of the 128 states. The cost of control is assumed to be 0.5 and the states are assigned a terminal penalty of 5 if WNT5A is 1 and 0 if WNT5A is 0. The control objective is to downregulate the WNT5A gene. From Figure 8.10, it is clear that the expected cost with control is much lower than without control, which agrees with our objective.

8.4.4 External Intervention with Infinite-Horizon Control

The goal of infinite-horizon control is to change the steady-state behavior of the PBN by employing external intervention over an infinite length of time. Let μ be the set of all admissible control policies $\mu = \{\mu_0, \mu_1, \ldots,\}$. A *stationary policy* is an admissible policy

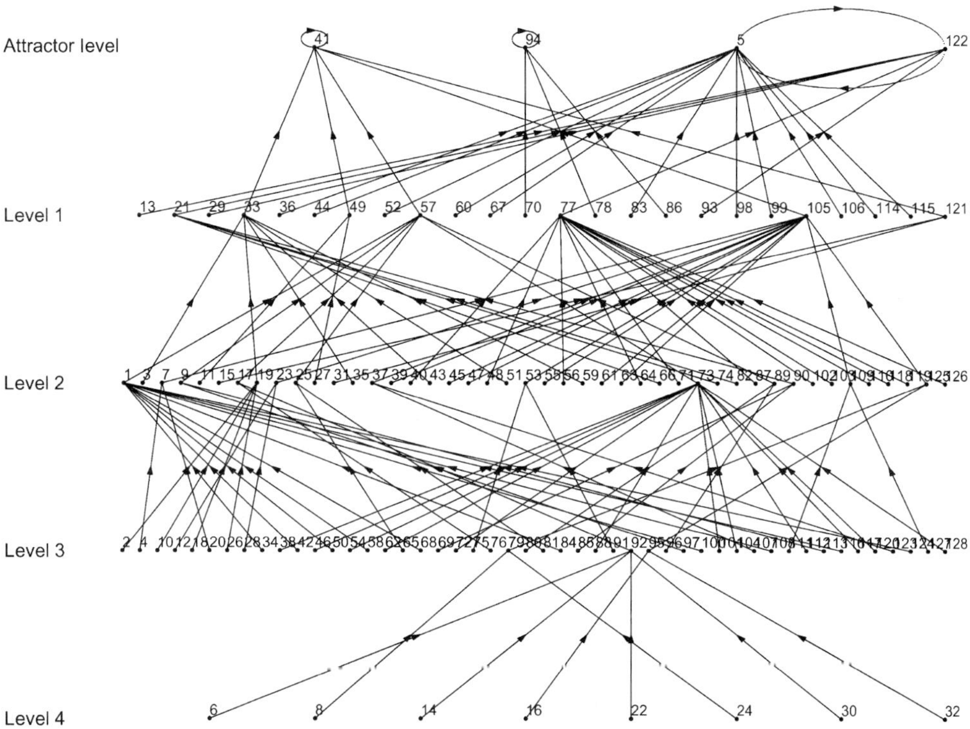

Figure 8.8 Network 3 [43].

of the form $\mu = \{\mu, \mu, \ldots,\}$. The optimal infinite-horizon control corresponds to finding a stationary policy, or μ, that minimizes the overall cost function, that is,

$$\mu^* = \arg\min_{\mu} J(\mu), \tag{8.34}$$

where μ^* is called the optimal policy. This setting seems similar to finite-horizon control and we may like to apply the formulation developed for finite-horizon control here for a solution of infinite-horizon control; however, two difficulties arise immediately. First, in the infinite-horizon control, the per-stage cost must also consider the destination in addition to the current state. That means the per-stage cost should be denoted as $C_t(z(t), v(t), z(t+1))$ instead of $C_t(z(t), v(t))$. In contrast to $C_t(z(t), v(t), z(t+1))$, we notice that $C_t(z(t), v(t))$ can be considered as the expected per-stage cost. To still be able to use $C_t(z(t), v(t))$ and sequentially the formulation for infinite-horizon control, we remove the dependency of the per-stage cost on the destination state at the next stage through marginalization. This leads to the expected per-stage cost calculated according to

$$C(z(t), v(t)) = \sum_{j=0}^{N_s-1} p_{z(t),j}(v(t))C_t(z(t), v(t), z(t+1) = j). \tag{8.35}$$

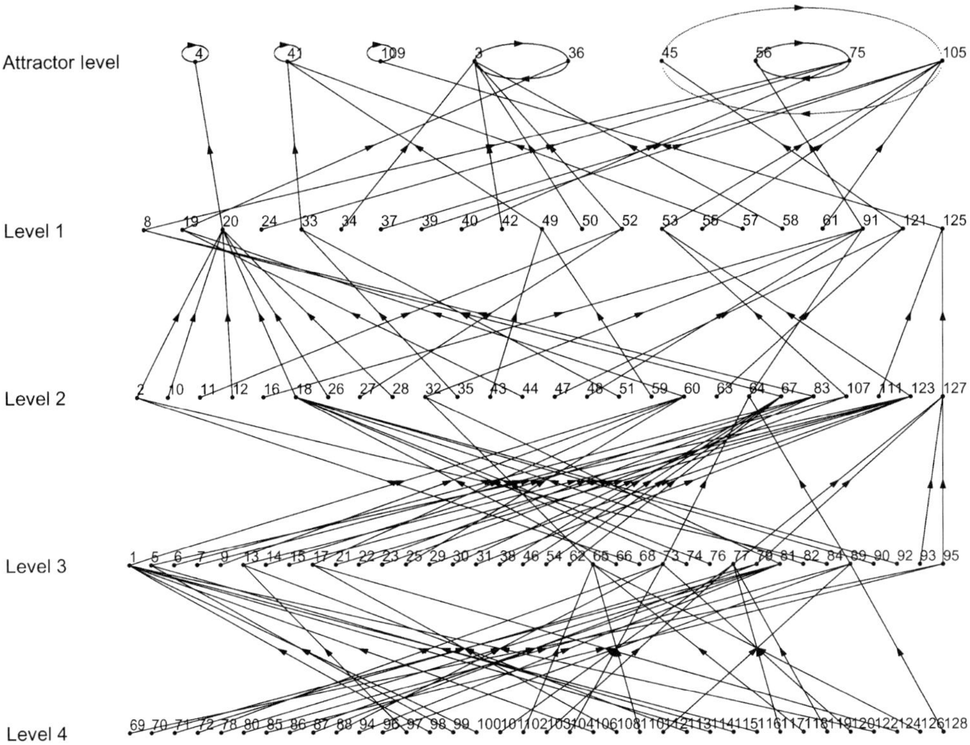

Figure 8.9 Network 4 [43].

Note that in (8.35) we assume that the per-stage cost is time invariant and thus we drop the subscript t from the cost functions. Also, note that the expected per-stage cost of (8.35) is bounded. This is because the control input and perturbation space are finite. However, even by using the expected per-stage cost, the cost function defined for finite-horizon control is still inapplicable. This is because, as there is no terminal stage, the sum of the infinitely many per-stage costs in the overall cost can lead to infinite cost, which cannot be minimized. To overcome this difficulty, two solutions have been proposed in [45].

8.4.4.1 The Discounted Approach

In the first case, we assume that the expected per-stage cost is bounded. A discounting factor $\alpha \in (0, 1)$ is introduced in the cost to make sure that the limit of the finite sums converges as the horizon length goes to infinity. More specifically, given a stationary control policy μ, the total cost with discounted and bounded cost per stage can be expressed as

$$J(\mu) = \lim_{M \to \infty} \left\{ \sum_{t=0}^{M-1} \alpha^t C(z(t), \mu) \right\}. \tag{8.36}$$

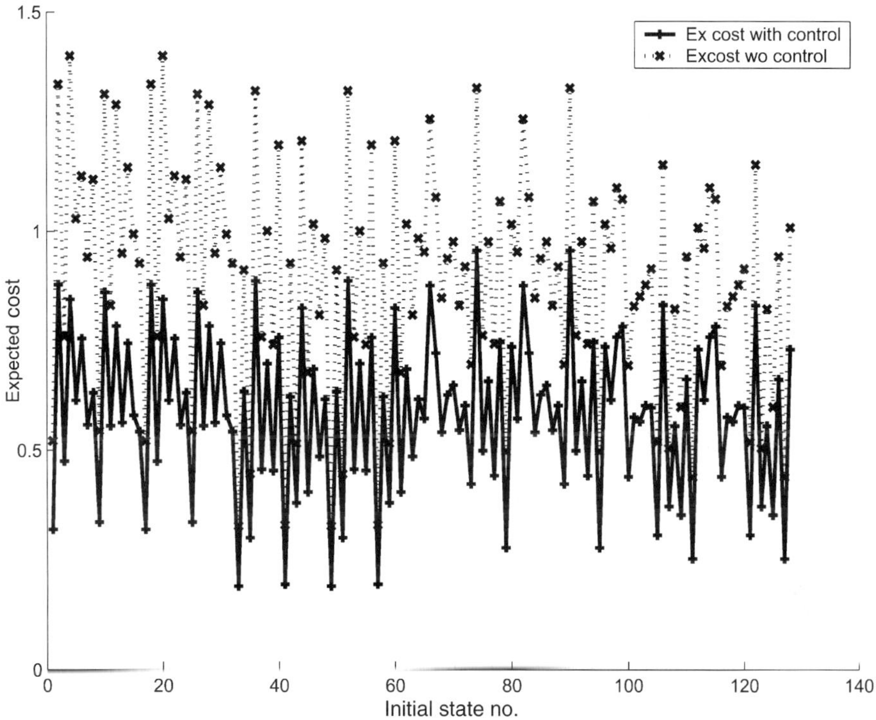

Figure 8.10 The expected cost for finite horizon of length 5 originated from different initial states [43].

In the general formulation, the inclusion of α in the cost captures the fact that costs incurred at a later time are less significant. In the case of cancer treatment, α signifies that the condition of the patient in the initial stages of treatment is more important than the condition at a later stage, or in other words, the reward for improving the condition of the patient in the present is more significant than the reward obtained from similar improvement at a later stage. This approach is reasonable if we keep in mind the expected life span of a cancer patient.

The optimal stationary control policy can be obtained by solving the optimization defined in (8.34). The dynamic programming solutions have been proposed in [45]. The algorithm is inspired by the connection between the cost (8.36) and the cost (8.27) of finite-horizon control. Note that if we define the costs (8.36) $C_t(z(t), \mu(t)) = \alpha^t C(z(t), \mu)$ and $C_M(z(M)) = 0 \forall z(M)$, then we obtain an expression of cost the same as the one for finite-horizon control. In light of this, the dynamic programming algorithm developed for finite-horizon control can be applied for the solution here by running it backward in time. In [45], two types of solutions, the *value iteration* procedure and the *policy iteration* procedure, are proposed. We focus here on the policy iteration procedure. To this end, we need to define the following two mappings:

1. For any cost function J, define the mapping TJ by

$$TJ(i) = \min_{\mu \in C} \left[C(i, \mu) + \alpha \sum_{j=0}^{N_s-1} p_{ij}(u) J(j) \right], \quad \forall i \in S. \tag{8.37}$$

   Note that TJ corresponds to the optimal cost function in the dynamic programming algorithm for finite-horizon control.

2. For any cost function J, define the mapping $T_\mu J$ by

$$(T_\mu J)(i) = C(i, \mu(i)) + \alpha \sum_{j=0}^{N_s-1} p_{ij}(u) J(j), \quad \forall i \in S. \tag{8.38}$$

   $T_\mu J$ can be considered as the cost function associated with the policy μ for the one-stage problem that has per-stage cost function C and terminal cost αJ.

In addition, we let $\mathbf{J} = [J(0), \ldots, J(N_s - 1)]^\top$, $\mathbf{J}_\mu = [J_\mu(0), \ldots, J_\mu(N_s - 1)]^\top$, $\mathbf{TJ} = [(TJ)(0), \ldots, (TJ)(N_s - 1)]$, and $\mathbf{T}_\mu\mathbf{J} = [(T_\mu J)(0), \ldots, (T_\mu J)(N_s - 1)]^\top$. We define a transition probability matrix corresponding to the stationary policy μ by

$$\mathbf{P}_\mu = \begin{pmatrix} p_{00}(\mu(0)) & \cdots & p_{0,N_s-1}(\mu(0)) \\ \vdots & \vdots & \vdots \\ p_{N_s-1,0}(\mu(N_s-1)) & \vdots & p_{N_s-1,N_s-1}(\mu(N_s-1)) \end{pmatrix}$$

and a cost vector $\mathbf{C}_\mu = [C(0, \mu(0)), \ldots, C(N_s - 1, \mu(N_s - 1))]^\top$. Then, the vector matrix form of mapping (8.38) can be expressed as

$$\mathbf{T}_\mu\mathbf{J} = \mathbf{C}_\mu + \alpha\mathbf{P}_\mu\mathbf{J}.$$

It has been shown in [47] that the cost $\mathbf{J}_\mu$ corresponding to the policy μ satisfies

$$\mathbf{J}_\mu = \mathbf{C}_\mu + \alpha\mathbf{P}_\mu\mathbf{J}_\mu$$

or

$$(\mathbf{I} - \alpha\mathbf{P}_\mu)\mathbf{J}_\mu = \mathbf{C}_\mu, \tag{8.39}$$

where $\mathbf{I}$ is the identity matrix. Note, for a given stationary policy μ, we can calculate $\mathbf{J}_\mu$ by solving a system of linear equations in (8.39). Then, if we can find a policy-improvement procedure that produces a new policy with smaller cost, then we can iterate between (8.39) and the policy-improvement procedure. The optimal policy is obtained when the minimum cost is reached. It is shown in [47] that an improved stationary policy $\hat{\mu}$ satisfies $\mathbf{T}_{\hat{\mu}}\mathbf{J}_{\mu} = \mathbf{T}\mathbf{J}_{\mu}$ and $\hat{\mu}$ will be an optimal policy if $\mathbf{J}_{\hat{\mu}} = \mathbf{T}\mathbf{J}_{\hat{\mu}}$. The facts are formalized by the following three theorems:

Theorem 1: Convergence of the discounted-cost algorithm: For any bounded cost function J,

$$J^*(i) = \lim_{M\to\infty} (T^M J)(i), \quad \forall i \in S, \tag{8.40}$$

where $(T^M J)(i) = (T(T^{M-1}J))(i)$ and $(T^0 J)(i) = J(i)$.

Theorem 2: Bellman's equation: The optimal cost function J^* satisfies

$$J^*(i) = \min_{\mu\in C}\left[C(i,\mu) + \alpha\sum_{j=0}^{N_s-1} p_{ij}(u)J^*(j)\right], i \in S \tag{8.41}$$

or, equivalently, $\mathbf{J}^* = \mathbf{T}\mathbf{J}^*$. Further, J^* is the unique solution of this equation within the class of bounded functions.

Theorem 3: Necessary and sufficient condition for optimality: A stationary policy μ is optimal if and only if $\mu(i)$ attains the minimum in Bellman's equation (8.41) for each $i \in S$; that is,

$$\mathbf{T}\mathbf{J}^* = \mathbf{T}_{\mu}\mathbf{J}^* \tag{8.42}$$

Based on the above discussion, an algorithm for finding the optimal control policy can be summarized in a chart.

1. Policy improvement: An improved stationary policy μ^{t+1} satisfying $\mathbf{T}_{\mu^{t+1}}\mathbf{J}_{\mu^t} = \mathbf{T}\mathbf{J}_{\mu^t}$ is obtained.
2. Stop if $\mathbf{J}_{\mu^t} = \mathbf{T}\mathbf{J}^{\mu^t+1}$, otherwise set $t = t+1$ and return to 1.

8.4.4.2 The Average-Cost-Per-Stage Approach

In the second case, we avoid the problem of a possibly infinite total cost by considering the average cost per stage, which is defined by

$$J(\mu) = \lim_{M\to\infty}\left\{\frac{1}{M}\sum_{t=0}^{M-1}\alpha^t C_t(z(t), v(t))\right\}. \tag{8.43}$$

The average-cost-per-stage formulation is essential when we are interested in the condition of the patient in the long run, and equal importance is given to the patient's condition in all stages.

The optimal policy is the stationary policy that minimizes (8.43) or the limit of J_M/M as $M \to \infty$, where J_M is the optimal finite-horizon cost over an interval of length M. The counterpart of the mapping defined for the discounted approach can be obtained as

$$TJ(i) = \min_{\mu \in C} \left[C(i, \mu) + \sum_{j=0}^{N_s - 1} p_{ij}(u) J(j) \right], \quad \forall i \in S \tag{8.44}$$

and

$$(T_\mu J)(i) = C(i, \mu(i)) + \sum_{j=0}^{N_s - 1} p_{ij}(u) J(j). \tag{8.45}$$

However, in the absence of the discounting factor, the iteration defined in Theorem 1 may diverge to infinity. Thus, calculating the average cost by taking $\lim_{M \to \infty}(J_M/M)$ is not feasible.

Instead, a *differential cost* h_t is introduced as

$$h(i) = J_t(i) - J_t(n_1), \quad \forall i \in S, \tag{8.46}$$

where n_1 is a fixed state. It is shown in [45] that the corresponding value iteration for the differential cost can be obtained as

$$\begin{aligned} \mathbf{h}_{t+1} &= \mathbf{Th}_t - Th_t(n_1)\mathbf{e} \\ &= \mathbf{Th}_t - \lambda e, \end{aligned} \tag{8.47}$$

where $\mathbf{h}_{t+1} = [\mathbf{h}_{t+1}(0), \ldots, h_{t+1}(N_s - 1)]^\top$, $\mathbf{Th}_t[Th_t(0), \ldots, Th_t(N_s - 1)$, $\mathbf{e} = [1, \ldots, 1]^\top$, and $\lambda = Th_t(n_1)$. The differential cost h plays a similar role as J in the discounted approach. Therefore, the counterpart of the Bellman equation (8.39) in the discounted approach is (8.47), from which the differential cost can be evaluated for a given policy. As a result, a similar algorithm as the discounted approach can now be constructed for the average-cost-per-stage approach. Before summarizing the algorithm, we state two theorems that form the basis of the algorithm.

Theorem 4: For ergodic Markov chains, the optimal average cost per stage is independent of the initial state.

Note the context-sensitive PBNs considered here satisfy the conditions for the ergodicity of the Markov chain.

Theorem 5: If a scalar λ and an N_s-dimensional vector h satisfy

$$\lambda + h(i) = \min_{\mu \in C} \left[C(i, \mu) + \sum_{j=0}^{N_s - 1} p_{ij}(u) h(j) \right], \quad i \in S \tag{8.48}$$

or, equivalently, $\lambda \mathbf{e} + \mathbf{h} = \mathbf{Th}$, then λ is the optimal average cost per stage $J^*(i) \forall i$, that is,

$$\lambda = \min_{\mu} J_{\mu}(i) = J^*(i). \tag{8.49}$$

Further, if μ^* attains the minimum in (8.48) for each i, then the stationary policy μ^* is optimal, that is, $J_{\mu^*} = \lambda$ for all $i \in S$.

Based on the above discussion and the theorems, the *policy iteration* algorithm for the average-cost-per-stage case can be summarized in the following chart.

1. Policy improvement: An improved stationary policy μ^{t+1} satisfying $\mathbf{T}_{\mu^{t+1}} \mathbf{h}_{\mu^t} = \mathbf{Th}_{\mu^t}$ is obtained.
2. Stop if $\mu^{t+1} = \mu^t$, otherwise set $t = t+1$ and return to 1.

8.4.5
Melanoma Application

To demonstrate the effectiveness of the infinite-horizon procedure, the seven-gene network containing the genes WNT5A, pirin, S100P, RET1, MART1, HADHB, and STC2 was again used in [45]. To obtain the PBN, the algorithm described in [47] was applied to construct four highly probable Boolean networks (Figures 11 to 14), which were used as the constituent Boolean networks in the PBN. Each constituent network was assumed to be derived from steady-state gene-expression data. The states are ordered as WNT5A, pirin, S100P, RET1, MART1, HADHB, and STC2, with WNT5A as the most significant bit (MSB) and STC2 as the least significant bit (LSB).

The control strategies of the previous section have been applied to the designed PBN with pirin chosen as the control gene (with $\mu = 1$ signifying the state of pirin is reversed and $\mu = 0$ signifying no intervention) and $\lambda = q = 0.01$. The cost of control is assumed to be 1 and the states are assigned penalties as follows:

$$C(i, \mu, j) = \begin{cases} 5 & \text{if } \mu = 0 \text{ and if and WNT5A is 1 for state } j \\ 6 & \text{if } \mu = 1 \text{ and if and WNT5A is 1 for state } j \\ 1 & \text{if } \mu = 1 \text{ and if and WNT5A is 1 for state } j \\ 1 & \text{if } \mu = 0 \text{ and if and WNT5A is 1 for state } j \\ 0 & \text{if } \mu = 0 \text{ and if and WNT5A is 0 for state } j \end{cases}$$

The penalty assignment is based on the fact that, for infinite-horizon problems, there is no terminal penalty; instead, the cost per stage contains the penalties of

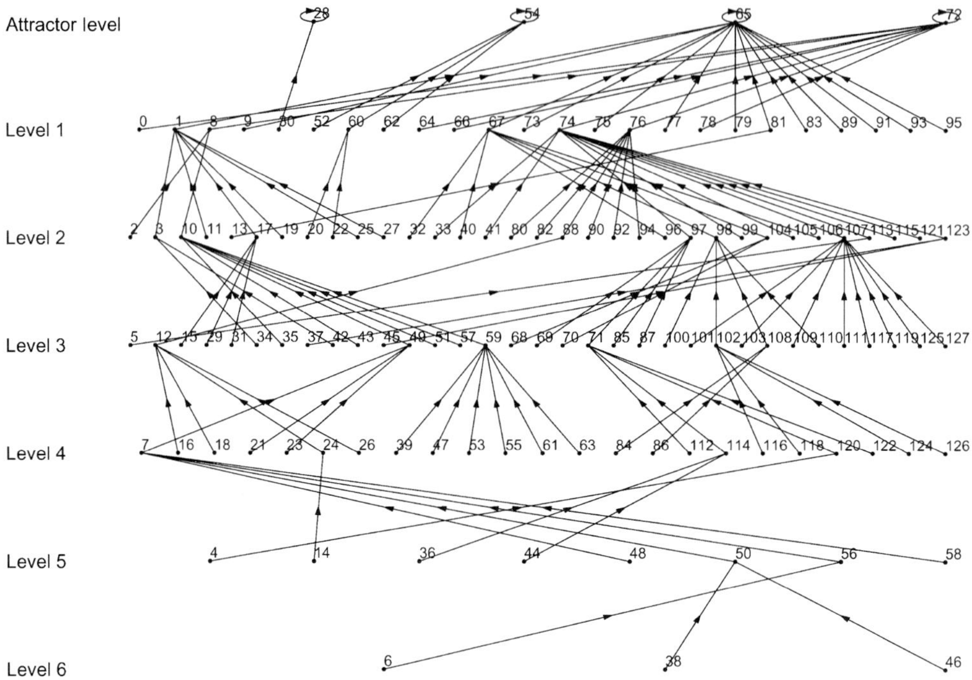

Figure 8.11 Network 1 [45].

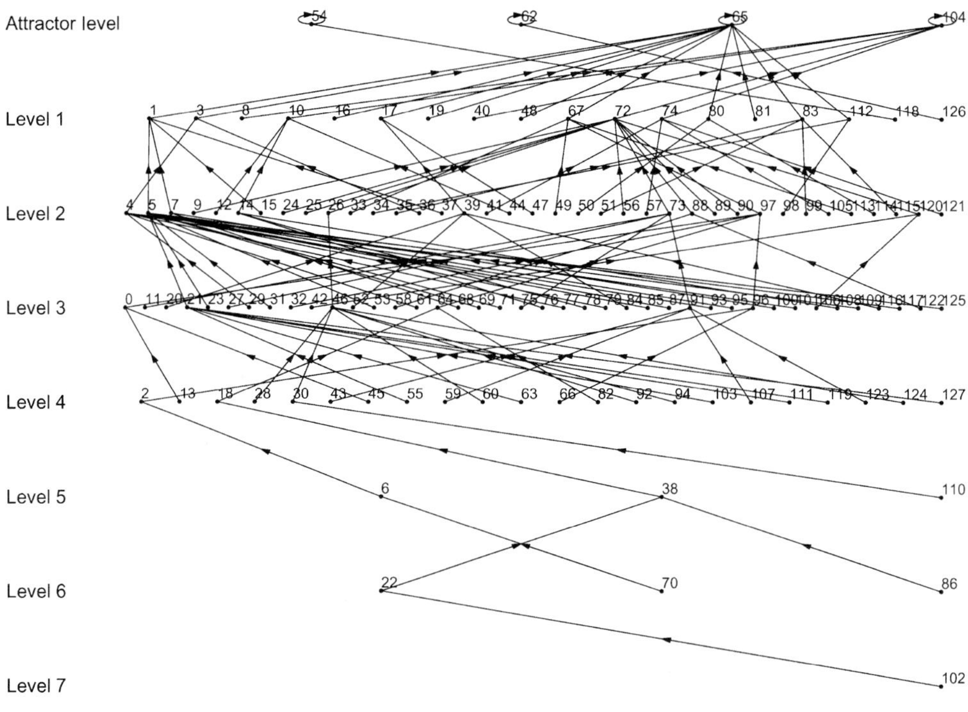

Figure 8.12 Network 2 [45].

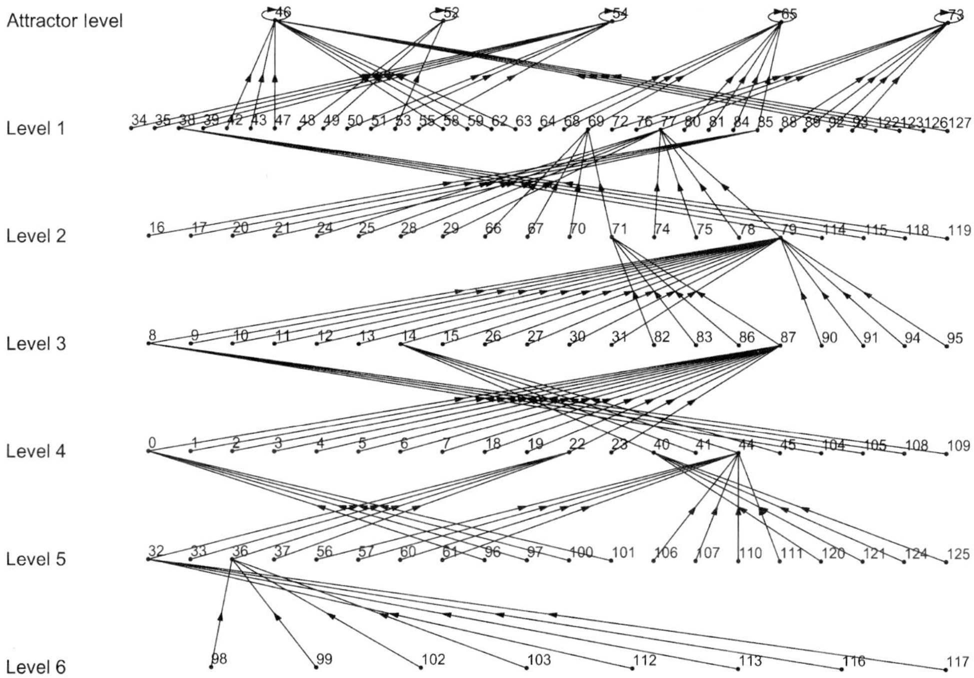

Figure 8.13 Network 3 [45].

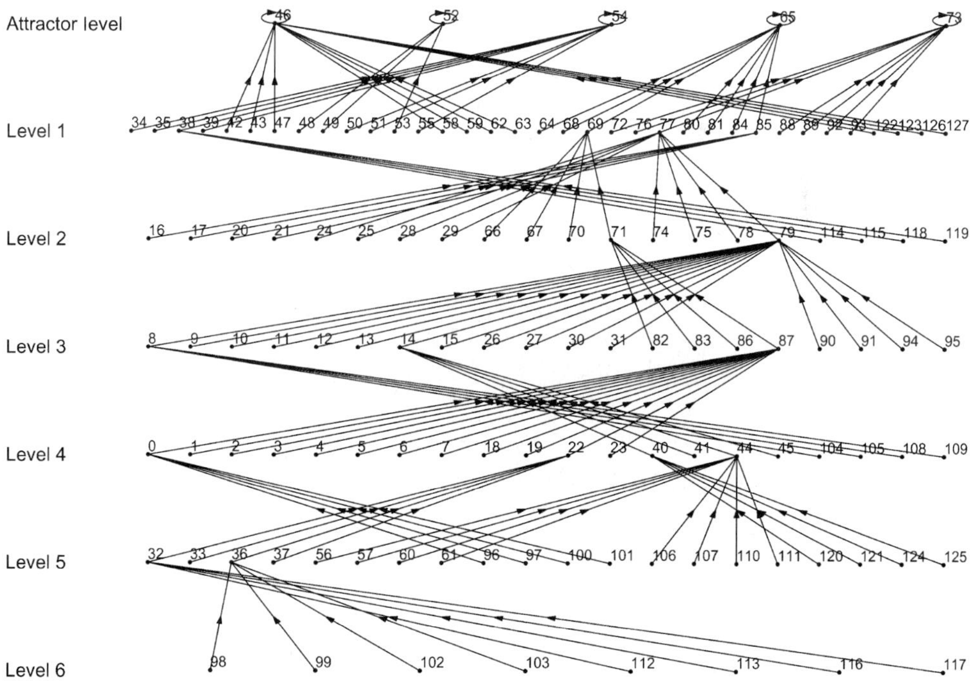

Figure 8.14 Network 4 [45].

each state. Since our objective is to downregulate the WNT5A gene, a higher penalty is assigned for destination states an upregulated WNT5A. Also, for a given WNT5A status for the destination state, a higher penalty is assigned when the control is active versus when it is not. First, the discounted approach is applied with the discount factor chosen to be 0.9. The control objective is to downregulate the WNT5A gene. Figure 8.15 shows the steady-state distributions of

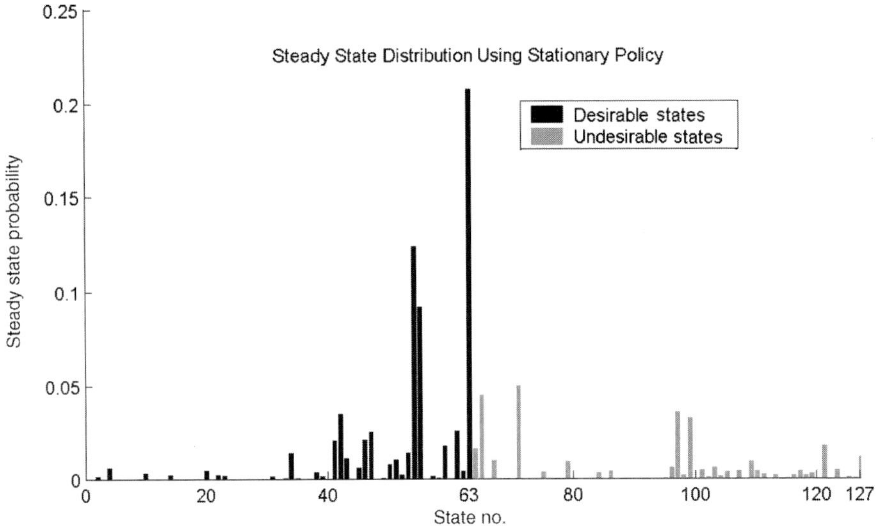

Figure 8.15 Steady state using discounted cost stationary policy [45].

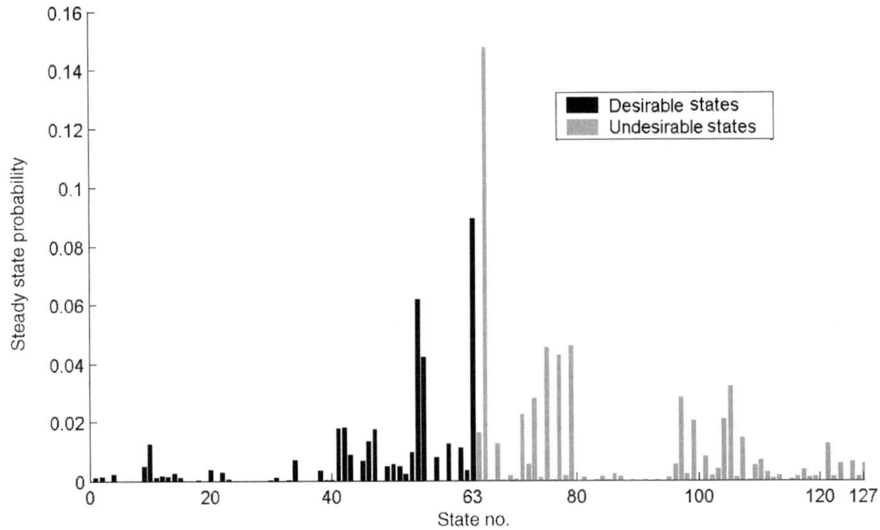

Figure 8.16 Original steady state [45].

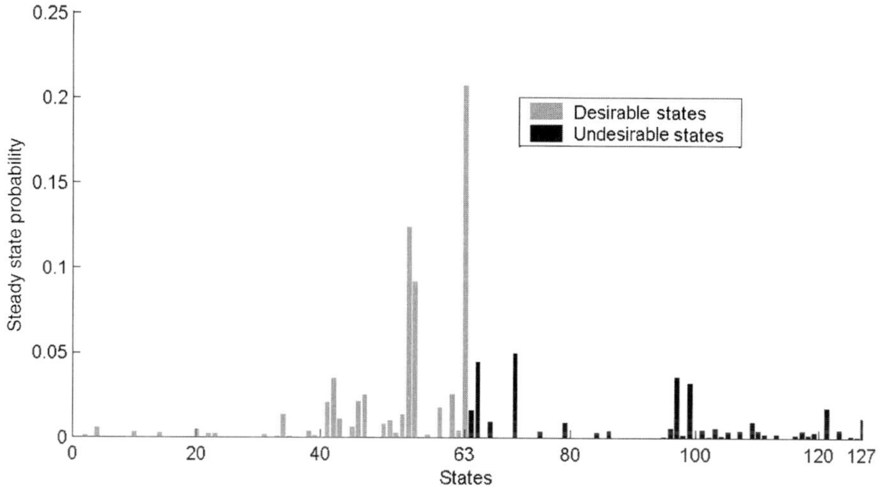

Figure 8.17 Steady state using average cost stationary policy [45].

the PBN using the obtained stationary policy, and Figure 8.16 shows the original PBN steady state for comparison. It should be noted that the desirable states are states from 0 to 63 since they have WNT5A 0, and in contrast the undesirable states are from 64 to 127 that have WNT5A 1. Comparing with Figure 8.16, we conclude that the stationary policy has enabled us to shift the probability mass from the bad states to states with lower metastatic competence. For example, state 65 (WNT5A is 1) has a high probability mass (0.15) in the original steady state, but stationary control has reduced its steady-state mass to 0.04. Next, the average-cost-per-stage approach is applied to design our optimal stationary policy. Both the value iteration and the policy iteration algorithms are used to calculate the optimal policy. The optimal policy obtained using the two iteration methods are the same. The steady-state distribution is shown in Figure 8.17 and is very similar to the steady-state distribution obtained using the previous total-cost formulation. Comparison of Figures 8.16 and 8.17 indicates that application of the stationary policy has been successful in shifting the steady-state distribution from undesirable to desirable states. The numerical value for the multiplication of the steady-state distribution with the cost vector is 1.7463 for the stationary policy, whereas it is 2.9830 for the uncontrolled PBN.

Acknowledgment

Y. Huang is supported by an NSF Grant CCF-0546345.

References

1 Alberts, B., Bray, D., Lewis, J., Raff, M., Roberts, K. and Watson, P. (2002) *Molecular Biology of the Cell*, 4th edn, Garland Publishing.

2 Lewin, B. (2003) *Genes VIII*, 1st edn, Prentice Hall.

3 Churchill, G.A. (2002) Fundamentals of experimental design for cDNA microarrays. *Nature Genetics*, **32** (Suppl.), 490–495.

4 Hernandez, P., Muller, M. and Appel, R.D. (2006) Automated protein identification by tandem mass spectrometry: issues and strategies. *Mass Spectrometry Reviews*, **25** (2), 235–254.

5 Lee, T.I., Rinaldi, N.J., Robert, F., Odom, D.T., Bar-Joseph, Z., Gerber, G.K., Hannett, N.M., Harbison, C.T., Thompson, C.M., Simon, I., Zeitlinger, J., Jennings, E.G., Murray, H.L., Gordon, D.B., Ren, B., Wyrick, J.J., Tagne, J.B., Volkert, T.L., Fraenkel, E., Gifford, D.K. and Young, R.A. (2002) Transcriptional regulatory networks in *Saccharomyces cerevisiae*. *Science*, **298**, 799–804.

6 MacBeath, G. and Schreiber, S.L. (2000) Printing proteins as microarrays for high-throughput function determination. *Science*, **289** (5485), 1760–1763.

7 Kitano, H. (ed.) (2001) *Foundations of System Biology*, The MIT Press.

8 Levchenko, A. (2001) Computational cell biology in the post-genomic era. *Molecular Biology Reports*, **28**, 83–89.

9 Kitano, H. (2002) Looking beyond the details: a rise in system-oriented approaches in genetics and molecular biology. *Current Genetics*, **41**, 1–10.

10 Kitano, H. (2002) Computational systems biology. *Nature*, **420**, 206–210.

11 Selinger, D.W., Wright, M.A. and Church, G.M. (2003) On the complete determination of biological systems. *Trends in Biotechnology*, **21** (6), 251–254.

12 de Jong, H. (2002) Modeling and simulation of genetic regulatory systems: a literature review. *Journal of Computational Biology*, **9** (1), 67–103.

13 Segal, E. (2004) Rich probabilistic models for genomic data. Ph.D. dissertation, Stanford University.

14 Friedman, N., Linial, M., Nachman, I. and Pe'er, D. (2000) Using Bayesian networks to analyze expression data. *Journal of Computational Biology*, **7** (3–4), 601–620.

15 Shmulevich, I., Dougherty, E.R., Kim, S. and Zhang, W. (2002) Probabilistic Boolean networks: a rule-based uncertainty model for gene regulatory networks. *Bioinformatics*, **18**, C261–C274.

16 Shmulevich, I., Dougherty, E.R. and Zhang, W. (2002) From Boolean to probabilistic Boolean networks as models of genetic regulatory networks. *Proceedings IEEE*, **90** (11), C1778–C1792.

17 Chen, T., He, H.L. and Church, G.M. (1999 January) Modeling gene expression with differential equations.Pacific Symposium on Biocomputing (PSB '99) vol. 4, Hawaii, USA, C29–C40.

18 Bower, J.M. and Bolouri, H. (eds) (2001) *Computational Modeling of Genetic and Biochemical Networks*, MIT Press, Cambridge, MA, USA.

19 Oppenheim, A.V. and Schafer, R.W. (1999) *Discrete-Time Signal Processing*, Prentice Hall.

20 Simon, I., Siegfried, Z., Ernst, J. and Bar-Joseph, Z. (2005) Combined static and dynamic analysis for determining the quality of time-series expression profiles. *Nature Biotechnology*, **23** (12), 1503–1508.

21 Chung, T., Brun, M. and Kim, S. (2006) Quantization of global gene expression data.5th International Conference on Machine Learning and Applications (ICMLA'06).

22 Di Camillo, B., Sanchez-Cabo, F., Toffolo, G., Nair, S.K., Trajanoski, Z. and Cobelli1, C. (2005) A quantization method based on threshold optimization for microarray

short time series. *BMC Bioinformatics*, **6** (Suppl. 4), S11.

23 Kauffman, S.A. (1993) *The Origins of Order: Self-organization and Selection in Evolution*, Oxford University Press, New York.

24 Dougherty, E.R., Kim, S. and Chen, Y. (2000) Coefficient of determination in nonlinear signal processing. *Signal Processing*, **80** (10), C2219–C2235.

25 Kim, S., Dougherty, E.R., Chen, Y., Sivakumar, K., Meltzer, P., Trent, J.M. and Bittner, M. (2000) Multivariate measurement of gene expression relationships. *Genomics*, **67** (2), C201–C209.

26 Kim, S., Dougherty, E.R., Bittner, M.L., Chen, Y., Sivakumar, K., Meltzer, P. and Trent, J.M. (2000) General nonlinear framework for the analysis of gene interaction via multivariate expression arrays. *Journal of Biomedical Optics*, **5** (4), C411–C424.

27 Shmulevich, I., Saarinen, A., Yli-Harja, O. and Astola, J. (2002) Inference of genetic regulatory networks via best-fit extensions in*Computational and Statistical Approaches to Genomics*, (eds W. Zhang and I. Shmulevich), Kluwer Academic Publishers, Boston, MA, USA.

28 Lähdesmäki, H., Shmulevich, I. and Yli-Harja, O. (2003) On learning gene regulatory networks under the Boolean network model. *Machine Learning*, **52** (1–2), C147–C167.

29 Zhou, X., Wang, X. and Dougherty, E.R. (2003) Construction of genomic networks using mutual information clustering and reversible-jump Markov-Chain-Monte-Carlo predictor design. *Signal Processing*, **83** (4), C745–C761.

30 Heyer, L.J., Kruglyak, S. and Yooseph, S. (1999) Exploying expression data: identification and analysis of coexpressed genes. *Genome Research*, **9**, C1106–C1115.

31 Zhou, X., Wang, X., Pal, R., Ivanov, I., Bittner, M.L. and Dougherty, E.R. (2004) A Bayesian connectivity-based approach to constructing probabilistic gene regulatory networks. *Bioinformatics*, **20** (17), C2918–C2927.

32 Robert, C.P. and Casella, G. (2004) *Monte Carlo Statistical Methods*, Springer-Verlag, New York.

33 Amundson, S.A., Bittner, M.L., Chen, Y., Trent, J.M., Meltzer, P. and Fornace, A.J. (1999) Fluorescent cDNA microarray hybridization reveals complexity and heterogeneity of cellular genotoxic stress response. *Oncogene*, **18**, C3666–C3672.

34 Kim, S., Dougherty, E.R., Bittner, M.L., Chen, Y., Sivakumar, K., Meltzer, P. and Trent, J.M. (2000) General nonlinear framework for the analysis of gene interaction via multivariate expression arrays. *Journal of Biomedical Optics*, **5** (4), C411–C424.

35 Bittner, M., Meltzer, P., Chen, Y., Jiang, Y., Seftor, E., Hendrix, M., Radmacher, M., Simon, R., Yakhini, Z., Ben-Dor, A., Sampas, N., Dougherty, E., Wang, E., Marincola, F., Gooden, C., Lueders, J., Glatfelter, A., Pollock, P., Carpten, J., Gillanders, E., Leja, D., Dietrich, K., Beaudry, C., Berens, M., Alberts, D. and Sondak, V. (2000) Molecular classification of cutaneous malignant melanoma by gene expression profiling. *Nature*, **406** (6795), 536–540.

36 Kim, S., Li, H., Dougherty, E.R., Cao, N., Chen, Y., Bittner, M. and Suh, E.B. (2002) Can Markov chain models mimic biological regulation. *Biological Systems*, **10** (4), 447–458.

37 Huang, S. (1999) Gene expression profiling, genetic networks, and cellular states: an integrating concept for tumorigenesis and drug discovery. *Journal of Molecular Medicine*, **77**, 469–480.

38 Shmulevich, I., Dougherty, E.R. and Zhang, W. (2002) Gene perturbation and intervention in probabilistic Boolean networks. *Bioinformatics*, **18**, 1319–1331.

39 Pal, R., Datta, A. and Dougherty, E.R. (2006) Intervention in probabilistic gene regulatory networks. *Current Bioinformatics*, **1**, 167–184.

40 Shmulevich, I., Dougherty, E.R. and Zhang, W. (2002) Control of stationary behavior in probabilistic Boolean networks by means of structural intervention. *Biological System*, **10**, 431–446.

41 Datta, A., Choudhary, A., Bittner, M.L. and Dougherty, E.R. (2003) External control in Markovian genetic regulatory networks. *Machine Learning*, **52**, 169–191.

42 Datta, A., Choudhary, A., Bittner, M.L. and Dougherty, E.R. (2004) External control in Markovian genetic regulatory networks: the imperfect information case. *Bioinformatics*, **20**, 924–930.

43 Pal, R., Datta, A., Bittner, M.L. and Dougherty, E.R. (2005) Intervention in context-sensitive probabilistic Boolean networks. *Bioinformatics*, **21**, 1211–1218.

44 Choudhary, A., Datta, A., Bittner, M. and Dougherty, E.R. (2005) Assignment of terminal penalties in controlling genetic regulatory networks. *American Control Conference*, 417–422.

45 Pal, R., Datta, A. and Dougherty, E.R. (2006) Optimal infinite-horizon control for probabilistic Boolean networks. *IEEE Transactions on Signal Processing*, **54** (6), 2375–2387.

46 Weeraratna, A.T., Jiang, Y., Hostetter, G., Rosenblatt, K., Duray, P., Bittner, M. and Trent, J. (2002) Wnt5a signalling directly affects cell motility and invasion of metastatic melanoma. *Cancer Cell*, **1**, 279–288.

47 Pal, R., Ivanov, I., Datta, A. and Dougherty, E.R. (2005) Generating Boolean networks with a prescribed attractor structure. *Bioinformatics*, **21**, C4021–C4025.

9
Structural Equation for Identification of Genetic Networks

Momiao Xiong

9.1
Introduction

Classical biology has mainly focused on studying the function of one gene or a few genes at a time [1]. However, the status of the cell and cellular processes are determined by large number of genes interwoven into networks, rather than a few genes [2]. A system-level understanding of a biological system holds a key to unraveling the relationship between genotype and phenotype.

Recent advances in genome sequencing and high-throughput technologies such as DNA chip and protein chip, which allow us to measure the spatiotemporal expression levels of thousands of genes or proteins [3–10], and the mass spectrometer [11,68] provide experimental tools to acquire knowledge of the systemic behavior of the biological processes. In the past several years, systems biology has evolved as a new discipline [12–17]. Systems biology investigates not only individual components of the biological system, but also their interaction and the behavior of the whole system. Investigation of the biological system involves (1) identification of individual components of the biological system; (2) discovery of the system structure that assembles individual components into a biological system; (3) revealing dynamic behavior of the biological system over time under various conditions; (4) studying the relationships between individual components when biological system is in equilibrium; (5) developing control mechanisms for optimally changing the biological systems into the desired systems; and (6) designing biological systems with good properties [12].

To develop statistical and computational methods for identification of genetic networks is one of the key issues in systems biology. Mathematical modeling and analysis have become central to unraveling genetic networks underlying biological processes. However, identification of large and complex genetic networks is difficult with direct intuitive approaches [18]. The widely used methods for construction of genetic networks have a bottom-up approach, that is, from submodels resulting from isolated and characterized components to a model of the integrated system. As

Analysis of Microarray Data: A Network-Based Approach. Edited by F. Emmert-Streib and M. Dehmer

ISBN: 978-3-527-31822-3

Alves and Savageau [18] point out, the fundamental problems inherent in the bottom-up approach are (1) the difficulty in identifying all the relevant components and relevant interactions and (2) the accurate estimation of all the relevant parameters in the models. In a top-down approach, the models that can reproduce the data are constructed based on the variables measured in the whole system. This approach is gaining increased interests due to microarray technology and large-scale gene expression data. Microarrays can be used to gain detailed information about transcriptional changes involved in a specific genetic network and global expression profiles assembled from diverse experiments, potentially leading to the identification of novel components of the signaling system. They can also be used to obtain a fingerprint of the transcriptional status of the cell under a given condition, which may be useful for characterizing the genetic networks in response to environmental stimuli [19].

One top-down approach is to infer regulatory networks by discovering distinct gene expression patterns (clusters) and identifying upstream transcriptional factor binding sites and coregulated genes [20,21]. This approach has some disadvantages. First, it is difficult to infer the detailed structure of genetic networks. Second, it has not provided tools for assessing how well the inferred networks explain the observed data. Third, this approach is unlikely to perform quantitative analysis of genetic networks. Several studies have used Boolean network [22,23] or differential equations [24,25], or neural networks [26] to model genetic networks.

These approaches, however, are in their infancy. The number of parameters in these models is extremely large. Since the size of tissue samples or the number of experiments is quite small, the observed data are often insufficient to estimate the parameters. The structures of the networks estimated by these methods are unable to reproduce the observed data. It is highly unlikely that the observed gene expression data set is large enough to support the parameter estimation of the full models in which all the genes in the networks are connected. To overcome this difficulty, several methods for the reconstruction of genetic networks have been developed. Woolf and Wang [27] proposed to use fuzzy logic to construct genetic networks. Several authors proposed to use Bayesian networks to estimate genetic networks and functional structure [28–30]. Bayesian networks can deal with stochastic feature of gene expression data and can use prior knowledge of genetic networks. Although Bayesian networks are powerful tools for modeling genetic networks, they are computationally intensive. Furthermore, discrete Bayesian networks do not explicitly provide quantitative relationship between genes in the networks.

Comprehensive understanding of the complex genetic networks will increasingly require a novel conceptual framework for a quantitative description of genetic networks and methods for integrating experimental and theoretical/computational approaches. To completely identify physically connected genetic networks and to infer true causal relationship between gene expressions using expression profiles may be too ambitious to be accomplished. Instead of reconstructing physically connected genetic networks, we attempt to model quasi-genetic networks that may not be physically connected but best fit the gene expression data. The identified quasi-genetic networks describe the quantitative relationship between genes in the networks. In this chapter, we propose to use structure equations [31] as a tool to identify and model quasi-genetic networks. We will provide (1) mathematical representation of genetic networks based on structural equations, (2) statistical methods for estimation and testing for the parameters in the models, (3) probabilistic criteria for assessing how well the models of the quasi-genetic networks explain observed data, and (4) optimization procedures for searching structure of the quasi-genetic networks.

Once a genetic network is identified, it is crucial to associate genetic networks with function and phenotypes of cells and tissues. We often use differentially expressed genes to discriminate phenotypes of cells. To associate genetic networks with function and phenotypes of cells, the notion of differentially expressed genes can be generalized to genetic networks and statistics to test for the differential expression of networks that need to be developed.

Coefficient parameters in the structural equations measure the regulatory effects of one gene on others or the strength of the gene–gene interactions. Functional mutations in the genes will often cause changes in regulatory effects. Thus, we expect that due to the accumulation of mutations in abnormal cells, the regulation of some genetic networks in abnormal cells will be significantly different from that in normal cells. Uncovering such differences may help us identify the causes of disease. To accomplish this task, several statistics to measure the differences in regulation between the genetic networks in normal and abnormal cells have been developed. By identifying differentially regulated genetic networks, a set of genes and genetic networks that influence the development of the diseases is likely to be discovered.

9.2 Models

Traditional regressions describe one-way or unidirectional cause-and-effect relations in which the variables on the left sides of the equations are dependent variables and the variables on the right sides of the equations are explanatory variables or independent variables. The explanatory variables are used to predict the outcomes of the dependent variables. However, in many cases, there are two-way, or simultaneous, relationships between the variables. Some variables are response variables in some equations but will be predictors in others. The variables in equations may influence each other. It is difficult to distinguish between dependent variables and explanatory variables. Therefore, we need to lump together a set of variables that can be simultaneously predicted by another set of variables. Complex cause–effect relationships require complex models. Structural equations, also called simultaneous equations, are such a powerful tool to represent two-way causal relationships among the variables [32].

For easy illustration of structural equation for modeling a general genetic network, we first consider an example. Figure 9.1 plots a part of TGFB pathway. A pathway can be easily described by a path diagram. Path diagrams are graphical

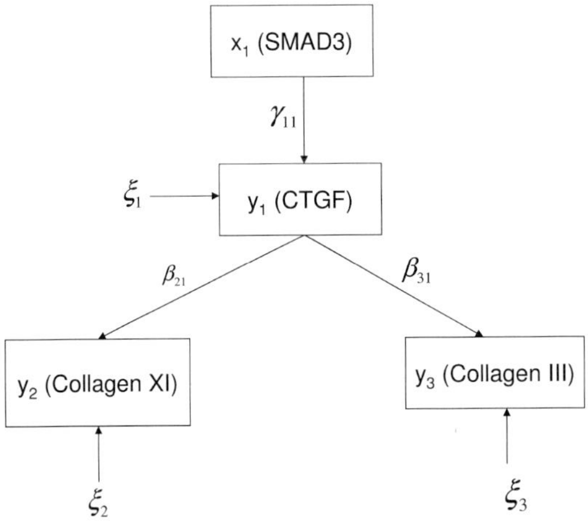

Figure 9.1 A path diagram for TGF-β pathways with 10 genes reconstructed from gene expression data of SSc studies of total 16 abnormal tissue samples and 32 normal tissue samples. The number along the edges is the estimated regulatory effect of one gene on another.

representation of pathway. They show variables interconnected with lines that are used to indicate regulation relationships or causal flow. We can think of a path diagram as a graph for showing which variables cause changes in other variables or which genes activate or inhibit expressions of other genes [31]. In the path diagram, the observed variables are enclosed in boxes. The unobserved variables or hidden variables are placed in the circles except for the disturbance terms, which are not enclosed. Straight single-headed arrows represent regulation relations or causal relations between the genes or variables.

In general, each gene expression is represented by a variable. The observed variables (gene expressions) are enclosed in boxes. The unobserved or latent variables are circled with the exception of the disturbance terms, which are not enclosed. Straight single-headed arrows represent causal relations between the variables connected by the arrows. Arrow can represent either activation (positive control) or inhibition (negative control), which will be described by a positive or a negative coefficient in the structural equations. Synergic action of genes will be represented by nonlinear terms in the nonlinear structural equations and will not be discussed in the linear structural equation models. The boxes and circles are also referred to as nodes. There may be intermediate steps between the nodes, which are unknown and omitted. A curved two-headed arrow indicates an association between two variables. The variables may be associated for any of a number of reasons. The association may be due to both variables depending on some other variable(s), or the variables may have a causal relationship, but this remains unspecified. Unenclosed variable signifies a disturbance term (error in either equation or measurement). Two straight single-headed arrows connecting two variables denote feedback relation.

The path diagram in Figure 9.1 can be described by the following structural equations:

$$\begin{aligned} y_1 &= \gamma_{11} x_1 + \xi_1, \\ y_2 &= \beta_{21} y_1 + \xi_2, \\ y_3 &= \beta_{31} y_1 + \beta_{32} y_2 + \xi_3. \end{aligned} \tag{9.1}$$

Structural equations classify variables into two class variables: endogenous and exogenous variables. The jointly dependent variables that are determined in the model are called endogenous variables. The explanatory variables that are determined outside the model or predetermined are called exogenous variables. The variable x_1 in Equation (9.1) is an exogenous variable. Although x_1 also represents the expression of the gene SMAD3, from the system point of view, it lies outside the model. The variables y_1, y_2, and y_3 are observed gene expressions or endogenous variables, which are determined inside the models. The coefficient γ_{11} represents the effect of the exogenous variable x_1 on the endogenous variable y_1, that is, the influence of the expression of the gene SMAD3 on the gene CTGF. The coefficients β_{21} and β_{31} denote the effects of the variable y_1 on the variables y_2 and y_3, respectively. The coefficient β_{32} denotes the effect of the variable y_2 on the variable y_3. Since these variables represent the influence of the expression of the gene CTGF on the expressions of the genes collagen XI and collagen III, they are called regulatory effects.

The regulatory effect of one gene on another consists of two parts: direct effect and indirect effect. The direct regulatory effect is the influence of one gene on another without mediation of other variables in a path diagram. The indirect regulatory effect of a gene is the effect that is mediated by at least one intervening gene. Therefore, the total regulatory effect is the sum of the direct regulatory effect and indirect regulatory effect.

For example, the total regulatory effect of the gene CTGF on the gene collagen III is

$$\text{total regulatory effect} = \beta_{32} + \beta_{21} + \beta_{23}.$$

Equation (9.1) can also be written in a matrix form:

$$\begin{bmatrix} y_1 \\ y_2 \\ y_3 \end{bmatrix} = \begin{bmatrix} 0 & 0 & 0 \\ \beta_{21} & 0 & 0 \\ \beta_{31} & \beta_{32} & 0 \end{bmatrix} \begin{bmatrix} y_1 \\ y_2 \\ y_3 \end{bmatrix} + \begin{bmatrix} \gamma_{11} \\ 0 \\ 0 \end{bmatrix} x_1 + \begin{bmatrix} \zeta_1 \\ \zeta_2 \\ \zeta_3 \end{bmatrix}$$

In general, the gene expressions in the genetic network can be described by the structural equations. For modeling a genetic network, we first select a set of genes that are of interests. The expressions of the selected genes in the network will be endogenous variables. The expressions of the genes that are connected to the selected genes, but outside the model, and environmental variables such as dosage of drugs and chemical compounds are exogenous variables. Assume that there are n endogenous variables and m exogenous variables. Let $y_1, y_2, \ldots, y_n$ be the endogenous variables and $x_1, x_2, \ldots, x_m$ be the exogenous variables. Assume that the errors (or noises) $\zeta_1, \zeta_2, \ldots, \zeta_n$ are independently and identically distributed to be normal with the mean of zeros and variances $\sigma_i^2, i = 1, 2, \ldots, n$.

We also assume that the errors are uncorrelated with the exogenous variables.

Then, the gene expressions in the network can be described by the following structural equations:

$$\begin{aligned} y_1 &= \beta_{12}y_2 + \beta_{13}y_3 + \cdots + \beta_{1n}y_n + \gamma_{11}x_1 + \gamma_{12}x_2 + \cdots + \gamma_{1m}x_m + \xi_1, \\ y_2 &= \beta_{21}y_1 + \beta_{23}y_3 + \cdots + \beta_{2n}y_n + \gamma_{21}x_1 + \gamma_{22}x_2 + \cdots + \gamma_{2m}x_m + \zeta_2, \\ y_n &= \beta_{n1}y_1 + \beta_{n2}y_2 + \beta_{n3}y_3 + \cdots + \beta_{nn}y_n + \gamma_{n1}x_1 + \gamma_{n2}x_2 + \cdots + \gamma_{nm}x_m + \zeta_n. \end{aligned} \tag{9.2}$$

In practice, not every variable will appear in the equations. Thus, many terms in the equations will disappear. Let $Y = [y_1, y_2, \ldots, y_n]^T$, $X = [x_1, x_2, \ldots, x_m]^T$, and $\zeta = [\zeta_1, \zeta_2, \ldots, \zeta_n]^T$. Let the coefficients in the equations be written in the following matrix forms:

$$B = \begin{bmatrix} 0 & \beta_{12} & \cdots & \beta_{1n} \\ \beta_{21} & 0 & \cdots & \beta_{2n} \\ \vdots & \vdots & \cdots & \vdots \\ \beta_{n1} & \beta_{n2} & \cdots & \beta_{nn} \end{bmatrix}, \qquad \Gamma = \begin{bmatrix} \gamma_{11} & \gamma_{12} & \cdots & \gamma_{1m} \\ \gamma_{21} & \gamma_{22} & \cdots & \gamma_{2m} \\ \vdots & \vdots & \cdots & \vdots \\ \gamma_{n1} & \gamma_{n2} & \cdots & \gamma_{nm} \end{bmatrix}.$$

Then, Equation (9.2) can be rewritten as

$$Y = BY + \Gamma X + \zeta. \tag{9.3}$$

The full model for gene expressions in a genetic network consists of a system of structural equations. The equations contain random variables, structural parameters, and, sometimes, nonrandom variables. The three types of random variables are latent, observed, and disturbance/error variables. The nonrandom variables (environment, e.g., drug doses) are explanatory variables whose values remain the same in repeated random sampling (fixed or nonstochastic variables). The links between the variables (e.g., the relationship between the gene expressions) are summarized in the structural parameters. The structural parameters are invariant constants that provide the activation or inhibition relation between the gene expressions, or generally, the "causal" relation between variables.

The structural parameters may describe the causal link between unobserved variables, between observed variables, or between unobserved and observed variables. The variables are classified into exogenous variables, which lie outside the model, and endogenous variables. Environmental stimuli such as drug dosage and some gene expressions can be viewed as exogenous variables. Most gene expressions are viewed as endogenous variables. The terms exogenous and endogenous are model specific. It may be that an exogenous variable in one model is endogenous in another.

9.3 Covariance Matrix

Covariance matrices are the fundamental quantities for the analysis of structural equations. Let Φ be the covariance matrix of exogenous variables and ψ be the covariance matrix of the errors. Covariance matrix of the endogenous variables is denoted by Σ_{yy}. For Equation (9.1), the covariance matrix Σ_{yy} can be calculated as follows:

$$\mathrm{VAR}(y_1) = \gamma_{11}^2\phi + \psi_{11},$$

$$\mathrm{COV}(y_1, y_2) = \beta_{21}(\gamma_{11}^2\phi + \psi_{11}), \quad \mathrm{VAR}(y_2) = \beta_{21}^2(\gamma_{11}^2\phi + \psi_{11}) + \psi_{22},$$

$$\begin{aligned}\mathrm{COV}(y_3, y_1) = &\beta_{31}(\gamma_{11}^2\phi + \psi_{11}) + \beta_{32}\beta_{21}(\gamma_{11}^2\phi + \psi_{11}) \\ &+ \beta_{32}[\beta_{21}^2(\gamma_{11}^2\phi + \psi_{11}) + \psi_{22}],\end{aligned}$$

$$\mathrm{COV}(y_3, y_2) = \beta_{31}\beta_{21}(\gamma_{11}^2\phi + \psi_{11}) + \beta_{32}(\beta_{21}^2[\gamma_{11}^2\phi + \psi_{11}) + \psi_{22}],$$

$$\begin{aligned}\mathrm{VAR}(y_3) = &\beta_{31}^2(\gamma_{11}^2\phi + \psi_{11}) + \beta_{32}^2[\beta_{21}^2(\gamma_{11}^2\phi + \psi_{11}) + \psi_{22}] \\ &+ \beta_{31}\beta_{32}[\beta_{21}^2(\gamma_{11}^2\phi + \psi_{11}) + \psi_{22}] + \psi_{33}.\end{aligned}$$

From these equations, we can see that the variance and covariance of the endogenous variables are the parameters of the model. This implies that from the knowledge of the variance and covariance of the endogenous variables we can estimate the parameters of the model. The variance and covariance of the endogenous variables are basic to an understanding of estimation and model fit. Now, we derive the general formula for the calculation of the endogenous and exogenous variables. It follows from Equation (9.3) that

$$Y = (I-B)^{-1}(\Gamma X + \zeta). \tag{9.4}$$

Thus, from the standard statistical theory of the variance and covariance of the composite function of random variables, we have

$$\begin{aligned}\Sigma_{yy} &= \mathrm{cov}(Y, Y) \\ &= (I-B)^{-1}[\Gamma \mathrm{cov}(X, X)\Gamma^{\mathrm{T}} + \mathrm{cov}(\zeta, \zeta)](I-B)^{-\mathrm{T}} \\ &= (I-B)^{-1}[\Gamma\Phi\Gamma^{\mathrm{T}} + \Psi](I-B)^{-\mathrm{T}}.\end{aligned} \tag{9.5}$$

Similarly, we can find the covariance matrix between the exogenous variables and endogenous variables:

$$\begin{aligned}\Sigma_{yx} &= \text{cov}(Y,X) = (I-B)^{-1}[\Gamma\text{cov}(X,X) + \text{cov}(\zeta,X)] \\ &= (I-B)^{-1}\Gamma\Phi.\end{aligned} \tag{9.6}$$

Thus, the covariance matrix of endogenous and exogenous variables is given by

$$\Sigma = \begin{bmatrix}\Sigma_{xx} & \Sigma_{xy} \\ \Sigma_{yx} & \Sigma_{yy}\end{bmatrix} = \begin{bmatrix}(I-B)^{-1}[\Gamma\Phi\Gamma^{\mathrm{T}} + \Psi](I-B)^{-\mathrm{T}} & (I-B)^{-1}\Gamma\Phi \\ \Phi\Gamma^{\mathrm{T}}(I-B)^{-\mathrm{T}} & \Phi\end{bmatrix}. \tag{9.7}$$

9.4 Estimation

Estimation of the parameters in the structural equations is rather complex. It involves many different estimation methods with varying statistical properties. In this chapter, we introduce two popular methods: maximum likelihood method and two-stage least square (2SLS) method.

9.4.1 Likelihood Function

We assume that both endogenous and exogenous variables follow normal distribution. Suppose that there are p endogenous variables and q exogenous variables. Let $Z = [Y, X]^{\mathrm{T}}$. Then, the density function of the vector of endogenous and exogenous variables is given by

$$f(Z,\Sigma) = (2\pi)^{-(p+q)/2}|\Sigma|^{-1/2}\exp\left[-\frac{1}{2}Z^{\mathrm{T}}\Sigma^{-1}Z\right]. \tag{9.8}$$

Then, the likelihood function of the n independent samples of observations is given by

$$f(Z_1, Z_2, \ldots, Z_n; \Sigma) = (2\pi)^{-n(p+q)/2}|\Sigma|^{-n/2}\exp\left[-\frac{1}{2}\sum_{i=1}^{n} Z_i^{\mathrm{T}}\Sigma^{-1}Z_i\right].$$

Let $S = 1/n\sum_{i=1}^{n} Z_iZ_i^{\mathrm{T}}$. The log-likelihood is then given by

$$L(\theta) = -\frac{n(p+q)}{2}\log(2\pi) - \frac{n}{2}\log|\Sigma(\theta)| - \frac{n}{2}\text{tr}\left[S\Sigma^{-1}(\theta)\right], \tag{9.9}$$

where tr denotes the trace of the matrix. Sine the constants will not affect the optimal values maximizing the likelihood function, Equation (9.9) can be reduced to

$$L(\theta) \approx -\log|\Sigma(\theta)| - \mathrm{tr}[S\Sigma^{-1}(\theta)]. \tag{9.10}$$

9.4.2
Maximum Likelihood Estimators

The estimators θ of the parameters in the structural equations can be found by maximizing the likelihood function or equivalently by minimizing the following fitting function [31]:

$$F_{\mathrm{ML}} = \log|\Sigma(\theta)| + \mathrm{tr}[S\Sigma^{-1}(\theta)] - \log|S| - (p+q). \tag{9.11}$$

When the estimated parameters satisfy $\Sigma(\hat{\theta}) = S$, then we have

$$F_{\mathrm{ML}} = \log|S| + (p+q) - \log|S| - (p+q) = 0.$$

Thus, when the estimated parameters perfectly predict the sampling covariance matrix, the fitting function F_{ML} will be equal to zero. We take the following simple example that consists of the first expression in Equation (9.1) to illustrate how to calculate the fitting function and find the maximum likelihood estimators:

$$y_1 = \gamma_{11}x_1 + \zeta_1. \tag{9.12}$$

Its covariance matrix is given by

$$\Sigma = \begin{bmatrix} \gamma_{11}^2\phi + \psi & \gamma_{11}\phi \\ \gamma_{11}\phi & \phi \end{bmatrix}.$$

Thus, its inverse matrix is

$$\Sigma^{-1} = \frac{1}{\phi\psi}\begin{bmatrix} \phi & -\gamma_{11}\phi \\ -\gamma_{11}\phi & \gamma_{11}^2\varphi + \psi \end{bmatrix}.$$

Let

$$S = \begin{bmatrix} \mathrm{var}(y_1) & \mathrm{cov}(y_1, x_1) \\ \mathrm{cov}(y_1, x_1) & \mathrm{var}(x_1) \end{bmatrix}$$

Then, we have the following fitting function:

$$F_{\mathrm{ML}} = \log|\psi\phi| + \frac{1}{\psi}\left[\mathrm{var}(y_1) - 2\gamma_{11}\mathrm{cov}(y_1, x_1) + \gamma_{11}^2\mathrm{var}(x_1)\right] + \frac{\mathrm{var}(x_1)}{\phi} - \log \times \left[\mathrm{var}(y_1)\mathrm{var}(x_1) - (\mathrm{cov}(y_1, x_1))^2\right] - 2. \quad (13)$$

For this simple problem, we can use the standard necessary condition for the minimization of the fitting function to find the maximum likelihood estimators:

$$\frac{\partial F_{\mathrm{ML}}}{\partial r_{11}} = -2\mathrm{cov}(y_1, x_1) + 2\gamma_{11}\mathrm{var}(x_1) = 0,$$

$$\frac{\partial F_{\mathrm{ML}}}{\partial \phi} = \frac{1}{\phi} - \frac{\mathrm{var}(x_1)}{\phi^2} = 0, \quad (9.14)$$

$$\frac{\partial F_{\mathrm{ML}}}{\partial \psi} = \frac{1}{\psi} - \frac{1}{\psi^2}\left[\mathrm{var}(y_1) - 2\gamma_{11}\mathrm{cov}(y_1, x_1) + \gamma_{11}^2\mathrm{var}(x_1)\right] = 0.$$

Solving Equation (9.14) yields

$$\gamma_{11} = \frac{\mathrm{cov}(y_1, x_1)}{\mathrm{var}(x_1)},$$

$$\phi = \mathrm{var}(x_1), \quad (9.15)$$

$$\psi = \mathrm{var}(y_1) - \frac{[\mathrm{cov}(y_1, x_1)]^2}{\mathrm{var}(x_1)}.$$

In general, fitting functions are complicated. It is infeasible to obtain the explicit solutions. Maximizing likelihood function is an unconstrained optimization problem. Many optimization methods can be used to solve this problem. Here, we introduce the widely used Newton–Raphson algorithm for minimizing the fitting function.

The iterative formula for updating the estimators is

$$\hat{\theta}_{k+1} = \hat{\theta}_k - \left[\frac{\partial F_{\mathrm{ML}}}{\partial\theta\partial\theta^{\mathrm{T}}}\right]^{-1} \frac{\partial F_{\mathrm{ML}}}{\partial\theta}\Big|_{\hat{\theta}_k}. \quad (9.16)$$

The procedures for Newton–Raphson algorithm are given by

1. start with initial value;
2. update the parameter by Equation (9.16);
3. stop when the criteria are satisfied.

Model (9.12) is used here to illustrate the iterative procedures. Assume that the sampling matrix S is given by

$$S = \begin{bmatrix} 2 & 1 \\ 1 & 1 \end{bmatrix}.$$

Then, the gradient and Hessian matrix of the fitting function F_{ML} is given by

$$\begin{bmatrix} \dfrac{\partial F_{\mathrm{ML}}}{\partial r_{11}} \\[2ex] \dfrac{\partial F_{\mathrm{ML}}}{\partial \phi} \\[2ex] \dfrac{\partial F_{\mathrm{ML}}}{\partial \psi} \end{bmatrix} = \begin{bmatrix} \dfrac{2(\gamma_{11}-1)}{\psi} \\[2ex] \dfrac{1}{\phi}-\dfrac{1}{\phi^2} \\[2ex] \dfrac{1}{\psi}-\dfrac{1}{\psi}(2-2\gamma_{11}+\gamma_{11}^2) \end{bmatrix},$$

$$\frac{\partial^2 F_{\mathrm{ML}}}{\partial\theta\partial\theta^{\mathrm{T}}} = \begin{bmatrix} \dfrac{2}{\psi} & 0 & -\dfrac{2(\gamma_{11}-1)}{\psi^2} \\[2ex] 0 & -\dfrac{1}{\phi^2}+\dfrac{2}{\phi^3} & 0 \\[2ex] -\dfrac{2(\gamma_{11}-1)}{\psi^2} & 0 & -\dfrac{1}{\psi^2}+\dfrac{2}{\psi^3}(2-2\gamma_{11}+\gamma_{11}^2) \end{bmatrix}.$$

Table 9.1 lists the intermediate estimated parameters and the values of the fitting function. Figure 9.2 plots the values of the fitting function as a function of the number of iterations. We can see that after six iterations, the fitting function reaches

Table 9.1 The estimated parameters and fitting function.

Number of iteration	$\hat{\gamma}_{11}$	$\hat{\phi}$	$\hat{\psi}$
0	0.5	0.5	0.5
1	0.5833	0.6667	0.6667
2	0.7750	0.8333	0.8044
3	0.9250	0.9524	0.9153
4	0.9881	0.9957	0.9828
5	0.9996	1.0000	0.9993
6	1.0000	1.0000	1.0000
7	1.0000	1.0000	1.0000
8	1.0000	1.0000	1.0000
9	1.0000	1.0000	1.0000
10	1.0000	1.0000	1.0000

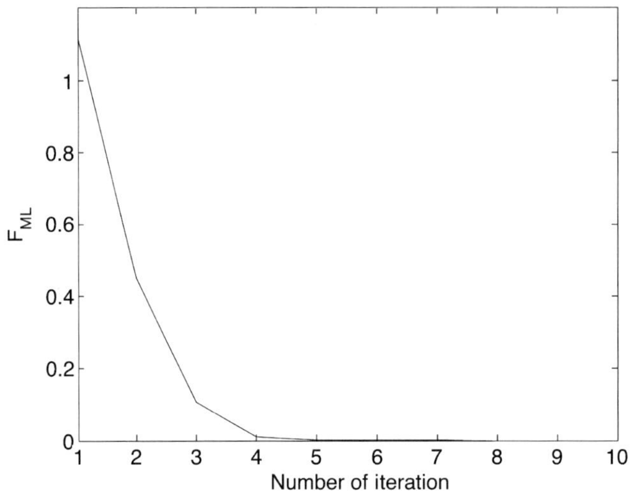

Figure 9.2 Values of the fitting function as a function of the number of iterations.

zero and the estimated parameters converge to the values obtained by explicit analytic solutions.

9.4.3
Asymptotic Properties of the Maximum Likelihood Estimators and Test Statistics

When sample size is large enough, the large sample theory ensures that the maximum likelihood estimators are consistent estimators and asymptotically distributed as a normal distribution [69]:

$$\hat{\theta}_n \sim N\left(\theta, \frac{1}{n} I^{-1}(\theta)\right),$$

where $I(\theta)$ is an information matrix defined as

$$I(\theta) = -E\left[\frac{\partial^2 L(\theta)}{\partial\theta\partial\theta^{\mathrm{T}}}\right].$$

If we use the fitting function to calculate the information matrix, then we have

$$I(\theta) = \frac{n-1}{2} E\left[\frac{\partial^2 F_{\mathrm{ML}}}{\partial\theta\partial\theta^{\mathrm{T}}}\right].$$

Thus, the estimators are asymptotically distributed as

$$\hat{\theta}_n \sim N\left(\theta, \frac{2}{n-1}\left\{E\left[\frac{\partial^2 F_{\mathrm{ML}}}{\partial\theta\partial\theta^{\mathrm{T}}}\right]\right\}^{-1}\right). \tag{9.17}$$

Consider model (9.12). Suppose that the sampling matrix

$$S = \begin{bmatrix} 2 & 1 \\ 1 & 1 \end{bmatrix}$$

is obtained for the sample size that is equal to 100. Then, the variance–covariance matrix is equal to

$$I^{-1}(\theta) = \begin{bmatrix} 0.0101 & 0 & 0 \\ 0 & 0.0202 & 0 \\ 0 & 0 & 0.0202 \end{bmatrix}.$$

When the variance–covariance matrix of the estimators is known, we can construct the statistic to test whether the gene is connected to another gene in the network. Suppose we want to test whether the ith gene influences the expression of the jth gene. We can construct the following statistic:

$$T = \frac{\hat{\beta}_{ij}^2}{\text{var}(\hat{\beta}_{ij})}.$$

The variance $\text{var}(\hat{\beta}_{ij})$ can be obtained from the inverse of the information matrix. The statistic will asymptotically follow a central $\chi^2_{(1)}$ distribution.

To test the overall model fit, we use the statistics $(n-1)F_{\text{ML}}$. It can be shown that $(n-1)F_{\text{ML}}$ follows a central χ^2 distribution with $1/2(p+q)(p+q+1)-t$ degrees of freedom [31], where t is the number of free parameters.

9.4.4 Two-Stage Least Square Method

The previous maximum likelihood method needs to assume the distribution of the endogenous and exogenous variables. However, in many cases, the exogenous variables are discrete variables. Normality assumptions on these variables will not hold. Therefore, we need to develop estimation methods that do not assume distributions of the endogenous and exogenous variables. Theoretic analysis indicates that the ordinary least square (OLS) methods for the estimation of the parameters in the structural equations, in general, will lead to inconsistent estimators. The statistical methods without assuming distribution of the endogenous and exogenous variables are either single-equation methods, which can be applied to each equation of the structural equations, or complete systems methods, which are applied to all structural equations. Here, the most important and widely used single-equation method is introduced: the 2SLS method [32,70].

9.4.4.1 Reduce Form

Any structural equations can be reduced to

$$Y = X\Pi + V. \tag{9.18}$$

In other words, the endogenous variables can be expressed as functions of exogenous variables. Regressing the endogenous variables on the exogenous variables, we have

$$\hat{\Pi} = (X^T X)^{-1} X^T Y. \tag{9.19}$$

The predicted value by regression is then given by

$$\hat{Y} = X(X^T X)^{-1} X^T Y. \tag{9.20}$$

9.4.4.2 Two-Stage Least Squares Estimation

The particular equation in the model can be written as

$$y = Y_1\beta + X_1\gamma + \xi, \tag{9.21}$$

where y is the $n\times 1$ vector of observation on the "dependent" variable of the particular equation of interest, Y_1 is the $n\times g$ matrix of observations on the other current endogenous variables included in the equation of interest, β is the $g\times 1$ structural regulatory coefficients associated with the variables in Y_1, X_1 is the $n\times k$ matrix of observations on the exogenous variables appearing in the equation, γ is the $k\times 1$ vector of coefficients associated with X_1, and ξ is the $n\times 1$ vector of the errors in the equation of interest.

Since the variables in Y_1 may be correlated with ξ, the OLS method cannot be applied to Equation (9.21). The basic idea behind 2SLS is to replace Y_1 by its approximate $\hat{Y}_1$, which is uncorrelated with X_1. This will be accomplished by applying OLS to the reduce form (9.18).

2SLS consists of the following two stages.

Stage 1: Regressing each variable in Y_1 on all the predetermined variables in the whole systems. Let $X = [X_1, X_2]$ be the $n\times K$ matrix of observations on all the predetermined variables in the whole system, and X_2 be the matrix of observations on those predetermined exogenous variables that are excluded from the equation under study. Then, the predicted value $\hat{Y}_1$ by applying OLS to the reduced form is then given by

$$\hat{Y}_1 = X(X^T X)^{-1} X^T Y_1. \tag{9.22}$$

Stage 2: In the second stage, we replace Y_1 by $\hat{Y}_1$. Then, regressing y on $\hat{Y}_1$ and X_1 leads to the following equation:

$$\begin{bmatrix} \hat{Y}_1^T \hat{Y}_1 & \hat{Y}_1^T X_1 \\ X_1^T \hat{Y}_1 & X_1^T X_1 \end{bmatrix} \begin{bmatrix} \hat{\beta} \\ \hat{\gamma} \end{bmatrix} = \begin{bmatrix} \hat{Y}_1^T y \\ X_1^T y \end{bmatrix}. \tag{9.23}$$

We can show that the estimators $\hat{\beta}$ and $\hat{\gamma}$ by 2SLS are consistent estimators. Let $Z = [\hat{Y}_1, X_1]$ and $\delta = [\beta^T, \gamma^T]^T$. Then, by large sample theory, the estimators $\hat{\beta}$ and $\hat{\gamma}$ will asymptotically be distributed as the following normal distribution:

$$\begin{bmatrix} \hat{\beta}_n \\ \hat{\gamma}_n \end{bmatrix} \sim N \begin{bmatrix} \begin{bmatrix} \beta \\ \gamma \end{bmatrix} & s^2 (Z^T Z)^{-1} \end{bmatrix},$$

where

$$s^2 = (y - Y_1 \hat{\beta} - X_1 \hat{\gamma})^T (y - Y_1 \hat{\beta} - X_1 \hat{\gamma}) / (n - g - k).$$

9.4.4.3 **Unweighted Least Squares (ULS)**

Another alternative method for parameter estimations without assuming distributions of the gene expressions, the ULS fitting function, is

$$F_{\text{ULS}} = \frac{1}{2} \text{Tr}\left[(S - \Sigma(\theta))^2 \right].$$

An analog to ordinary least square is evident. In OLS, the sum of squares of the residual term is minimized. The error is the discrepancy between the observed dependent variable and the one predicted by the model. With F_{ULS}, we minimize the sum of squares of each element in the residual matrix $S - \Sigma(\theta)$. The residual matrix in this case consists of the differences between the sample variances and covariances and the corresponding ones predicted by the model.

9.4.4.4 **Generalized Least Squares (GLS)**

A problem with F_{ULS} is that it implicitly weights all elements of $S - \Sigma(\theta)$ as if they have the same variances and covariances. This is similar to the inappropriate application of OLS when the disturbances from a regression equation are heteroskedastic or autocorrelated. It would seem reasonable to apply a GLS fitting function that weights the elements of $S - \Sigma(\theta)$ according to their variances and covariances. A general form of the GLS fitting function is

$$F_{\text{GLS}} = \frac{1}{2} \text{Tr}\left\{ [(S - \Sigma(\theta)) W^{-1}]^2 \right\},$$

where W^{-1} is a weight matrix for the residual matrix. The weight matrix W^{-1} is either a random matrix that converges in probability to a positive matrix or a positive definite matrix of constants.

9.5 Model Selection

9.5.1 Model Selection Criterion

Learning about genetic networks consists of two parts: parameter learning and structure learning. For parameter learning, in the previous section, we assume that the network structure is known. However, in most cases, the network structure is unknown and needs to be identified. Learning network structure from genome-wide gene expression profiles consists of two steps. The first step is to select the set of genes whose reconstructed network best fits the gene expression data. The second step is to learn the structure of the networks for a set of selected genes, which provides the best fit to the gene expression data.

To identify the structure of the network, an overall model fit measure is needed to asses how well a genetic network fits the data and to compare the merits of alternative network structure [33]. The overall model fit measure is to calculate the difference between the covariance matrix predicted by the model and the sample covariance matrix from the observed data. Those differences measure how similar the hypothesized genetic network model is. The model fit measure allows us to rank genetic networks according to their ability to fit the observed data. A widely used model fit measure is Akaike information criterion (AIC) [31,34], which is defined as

$$(N-1)F_{\mathrm{ML}}-2d,$$

where N is the number of samples, F_{ML} is the fitting function, and $d=1/2(p+q)(p+q+1)-t$ is degrees of freedom, where t is the number of free parameters in the model. The AIC value provides a relative ordering of different models fitting the data. The smaller the AIC value, the better the model fits the data.

However, AIC information cannot be employed to test whether the identified genetic network is valid. Fortunately, the statistic $(N-1)F_{\mathrm{ML}}$ is asymptotically distributed as a $\chi^2_{(d)}$ distribution under the null hypothesis H_0: $\Sigma=\Sigma(\theta)$. It should be noted that the null hypothesis means that the constraints on Σ imposed by the genetic network model are valid. In contrast to ordinary tests where the probability of obtaining a χ^2 value larger than a prespecified value is the probability of committing error for the rejection of null hypothesis, in the model selection test here, the probability of obtaining a χ^2 value larger than a prespecified value is the probability of ensuring that the fitted model is correct and is referred to as the fitting probability. Therefore, the higher the probability of the χ^2, the closer is the fitted model for the genetic network to the true genetic network.

9.5.2
Genetic Algorithms (GAs)

Second step for model selection is to search the genetic networks that best fit the observed data. Searching genetic network is a very difficult problem because of the large number of possible networks. To exhaustively search all possible networks is infeasible, in practice, even with high-performance computers. Approximation methods should be developed. GAs can be used for searching networks [35]. Network search consists of two parts. First, we need to search a set of genes that are included in the network from the total available genes. Then, for the fixed set of genes, we search the structures of the network that specify how the genes in the network are connected. We developed a new type of GA that accomplishes these two tasks simultaneously.

We index all genes in the network. Thus, an integer number is assigned to each gene. Assume a string with length k: $g_1, g_2, \ldots, g_k$. This string will denote what genes will be included in the network.

We use a $k \times k$ connective matrix C to represent the structure of a network with k genes. The elements of C are given by

$$c_{ij} = \begin{cases} 1 & \text{if node } j \text{ is directed to node } i, \\ 0 & \text{otherwise.} \end{cases}$$

GAs begin with a population that consists of a large number of individuals. In our genetic algorithm, individuals of the population represent selected genes and network structures. This type of individual is denoted by a string:

$$g_1 g_2 \cdots g_k c_{11} c_{21} \cdots c_{k1} \cdots c_{1k} c_{2k} \cdots c_{kk},$$

which is usually referred to as a chromosome in the GA literature (as opposed to a real chromosome). The first part of the chromosome $g_1 g_2 \cdots g_k$ is a set of integer numbers representing selected genes that will be included in the network. The second part $c_{11} c_{21} \cdots c_{k1} \cdots c_{1k} c_{2k} \cdots c_{kk}$ is a binary string indicating the network structure. For Figure 9.1, if we index the genes CTGF, collagen XI, collagen III, and SMAD3 by 1, 2, 3, and 4, then the string representing the genetic network in Figure 9.1 will be $S = 1234011000100$. GAs attempt to find individuals from the population with the best fitness (e.g., smallest AIC value). The searching procedure of GAs can be briefly described as follows. First, the initial population is randomly generated, and the fitness of each individual is calculated. Second, individuals with good fitness are selected as parents. These parents produce children by the operations of crossover and mutation. A crossover operation in a GA algorithm produces two children by an exchange of chromosome segments between two parents. The mutation operation creates children by changing their own chromosome. All new produced children are added to the population. Some individuals with worse fitness (e.g., higher AIC values) are removed from the extended population (including both parents and children) to generate a new population with its initial size but with better fitness. Crossover and mutation play different roles in the genetic algorithm. Crossover increases the average fitness of the population. Mutation can help the algorithm to avoid local optima by exploring new states. After numerous iterations of GAs, most likely or near most likely networks to fit the data can be found. When the difference between AIC values of two successive iterations is less than a prespecified threshold, the iteration of GAs will be stopped.

9.5.3 Illustration of Structural Equations for Modeling Genetic Networks

To illustrate the use of structural equations for modeling genetic networks, we first analyze the time course expression profiles of 6220 genes using oligonucleotide arrays in synchronized yeast cells during the cell cycle [36]. Data were collected at 17 points that included nearly two full cell cycles. The ideal method for reconstruction of genetic networks for time course data is dynamic (rather than ordinary) structural equations. However, here for the purpose of illustration, we use gene expression time course data of yeast cell cycle as an example to explain how to model genetic networks using structural equations. Although gene expression data from yeast cell cycle are time course data, their dynamics are stable. When the time intervals at which gene expressions are measured are not small, the observed expression can be viewed as being sampled from near steady state of yeast cell cycle dynamic system. Therefore, it is possible to use ordinary structural equations to model such systems. This example also serves to investigate how well ordinary structural equations are used to approximate stable dynamic systems.

The genetic network shown in Figure 9.3 [37] was reconstructed by applying the structural equations to the time course expression profiles of the genes Cdc28, Clb1, Clb3, Mcm1, Mcm2, Swi4, Cln3, Cdc47, and Cdc6, which play an important role in the M/G1 phase of the cell cycle. The regulatory structure of the network can be confirmed by the experiments [38,39]. Zhang [40] investigated transcriptional regulation of M/G1 phase in budding yeast using the same gene expression data and found that the genes Clb1, Clb2, Clb3, and Cdc 28 regulated the expressions of the genes Cln3, Swi4, Cdc6, and Cdc47. However, he could not further infer the regulatory relations either among the genes Clb1, Clb2, Clb 3, and Cdc28 or among the genes Cln3, Swi4, Cdc6, and Cdc47. Therefore, our results give a more detailed structure of the network than that of Zhang [40].

A common approach for identifying networks is to use model selection to choose those networks that best fit the data. We use both AIC values and fitting probability to score the model. AIC values, which have a close relationship with the likelihood function, are widely used model selection criteria. However, AIC values measure only the relative goodness of fit. On the contrary, the fitting probability quantifies how well the model explains the observed data. Therefore, we not only use AIC values to select the model, but we also report the fitting probability of the selected models to indicate how reliable the selected models are.

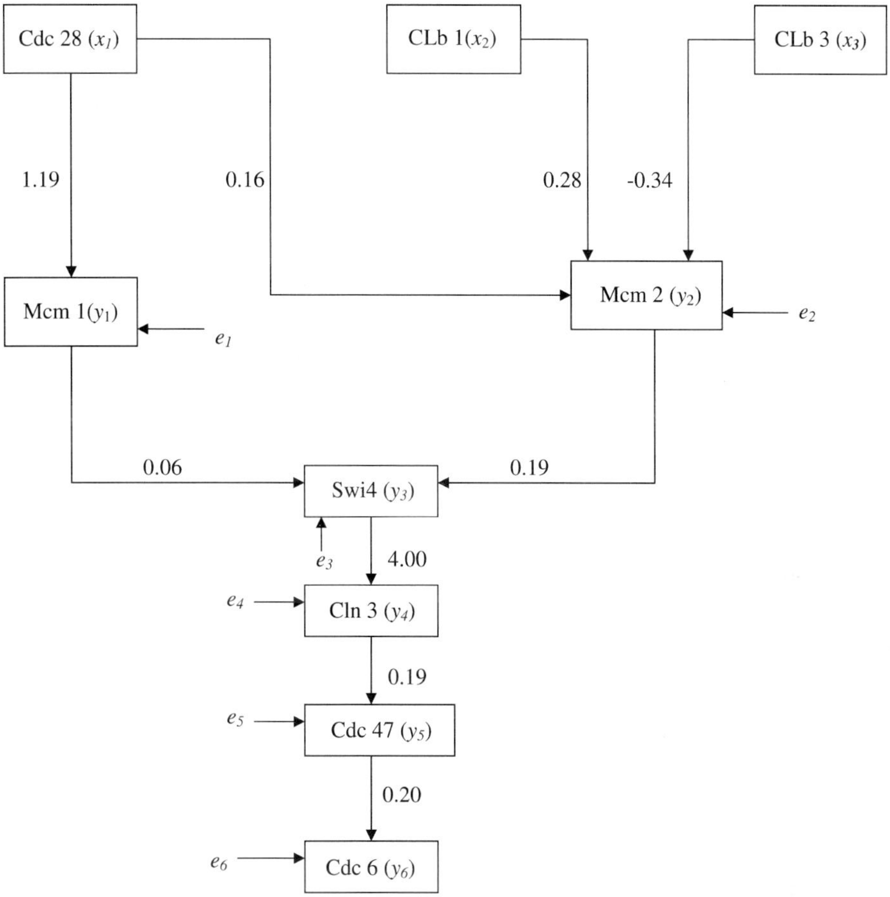

Figure 9.3 A path diagram for a genetic network of nine genes reconstructed from yeast cell cycle data.

The gene Swi4 plays an important role in cell cycle progression [39,40]. As an example, we searched for genetic networks with 10 genes, including Swi4. We use genetic algorithms to search for optimal subsets of genetic networks with the smallest AIC values or the largest fitting probability. There may be a large number of the genetic networks with similar score values. But, for each implementation of GAs, we obtain one set of genetic network that fits the data well. Therefore, in searching for genetic networks with Swi4 and other nine genes, we repeatedly applied genetic algorithms to the data set 500 times, which yielded 500 genetic networks with AIC values ranging from −70.50 to −55.02 and fitting probabilities ranging from 1 to 0.997. It was interesting to observe that of these 500 genetic networks, 371 were partitioned into more than two disconnected networks, while the remaining 129 genetic networks were fully connected. We ranked the fully

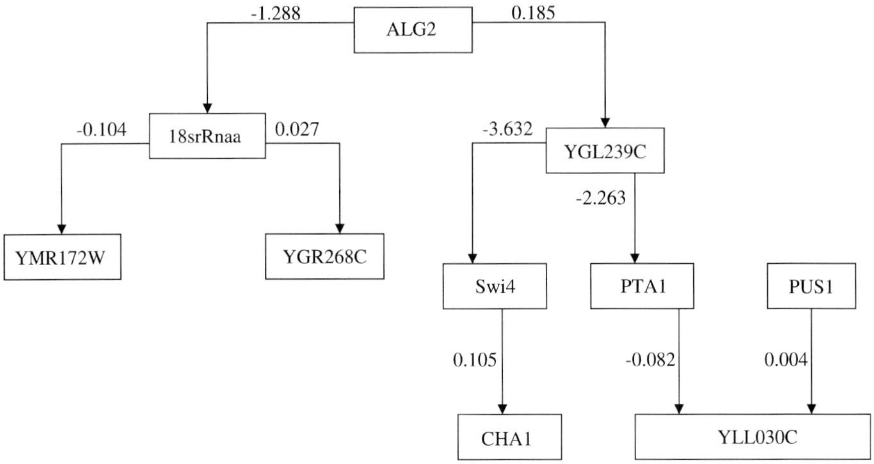

Figure 9.4 A fully connected genetic network with the smallest AIC value (AIC = −64.02, fitting probability = 1.0000) among the 129 fully connected genetic networks with 10 genes for yeast cell cycle gene expression data reconstructed from yeast cell cycle data.

connected genetic networks according to their AIC values. The highest scoring fully connected genetic network with the smallest AIC value is plotted in Figure 9.4 [37]. Since the coefficients in the equations measure the strength of regulation of one gene on the expression of another gene, they were referred to as the regulatory effects of the genes. Since the connected genes in a reconstructed network may or may not be physically connected in reality, the regulatory effect may be a direct effect that is unmediated by other genes or may be an indirect effect that is mediated by other genes that do not appear in the reconstructed networks. Unfortunately, the currently proposed method cannot distinguish direct and indirect effects of the genes. The genetic network in Figure 9.4 has 10 genes: cell cycle gene Swi4; a glycosyltransferase gene ALG2 involved in dolichol pathway, which was regulated at two critical control points in the G1 phase of the cell cycle, G0/G1 and START [41]; an essential gene of *Saccharomyces cerevisiae* affecting pre-tRNA processing, PTA1 [42]; a pseudouridine synthetase gene PUS1, which catalyzes the formation of pseudouridines in tRNAs [43]; a serine and threonine catabolism gene CHA1 [44]; and five other unknown genes.

To investigate the effect of removing a gene from the genetic network, we plotted Figure 9.5 [37] in which the gene Swi4 was removed from the genetic network shown in Figure 9.4. Surprisingly, most of the regulatory effects in the genetic network were not changed except for the regulatory effect of YGL239C on CHA1. This had an important implication: removing a gene will only affect the effects of the genes that were directly connected with the removed gene, but did not have a significant impact on other parts of the genetic network.

As the number of genes in gene expression profiles increases, the total number of all the possible genetic networks exponentially increases. This number of possible

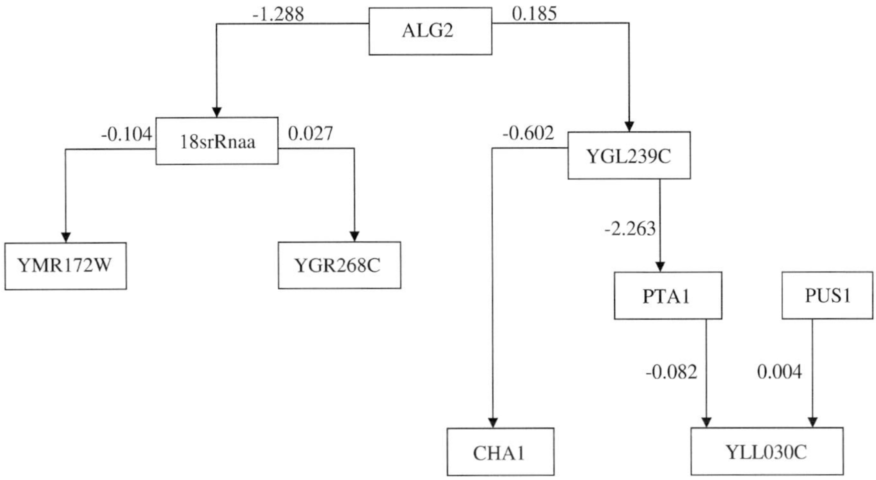

Figure 9.5 A genetic network in which the gene Swi4 was removed from the genetic network shown in Figure 9.2.

genetic networks is too large to be exhaustively searched. There are two approaches to deal with this problem. One approach is an ensemble method for identifying genetic networks that are consistent with the existing gene expression profiling data [45]. The second approach is to use genetic algorithms for searching genetic networks with smallest AIC values. The AIC value and fitting probability are referred to as the score of the genetic networks. To investigate whether genetic algorithms can identify the networks with the highest score and how many iterations are required to achieve the highest score, genetic algorithms were applied to yeast cell cycle data [36] to search networks with 12 genes. Figures 9.6 and 9.7 plot the AIC value and the fitting probability against the number of iterations, respectively. From Figure 9.6, we can see fluctuations in AIC values, but we still can observe the decreasing trend of AIC values. Figure 9.7 shows that after 80 iterations, these runs reach a fitting probability 1.

The fitting probability is a function of the number of genes in the network. The fitting probability will decrease when the number of genes in the network increases. To demonstrate this, we first fix the number of genes in the network and run 100 genetic algorithms to search for networks with the fixed number of genes. For each fixed number of genes in the network, we can obtain the largest fitting probability by repeating genetic algorithms 100 times. From Figure 9.8, we can observe that when the number of genes in the network was more than 14 the fitting probability became small, which implied that the genetic network did not fit the data well. The size of the genetic network (i.e., the number of genes in the network) is limited by the number of tissue samples in microarray experiments.

To further evaluate the performance of the proposed model for reconstructing genetic networks, we take 85 regulators of yeast listed in Lee *et al.* [46], where the remaining 21 of 106 regulators cannot be found in the data set of Cho *et al.* [36], as

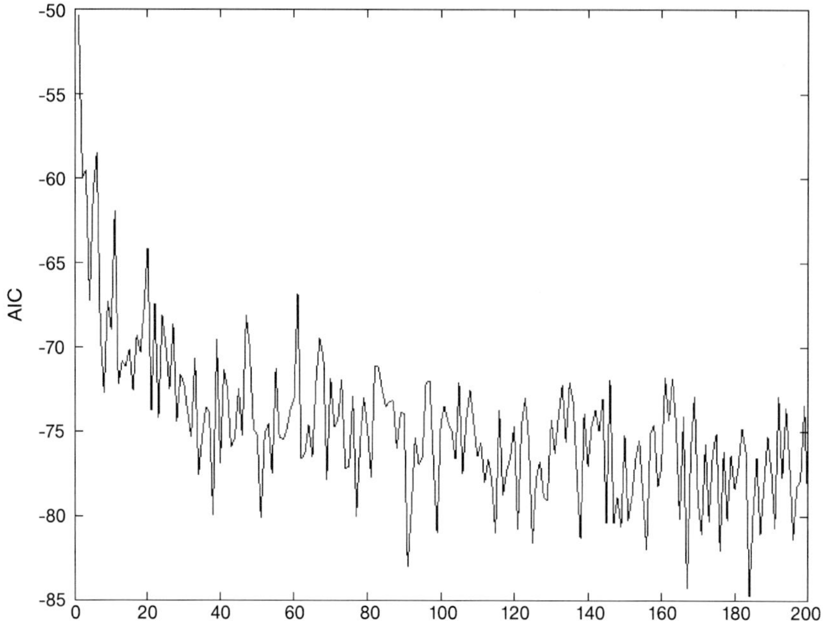

Figure 9.6 AIC values of 200 genetic networks with 12 genes as a function of the number of running genetic algorithms that were applied to yeast cell cycle gene expression data.

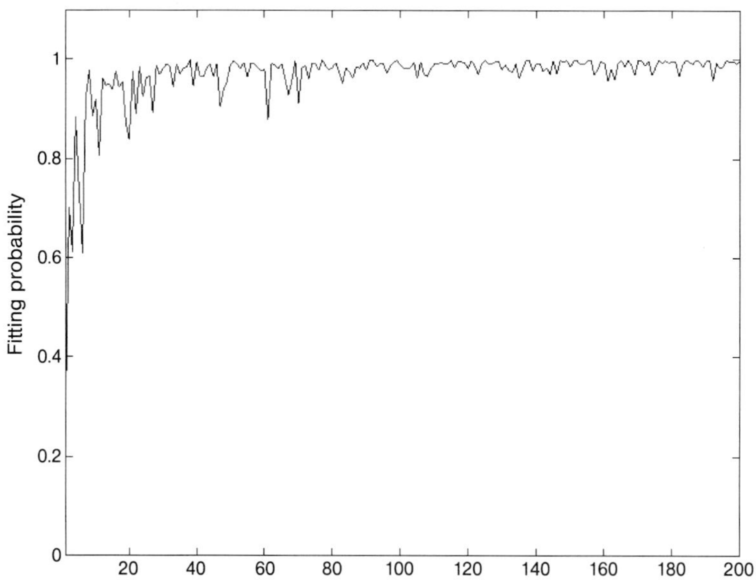

Figure 9.7 A fitting probability of 200 genetic networks with 12 genes as a function of the number of running genetic algorithms that were applied to yeast cell cycle gene expression data.

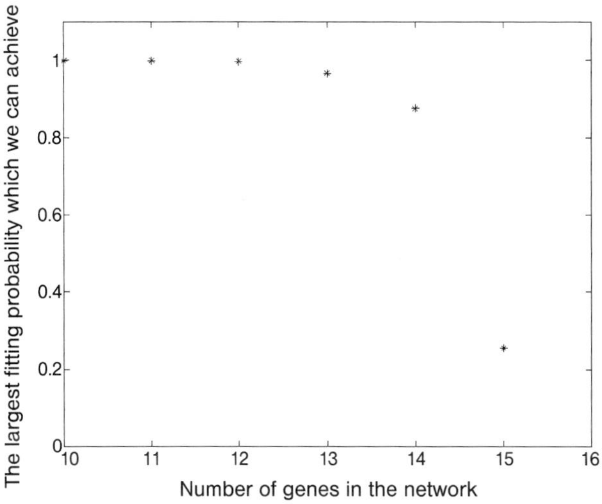

Figure 9.8 Largest fitting probability that can be achieved after 100 iteration of genetic algorithms as a function of number of genes in the genetic networks.

the primary genes for the reconstruction of genetic networks with six genes. For each regulator that was used as an exogenous variable in the structural equations or a primary gene of the genetic network being reconstructed, genetic algorithms were applied 300 times to the yeast cell cycle data for searching genetic networks with six genes. From the reconstructed genetic networks with the regulators as primary genes of the genetic networks, we can find pairs of the regulator-regulated target gene. The identified regulator-regulated gene interactions are compiled in Table 9.2 (http://www.genetics.org/supplemental/), where P-values were given by location analysis in Lee *et al.* [46]. From Table 9.2 we can see that the regulator-regulated gene interactions predicted by the proposed structural equation model had small P-values in genome-wide location analysis [46], which indicated that those predicted regulator-regulated gene interactions were confirmed by experimental results of genome-wide location analysis.

As another example, a linear structural equation model was applied to analyze the expression profiles of 10 genes in twin studies. The tissue samples include lesional and nonlesional skin of 11 monozygotic and 5 dizygotic patients, and unaffected skin of their 16 normal pairs [47]. Since we assume that the structure of the networks

Table 9.2 Some of the nonlinear transformations.

Function	Derivative	
Polynomial	x^2+x+1	$2x+1$
Gaussian	$e^{-\frac{(x-c)^2}{2\sigma^2}}$	$\frac{c-x}{\sigma^2}\times e^{-\frac{(x-c)^2}{2\sigma^2}}$

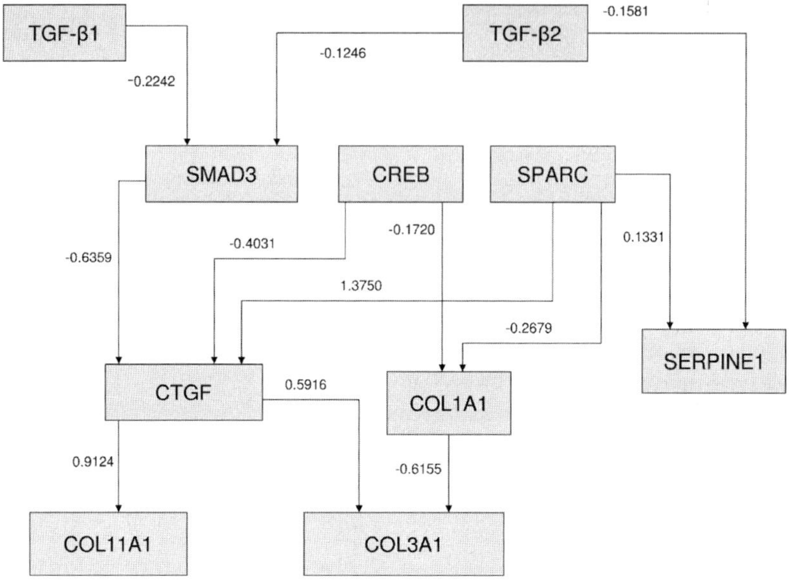

Figure 9.9 A path diagram for TGF-β pathways with 10 genes reconstructed from gene expression data of SSc studies of total 16 abnormal tissue samples and 32 normal tissue samples. The number along the edges is the estimated regulatory effect of one gene on another.

with 10 candidate genes is unknown, we repeatedly applied genetic algorithms to the data set 200 times. The path diagram of the network with the largest fitting probability $P = 0.8864$ is shown in Figure 9.9. The structural equations for the network are given by

$$\begin{aligned}
&\mathrm{SMAD3} = -0.2242\mathrm{TGF}\beta1 - 0.1242\mathrm{TGF}\beta2,\\
&\mathrm{CTGF} = -0.6359\mathrm{SMAD3} - 0.4031\mathrm{CREB} + 1.3750\mathrm{SPARC},\\
&\mathrm{COL1A1} = -0.1720\mathrm{CREB} - 0.2679\mathrm{SPARC},\\
&\mathrm{COL3A1} = 0.5916\mathrm{CTGF} - 0.6155\mathrm{COL1A1},\\
&\mathrm{COL11A1} = 0.9124\mathrm{CTGF},\\
&\mathrm{SERPINE1} = 0.1331\mathrm{SPARC} - 0.1581\mathrm{TGF}\beta2.
\end{aligned}$$

The coefficients in the equations measure the magnitude of impact of one gene on the expression of another gene and hence are referred to as the regulatory effects of the genes. The positive and negative regulatory effects of the gene indicate increasing and decreasing expression levels of the regulated genes, respectively.

The proposed algorithm correctly identified the structure of the network. The regulatory relations between the genes in the reconstructed network can be confirmed by the experiments. Numerous studies have shown that TGF-β families initiate activation and the transduction of SMAD3 protein by binding to type I and type II TGF-β receptors [48]. Since the regulatory effects of TGF-β1 and TGF-β2 on SMAD3 and the regulatory effect of SMAD3 on CTGF were negative, increasing expressions of TGF-β1 and TGF-β2 will decrease the expression of SMAD3, which in turn will increase the expression of CTGF. This is supported by the report that TGF-β markedly increased the expression of CTGF in human fibroblasts [49]. CTGF was reported as a downstream mediator of bioactivities of TGF-β [50]. CTGF also was reported to enhance expression of collagen [51]. SPARC was shown to be able to regulate the expression of collagen type I (COL1A1) in mesangial cells [52]. The experiments also confirmed that CREB blocked expression of CTGF and collagen type I (COL1A1) [53].

9.6 Identification of Differentially Expressed Genetic Networks

Differentially expressed genes are tools widely used for studying functions of genes. Similar to differentially expressed gene, we can extend the concept of differentially expressed genes to differentially expressed genetic networks. Differentially expressed genetic networks are a property of the network as a whole. The differential expression of the genetic network may be due to the differential expression of some individual genes in the network or other factors such as the gene–gene interaction. The concept of differentially expressed genetic networks will play an important role in functional studies of cells.

Recently, several statistics have been proposed to test the differential expression of the genetic networks. In this chapter, we introduce two typical statistics to identify differentially expressed genetic networks.

9.6.1 The Generalized T^2 Statistic for Testing the Differential Expression of Genetic Networks

Let $\bar{X}_1$ and $\bar{X}_2$ be the mean values of expression of the all genes in the network from normal and abnormal tissues, respectively. Let S_{pool} be the pooled estimate of common covariance matrix between gene expressions. It can be shown that [54]

$$T^2 = \left[\frac{n_1 + n_2 - p - 1}{(n_1 + n_2 - 2)p}\right]\left[\frac{n_1 n_2}{n_1 + n_2}\right] D^2,$$

follows an F-distribution with $v_1 = p$ and $v_2 = n_1 + n_2 - p - 1$ degrees of freedom, where

$$D^2 = (\bar{X}_1 - \bar{X}_2)^{\mathrm{T}} S_{\text{pool}}^{-1} (\bar{X}_1 - \bar{X}_2),$$

n_1 and n_2 are the sample sizes of normal and abnormal tissues, respectively, and p is the number of genes selected in the test statistic. Consequently, T^2 can be used to test whether the population means, μ_1 and μ_2, differ significantly and to test for the significance of separation of two populations (normal and abnormal tissues). Formally, the null hypothesis H_0: $\mu_1 = \mu_2$ versus the alternative hypothesis H_a: $\mu_1 \neq \mu_2$ will be assumed. If H_0 is rejected based on T^2 test, we can conclude that the separation between normal and abnormal tissue populations is significant and the genetic network is differentially expressed.

9.6.2
Nonlinear Tests for Identifying Differentially Expressed Genetic Networks

The carefully designed nonlinear transformation of the statistics may only amplify signal rather than noise. Thus, statistics based on nonlinear transformation that will be more powerful than the originally statistics is likely to develop. The results of nonlinear functions of the mean values of expression of all the genes in the network from normal and abnormal tissues can be used to construct nonlinear test statistics for testing differential expressions of genes or genetic networks [71]. The quadratic form $X^T CX$ of asymptotically normal random vectors provides a statistical framework for construction of test statistics. Let $\hat{\Lambda}$ be an estimator of matrix Λ. The nonlinear statistics can be defined as

$$T_N = [f(\bar{Y}) - f(\bar{X})]^T (\hat{\Lambda})^- [f(\bar{Y} - f(\bar{X})], \tag{9.24}$$

where $(\hat{\Lambda})^-$ is the generalized inverse of the matrix $\hat{\Lambda}$. Let $r = \text{rank}(\hat{\Lambda})$. It can be shown [55] that under the null hypothesis of no differential expressions of the gene or genetic network, that is, H_0: $\mu_x = \mu_y$, the statistic T_N is asymptotically distributed as a central $\chi^2_{(r)}$ distribution, and under the alternative hypothesis H_a: $\mu_x \neq \mu$, the statistic T_N is asymptotically distributed as a noncentral $\chi^2_{(r)}$ distribution with the following noncentrality parameter:

$$\lambda_N = \left[f(\mu_y) - f(\mu_x)\right]^T \Lambda^- \left[f(\mu_y) - f(\mu_x)\right]. \tag{9.25}$$

Now we consider a special vector-valued nonlinear function. Let $g(x)$ be a real-valued nonlinear function that has a nonzero derivative at its mean $E[x] = \mu$. Defining

$$f(\mu_y) = [g(\mu_{1y}), \ldots, g(\mu_{ky})]^T, \qquad f(\mu_x) = [g(\mu_{1x}), \ldots, g(\mu_{kx})]^T,$$

where $\mu_x = [\mu_{1x}, \ldots, \mu_{kx}]^T$ and $\mu_y = [\mu_{1y}, \ldots, \mu_{ky}]^T$. According to this definition, Jacobian matrices B and C have the following simple forms:

$$b_{ii} = \frac{\partial g(\mu_{ix})}{\partial \mu_{ix}}, \qquad b_{ij} = 0 \quad \forall i \neq j \quad B = (b_{ij})_{k \times k},$$

$$c_{ii} = \frac{\partial g(\mu_{iy})}{\partial \mu_{iy}}, \qquad c_{ij} = 0 \quad \forall i \neq j \quad C = (c_{ij})_{k \times k}.$$

Test statistic T_N in Equation (9.24) defines a class of nonlinear tests. Various nonlinear functions satisfying some regularity conditions can be used to construct the test statistics. Table 9.2 lists some of the nonlinear functions and their corresponding derivatives.

9.6.3 Examples

Below, examples to show how to use the above statistics for testing the differential expressions of the genetic networks are given [37,56]. The differential expression of the genetic network may be due to the differential expression of some individual genes in the network or other factors such as the gene–gene interaction. To show that highly differentially expressed genetic networks may contain highly differentially expressed genes, we analyzed the expression profiles of 5483 genes using oligonucleotide arrays in 74 multiple myeloma (MM) tissue samples and 31 normal tissue samples [37,57]. These data were transformed by a natural logarithm and normalized by subtracting the mean of each gene and then divided by its standard deviation. Genetic algorithms were repeatedly applied to the data set 200 times to search for the most likely genetic networks with 10 genes that best fit the data. The AIC values for the resulting 200 genetic networks ranged from −56.31 to −67.55 and the fitting probability ranged from 0.9982 to 1. For each resulting network, we calculated the test statistic T^2 and P-values for testing the difference in expression of the genetic network between normal and abnormal samples. A specific AIC value was taken as a threshold and all genetic networks with AIC values larger than the threshold value were discarded. We then ranked the genetic networks according to their T^2 values. The genetic network representing the most significant difference in the T^2 test had a P-value $<10^{-16}$, as shown in Figure 9.10. We can see several remarkable features from Figure 9.8. First, this network consisted of two subnetworks. The P-value of one subnetwork was $<10^{-16}$ and that of another subnetwork was 0.043. Second, the gene HG4462-HT4736 (immunoglobulin heavy chain) that was most significantly differentially expressed gene in the set of total genes (P-value $< 10^{-16}$) was included in the subnetwork with a significantly differentially expressed subnetwork. Third, in the network, we observed another differentially expressed gene DRIL1 (P-value $= 0.000\,817$), which was directly linked to the gene HG4462-HT4736. The gene DRIL1 is a dead ringer transcription regulator and was recently identified as an oncogene [58]. Fourth, the remaining genes in the network were not significantly differentially expressed. This demonstrated that the differential expression of the genetic network is a systems property, which does not imply that all genes in the network are differentially expressed, but that the genetic network as a whole is differentially expressed between the normal and tumor tissues.

The differential expression of the genetic network may be largely due to the differential expression of some genes in the network. However, this need not always be the case. For example, it is possible that all genes in a genetic network are not highly differentially expressed, but the network as a whole is highly differentially

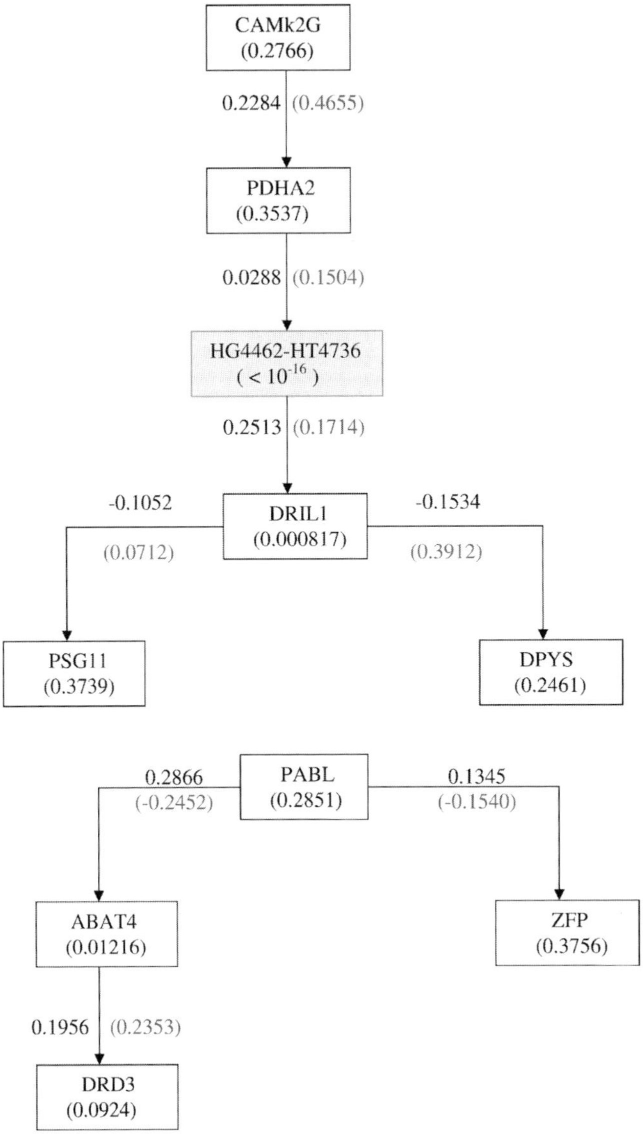

Figure 9.10 The most significantly differentially expressed genetic network among the 200 genetic networks with 10 genes for MM data set where AIC is −65.70 and the fitting probability is 1.000. The numbers in parenthesis below the name of the gene denote the *P*-value of the differential expression of individual gene, the numbers along the edge denote the gene regulatory effect in the tumor tissues, and the numbers in parenthesis along the edge denote the gene regulatory effect in the normal tissues.

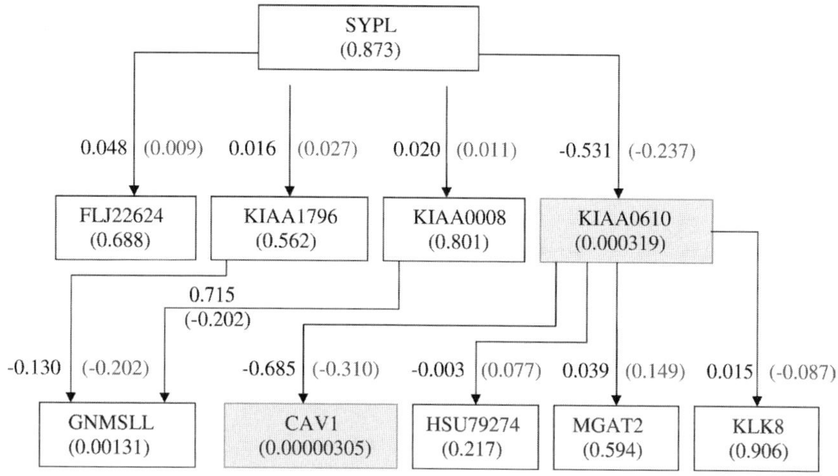

Figure 9.11 The second most significantly differentially expressed genetic network among the 200 genetic networks with 10 genes for prostate data set where AIC $= -52.93$ and the fitting probability is 0.9953. The numbers in parenthesis below the name of the gene denote the P-value of the differential expression of individual gene, the numbers along the edge denote the gene regulatory effect in the tumor tissues, and the numbers in parenthesis along the edge denote the gene regulatory effect in the normal tissues.

expressed. To show this, we analyzed the expression profiles for 12 531 genes using Affymetrix oligonucleotide array in 50 normal and 52 tumor prostate tissues [59]. Again, genetic algorithms were repeatedly applied to this data set 200 times to search for the most likely genetic networks with 10 genes that best fit the data. The AIC values for the resulting 200 genetic networks ranged from -36.31 to -61.84 and the fitting probabilities ranged from 0.53 132 to 0.99 999. The second most significantly differentially expressed genetic network for prostate data set, which is shown in Figure 9.11, had an AIC value of -52.93, fitting probability of 0.9953, and P-value for testing significance of differential expression of the genetic network of 2.93×10^{-11}. However, the P-value of the most significantly differentially expressed gene in the network (CAV1) was equal to 3.047×10^{-6} and was much larger than that of the genetic network as a whole. In addition, although two genes, GNMSLL (P-value $= 0.001\,308$) and KIAA0610 (P-value $= 0.0\,003\,191$), showed mild evidence of the significance of difference expressions, the remaining seven genes in the network did not show any evidence of differential expressions. This example demonstrated that the genetic network as a whole might be more significantly differentially expressed than the individual genes in the network. It was reported that the gene CAV1 was involved in breast cancer [46,60] and ovarian carcinoma [61].

To compare nonlinear test statistics with the generalized T^2, we study gene expression profiles of invasive lobular and ductal carcinomas of breast [62]. There are 38 invasive ductal carcinoma (IDC) and 21 invasive lobular carcinoma (ILC) patients with over 42 000 genes profiled using cDNA. Specimens are separated by the modified Scarff–Bloom–Richardson method. The nonlinear statistics and

Table 9.3 *P*-values of T^2 and nonlinear statistics for testing differential expressions of the cell cycle regulation and MKKK pathways between IDC samples and ILC samples.

Pathway	T^2	Quadratic	Gaussian
Cell cycle regulation	0.001 706	2.9896e−006	1.5848e−006
MKKK	0.001 310	2.5812e−005	2.4380e−006

the Hotelling's T^2 statistic were applied to this data set for testing differential expressions of two pathways: cell cycle regulation pathway and MKKK pathway between IDC samples and ILC samples. Cell cycle regulation pathway includes 18 genes: BRCA2, CDC2, CDC25A, CDK4, CDK7, E2F1, FRAP1, GAK, HK2, IGF1R, KRAS2, PLK1, PPM1D, PRKCA, SFN, STAT1, TADA3L, and TP53BP2; and MKKK pathway includes 16 genes: BRAF, DLK1, MAP3K1, MAP3K11, MAP3K14, MAP3K2, MAP3K3, MAP3K4, MAP3K5, MAP3K7, MAP4K1, MAP4K3, MST1, PAK1, PAK3, and RAF1. Table 9.3 lists the *P*-values of T^2, quadratic, and Gaussian nonlinear statistics for testing differential expressions of cell cycle regulation and MKKK pathways. In general, nonlinear statistics show stronger significance than the generalized T^2 statistic.

9.7 Differentially Regulated Genetic Networks

In the previous section, we studied the differentially expressed genetic networks. In this section, we introduce another important concept: differentially regulated genetic networks that study the differences in the gene regulatory effects between the normal and abnormal tissues. Identification of differentially regulated genetic networks consists of three steps. First, we reconstruct genetic networks using structural equations and gene expression data in all available samples. Second, we fix the structure of the genetic networks and then estimate network parameters by using gene expression data of normal and abnormal samples. Third, we rank the genetic networks according to some statistics, which measure the extent of the difference in regulatory effects of the genetic networks between normal and abnormal tissue samples.

9.7.1 Index for Measuring Difference in Regulation of Genetic Networks

Let $A=[B\Gamma]$ be a coefficient matrix of structural equations for modeling a genetic network. Let A_1 and A_2 be its corresponding coefficient matrices in the normal and abnormal tissue samples, respectively. Let $W=A_1-A_2$ and w_{ij} be an element of the matrix W. Since w_{ij} is a parameter in the network, its asymptotic standard deviation can be calculated from the square root of the main diagonal of the asymptotic covariance matrix of the estimated parameters in the network and denoted by

$S_{W_{ij}}$. We define the test statistic T_G as follows:

$$T_G = \frac{W_{ij}}{S_{W_{ij}}}.$$

Although the exact distribution of T_G is unknown, its asymptotical distribution can be approximated by a t distribution with $N-2$ degrees of freedom. This statistic can be used to test the difference in the regulatory effect of one gene on another between normal and abnormal tissues.

The difference in the regulatory effect of one gene on another cannot measure the difference in the global behavior of the genetic networks between normal and abnormal tissues. A simple quantity to measure the difference in global behavior of genetic networks between the normal and abnormal tissues is the largest absolute value of the difference of the regulatory effect of one gene on another in the network between the normal and abnormal tissues, that is, $w_0 = \max_{i,j} |w_{ij}| = |w_{i_0 j_0}|$. The statistic T_G for testing the difference of individual regulatory effect can be used to test the difference in global behavior of genetic networks. Specifically, the statistic for testing the differential regulation of the genetic networks is given by

$$T_{G_0} = \frac{w_{i_0 j_0}}{S_{w_{i_0 j_0}}}.$$

The P-value is calculated by a permutation test. The gene expression profile matrix is randomly permuted, and the structural equation model and genetic algorithms are applied to randomly permutated gene expression data to reconstruct the genetic network hundreds or thousands of times. Then, we calculate T_{G_0} and obtain an empirical distribution of T_{G_0}. The P-value of the test is then defined as the probability that T_{G_0} exceeds its observed value. The statistic T_{G_0} can be used to measure the difference in regulation of the genetic network.

The difference in global behavior of genetic networks between the normal and abnormal tissues depends on the whole regulatory coefficient matrix. A scalar associated with a matrix W is a norm of the matrix W that denotes a real-valued function of W (of the elements w_{ij} of W). The norm is relevant with all elements of the matrix and hence can be used to measure the difference in regulation of the whole genetic networks. Four metrics borrowed from the norms of the matrix for measuring the difference in regulation of the genetic networks are defined as follows [63]:

1. $||W||_1 = \max_j \left(\sum_{i=1}^{p+q} |w_{ij}|^{ijij} \right)$ = maximum of sums of absolute value of column elements of the matrix.
2. $||W||_\infty = \max_i \left(\sum_{j=1}^{p+q} |w_{ij}| \right)$ = maximum of sums of absolute values of row elements of the matrix.
3. $||W||_2$ = square root of the maximum eigenvalues of the matrix $W^T W$, a spectral norm.
4. $||W||_E = [\sum_i \sum_j (|w_{ij}|)^2]^{1/2}$, an Euclidean norm.

9.7.2 Examples

There are three important cases: (i) the genetic network is differentially regulated but not differentially expressed; (ii) the genetic network is differentially expressed but not differentially regulated; or (iii) the genetic network is both differentially regulated and expressed. We first use the largest difference of the gene regulatory effect in the network between normal and abnormal samples as a measure to quantify the difference in regulation of the network. Then we will compare all five measures. The most differentially regulated genetic network for the multiple myeloma data set studied in Section 9.6.3, which had an AIC value of −65.45 and fitting probability of 1, is plotted in Figure 9.12. The network with 10 genes was partitioned into two

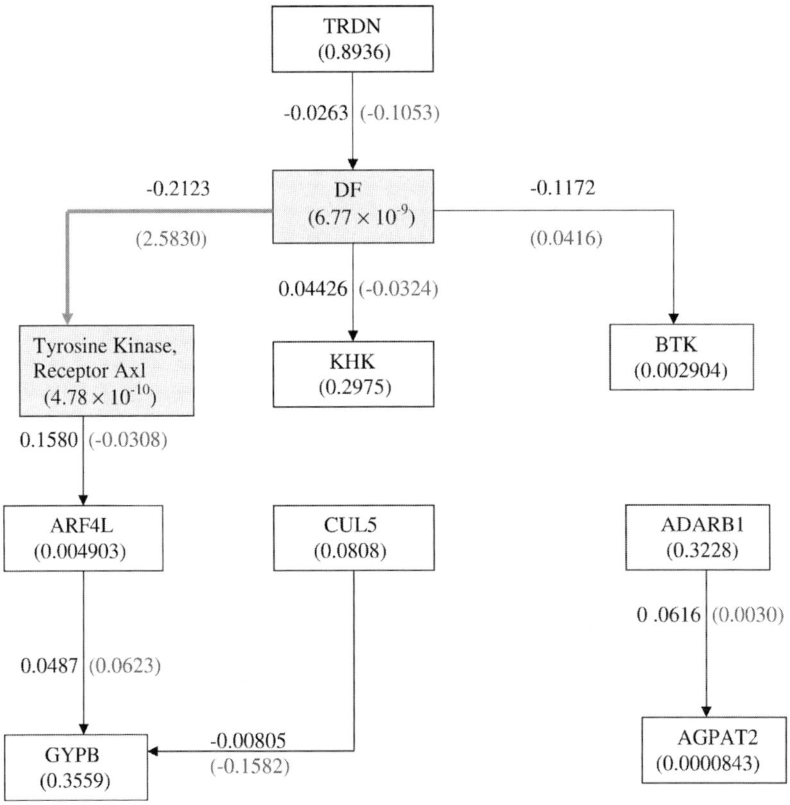

Figure 9.12 The most significantly differentially regulated genetic network among the 200 genetic networks with 10 genes for MM data set where AIC = −65.45 and the fitting probability is 1.0000. The numbers in parenthesis below the name of the gene denote the *P*-value of the differential expression of individual gene, the numbers along the edge denote the gene regulatory effect in the tumor tissues, and the numbers in parenthesis along the edge denote the gene regulatory effect in the normal tissues.

subnetworks: one subnetwork with eight genes and one subnetwork with two genes. The largest difference of the gene regulatory effect was 2.7953 ($T_{G_0} = 25.68$, P-value = 0.00 062), which was associated with the regulation of the gene DF (D component of complement adipsin) on Ax1, where the P-value was obtained by a permutation test. From Figure 9.12, we can also observe that other regulatory effects in the network for the tumor and normal samples are not significantly different. It was interesting to note that both genes DF (P-value = 6.77×10^{-9}) and receptor Ax1 (P-value = 4.78×10^{-10}) as well as the network (P-value = 3.87×10^{-14}) were differentially expressed. The expression of the genes Ax1 and DF and fitted structural equation line of the expression of gene DF as a function of the expression of the gene Ax1 in tumor and normal samples are shown in Figure 9.13. The slope of the line represented the regulatory effect of the gene DF on the gene Ax1. We can clearly see the different regulatory effects in the tumor and normal samples from Figure 9.13. It was reported that the gene DF was a novel serine protease [64] and involved in myeloid cell differentiation [65]. The gene Ax1 was a tyrosine kinase receptor and was recently found to be downregulated in mature bone marrow derived dendritic cells [66].

Now we present a data set in which the genetic network is differentially expressed but not differentially regulated. The most significantly differentially expressed

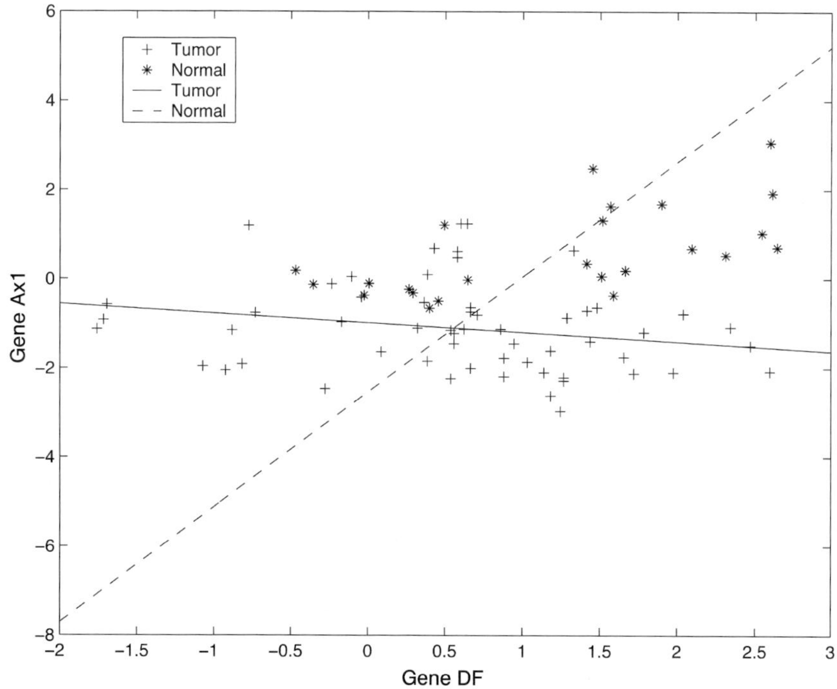

Figure 9.13 Fitted structural equation line of the expression of the gene DF as a function of the expression of the gene Ax1.

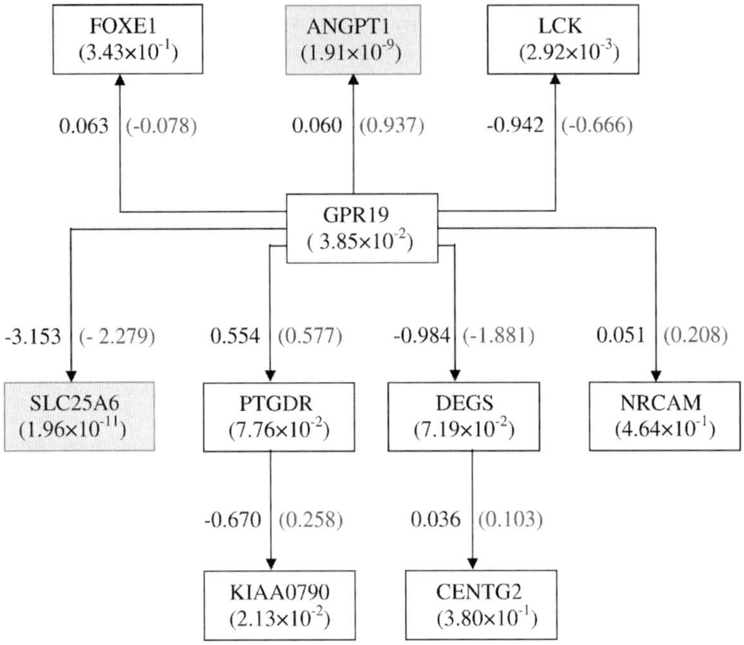

Figure 9.14 The most significantly differentially expressed genetic network among the 200 genetic networks with 10 genes for the prostate data set where AIC = −51.56 and the fitting probability is 0.9827. The numbers in parenthesis below the name of the gene denote the *P*-value of the differential expression of individual gene, the numbers along the edge denote the gene regulatory effect in the normal tissues, and the numbers in parenthesis along the edge denote the gene regulatory effect in the tumor tissues.

genetic network for prostate data set [59], which is shown in Figure 9.14, had an AIC value of −51.56, fitting probability of 0.9827, and *P*-value for testing the significance of differential expression of 2.47×10^{-12}. The gene SLC25A6 (*P*-value = 1.96×10^{-11}) and the gene ANGPT1 (*P*-value = 1.91×10^{-9}) in the network showed significant evidence of differential expressions. However, the rank of the genetic network in the differentially regulated genetic network for the prostate data was 122. The largest difference in the gene regulatory effects of the network was 0.877 ($T_G = 8.5253$, *P*-value = 0.15). This demonstrated that although this genetic network was highly differentially expressed, it was not differentially regulated.

Now we give an example showing that differential regulation is a useful concept of genetic networks for identifying mutations causing diseases. Recall twin studies presented in the previous section [47]. The reconstructed genetic network and estimated regulatory effects of the genes are shown in Figure 9.15. The number in parenthesis along the edge is the regulatory effect of the genes in the normal tissue samples. The largest difference of the regulatory effects of the genes was 1.3553 ($T_{G_0} = 12.4387$, *P*-value = 1.1102×10^{-16}), which was associated with the regulation of CREB on CTGF, where the *P*-value was obtained by a permutation test. In the normal tissues, the function of CREB is to inhibit the expression of

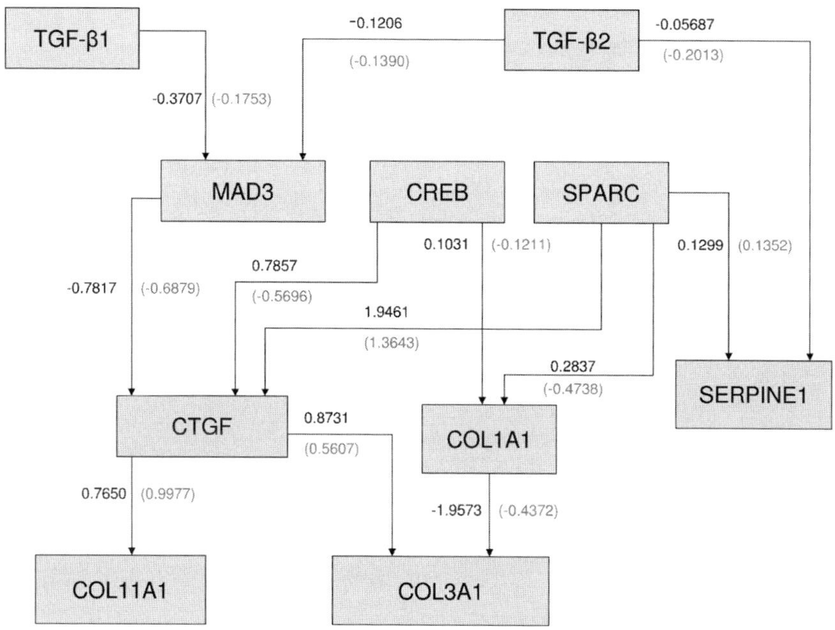

Figure 9.15 A path diagram for TGF-β pathways with 10 genes reconstructed from gene expression data of total 32 normal tissue samples and 16 abnormal tissue samples. The number along the edges is the estimated regulatory effect of one gene on another from abnormal tissue samples. The number in parenthesis along the edges is the estimated regulatory effect of one gene on another from normal tissue samples.

collagen type I and CTGF, which in turn regulates the expression of collagen III. As shown in Figure 9.15, in the abnormal tissues, CREB increased the expression of collagen type I and CTGF, which in turn increased the expression of collagen type III. From Figure 9.15, we can also see that SPARC changed regulatory roles from negative control to positive control when the normal skin tissues became lesional. In the normal tissues, SPARC inhibited expression of collagen type I but increased production of collagen type I.

The inferred regulatory relations between the genes in the network should be validated by perturbing network and analyzing its response to perturbation. The changes in regulatory roles of genes from activation to inhibition or vice versa due to the effect of tissues should also be validated by perturbing network. Several methods, for example, antisense RNA and RNA interference, can be used to perturb networks.

SPARC siRNA and CTGF siRNA were used to suppress the expression of SPARC and CTGF. RT-PCR was used to measure changes of expressions of collagen type I and type III for investigations of regulatory relations between genes in the network [47]. Figure 9.16 shows the effects of SPARC siRNA, CTGF siRNA, and TGF-β1 in

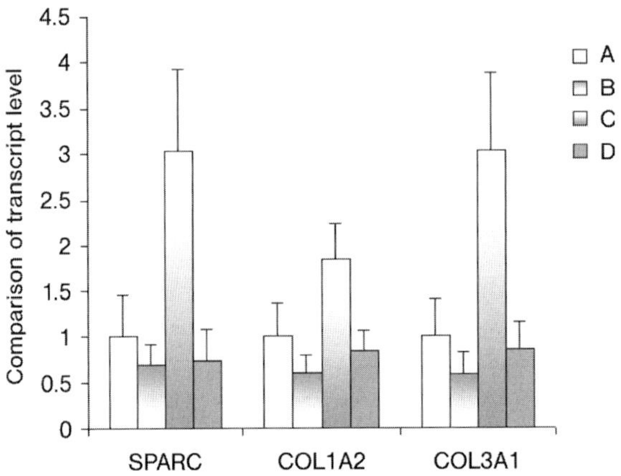

Figure 9.16 Comparison of transcript levels of SPARC, COL1A2, and COL3A1 in different conditions: (a) cultured fibroblasts with transfection media, (b) cultured fibroblasts with SPARC siRNA transfection 10 μg/ml for 72 hours, (c) cultured fibroblasts treated with TGF-β1 10 ng/ml, (d) cultured fibroblasts with SPARC siRNA transfection 10 μg/ml for 24 hours before addition of TGF-β1 10 ng/ml. Error bars represent standard deviation in four assays.

transfected and normal fibroblasts. Several features emerged from Figure 9.16. First, it is shown that in normal fibroblasts TGF-β1 increased the expression of CTGF and collagen type III. This observation was consistent with the prediction of the structural equation model for the network. In both lesional and nonlesional skin tissues, regulation of TGF-β1 on CTGF was through negative control of TGF-β1 on MAD3 and MAD3 on CTGF. Therefore, the structural equation model predicted that addition of TGF-β1 would increase the expression of CTGF, which in turn would increase the expression of collagen type III.

In fibroblasts transfected with SPARC siRNA, it was observed that expression of both collagen type I and type III were decreased. The structural equation model precisely predicted a large reduction in the expression of collagen type I and type III by inhibition of the SPARC gene. Significant changes in regulation of a gene may imply the occurrence of a mutation or gene variations within the gene, which provides information for identifying disease genes. This was supported by studies associated with SPARC in SSc. Our previous genetic studies associated with an isolated population with a high prevalence of SSc (Choctaw Indians), as well as a multiethnic cohort of SSc patients, strongly suggest that the SPARC gene influenced disease susceptibility [67].

To compare the five metrics for characterizing the difference in regulation of the genetic networks under different conditions, we present Table 9.4 showing the correlation coefficients between rakings of the genetic networks made by the five metrics. We can see that the correlation coefficients between rankings of the five metrics were very high. This suggested that the five metrics can provide similar evidence showing differential regulation of the genetic networks in normal and abnormal tissues.

Table 9.4 Correlation coefficients between rankings of the genetic networks for sample gene expression profiles of prostate normal and tumor tissue made by the five metrics.

	$\max_{i,j}\lvert w_{ij}\rvert$	$\lVert W\rVert_1$	$\lVert W\rVert_\infty$	$\lVert W\rVert_2$	$\lVert W\rVert_E$
$\max_{i,j}\lvert w_{ij}\rvert$	1	0.9418	0.9864	0.9921	0.9839
$\lVert W\rVert_1$		1	0.9234	0.9676	0.9509
$\lVert W\rVert_\infty$			1	0.9841	0.9773
$\lVert W\rVert_2$				1	0.9885
$\lVert W\rVert_E$					1

9.8 Conclusions

Genetic networks have two aspects: structure of the networks and strength of the interaction between the genes in the networks. To understand genetic networks comprehensively, in addition to studying structure nature, we also need to quantify the strength of the interaction between the genes. Because of the large variation in observed gene expression profiles, quantitative models for genetic networks may not be accurate, but they will still be a useful tool for guiding experiments and understanding complex biological systems, particularly, when advances in experimental technologies are made and the precision of experimental data is improved.

Identification of genetic networks consists of two steps: parameter estimation and structure discovery. In the first step, we assume that the structure of the network is known. A remarkable feature of the regulatory relation among genes in the network is that the expression levels of the genes are determined by the simultaneous interaction of the regulatory relations in the network. Using ordinary regression and least square method for the estimation of the parameters will result in inconsistent estimates of the parameters in the network. The proposed structural equation models and several estimation procedures based on covariance analysis can avoid this problem and lead to consistent estimates of the parameters in the networks.

The second step is to identify the structure of the networks when it is unknown. The genetic networks that best fit the data may not be truly physically connected, but can reveal causal relations between variables in the network and predict behavior of biological systems. Model selection can be used to accomplish this task. Since searching optimal models from extremely large number of potential networks is computationally expensive, genetic algorithms are a powerful tool for searching most likely genetic networks fitting the data.

Structure discovery is, essentially, to identify causal relations between variables in the model. The definition of cause has three crucial components: isolation, association, and direction of influence [31]. Much of the debate about causal relations comes from inability of complete isolation of the variables. Although structural equation models make various assumptions to approximate isolation, it is impossible to achieve perfect isolation in practice. Therefore, limitation of structural equation models for genetic networks is that they may not reveal true causal relations of the variables in the models, which will affect their precision in predicting behavior of biological systems.

Gene regulation is a complex phenomenon. The expression level of each gene is a nonlinear function of its regulatory input. Linear tools for reconstruction of genetic networks can only approximately model the response of the gene to the regulatory input. Like other mathematical models, the structural equations for modeling genetic networks have limitations. The identified genetic networks based on the linear structural equations may not represent the true biological systems.

References

1 Davidson, E.H., Rast, J.P., Oliveri, P., Ransick, A., Calestani, C., Yuh, C.H., Minokawa, T., Amore, G., Hinman, V., Arenas-Mena, C., Otim, O., Brown, C.T., Livi, C.B., Lee, P.Y., Revilla, R., Rust, A.G., Pan, Z., Schilstra, M.J., Clarke, P.J., Arnone, M.I., Rowen, L., Cameron, R.A., McClay, D.R., Hood, L. and Bolouri, H. (2002) A genomic regulatory network for development. *Science*, **295**, 1669–1678.

2 Strohman, R. (2002) Maneuvering in the complex path from genotypes to phenotype. *Science*, **296**, 701–703.

3 Jong, H.D. (2002) Modeling and simulation of genetic regulatory systems: a literature review. *Journal of Computational Biology*, **9**, 67–103.

4 Brown, P.A. and Botstein, D. (1999) Exploring the new world of the genome with DNA microarrays. *Nature Genetics*, **21** (Suppl.), 33–37.

5 Lipschutz, R.J., Fodor, S.P.A., Gingeras, T.R. and Lockhardt, D.J. (1999) High density synthetic oligonucleotide arrays. *Nature Genetics*, **21** (Suppl.), 20–24.

6 Lockhart, D.J. and Winzeler, E.A. (2000) Genomics gene expression and DNA arrays. *Nature*, **405**, 827–836 (review).

7 Hughes, T.R. and Shoemaker, D.D. (2001) DNA microarrays for expression profiling. *Current Opinion in Chemical Biology*, **5**, 21–25.

8 Young, R.A. (2000) Biomedical discovery with DNA arrays. *Cell*, **102**, 9–15 (review).

9 Figeys, D. and Pinto, D. (2001) Proteomics on a chip: promising developments. *Electrophoresis*, **22**, 208–216.

10 Houseman, B.T., Huh, J.H., Kron, S.J. and Mrksich, M. (2002) Peptide chips for the quantitative evaluation of protein kinase activity. *Nature Biotechnology*, **20**, 270–274.

11 Mann, M. (1999) Quantitative proteomics. *Nature Biotechnology*, **17**, 954–955.

12 Kitano, H. (2002) System biology: a brief overview. *Science*, **295**, 1662–1664.

13 Csets, M.E. and Doyle, J.C. (2002) Reverse engineering of biological complexity. *Science*, **295**, 1664–1669.

14 Guet, C.C., Elowitz, M.B., Hsing, W. and Leibler, S. (2002) Combinatorial synthesis of genetic networks. *Science*, **296**, 1466–1470.

15 Maslov, S. and Sneppen, K. (2002) Specificity and stability in topology of protein networks. *Science*, **296**, 910–913.

16 Noble, D. (2002) Modeling the heart – from genes to cells to the whole organ. *Science*, **295**, 1678–1682.

17 Ideker, T., Thorsson, V., Ranish, J.A., Christmas, R., Buhler, J., Eng, J.K., Bumgarner, R., Goodlett, D.R., Aebersold, R. and Hood, L. (2001) Integrated genomic and proteomic analyses of a systematically perturbed metabolic network. *Science*, **292**, 929–934.

18 Alves, R. and Savageau, M.A. (2000) Extending the method of mathematically controlled comparison to include numerical comparisons. *Bioinformatics*, **16**, 786–798.

19 Schulze, A. and Downward, J. (2001) Navigating gene expression using microarrays – a technology review. *Nature Cell Biology*, **3**, E190–E195.

20 Banerjee, N. and Zhang, M.Q. (2002) Functional genomics as applied to mapping transcription regulatory networks. *Current Opinion in Microbiology*, **5**, 313–317.

21 Bolouri, H. and and Davidson, E.H. (2002) Modeling DNA sequence-based *cis*-regulatory gene networks. *Developmental Biology*, **246**, 2–13.

22 Liang, S., Fuhrman, S. and Somogyi, R. (1998) Reveal a general reverse engineering algorithm for inference of genetic network architectures. *Pacific Symposium on Biocomputing*, **3**, 18–29.

23 Akutsu, T., Miyano, S. and Kuhara, S. (2000) Algorithms for identifying Boolean networks and related biological networks based on matrix multiplication and fingerprint function. *Journal of Computational Biology*, **7**, 331–343.

24 Chen, T., He, H.L. and Church, G.M. (1999) Modeling gene expression with differential equations. *Pacific Symposium on Biocomputing*, **4**, 29–40.

25 von Dassow, G., Meir, E., Munro, E.M. and Odell, G.M. (2000) The segment polarity network is a robust developmental module. *Nature*, **406**, 188–192.

26 Wahde, M. and Hertz, J. (2000) Coarse-grained reverse engineering of genetic regulatory networks. *Biosystems*, **55**, 129–136.

27 Woolf, P.J. and Wang, Y. (2000) A fuzzy logic approach to analyzing gene expression data. *Physiological Genomics*, **3**, 9–15.

28 Friedman, N., Linial, M., Nachman, I. and Pe'er, D. (2000) Using Bayesian networks to analyze expression data. *Journal of Computational Biology*, **7**, 601–620.

29 Hartemink, A.J., Gifford, D.K., Jaakkola, T.S. and Young, R.A. (2001) Using graphical models and genomic expression data to statistically validate models of genetic regulatory networks. *Pacific Symposium on Biocomputing*, 422–433.

30 Imoto, S., Goto, T. and Miyano, S. (2002) Estimation of genetic networks and functional structures between genes by using Bayesian networks and nonparametric regression. *Pacific Symposium on Biocomputing*, 175–186.

31 Bollen, K.A. (1989) *Structural Equations with Latent Variables*, John Wiley & Sons, Ltd, New York.

32 Gujarati, D.N. (2003) *Basic Econometrics*, 4th edn, McGraw Hill, New York.

33 Jordan, M.I. (1999) *Learning in Graphical Models*, The MIT Press, Cambridge, MA.

34 Maruyama, G.M. (1998) *Basics of Structural Equation Modeling*, Sage Publications, Thousand Oaks.

35 Larranaga, P., Poza, P., Yurramendi, Y., Marga, R.H. and Kuijpers, C.M.H. (1996) Structure learning of Bayesian networks by genetic algorithms: a performance analysis of control parameters. *IEEE Transaction on Pattern Analysis and Machine Intelligence*, **18**, 912–926.

36 Cho, R.J., Campbell, M.J., Winzeler, E.A., Steinmetz, L., Conway, A., Wodicka, L., Wolfsberg, T.G., Gabrielian, A.E., Landsman, D., Lockhart, D.J. and Davis, R.W. (1998) A genome-wide transcriptional analysis of the mitotic cell cycle. *Molecules and Cells*, **2**, 65–73.

37 Xiong, M., Li, J. and Fang, X. (2004) Identification of genetic networks. *Genetics*, **166**, 1037–1052.

38 Koch, C. and Nasmuth, K. (1994) Cell cycle regulated transcription in yeast. *Current Opinion in Cell Biology*, **6**, 451–459.

39 McInerny, C.J., Partridge, J.F., Mikesell, G.E., Creemer, D.P. and Breeden, L.L. (1997) A novel Mcm1-dependent

element in the SWI4 CLN3, CDC6, and CDC47 promoters activates M/G1-specific transcription. *Genes and Development*, **11**, 1277–1288.

40 Zhang, M.Q. (1999) Large-scale gene expression data analysis: a new challenge to computational biologists. *Genome Research*, **9**, 681–688.

41 Lennon, K., Pretel, R., Kesselheim, J., Heesen, S. and Kukuruzinska, S.M.A. (1995) Proliferation-dependent differential regulation of the dolichol pathway genes in *Saccharomyces cerevisiae*. *Glycobiology*, **5**, 633–642.

42 O'commor, J.P. and Peebles, C.L. (1992) PTA1 an essential gene of *Saccharomyces cerevisiae* affecting pre-tRNA processing. *Molecular and Cellular Biology*, **12**, 3843–3856.

43 Arluison, V., Batellier, G., Ries-Kautt, M. and Grosjean, H. (1999) RNA: pseudouridine synthetase Pus1 from *Saccharomyces cerevisiae*: oligomerization property and stoichiometry of the complex with yeast tRNA(Phe). *Biochimie*, **81**, 751–756.

44 Bornaes, C., Petersen, J.G. and Holmberg, S. (1992) Serine and threonine catabolism in *Saccharomyces cerevisiae*: the CHA1 polypeptide is homologous with other serine and threonine dehydratases. *Genetics*, **131**, 531–539.

45 Battogtokh, D., Asch, D.K., Case, M.E., Arnold, J. and Schuttler, H.-B. (2002) An ensemble method for identifying regulatory circuits with special reference to the *qa* gene cluster of *Neuropora crassa*. *Proceedings of the National Academy of Science of the United States of America*, **99**, 16904–16909.

46 Lee, T.I., Rinaldi, N.J., Robert, F., Odom, D.T., Bar-Joseph, Z., Gerber, G.K., Hannett, N.M., Harbison, C.T., Thompson, C.M., Simon, I., Zeitlinger, J., Jennings, E.G., Murray, H.L., Gordon, D.B., Ren, B., Wyrick, J.J., Tagne, J.B., Volkert, T.L., Fraenkel, E., Gifford, D.K. and Young, R.A. (2002) Transcriptional regulatory networks in *Saccharomyces cerevisiae*. *Science*, **298**, 799–804.

47 Xiong, M., Feghali-Bostwick, C.A., Arnett, F.C. and Zhou, X. (2005) A systems biology approach to genetic studies of complex diseases. *FEBS Letters*, **579**, 5325–5332.

48 Zimmerman, C.M. and Padgett, R.W. (2000) Transforming growth factor beta signaling mediators and modulators. *Gene*, **249**, 17–30.

49 Igarashi, A., Okochi, H., Bradham, D.M. and Grotendorst, G.R. (1993) Regulation of connective tissue growth factor gene expression in human skin fibroblasts and during wound repair. *Molecular Biology of the Cell*, **4**, 637–645.

50 Dammeier, J., Brauchle, M., Falk, W., Grotendorst, G.R. and Werner, S. (1998) Connective tissue growth factor: a novel regulator of mucosal repair and fibrosis in infla- mmatory bowel disease? *The International Journal of Biochemistry and Cell Biology*, **30**, 909–922.

51 Frazer, K., Nahman, N.S., Jr, Rothe, K.L., Falkenhain, M.E., Frazer, K.M., Dacio, L.E., Madia, J.D., Leonhart, K.L., Kronenberger, J.C. and Stauch, D.A. (1996) Simulation of fibroblast cell growth matrix production, and granulation tissue formation by connective tissue growth factor. *The Journal of Investigative Dermatology*, **107**, 404–411.

52 Francki, A. (1999) SPARC regulates the expression of collagen type I and transforming growth factor-beta1 in mesangial cells. *Journal of Biological Chemistry*, **274**, 32145–32152.

53 Houglum, K. Lee, K.S. and Chojkier, M. (1997) Proliferation of hepatic stellate cells is inhibited by phosphorylation of CREB on serine 133. *European Journal of Clinical Investigation*, **99**, 1322–1328.

54 Anderson, T.W. (1984) *An Introduction to Multivariate Statistical Analysis*, 2nd edn, John Wiley & Sons, Ltd, New York.

55 Greenwood, P.E. and Nikulin, M.S. (1996) *A Guide to Chi-Squared Testing*, John Wiley & Sons, Ltd, New York.

56 Xiong, H. (2005) Non-linear tests for identifying differentially expressed genes or genetic networks. *Bioinformatics*, **22**, 919–923.

57 Zhan, F., Hardin, J., Kordsmeier, B., Bumm, K., Zheng, M., Tian, E., Sanderson, R., Yang, Y., Wilson, C., Zangari, M., Anaissie, E., Morris, C., Muwalla, F., van Rhee, F., Fassas, A., Crowley, J., Tricot, G., Barlogie, B. and Shaughnessy, J., Jr. (2002) Globalgene expression profiling of multiple myeloma monoclonal gammopathy of undetermined significance, and normal bone marrow plasma cells. *Blood*, **99**, 1745–1757.

58 Peeper, D.S., Shvarts, A., Brummelkamp, T., Douma, S., Koh, E.Y., Daley, G.Q. and Bernards, R. (2002) A functional screen identifies hDRIL1 as an oncogene that rescues RAS-induced senescence. *Nature Cell Biology*, **4**, 148–153.

59 Singh, D., Febbo, P.G., Ross, K., Jackson, D.G., Manola, J., Ladd, C., Tamayo, P., Renshaw, A.A., D'Amico, A.V., Richie, J.P., Lander, E.S., Loda, M., Kantoff, P.W., Golub, T.R. and Sellers, W.R. (2002) Gene expression correlates of clinical prostate cancer behavior. *Cancer Cell*, **1**, 203–209.

60 Fiucci, G., Ravid, D., Reich, R. and Liscovotch, M. (2002) Caveolin-1 inhibits anchorage-independent growth anoikis and invasiveness in MCF-7 human breast cancer cells. *Oncogene*, **21**, 2365–2375.

61 Wiechen, K., Diatchenko, L., Agoulnik, A., Scharff, K.M., Schober, H., Arlt, K., Zhumabayeva, B., Siebert, P.D., Dietel, M., Schafer, R. and Sers, C. (2001) Caveolin-1 is down-regulated in human ovarian carcinoma and acts as a candidate tumor suppressor gene. *The American Journal of Pathology*, **159**, 1635–1643.

62 Zhao, H., Langerød, A., Youngran, J.I., Kent, W., Jahn, N., Nesland, M., Tibshirani, R., Bukholm, I.K. and Kåresen, R. (2004) http://www.ncbi.nlm.nih.gov/geo/gds/gds_browse.cgi?gds=619.

63 Graybill, A. (1976) *Matrices with Applications in Statistics*, 2nd edn, Wadsworth International Group, Belmont, CA.

64 Volanakis, J.E. and Narayana, S.V. (1996) Complement factor D a novel serine protease. *Protein Science*, **5**, 553–564.

65 Wong, E.T., Jenne, D.E., Zimmer, M., Porter, S.D. and Gilks, C.B. (1999) Changes in chromatin organization at the neutrophil elastase locus associated with myeloid cell differentiation. *Blood*, **94**, 3730–3736.

66 Chen, Z., Gordon, J.R., Zhang, X. and Xiang, J. (2002) Analysis of the gene expression profiles of immature versus mature bone marrow-derived dendritic cells using DNA arrays. *Biochemical and Biophysical Research Communications*, **290**, 66–72.

67 Zhou, X., Tan, F.K., Reveille, J.D., Wallis, D., Milewicz, D.M., Ahn, C., Wang, A. and Arnett, F.C. (2002) Association of novel polymorphisms with the expression of SPARC in normal fibroblasts and with susceptibility to scleroderma. *Arthritis and Rheumatism*, **46**, 2990–2999.

68 McLuckey, S.A. and Wells, J.M. (2001) Mass analysis at the advent of the 21st century. *Chemical Reviews*, **101**, 2571–2606.

69 Serfling, R.J. (1980) *Approximating Theorems of Mathematical Statistics*. John Wiley & Sons, New York.

70 Johnston, J. (1980) *Econometric Methods*, 2nd edn. McGraw-Hill, New York.

71 Xiong, H. (2006) Non-linear test for identifying differentially expressed genes or genetic networks. *Bioinformatics*, **22**, 919–923.

10
Detecting Pathological Pathways of a Complex Disease by a Comparative Analysis of Networks

Frank Emmert-Streib and Matthias Dehmer

10.1
Introduction

With his studies of pea plants, Gregor Mendel, often called the "father of modern genetics," paved the way for a genetical understanding of heredity and phenotypic variations of plants and animals. He demonstrated that genes are responsible for the phenotypic appearance of organisms or, in its simplest form, that one gene determines a trait, for example, the color of a seed, completely. Such genes are called Mendelian genes. Despite explaining many observable phenotypes successfully with Mendelian genes, our understanding of diseases, which can be seen as phenotype of organisms, is limited. It has been realized that many diseases, such as diabetes, heart disease or cancer, depend on altered interactions between multiple genes, rather than changes in a single gene. This means that groups of interacting genes, frequently called pathways, seem to be more appropriate for a systematic analysis, and hence, understanding of such complex diseases. In this chapter, we focus on one such complex disease – the chronic fatigue syndrome (CFS) – and introduce a method that allows us to detect pathways that are changed during pathogenesis.

The chronic fatigue syndrome affects approximately one million people in the United States alone [1]. So far, the diagnosis and evaluation of CFS has been primarily based on the presence or absence of various symptoms over time [2] rather than, for example, clear clinical signals [3]. Patients suffering from CFS experience persisting or relapsing fatigue of at least a 6-month duration that is not substantially reduced by rest and causes a significant reduction in activities [4]. Currently, the definition of CFS is largely based on case studies such as Fukuda *et al.* [5].

In this chapter, we aim to classify neither patients as sick (chronic fatigue) or nonsick nor genes based on experimental data from various methodologies. Instead, we use this information as prior information to constrain our analysis, because we expect that the signature of this disease in experimental data of any kind will be vanishingly small due to our lack of understanding of this disease,

Analysis of Microarray Data: A Network-Based Approach. Edited by F. Emmert-Streib and M. Dehmer

ISBN: 978-3-527-31822-3

preventing an efficient design of experiments. Previous studies of the CFS confirm this observation. Whistler *et al.* [3] found that detecting differentially expressed genes from microarray experiments is not possible if the patients are not grouped, prior to the analysis, according to their medical state. Also, the microarray data could be clustered meaningfully only if the data of the patients were classified beforehand based on clinical data and the grouping of genes according to their biological pathways. The implications thereof are that this disease is likely not to be characterized by single gene biomarkers. In this chapter, we pick up these results and use clinical data that classify patients in various groups [4]. From these groups we use only the group of sick (CF) and nonsick people, because we think that this distinction can be done reliably. We are aware of the speculation that the chronic fatigue disease is a heterogeneous, rather than a homogeneous, illness, which means that it is likely that the overall class of chronic fatigue patients can be subdivided into smaller classes corresponding to different representatives of the CFS [3]. For our study, however, this coarse classification is sufficient, because we do also not aim to detect these subclasses. In addition to the clinical data, we use the gene ontology (GO) database [6] as prior information to group genes according to their GO terms associating them to biological processes. For the resulting groups of genes, we construct an undirected dependency graph (UDG) from expression data of peripheral blood. This gives us, for example, one UDG representing the stage "sick" (CF) and one representing "nonsick" for the biological process of humoral immune response (GO:0006959). Our central hypothesis here is that the CFS has a signature detectable on a pathway level rather than on the expression level of single genes [7]. More precisely, we suggest an extension of the concept of differentially expressed genes at a pathway level, which means that the orchestra of expressed genes belonging all to a certain pathway is modified in response to the presence of a disease. Mathematically, we detect this modification by evaluating the similarity of the undirected dependency graphs representing pathways. We are aware of the fact that the available data from microarray experiments may not be sufficient to fully reconstruct the molecular interactions of genes and their products. For this reason, the UDGs are an approximation of the true pathways capturing some, but certainly not all, structural information of the underlying interactions. Following Xiong *et al.* [8], we name these approximations quasi-pathways. The assumption behind this hypothesis is that in the simplest case the phenotype of an organism is affected by a single gene mutation that means there is a one-to-one correspondence between phenotype and a single gene. In general, however, this oversimplified view is wrong, because otherwise we can only observe N different phenotypes, whereas N corresponds to the number of genes of the genome of an organism. This implies that combinations of genes and their controlled expression determine a phenotype. Since biological processes in general have a high intrinsic variability – not to be confused with a measurement error [9,10] – for example, in contrast to technical processes, it is plausible that not only one but multiple expression patterns of genes result in a specific phenotype. This makes biological processes robust and stable against perturbations [9,10] despite the sloppy functioning of their parts. This short discussion makes it clear that the concept of differentially expressed genes loses its significance in the context of injective phenotypes, which represent in our case the symptoms of CFS.

Before we present the mathematical details of our method, we want to visualize its underlying idea in the following section.

10.2 Outline of Our Method

As described in the previous section, we do not aim to classify patients according to the CFS stages, but group patients based on expert knowledge and clinical data [11] prior to our analysis [7]. From the resulting nine categories [4], we use data only from patients suffering from chronic fatigue syndrome in the worst stage and nonfatigue (NF) in the least stage. This corresponds to two patient classes – sick versus nonsick. In principle, we would like to find a way to compare these patient classes directly to identify marker of any similar characteristic for the sick or the nonsick class. However, this seems to be not possible. For this reason, we work around this obstacle by comparing patient classes not directly but indirectly. More precisely, from the microarray data of peripheral blood we extract groups of genes that participate in the same biological pathway according to the gene ontology database [6] and reconstruct their underlying network structure. This results in two sets of networks $\{G_i\}$ and $\{G'_i\}$, each corresponding to a biological pathway i for the sick G_i and nonsick G'_i class. These two classes are then analyzed comparatively, pathway by pathway, by calculating the distance between the corresponding graphs with an appropriate measure. This leads us, in the simplest case, to a binary decision about this comparison. Both pathway graphs are either similar ($d(G_i, G'_i) = 0$) or dissimilar ($d(G_i, G'_i) = 1$). From

$$\begin{aligned} d(G_1, G_1) &= 1 \\ d(G_2, G_2) &= 0 \\ \vdots & \quad \vdots \\ d(G_n, G_n) &= 0 \end{aligned} \tag{10.1}$$

we obtain a set of pathways $CP = \{i | d(G_i, G_i) = 1,\ i \in \{1, \ldots, n\}\}$ corresponding to candidate pathways that are changed due to the influence of disease progression.

Schematically, our approach is visualized in Figure 10.1. In the following sections, we will present our approach in more detail. Despite the fact that we are not able to estimate the network structure of the biological pathways with infinite precision (perfectly) – for this reason we speak of pseudo-pathways rather than pathways – our approach can be translated into a statistical framework.

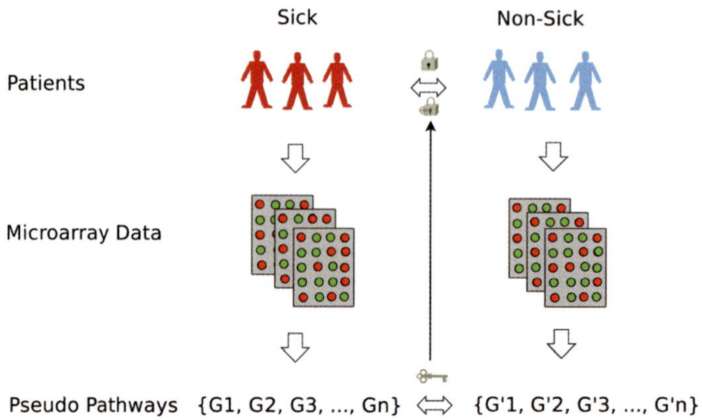

Figure 10.1 Visualization of our approach to detect differences in biological pathways due to pathogenesis of the CFS.

In Section 10.3, we introduce the statistical method to estimate the networks, and discuss different measures to compare graphs in Section 10.4. Then in Section 10.5, we present results of the application of our method to data from the CFS, and in Section 10.5.1 we discuss the influence of measurement errors. Finally, we discuss the obtained results and finish with an outlook in Section 10.6.

10.3
Detecting Pathological Pathways

Our approach is based on three different sources of information – clinical data [11], gene expression data of peripheral blood mononuclear cells [11], and the gene ontology database [6].

The clinical data provide us with a classification of the 227 patients, who participated in the study, into nine categories [4], from which we use data only from patients suffering from chronic fatigue syndrome in the worst stage and nonfatigue in the least stage. This corresponds to two patient classes – sick versus nonsick. Henceforth, we use this abbreviation to simplify the communication. We assume that people from either of these two classes can be distinguished reliably provided they respondents participated "fair" in the sense that they correctly answered the questionnaire on which the classification was based.

We obtain a meaningful grouping of genes according to biological processes they participate in the GO database. The GO database is hierarchically organized in three major groups – biological process, molecular function or cellular component. Since we hypothesize that pathways should change significantly due to the influence of the disease, we grouped genes according to biological processes they participated in. Practically, however, this grouping was constrained by two factors. First, we were only looking for groups consisting of 20–50 genes, because inferring meaningful

Table 10.1 Biological processes used in our analysis.

Number	GO term	Description
1	GO:0006471	Protein amino acid ADP-ribosylation
2	GO:0007219	Notch signaling pathway
3	GO:0008360	Regulation of cell shape
4	GO:0042157	Lipoprotein metabolism
5	GO:0007126	Meiosis
6	GO:0006958	Complement activation, classical pathway
7	GO:0007222	Frizzled signaling pathway
8	GO:0006633	Fatty acid biosynthesis
9	GO:0043087	Regulation of GTPase activity
10	GO:0042742	Defense response to bacteria
11	GO:0001525	Angiogenesis
12	GO:0045892	Negative regulation of transcription, DNA dependent

connections between genes becomes more complex for larger groups of genes [12]. Second, we were searching for gene groups that were present to a large extent on the microarray chip used. More precisely, the selected genes corresponding to a certain pathways could be found with 75% or more on the chip. Considering these constraints, we selected 12 biological processes fulfilling our conditions. These pathways are given in Table 10.1.

10.3.1 Pathways as Undirected Dependency Graphs

For the pathways given in Table 10.1, we calculated undirected dependency graphs. An UDG is an undirected, unweighted graph, whose edges indicate that the connected nodes depend in same way on each other. In our case, the nodes of the graph correspond to genes and an edge might be interpreted as, for example, gene A regulating the transcription of gene B, gene (protein) A interacting with gene (protein) B, or gene (protein) A phosphorylating gene (protein) B. Because we have no explicit information about the physical interactions between two genes or products thereof, the edges in the UDGs should be abstractly viewed. For this reason we speak about UDG and not about, for example, signaling or transcription regulatory networks. As a measure of dependency between two genes, we used the correlation and partial correlation coefficient calculated from the expression profiles, $\mathbf{x}_i$ for $i \in \{1, \ldots, N\}$, of the genes from the microarray data. The components of the profiles correspond to patients. For example, we used the expression profiles of the genes from *regulation of GTPase activity* (GO:0043087) from the sick and nonsick patients according to their classification described in the previous section and calculated two UDGs, one representing the regulation of GTPase activity pathway for the sick patients and one for the nonsick patients. More formally, an UDG is obtained via algorithm 1. An UDG G of first order is an undirected, unweighted graph with N nodes (number of genes). The thresholds Θ_c and Θ_{pc} were obtained from randomization tests by generating random profiles $\hat{\mathbf{x}}_i$ by randomly assigning expression values to the components of $\hat{\mathbf{x}}_i$ from all available expression data. These thresholds correspond to p-values with $p = \alpha$. The significance level α was set to 0.05. For the correlation we used Pearson correlation

$$\rho_{\mathrm{x}_i\mathrm{x}_j} = \frac{\sigma_{\mathrm{x}_i\mathrm{x}_j}}{\sigma_{\mathrm{x}_i}\sigma_{\mathrm{x}_j}}, \tag{10.2}$$

whereas the estimator for the variance $\sigma^2_{\mathrm{x}_i}$ and covariance $\sigma_{\mathrm{x}_i\mathrm{x}_j}$ is given by

$$s^2_{\mathrm{x}_i} = \frac{\sum_m (x_i^m - \mu_{\mathrm{x}_i})^2}{N-1}, \tag{10.3}$$

$$s_{\mathrm{x}_i\mathrm{x}_j} = \frac{\sum_m (\mathrm{x}_i^m - \mu_{\mathrm{x}_i})(\mathrm{x}_j^m - \mu_{\mathrm{x}_j})}{N-1}, \tag{10.4}$$

resulting in the estimator for the correlation

$$r_{\mathrm{x}_i\mathrm{x}_j} = \frac{s_{\mathrm{x}_i\mathrm{x}_j}}{s_{\mathrm{x}_i}s_{\mathrm{x}_j}}. \tag{10.5}$$

Algorithm 1: Undirected dependency graph of first order

connect all N nodes with an edge with each other
for all (i, j) **do**
  calculate the correlation $r_{\mathbf{x}_i\mathbf{x}_j}$
end for
for all $e_{\mathbf{x_i},\mathbf{x_j}}$ **do**
  if $r_{\mathbf{x}_i\mathbf{x}_j} < \Theta_c$ **then**
    delete the edges $e_{\mathbf{x_i},\mathbf{x_j}}$ connecting node $\mathbf{x}_i$ and $\mathbf{x}_j$
  end if
end for
for all (i, j, k) with $e_{\mathbf{x_i},\mathbf{x_j}}$ **do**
  calculate the partial correlation of first order $r_{\mathbf{x}_i\mathbf{x}_j|\mathbf{x}_k}$
end for
for all $e_{\mathbf{x_i},\mathbf{x_j}}$ **do**
  if $r_{\mathbf{x}_i\mathbf{x}_j|\mathbf{x}_k} < \Theta_{pc}$ **then**
    delete the edge connecting node $\mathbf{x}_i$ with $\mathbf{x}_j$
  end if
end for

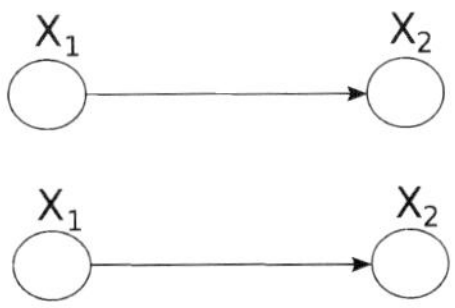

Figure 10.2 Top: True network structure. Bottom: Estimated network structure of an algorithm purely based on correlations.

The partial Pearson correlation of first order is obtained by

$$r_{x_i x_j | x_k} = \frac{r_{x_i x_j} - r_{x_i x_k} r_{x_j x_k}}{\sqrt{(1-r^2_{x_i x_k})}\sqrt{(1-r^2_{x_j x_k})}}. \tag{10.6}$$

The reason why we used partial correlation in addition to correlation is exemplified in Figures 10.2 and 10.3. Suppose the upper graph in Figure 10.2, represents the true relation between gene A and gene B. An inference algorithm purely based on correlation coefficients would estimate the lower graph in Figure 10.2 because the correlation between gene (profile) A and gene (profile) B would be high. Suppose now the upper graph in Figure 10.3 represents the true relation between genes A, B, and C. An inference algorithm purely based on correlations would estimate now a network structure including one additional edge from gene A to C as shown in the lower Figure 10.3. The reason therefore is clear because gene A affects gene C and, hence, there is a correlation between both variables. However, apparently, the edge between gene A and gene C is wrong. To prevent such false positive edges in the inferred network due to indirect rather than direct effects between genes, we calculated partial correlation coefficients of first order as well. For the situation depicted in Figure 10.3, $r_{\{x_1, x_3 | x_2\}}$ would be zero because gene 2 blocks the information flow between genes 1 and 3. Hence, using partial correlation of first order is the first step to estimate the underlying causal structure of interactions [13]. We want to remark that calculating partial correlation

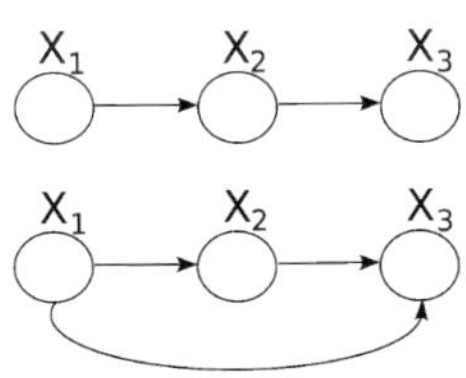

Figure 10.3 Top: True network structure. Bottom: Estimated network structure of an algorithm purely based on correlations. Since gene 1 affects indirectly gene 3, the correlation between both variables will be high.

coefficients up to order n would transfer our approach to the PC-algorithm [14,15], whose theoretical fundament is based on the graph theoretical entity *d-separation* and its correspondence to partial correlations [16–18]. However, from a practical side, there were three reasons that made us restrict the order to one. First, the computational effort calculating higher order partial correlation coefficients is NP-hard. Second, simulation studies show that the percentage reduction of false positive edges is largest for $n = 1$. Increasing the order beyond 1 has only a minor effect [19,20]. Third, the major objective of this study is not to infer causal molecular network structures but to detect significant differences between network structures due to disease pathogenesis.

10.4
Measuring the Similarity of Pathways

In this section, we define a similarity measure to evaluate the modifications of UDGs sick versus a nonsick pathway. For this, we first give a brief overview about methods to determine the structural similarity of graphs.

Graph similarity or distance measures have been frequently investigated [21–30]. To construct a meaningful graph similarity measure, we have to take into account certain structural characteristics and parameters of the graphs under consideration, for example, degree sequences and measures based on metrical graph properties [31]. However, selecting such structural characteristics in practice is difficult. In the literature, there are two major paradigms for matching graphs structurally: *exact graph matching* and *inexact graph matching* [24,32]. An exact graph matching technique is based mainly on the principle that we are looking for a graph or subgraph structure of a graph that matches exactly a graph or subgraph structure of another graph. To express the structural equivalence of graphs, we use the known definition of a graph isomorphism [33–35]. We notice that most of the graph metrics or graph similarity measures of the exact graph-matching paradigm are based on isomorphic and subgraph isomorphic relations [21–25,28–30,36,37], respectively. An example of such a classical graph metric that is based on subgraph–isomorphic relations is the well-known Zelinka distance [30]. This graph metric is based mainly on the principle that two graphs are more similar, the bigger the common induced (isomorphic) subgraphs. Zelinka was the first to introduce this measure for unlabeled graphs with an equal number of vertices [25,28–30]. This result can be stated as follows [30].

Theorem 1 Let $H = (V_H, E_H)$ and $G = (V_G, E_G)$ be unlabeled graphs without reflexive and multiple edges and it holds $|V_H| = |V_G| = n$. $\overline{\text{SUB}}_m(H)$ denotes the set of induced subgraphs of order m. H^* denotes the isomorphism classes of such graphs in which H lies and let

$$\text{SUB}_m(H) := \{H^* | H \in \overline{\text{SUB}}_m(H)\}. \tag{10.7}$$

$SUB_m(H)$ is the set of isomorphism classes in which the induced subgraphs of H with order m lie. Then,

$$d_Z(H,G):=n-\mathrm{SIM}(H,G), \tag{10.8}$$

is a graph metric, where

$$\mathrm{SIM}(H,G):=\max\{m|\mathrm{SUB}_m(H)\cap \mathrm{SUB}_m(G)\neq\varnothing\}. \tag{10.9}$$

Generally, if we assume a graph $G=(V, E)$, the graph $G?=(V?, E?)$ is called a subgraph iff $V?\subseteq V$ and $E?\subseteq E$. Moreover, if it holds $E?=E\cap(V?\times V?)$, then we call $G?$ the induced subgraph of G [33,38]. An isomorphism class generally denotes the set of graphs that is isomorphic to a given graph G. As an example we look at Figure 10.4.

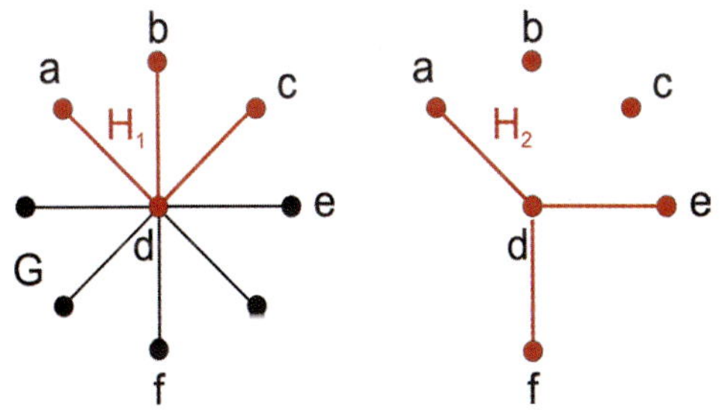

Figure 10.4 Two undirected graphs G and H_2. The graph H_1 (indicated by red vertices and edges) is embedded in G.

The graph G is an undirected star-like graph with nine vertices. The embedded graph H_1 (indicated by red vertices and edges) is an induced subgraph of G and has four vertices. In contrast to this, H_2 has six vertices. We conclude that H_2 is a subgraph of G, but not an induced subgraph. We clearly see that an induced subgraph H of a graph G contains all edges of E regarding the vertex set of H. That also means that an induced subgraph H covers more structural information of G than a simple subgraph. Therefore, most of the classical graph metrics or graph similarity measures are based on induced subgraph relations. It is evident that Theorem originally stated by Zelinka is not applicable for practical applications because the graphs under consideration mostly possess different orders and vertex or edge labels. Therefore, a more general graph metric has been introduced by Sobik [28,29] and Kaden [25], who generalized this measure for arbitrary and also labeled graphs of different orders [25,28–30]. To demonstrate the exact graph matching paradigm by an example, we now present the known generalization of the Zelinka distance due to Sobik [28,29].

Theorem 2 Let H: $(V, E, f_V, f_E, A_V, A_E)$ be a finite and labeled graph. A_V, A_E denote finite, nonempty vertex and edge alphabets and $f_V: V\rightarrow A_V, f_E: E\rightarrow A_E$ the associated vertex and edge labeling functions. Now, let H and G be finite, labeled graphs of arbitrary orders. Then,

$$d_S(H,G) := \max\{|H|,|G|\} - \mathrm{SIM}(H,G)\} \tag{10.10}$$

is a graph metric.

In Figure 10.5 we give an example for the application of Theorem 2 and compute the pairwise (generalized) Zelinka distances d_S for the shown graphs. However, regarding the practical application it is known that the subgraph isomorphism problem is NP-complete [35]. For this reason, such measures are not suitable to classify large sets of graphs, especially graphs with many vertices.

In contrast to methods from the exact graph-matching paradigm that are based on isomorphic relations, there exists an important class of graph similarity or distance measures based on the so-called *graph edit distance* (GED) belonging to the inexact graph-matching paradigm. The graph edit distance can also be viewed as a powerful extension of the well-known *string edit distance* originally developed by Levenstein [39]. We want to remark that the graph edit distance, introduced by Bunke *et al.* [21,40], is a graph distance metric that naturally belongs in the subclass of so-called *error-correcting graph matching* [21,40–42]. The basic idea of the graph edit distance is to define certain graph edit operations such as insertion or deletion of an edge/vertex or relabeling of a vertex along with costs associated with each such operation. Bunke *et al.* [21,40] denote as an optimal inexact match a sequence of edit operations that transforms a graph G into H by producing minimal transformation costs. If we

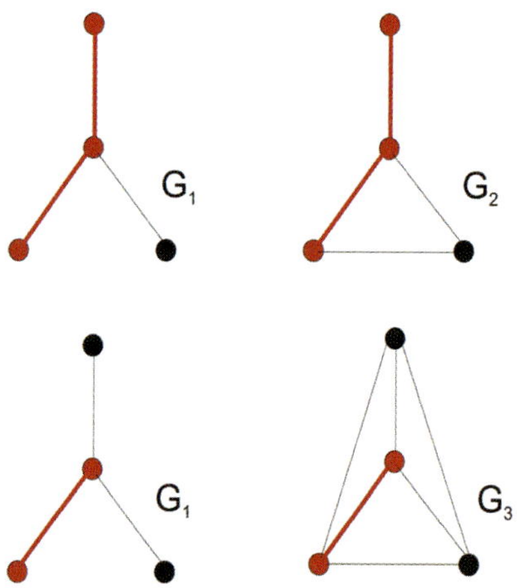

Figure 10.5 The undirected graphs G_1, G_2, and G_3. The pairwise computation of the generalized Zelinka distance leads us to $d_S(G_1, G_2) = 4 - 3 = 1$ and $d_S(G_1, G_3) = 4 - 2 = 2$. The subgraphs indicated by red vertices and edges are possible maximal induced subgraphs.

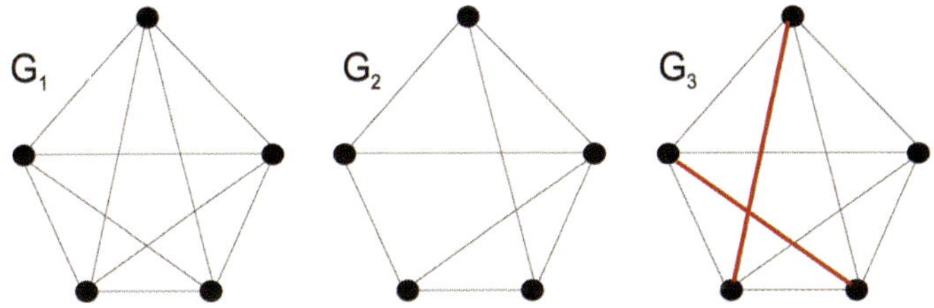

Figure 10.6 Two simple unlabeled, undirected graphs $G_1 = (V_1, E_1)$ and $G_2 = (V_2, E_2)$ we compared structurally. G_1 is the so-called complete graph K_5 with five vertices. It holds $|V_1| = |V_2| = 5$ and $|E_1| > |E_2|$. By inserting two edges (indicated by red color) in G_2, we get the alignment graph G_3. After performing this edit operation, it holds that $G_1 \cong G_3$, that is, G_1 and G_3 are isomorphic.

assume that $m_1, m_2, \ldots, m_n$ are all possible transformations mapping G to H, then the optimal inexact match [21] m' is defined by

$$c(m') = \min\{c(m_i)|1 \leq i \leq n\}. \tag{10.11}$$

This means that the graph edit distance between two graphs is the minimum cost associated with a sequence of edit operations. The optimal error-correcting graph isomorphism is then defined as the resulting isomorphism, after obtaining this optimal sequence of edit operations [42]. We now state the original result of Bunke *et al.* [21,40].

Theorem 3 Let $d(H, G)$ be the costs for determining the optimal inexact match between H and G. Then, $d(H, G)$ is a graph metric.

As an example, we exemplarily determine the graph edit distance of two simple unlabeled graphs in Figure 10.6. If we set all edit costs equal to 1, we observe that by inserting two edges into G_2, we get G_1. Thus, we transform G_2 to G_1 and receive the final alignment graph G_3. For this example, we obtain $d(G_1, G_2) = 2$. Bunke [43,44] furthermore explored relations between the edit distance of two graphs and their maximum common (isomorphic) subgraph. Particularly, it has been shown that the optimal error-correcting graph isomorphism for a certain graph-edit cost function is the maximum common subgraph [43,44]. Regarding the computational complexity of the GED, it has been shown that for general (unlabeled) graphs the computational complexity grows exponentially with the size of the problem. Thus, the problem to determine the similarity between two graphs is NP-complete [35,42,45].

10.4.1 GED for Graphs With Unique Vertex Labels

The graph similarity measure we use to compare the networks corresponding to pathways of the CFS is the graph edit distance regarding a special class of graphs.

In our case, the comparison of two pathways representing graphs was constrained by two important properties that made the comparison using the graph edit distance tractable:

- The compared graphs have always the same number of vertices because graphs correspond to pathways and the vertices correspond to genes. However, the definition of a pathway does not change between sick and nonsick people.
- The vertices of our graphs are uniquely labeled by the names of the genes.

Especially, the second condition implies that we did not need to find an optimal alignment between the vertices of both graphs for which the optimal edge manipulations have to be found, but it is clear that vertices with the same gene name should be matched if we want to determine the overall structural deformation between both graphs. It is interesting to note that for graphs with unique vertex labels, some already mentioned graph problems are efficiently computable. For example, Dickinson *et al.* [46] showed that the computational complexity for this case of the graph edit distance equals $O(n^2)$, where n denotes the number of vertices of the underlying graph. Moreover, Dickinson *et al.* [46] proved a useful theorem, which states that some computational graph procedures require quadratic time complexity.

Theorem 4 Let U be the class of graphs having unique node labels. Then, there exist computational procedures that solve the following problems in quadratic time with respect to the number of vertices in an underlying graph:

1. Graph edit distance.
2. Graph isomorphism problem.
3. Subgraph isomorphism problem.
4. Maximum common subgraph problem (e.g., to compute the generalized Zelinka distance, see Theorem 2).

Now, we answer the question how the graph edit distance for graphs $G \in U|\{n\}$can be easily computed, where $U|\{n\}$ denotes the class of graphs having unique vertex labels and n vertices. To determine the graph edit distance of two such graphs G and H, we are looking for sequences of edit operations that transform G to H. This leads us to the following definition.

Definition 1 Let $G, H \in U|\{n\}$. We call $n_{\text{GED}}(G, H)$ the graph edit distance between G and H, where $n_{\text{GED}}(G, H)$ counts the number of edge deletions and insertions we need to transform graph G to H. By dividing $n_{\text{GED}}(G, H)$ with the total number of edges possible in such a graph with n vertices, we obtain the measure

$$p_{\text{GED}}(G, H) = \frac{2n_{\text{GED}}(G, H)}{n(n-1)} \in [0, 1]. \tag{10.12}$$

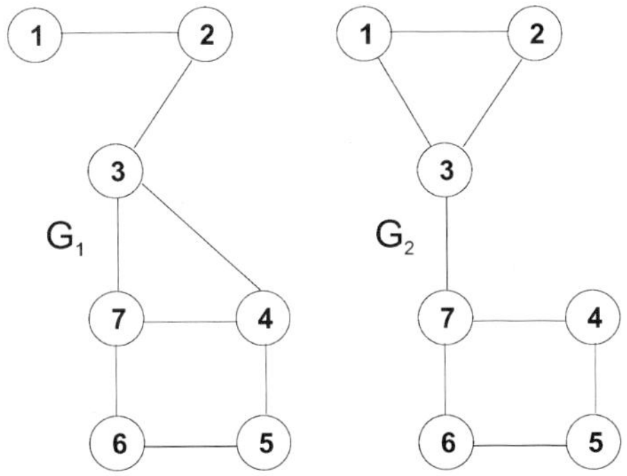

Figure 10.7 Two undirected graphs G_1 and G_2 with seven vertices to determine the graph edit distance. We see that G_1 and G_2 have unique vertex labels.

We note that Equation 10.15 is simply the percentage graph edit distance between graph G and H. The factor "2" indicates that we deal with undirected networks. By applying Theorem 4, we get that the computational complexity of $p_{\mathrm{GED}}(G, H)$ is $O(n)^2$. In Figures 10.7 and 10.8, we depict the procedure to determine the graph edit distance between two undirected uniquely vertex labeled graphs. Figure 10.7 shows such two graphs G_1 and G_2 we wanted to compare. To get the alignment graph G_3 depicted in Figure 10.8, we applied an edge insertion (indicated by a dashed line) and an edge deletion (indicated by a dotted line) to G_1. This sequence of edit operations transforms G_1 to G_2. By assuming that all edit costs are equal to 1, we get $n_{\mathrm{GED}}(G_1, G_2) = 2$.

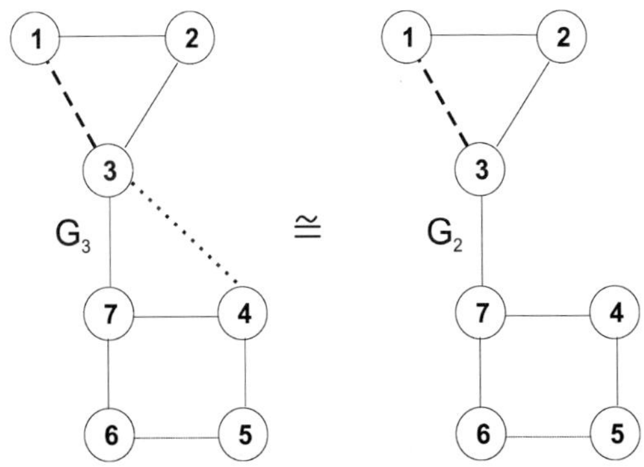

Figure 10.8 To transform the graph G_1 to G_2 (see Figure 10.7), we perform one edge insertion (dashed line) and one edge deletion (dotted line) in G_1. Then, we obtain the resulting alignment graph G_3 and it holds $G_3 \cong G_2$.

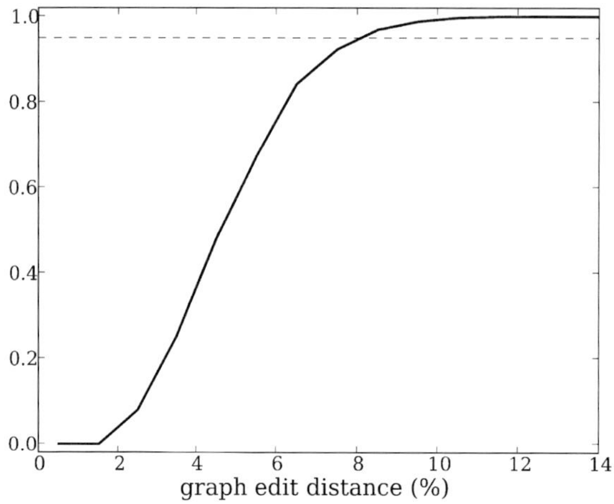

Figure 10.9 Cumulative distribution of graph edit distances of random networks. The dashed line corresponds to 0.95.

10.4.2
Statistical Significance of the GED

Finally, in order to decide if the obtained graph edit distance between two graphs is similar (the disease had no influence) or unsimilar (the disease had an influence), we needed to find a threshold Θ_{GED} to separate similar ($p_{\text{GED}}(G_A, G_B) \leq \Theta_{\text{GED}}$) from unsimilar ($p_{\text{GED}}(G_A, G_B) > \Theta_{\text{GED}}$) pathways. Naturally, this threshold was found via randomization tests by generating an ensemble of random networks with the same number of nodes and the same mean number of edges as the inferred networks from the microarray data. In Figure 10.9 we show the cumulative distribution of the percentage graph edit distances of these random networks. Choosing a significance level of $\alpha = 0.05$ leads to $\Theta_{\text{GED}} = 0.08$ above which we view two graphs as statistically unsimilar and, hence, as indicator that the CFS might be responsible for this modification.

10.5
Results for the Chronic Fatigue Syndrome

Now we present the results of our method described in the previous section applied to the data from the CFS study [11]. The graph edit distances for the biological processes in Table 10.1 are shown in Figure 10.10. The results are ranked in descending order of the graph edit distances. The horizontal dashed line corresponds to a p-value of $\alpha = 0.05$. Graph edit distances above this line are statistically significant with a p-value less than α. One can clearly see that the first bar is much

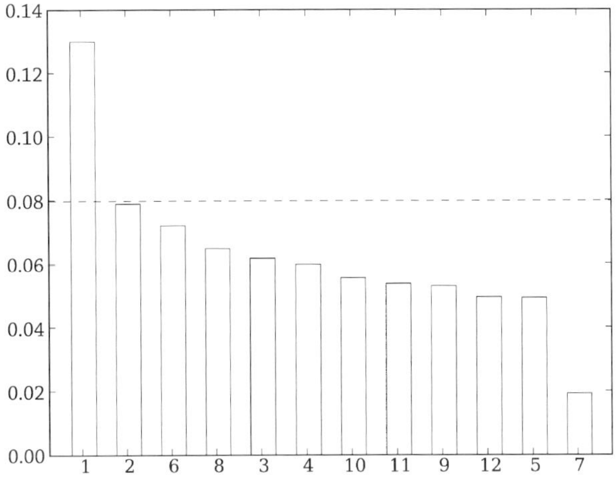

Figure 10.10 Similarity values of networks obtained by the comparison of the same pathway from sick versus nonsick patients. The similarity between the networks was calculated with the graph edit distance of Bunke. The dashed line corresponds to a p-value of 0.05.

higher than all other bars. The biological process of this pathway is protein amino acid ADP-ribosylation (GO:0006471).

The second highest graph edit distance is from the Notch signaling pathway (GO:0007219) that is only slightly below Θ_{GED}. All other graph edit distances are clearly below the significance threshold indicating that our method is rather conservative. This is a desirable feature because it allows follow-up experiments to focus on a very restricted subset of all possible configurations.

The biological function of ADP-ribosylation is related to DNA-repair, apoptosis, and disease response [47,48]. A possible connection to the CFS is provided by a hypothesis that oxidative stress might play a major role as possible cause for the CFS. There are several experimental studies available supporting this hypothesis [49,50]. Since oxidants and free radicals damage DNA, the oxidative stress could affect the ADP-ribosylation pathway to initiate DNA-repair, and, hence, to counterbalance the destructive influence of oxidative stress. This argument is plausible but, of course, needs to be studied experimentally in more detail.

A natural question arising now is how stable are the obtained results against the influence of noise. Since the amount of the available data is finite and measurement errors are present, it is clear that the reconstruction of the UDGs, as described in Section 10.3.1, is erroneous. For microarray experiments, it is reasonable to assume an error of about 20% [51]. To gain a better understanding of the obtained results, we investigate now the stability of the results against reconstruction errors of the underlying networks.

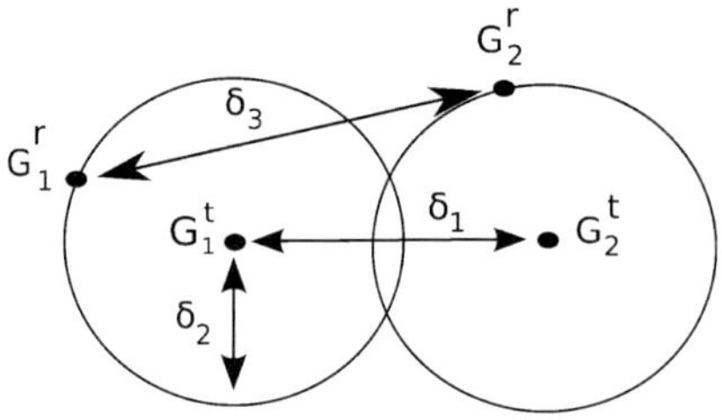

Figure 10.11 Visualization of the influence of a measurement error leading to a reconstruction error δ_2 for the true networks G_1^t and G_2^t. The true graph edit distance between the true networks is $d_{GED}(G_1^t, G_2^t) = \delta_1$ and δ_3 is the graph edit distance between the reconstructed networks G_1^r and G_2^r.

10.5.1
Influence of Measurement Errors

We analyze this influence in the following way. First, we generate two networks with N nodes and an edge density of p_e drawn from a Gaussian distribution Gauss(p_0, σ) with a graph edit distance of δ_1. We call these graphs the true networks, G_1^t and G_2^t. Then we generate two graph G_1^r and G_2^r with $d_{\text{GED}}(G_1^t, G_1^r) = d_{\text{GED}}(G_1^t, G_1^r) = \delta_2$. We call these graphs reconstructed networks, because the graph edit distance δ_2 mimics the reconstruction error of the corresponding network. Schematically, this procedure is depicted in Figure 10.11.

Calculating the graph edit distance $d_{\text{GED}}(G_1^r, G_2^r) = \delta_3$ provides now the desired information regarding the stability of the true graph edit distance $d_{\text{GED}}(G_1^t, G_2^t)$. Here, we do this calculation for an ensemble of networks, as described by algorithm 2, resulting in a distribution of $d_{\text{GED}}(G_1^r, G_2^r)$ values in dependence on δ_1 and δ_2. Because δ_2 represents the distance of the true from the reconstructed graph we keep it constant to $0.2\delta_1$ corresponding to a typical error level of 20% of biological experiments. The variation of the parameter δ_1 allows us to study the robustness as function of the distance to Θ_{GED}. Statistically, we measure the robustness as the number of true positive distances (TP) if $\delta_1 > \Theta_{\text{GED}}$ and the number of true negative distances (TN) if $\delta_1 < \Theta_{\text{GED}}$. In Figure 10.12, we show the true negative/positive rate (TNR/TPR) as function of the distance of δ_1 from Θ_{GED} in percentage. We again emphasize that δ_1 is the true graph edit distance and Figure 10.12 gives the probabilities that a true graph edit distance, larger/smaller than Θ_{GED}, is estimated as δ_3 with δ_3 larger/smaller than Θ_{GED} if the reconstruction error is δ_2 or formally

$$\text{TPR}(\delta_1) = p(\delta_3 > \Theta_{\text{GED}} | \delta_1 > \Theta_{\text{GED}}), \tag{10.13}$$

$$\text{TNR}(\delta_1) = p(\delta_3 < \Theta_{\text{GED}}) | \delta_1 < \Theta_{\text{GED}}). \tag{10.14}$$

From these results we can conclude that the protein amino acid ADP-ribosylation pathway (GO:0006471), which has 62% distance from Θ_{GED}, receives very strong confirmation from our stability analysis if we assume that the true graph edit distance is larger than Θ_{GED}. However, if we assume that the true graph edit distance

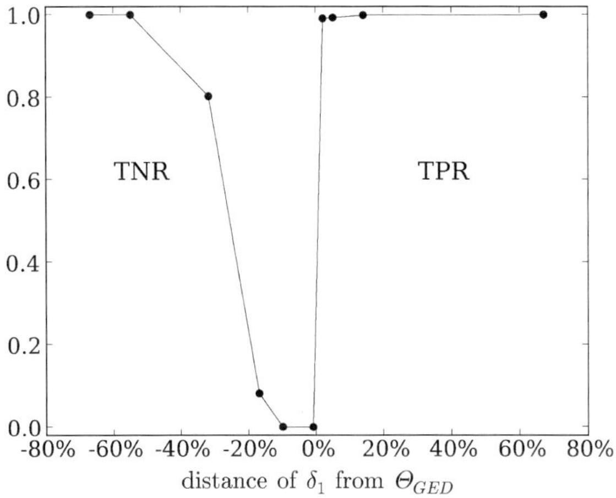

Figure 10.12 Influence of a reconstruction error δ_2 on the stability of the solution.

is less than Θ_{GED}, the probability to obtain a graph edit distance $\delta_3 \geq 0.63$ is vanishingly small. This is revealed by additional simulations similar to those described by algorithm 2. In contrast, the Notch signaling pathway (GO:0007219), which has -3% distance from Θ_{GED}, is unlikely to be a false negative (FN), because the probability that any true graph edit distance with $\delta_1 > \Theta_{\mathrm{GED}}$ results in $\delta_3 < \Theta_{\mathrm{GED}}$ is very low as can be seen directly from

$$\mathrm{FNR}(\delta_1) = p(\delta_3 < \Theta_{\mathrm{GED}} | \delta_1 > \Theta_{\mathrm{GED}}) = 1 - \mathrm{TPR}(\delta_1) \tag{10.15}$$

and Figure 10.12. A similar argument holds for all other pathways that were detected as similar, that is, with a graph edit distance less than Θ_{GED}. Hence, our method identified exactly one pathway as statistically significant – the ADP-ribosylation pathway (GO:0006471).

Algorithm 2: Graph edit distances between reconstructed graphs.

δ_1: distance between the true graphs
δ_2: distance between a true and reconstructed graph
p_0: mean edge density (sparse networks)
N number of nodes
repeat
  draw a random variable p_e (edge density) from $Gauss(p_0, \sigma)$
  construct G_1^t (true network)
  construct G_2^t (true network) with $d_{GED}(G_1^t, G_2^t) = \delta_1$
  for all $i = 1$ to 2 **do**
  construct G_i^r('reconstructed'network)with $d_{GED}(G_i^t, G_i^r) = \delta_2$
  end for
  calculate $\delta_3 = d_{GED}(G_1^r, G_2^r)$
  save δ_3
until ensemble is complete

10.6
Discussions and Conclusions

In this chapter, we described a novel method, introduced recently, to detect differences in biological pathways due to pathogenesis [7]. Our method can be summarized by the following steps: First, we represent pathways as networks we call undirected dependency graphs. Second, we compare networks corresponding to the same pathway from sick and nonsick patients graph theoretically. Third, we decide statistically if the compared networks are similar or different. We exemplified our method on expression data from peripheral blood mononuclear cells from patients suffering from the chronic fatigue syndrome and control patients who were healthy. Application of our method to these data revealed that the pathway of protein amino acid ADP-ribosylation (GO:0006471) is significantly changed due to the influence of the disease. Additional simulation studies clearly indicate that even in the presence of a network reconstruction error of about 20%, which seems to be a reasonable assumption of the noise level present in data from microarray experiments [51], this result is highly reliable.

Biologically, it is known that ADP-ribosylation is involved in DNA-repair, apoptosis, and disease response [47,48]. It has been hypothesized that oxidative stress might play a major role as possible cause for CFS. Several studies supporting this hypothesis [49,50] are now available. Since oxidants and free radicals damage the DNA, oxidative stress could affect the ADP-ribosylation pathway to initiate DNA-repair to counterbalance the destructive influence of oxidative stress. This gives a plausible involvement of the ADP-ribosylation pathway in the context of the aforementioned hypothesis.

We emphasize that we did not aim to detect the differential expression of single genes between chronic fatigue and healthy people, because previous studies already demonstrated the limitations of this approach to the CFS [3], but to detect modifications due to the pathogenesis of complete pathways the genes participate in. This conceptual understanding was already introduced in a similar form by Xiong *et al.* [8], however, by using a different mathematical framework. Our approach, which favors groups of interacting genes rather than single genes, is justified by recent insights into the working mechanisms of complex diseases in general [52]. We believe that our understanding of complex diseases, such as the chronic fatigue syndrome, will tremendously improve if we continue developing methods on a pathway and, hence, systems level.

Acknowledgments

We would like to thank Chris Bauch, Jie Chen, Alberto de la Fuente, Earl F. Glynn, Arcady Mushegian, Chris Seidel, Bill Shipley, and Korbinian Strimmer for fruitful discussions, and Mike Coleman and Daniel Thomasset for administrating the computer system.

References

1 Reynolds, K.J., Vernon, S.D., Bouchery, E. and Reeves, W.C. (2004) The economic impact of chronic fatigue syndrome. *Cost Effectiveness and Resource Allocation*, **2**, 4.

2 Jason, L.A., King, C.P., Frankenberry, E.L. and Jordan, K.M. (1999) Chronic fatigue syndrome: assessing symptoms and activity level. *Journal of Clinical Psychology*, **55** (44), 411–424.

3 Whistler, T., Unger, E.R., Nisenbaum, R. and Vernon, S.D. (2003) Integration of gene expression clinical and, epidemiologic data to characterize chronic fatigue syndrome. *Journal of Transnational Medicine*, **1**, 10.

4 Reeves, W.C., Wagner, D., Nisenbaum, R., Jones, J.F., Gurbaxani, B., Solomon, L., Papanicolaou, D.A., Unger, E.R., Vernon, S.D. and Heim, C. (2005) Chronic fatigue syndrome – a clinically empirical approach to its definition and study. *BMC Medicine*, **3**, 19.

5 Fukuda, K., Straus, S.E., Hickie, I., Sharpe, M.C., Dobbins, J.G. and Komaroff, A. (1994) The chronic fatigue syndrome: a comprehensive approach to its definition and study. *Annals of Internal Medicine*, **121**, 953–959.

6 Ashburner, M., Ball, C.A., Blake, J.A., Botstein, D., Butler, H., Cherry, J.M., Davis, A.P., Dolinski, K., Dwight, S.S., Eppig, J.T., Harris, M.A., Hill, D.P., Issel-Tarver, L., Kasarskis, A., Lewis, S., Matese, J.C., Richardson, J.E., Ringwald, M., Rubin, G.M. and Sherlock, G. (2000) Gene ontology: tool for the unification of biology. The gene ontology consortium. *Nature Genetics*, **25** (1), 25–29.

7 Emmert-Streib, F. (2007) The chronic fatigue syndrome: a comparative pathway analysis. *Journal of Computational Biology*, **14** (7), 961–972.

8 Xiong, M., Li, J. and Fang, X. (2004) Identification of genetic networks. *Genetics*, **166**, 1037–1052.

9 Fraser, H.B., Hirsh, A.E., Glaever, G., Kumm, J. and Eisen, M.B. (2004) Noise minimization in eukaryotic gene expression. *PLOS Biology*, **2** (6), e137.

10 McAdams, H.H. and Arkin, A. (1997) Stochastic mechanisms in gene expression. *Proceedings of the National Academy of Sciences of the United States of America*, **94**, 814–819.

11 Dataset provided by CAMDA, (2006).

12 Waddell, P.J. and Kishino, H. (2000) Cluster inference methods and graphical models evaluated on nci60 microarray gene expression data. *Genome Informatics*, **11**, 129–140.

13 Shipley, B. (2000) *Cause and Correlation in Biology*, Cambridge University Press.

14 Spirtes, P. and Glymour, C. (1991) A algorithm for fast recovery of sparse causal graphs. *Social Science Computer Review*, **9**, 62–72.

15 Spirtes, P., Glymour, C. and Scheines, R. (2000) *Causation Prediction, and Search*, MIT Press.

16 Pearl, J. (1988) *Probabilistic Reasoning in Intelligent Systems*, Morgan Kaufmann.

17 Verma, T. and Pearl, J. (1988) Causal networks: semantics and expressiveness. Proceedings of the 4th Workshop on Uncertainly in Artificial Intelligence Mountain View, CA, pp. 352–359.

18 Geiger, D., Verma, T. and Pearl, P. (1989) D-separation: from theorems to algorithms. *Workshop on Uncertainty in AI*.

19 Emmert-Streib, F. and Dehmer, M. (2006) Theoretical bounds for the number of inferable edges in sparse random networks. BIOCOMP'06 – The 2006 International Conference on Bioinformatics and Computational Biology, 472–476.

20 de la Fuente, A., Bing, N., Hoeschele, I. and Mendes, P. (2004) Discovery of meaningful associations in genomic data using partial correlation coefficients. *Bioinformatics*, **20** (18), 3565–3574.

21 Bunke, H. (1983) What is the distance between graphs? *Bulletin of the EATCS*, **20**, 35–39.

22 Dehmer, M., Emmert-Streib, F. and Kilian, J. (2006) A similarity measure for graphs with low computational complexity. *Applied Mathematics and Computation*, **182** (1), 447–459.

23 Dehmer, M., Emmert-Streib, F., Mehler, A. and Kilian, J. (2006) Measuring the structural similarity of web-based documents: a novel approach. *International Journal of Computational Intelligence*, **3** (1), 1–7.

24 Dehmer, M. (2006) Strukturelle Analyse Web-basierter Dokumente, Gabler Edition Wissenschaft – Deutscher Universitätsverlag (eds F. Lehner and F. Bodendorf) Series: Multimedia und Telekooperation.

25 Kaden, F. (1982) Graphmetriken und Distanzgraphen. ZKI-Informationen, Akademie der Wissenschaften der DDR, pp. 1–63.

26 Schädler, C. (1999) Die Ermittlung struktureller Ähnlichkeit und struktureller Merkmale bei komplexen Objekten: Ein konnektionistischer Ansatz und seine Anwendungen PhD thesis,Technische Universität, Berlin.

27 Schenker, A., Bunke, H., Last, M. and Kandel, A. (2005) *Graph-Theoretic Techniques for Web Content Mining*, World Scientific Publishing Company.

28 Sobik, F. (1982) Graphmetriken und Klassifikation strukturierter Objekte. ZKI-Informationen, Akademie der Wissenschaften der DDR, **2** (82), 63–122.

29 Sobik, F. (1984) A graph theoretic approach to the characterization of classes of structured objects. *Computers and Artificial Intelligence*, **3**, 235–247.

30 Zelinka, B. (1975) On a certain distance between isomorphism classes of graph. *Matematiky*, **100**, 371–373.

31 Skorobogatov, V.A. and Dobrynin, A.A. (1988) Metrical analysis of graphs. *MATCH*, **23**, 105–155.

32 Bunke, H. (2000) Graph matching: theoretical foundations, algorithms, and applications. Proceedings of Vision Interface 2000, pp. 82–88.

33 Bang-Jensen, J. and Gutin, G. (2000) *Digraphs. Theory Algorithms and Applications*, Springer Verlag, London, Berlin, Heidelberg.

34 Halin, R. (1989) *Graphentheorie*, Akademie Verlag, Berlin.

35 Ullman, J.R. (1976) An algorithm for subgraph isomorphism. *Journal of ACM*, **23** (1), 31–42.

36 Dehmer, M. and Mehler, A. (2006) *A New Method of Measuring Similarity for a Special Class of Directed Graphs*, Tatra Mountains Mathematical Publications, Slovakia, in press.

37 Dehmer, M., Emmert-Streib, F. and Wolkenhauer, O. (2006) Perspectives of graph mining techniques. *Rostocker Informatik Berichte*, **30** (2), 47–57.

38 Berge, C. (1973) *Graphs and Hypergraphs*, American Elsevier Publishers.

39 Levenstein, V.I. (1966) Binary codes capable of correcting deletions, insertions and reversals. *Soviet Physics Doklady*, **10**, 707–710.

40 Bunke, H. and Allermann, G. (1983) Metric on graphs for structural pattern recognition.*Signal Processing II: Theory and Applications*, Chapter A, 257–260.

41 Irniger, C.A.M. (2005) Graph matching. Filtering databases of graphs using machine learning techniques. PhD thesis, Philosophisch-naturwissenschaftliche Fakultät,Universität Bern.

42 Messmer, B. and Bunke, H. (1998) A new algorithm for error-tolerant subgraph isomorphism detection. *IEEE Transactions on Pattern Analysis and Machine Intelligence*, **20** (5), 493–504.

43 Bunke, H. (1997) On a relation between graph edit distance and maximum common subgraph. *Pattern Recognition Letters*, **18** (8), 689–694.

44 Bunke, H. (1999) Error correcting graph matching: on the influence of the underlying cost function. *IEEE*

Transactions on Pattern Analysis and Machine Intelligence, **21** (9), 917–922.

45 Zhang, K., Statman, R. and Shasha, D. (1992) On the editing distance between unordered labeled trees. *International Information Processing Letters* **42** (3), 133–139.

46 Dickinson, P.J., Bunke, H., Dadej, A. and Kraetzl, M. (2004) Matching graphs with unique node labels. *Pattern Analysis Applications*, **7**, 243–266.

47 Oliver, F.J., Menissier-de Murcia, J. and de Murcia, G. (1999) Poly(ADP-ribose) polymerase in the cellular response to DNA damage apoptosis, and disease. *The American Journal of Human Genetics*, **64**, 1282–1288.

48 Vispe, S., Yung, T.M.C., Ritchot, J., Serizawadagger, H. and Satoh, M.S. (2000) A cellular defense pathway regulating transcription through poly(ADP-ribosyl)ation in response to DNA damage. *Proceedings of the National Academy of Sciences of the United States of America*, **29** (97), 9886–9891.

49 Jammes, Y., Steinberg, J.G., Mambrini, O., Bregeon, F. and Delliaux, S. (2005) Chronic fatigue syndrome: assessment of increased oxidative stress and altered muscle excitability in response to incremental exercise. *Journal of Internal Medicine*, **257** (3), 299–310.

50 Kennedy, G., Spence, V.A., McLaren, M., Hill, A., Underwood, C. and Belch, J.J. (2000) Oxidative stress levels are raised in chronic fatigue syndrome and are associated with clinical symptoms. *Free Radical Biology and Medicine*, **39** (5), 584–589.

51 Tu, Y., Stolovitzky, G. and Klein, U. (2002) Quantitative noise analysis for gene expression microarray experiments. *Proceedings of the National Academy of Sciences of the United States of America*.

52 Strohman, R. (2002) Maneuvering in the complex path from genotype to phenotype. *Nature*, **296** (26), 701–703.

11
Predicting Functional Modules Using Microarray and Protein Interaction Data

Yu Chen and Dong Xu

11.1
Introduction

As we move into the postgenomic era, systems-level understanding of biological objects has become a central theme of biomedical sciences [1]. It is recognized that thousands of components of a living cell are interconnected, so that dynamic and complex networks such as protein interaction network, regulatory network, and metabolic network govern its fundamental properties. To understand organizations of these networks, their substructures can be defined as functional units performing specific tasks, since cellular processes are often determined by functional modules such as operons, regulons, interaction cascades, molecular complexes, and signaling pathways [2]. Recently, biological networks have been investigated in terms of modularity by treating substructures of complex networks as functional units [3,4]. A functional module [2] is defined as a separate substructure of a network involving a group of genes or their products that are related by physical or genetic interactions, for example, coregulation, in the same protein complex or in the same metabolic or signaling pathway [5,6]. The idea of functional modules provides an effective approach to understanding and model biomolecular network systematically. Studies in this area offer unprecedented opportunities to derive underlying principles of how intricate cellular networks are constructed and how network architectures contribute to biological functionality. A series of important questions need to be addressed. For example, what are the characteristics of network architectures that distinguish them from randomly generated networks? How the network structures are related to biological functions? Are some topological patterns preferred at certain times or conditions?

The high-throughput data such as gene expression microarray data and protein interaction data are becoming an essential tool for systems-level understanding of biological networks. Protein–protein interactions refer to the association of protein molecules and are important for many biological functions. Proteins might interact for a long time to form part of a protein complex; a protein may be carrying another protein or a protein may interact briefly with another protein just to modify it.

Analysis of Microarray Data: A Network-Based Approach. Edited by F. Emmert-Streib and M. Dehmer

ISBN: 978-3-527-31822-3

Modules are found in protein–protein interaction networks and are defined as subnetworks whose components are more likely to connect to each other, rather than to entities outside the subnetworks. Recent analyses of yeast proteome identified several hundred novel core complexes and conditional binding modules [7]. Gene expression experiments often do not reveal direct regulations. However, the expression profiles of regulators provide information about their activities and expression data contain information on regulation between regulators and their targets. A modular map of yeast transcriptional regulatory networks is obtained on the basis of multiple microarray data [8]. These high-throughput data can characterize biological systems at the genome scale, enabling us to collect comprehensive information for the control mechanism and organization of the underlying molecules. For example, a stimulus introduced may lead to the expression level changes in a large number of genes potentially involved in the corresponding modules or regulons [9]. Although a number of studies addressed identification of functional modules [10–14], they typically used single type of data and focused on a particular aspect of module, for example, protein–protein interaction. There is room for significant improvement of module identification, especially on how to use diverse types of data in constructing modules. Each type of high-throughput data may reflect different aspects of gene/protein associations. For example, yeast two-hybrid data indicate direct binary interactions between proteins, while protein complexes identified by mass spectrometry detect the components of the same physical machinery, although the components may not all interact directly with each other. Moreover, proteins interact with one another with a wide range of affinities and timescales. Consequently, each measurement of protein–protein interactions can only capture a snapshot of the dynamic protein interaction map under a specific condition. How to effectively derive biomolecular network by integrating heterogeneous high-throughput data is a major challenge [15,16].

To address this challenge, Bayesian approaches are widely used to integrate the evidences from multiple sources into a single probabilistic framework where the functional linkages are measured by combing the similarity relationships from gene expression profiles, protein interactions, gene genomic sequence data, and phenotype profiling data [16–19]. We constructed a functional gene/protein network based on a Bayesian probability approach. In the constructed functional linkage network, each node represents one protein and the weight of each edge is characterized by the Bayesian probability of function similarity between the two connected proteins. In our study, protein function is defined by the GO (Gene Ontology) biological process [20]. We quantitatively measured functional relationships between genes underlying each type of high-throughput data (protein binary interactions, protein complexes, and microarray gene expression profiles) and coded them into "functional linkage graph" (interaction network), where each node represented one protein and the weight of each edge was characterized by the Bayesian probability of function similarity between the two connected proteins. Each type of data was integrated into a probabilistic functional network by measuring the likelihood of genes in the same functional module. The constructed functional linkage graph provided rich information on the organizations and mechanisms of networks. We then clustered the constructed functional linkage graph to detect putative functional modules. The outcome of our analysis is a set of functional modules, defined as clusters of the graph, where the genes/proteins within an individual cluster typically have a consistent biological role.

A unique feature of our method is that the modules are identified through an integrated probabilistic functional network enriched in characterizing functional and topological properties in a systematic and quantitative way. The discovered modules are validated by two quantitative measures, that is, functional homogeneity and interconnectivity of modules. The results are compared with modules decomposed from controlled networks and random networks. The identified modules show superior performance both functionally and topologically. Furthermore, we also investigated the properties (e.g., dispensability) of nodes that were classified according to their patterns of intra- and intermodule connections. Our study provides some useful information on the architecture and mechanism of modularity and sheds some light on the design principles of biological networks in general.

11.2 Materials and Methods

11.2.1 Data sets

All the data used in this study were from the yeast *Saccharomyces cerevisiae*. The protein–protein interaction data of yeast two-hybrid experiments were from Uetz *et al.* [21] and Ito *et al.* [22], together with 5075 unique interactions among 3567 proteins. The protein complex data were obtained from Gavin *et al.* [23] and Ho *et al.* [24]. We assumed that any two proteins in a complex had an association with each other, and the complexes yielded 49 313 associations. The gene expression profiles of microarray data were from Gasch *et al.* [25], which included 174 experimental conditions for all genes in yeast. For each experiment, if any data were missing, we substituted gene expression ratio to the reference state with the average ratio of all genes under that specific experimental condition. A Pearson correlation coefficient was calculated for each possible gene pair to quantify possible coexpression between the two genes. The dispensability of a protein can be quantified by its contribution to survival and reproduction of the carrier upon gene deletion. This contribution can be experimentally measured by the growth rate of the carrier. The growth rates of gene deletion mutants in the yeast *S. cerevisiae* were measured in the genome scale, where 4706 homozygous diploid deletion strains were monitored in parallel in nine different medium conditions [26]. For each mutant, we estimated the deleted gene's fitness f_i as $1 - r_i/r_{max}$, where r_i is the growth rate of the strain with the gene deleted and r_{max} is the maximal growth rate [27]. The fitness values of essential genes (lethal upon deletion) are 1.

11.2.2
Protein Function Annotation and GO Index

In our study, function annotation of protein is defined by the GO biological process [20]. The GO biological process ontology is available at http://www.geneontology.org. It has a hierarchical structure with multiple inheritances. After acquiring the biological process functional annotation for the known proteins along with their GO identification (ID), we generated a numerical GO index, which represented the hierarchical structure of the classification. The more detailed the level of the GO index, the more specific the function assigned to a protein. The following shows an example of a GO index hierarchy, with the numbers on the left giving the GO indices and the numbers in the parentheses indicating the GO IDs:

2	Cellular process (GO:0009987)
2-1	Cell communication (GO:0007154)
2-1-8	Signal transduction (GO:0007165)
2-1-8-1	Cell surface receptor linked signal transduction (GO:0007166)
2-1-8-1-4	G-protein coupled receptor protein signaling pathway (GO:0030454)
2-1-8-4-4-12	Signal transduction during conjugation with cellular fusion (GO:0000750)

The SGD database (http://www.yeastgenome.org/) has annotated 4044 proteins with GO biological process IDs and we converted them into GO indices.

We calculated protein function similarity by comparing the level of similarity that the two proteins share in terms of their GO indices. For example, if both gene 1 and gene 2 have annotated functions, assume gene 1 has a function represented by GO index 2-1-8-1 and gene 2 has a function represented by GO index 2-1-8. When compared with each other for the level of matching GO index, they match with each other through 2-1-8, that is, index level 1 (2), index level 2 (2-1), and index level 3 (2-1-8). In general, the function similarity between proteins X and Y is defined by the maximum number of index levels from the top shared by X and Y. In this example, the function similarity between gene 1 and gene 2 is 3.

11.2.3
Construction of Probabilistic Functional Network

Following our early work [27], we calculated the probabilities for two genes sharing the same function based on different types of high-throughput data, that is, microarray data, protein binary interaction data, and protein complex data. Assuming $H=\{M, B, C\}$ denotes the interaction events in different types of high-throughput data, where M represents two genes correlated in gene expression profiles with a certain Pearson correlation coefficient in microarray gene expression data,

B represents a protein binary interaction, and C represents a protein complex association, the *posterior* probability that two proteins have the same function at a given level, $p(S|H)$, is computed using the Bayes' formula

$$p(S|H) = \frac{p(H|S)P(S)}{P(H)}, \tag{11.1}$$

where S represents the event that two genes/proteins have the same function at a given level of GO index. The probability $p(S)$ is the relative frequency of genes/proteins whose functions are same at the given level of GO index by chance. $p(H|S)$ is the conditional (a priori) probability that two genes/proteins have the event H given that they have the same function at a given level of GO index. The probability $p(H)$ is the frequency of H in the entire data set, for example, the frequency of a certain correlation coefficient over all gene pairs in yeast, which was calculated from the genome-wide gene expression profiles ($H = M$) or the relative frequency of two proteins having a known binary interaction over all possible pairs in yeast, is estimated from the known protein interaction data set ($H = B$). The probabilities $p(H|S)$ and $p(S)$ were computed based on a set of proteins whose functions were annotated in the GO biological process.

Thus, the high-throughput data are encoded into a graph of "functional linkage" network: $G = \langle V, E \rangle$, where the vertices V of the graph are connected through the edges E, as illustrated in Figure 11.7. Each vertex represents a protein, and an edge represents a detectable association between the two connected genes in the high-throughput data. The weight of an edge W_{ij} reflects the functional similarity between the connected proteins i and j, and it is calculated as

$$W_{ij} = \sum_{k=1}^{12} -\log(1-(1-P_k(S|M))(1-P_k(S|B))(1-P_k(S|C))), \tag{11.2}$$

where k is the GO index similarity level according to the definition in Section 11.2.2. The maximum level of GO biological process index is 12. $P_k(S|M)$, $P_k(S|C)$, and $P_k(S|B)$ are available from the probability table calculated in our previous work [27]. The way in which Equation 11.2 combines the three types of the high-throughput data assumes that each type of data can be integrated independently. Although such an assumption may not be true for accurately calculating the probability, it is useful to assess the reliability of functional similarity. The smaller the value of W_{ij}, the more functional similarity proteins i and j have.

11.2.4
Identification of Functional Modules by Clustering the Network

We applied the Markov cluster (MCL) algorithm [28] to identify functional modules by clustering the constructed functional linkage graph. Recently, the MCL algorithm is being widely applied in clustering biological data, ranging from protein/gene sequence data to protein interaction data [10,29,30]. In a biological network, the members of a module have more functional similarity or linkage to each other than to proteins in other modules [5,25]. These properties of functional modules make them ideally suited to the MCL algorithm. The basic idea underlying the MCL algorithm is that a "dense" region with many edges in a graph allows random walks to have a high probability for paths that start and end in the same dense region. This is especially true if one looks at a subset of all random walks departing from a specific node. If this node is situated in a dense region, random walks departing from it will generally have a tendency to stay in the same region. The difference between the MCL algorithm and other partition clustering methods is that the former allows some nodes not in any clusters. An implementation of the MCL algorithm is available at http://micans.org/mcl/ with additional information.

The flowchart provided in Figure 11.1 illustrates how we used the MCL algorithm for identifying tight clusters. The functional linkage graph (Figure 11.1a) is constructed from protein binary interaction data, protein complex data, and microarray data. Figure 11.1b is the weight matrix used as the input for clustering, where each row or column of the matrix represents a given protein, and each entry in a matrix cell S_{ij} represents a similarity between protein i and protein j. Its value is calculated as

$$S_{ij} = \frac{1}{W_{ij}}, \tag{11.3}$$

where W_{ij} is the weight of edge as described in Section 11.2.3. This matrix is supplied to the MCL algorithm. The number of clusters (modules) is determined by one main control parameter I that affects cluster granularity. $I=2.0$, as suggested in [28], was used in all our clustering analyses. We used the default values in the software package for other parameters.

11.2.5
Evaluation of Topological and Functional Properties of Modules

We used several parameters to characterize the properties of identified modules. The *degree* of a node i is the number of edges between node i and its nearest neighbors. It is denoted by k_i. The clustering coefficient, that is, the number of observed edges among k_i divided by the maximum number of all possible edges among k_i, measures the local cohesiveness around of the node i. By averaging the clustering coefficient over all nodes in a cluster, a global parameter C_A (the average clustering coefficient of a cluster) is used to measure the "tightness" of the cluster:

$$C_A = \left[\sum_{i=1}^{n} \frac{2E_i}{k_i(k_i-1)}\right] / n, \tag{11.4}$$

where n is the number of nodes in the module, E_i is the number of edges between the nearest neighbors of node i, and k_i is the degree of node i. The highest value of C_A is 1 when the module is a complete graph.

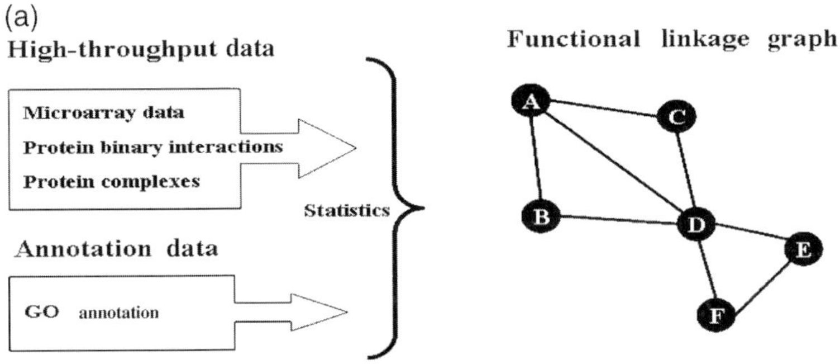

(b)

Weighted matrix

	A	B	C	D	E	F
A	1.0	0.8	0.3	0.5	—	—
B	0.8	1.0	—	0.6	—	—
C	0.3	—	1.0	0.7	—	—
D	0.5	0.6	0.7	1.0	0.9	0.2
E	—	—	—	0.9	1.0	0.4
F	—	—	—	0.2	0.4	1.0

(C)

MCL clustering

Functional module identification

Figure 11.1 A flowchart of network clustering. (a) Coding high-throughput data and biological annotation data into a functional linkage graph with an example of functional linkage graph for six proteins (A–F). (b) The weighted matrix as the input of MCL clustering. (c) Identification of functional modules.

Since it is insufficient to evaluate the module properties based solely on topology, we also proposed a measure to evaluate the biological properties of modules. To examine biological consistency of an identified module, we analyzed the GO biological process annotations of proteins within an individual cluster/module. We assign a *functional significance score* (FSS) to each module, which indicates the homogeneity of functional annotation within that cluster and reflects the biological role of the module. FSS is calculated as

$$\mathrm{FSS} = \left[\sum_{i}^{n}\sum_{j}^{n}\left(\frac{S_{ij}}{12}\right)\right] \Big/ \left[\frac{n(n-1)}{2}\right], \tag{11.5}$$

where n represents the number of nodes within a cluster and S_{ij} represents function similarity between nodes i and j in terms of their GO index as defined in Section 11.2.2. The number 12 is a normalization factor, which represents the maximum levels of GO biological process index. FSS ranges between 0 and 1. The higher the value of FSS (closer to 1), the more homogeneous the function annotation of a cluster.

11.3 Results

11.3.1 Modules Discovered from the Probabilistic Functional Network

In this study, we focused on highly significant modules. Hence, for the construction of functional network, we selected only confident edges between two proteins if they satisfied the following stringent criteria: (1) the interaction between two proteins is found in both yeast two-hybrid interaction and protein complex association; and (2) gene expression profile correlation coefficient is $\geq$0.9. Other researchers also used the same criteria for selecting the high-quality interaction data to minimize false positives [17,31]. The criteria resulted in 3990 interactions using which we calculated the edges and their weights, and then applied the MCL algorithm. Overall, 86 clusters were identified each containing at least four proteins. We found that genes belonging to the same cluster typically had a consistent biological process annotation, indicating the modularity of cellular networks. Figure 11.2 presents three examples of functional modules. Figure 11.2a shows the AP-1 adaptor complex. It represents a protein complex or a molecular machine where several protein components stably bind each other to perform a function (in a similar way as ribosome, histones, polymerases, etc. do) [32]. Figure 11.2b shows a functional module with a few proteins that control or perform a particular cellular function through temporal/transient interactions among themselves. The associations among genes are from gene coexpressions that do not form a macromolecular physical complex and they may be dynamically changed with the changes of location and time [18,33,34]. Our method successfully discovers both stable complexes and dynamic modules but does not distinguish between the two types. Figure 11.2c shows a combined type of module consisting of protein complex and dynamic associations. Gene products of YPR037C, YLR321C, YCR052W, and YMR091C are members of the complex with the capacity to remodel the structure of chromatin, termed RSC [35]. This complex catalyzes the transfer of a histone octamer from a nucleosome core particle to naked DNA, involving many essential cellular processes. The protein YNR032W is serine/threonine protein phosphatase, forming a dynamic interaction with YFR037C. This module is detected by integrating protein interaction data and microarray coexpression data, as either type of high throughput does not suffice to identify such modules. The different categories of modules in Figure 11.3 provide diverse coverage of biological processes on various types of

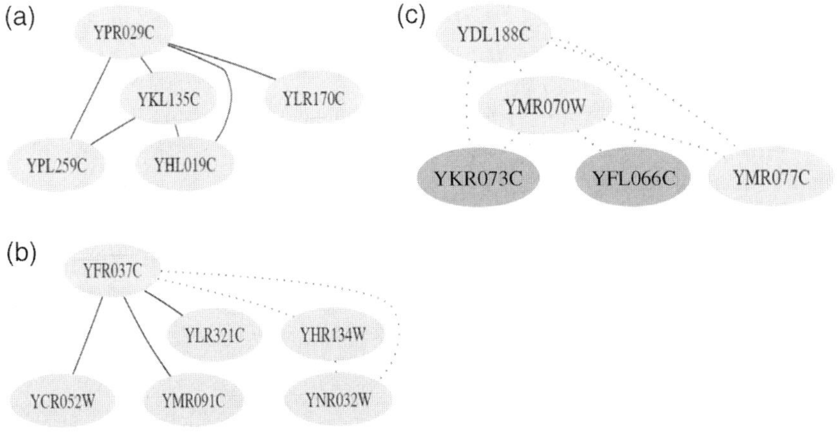

Figure 11.2 Examples of predicted functional modules in different categories. (a) AP-1 adaptor complex that links clathrin to the membrane surface of a vesicle. The five proteins are different subunits of the complex. (b) A dynamic functional module in terms of a signal transduction cascade that includes a transcriptional activator, a protein phosphatase, a vacuolar sorting protein, and two unannotated proteins. (c) The combined type of functional module with the RSC complex and transient interactions. In the figure, dark gray represents the unannotated proteins and light gray represents annotated proteins. The patterns of lines represent different types of associations: a solid line represents a protein interaction and a dotted line represents coexpression. Detailed information on genes in each module is available at http://digbio.missouri.edu/NetCluster/. ("A" came from cluster 42, "B" from cluster 48, and "C" from cluster 36.) The plots are generated using Graphviz, an open-source graph visualization software package available at http://www.graphviz.org.

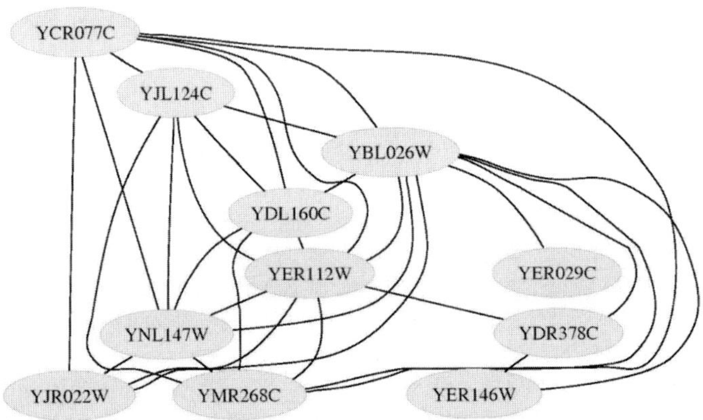

Figure 11.3 The identified module involved in the RNA processing, splicing, and decay. Detailed information on 11 genes within this module is available as the cluster 8 at http://digbio.missouri.edu/NetCluster/. The lines denote protein–protein interactions.

responses to the environment. It also confirms the power of data integration in characterizing the nature of biomolecular network. Figure 11.3 shows one identified module with more complicated topology that includes 11 components with known functions. It is a nuclear complex involving RNA processing, splicing, and decay [36]. In this module, the average node degree is 5.4 and the average clustering coefficient (C_A) is 0.85, which indicates a cluster with dense edges.

11.3.2 Evaluation of Modules

To identify modules from a network graph, several network clustering approaches based on local or global topological properties of networks using protein–protein interaction data or gene expression microarray data [37–39] have recently been developed. Compared with existing approaches, a unique feature of our method is that modules were identified in the context of functionality at the genome scale. We integrated protein–protein interaction data and gene expression profiles into a probabilistic functional network where a probabilistic weight was assigned to each edge based on the projected functional linkage. To investigate the properties of the discovered modules, we measured the functional homogeneity and the topological property of each module. The results are also compared with modules identified from a constructed connection network and the random networks.

We compared the FSSs of modules in the following three classes. (I) The modules are identified from the probabilistic functional network, where the weight of an edge presents the likelihood of functional similarity. (II) The modules are identified from the connection network that is constructed solely based on topological connectivity of interactions without adding the functional linkage information. We combine protein–protein interaction network and the gene coexpression network into a graph where the weight of an edge is binary: 1 means that the two nodes have an association (protein–protein interaction or gene coexpression, or both) and 0 means no association. We obtained 88 clusters that have at least four components. (III) The modules are identified from the random network. The random networks are generated by randomly rewiring the edges of the constructed functional linkage graphs while keeping the same edge weights [40,41]. This method guarantees the preservation of the distribution of connectivity. The MCL method was applied to each of the 10 random networks. We then calculated the FSS for the modules that have more than four components.

Figure 11.4a shows the FSSs of modules in classes I, II, and III. The average values of classes I, II, and III are 0.44, 0.38, and 0.18, respectively. We also compared the average clustering coefficient of a cluster (C_A) for the modules of classes I, II, and III as shown in Figure 11.4b, where the average values of classes I, II, and III are 0.65, 0.64, and 0.03, respectively. We can find that the modules of class I not only have the highest functional consistency but also indicate the similar tightness as modules of class II. Moreover, some of the modules in class I include the components that have consistent functions, but these components could not be clustered into any individual module in class II (see Table 11.1). This means that the

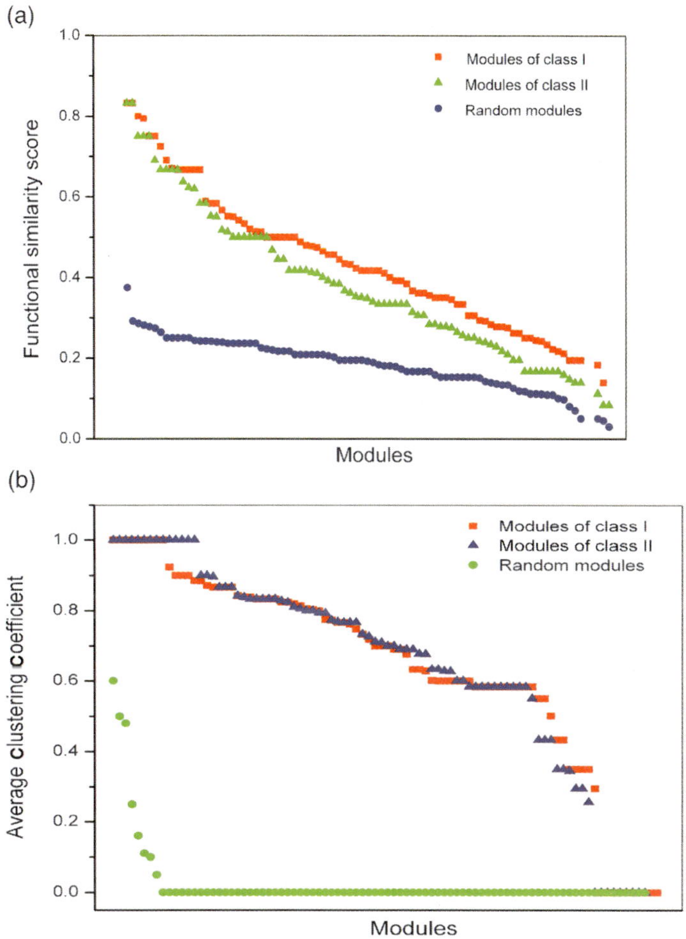

Figure 11.4 (a) Scatter plot of FSSs for modules identified from the probabilistic functional network (class I), a constructed connectivity network (class II), and random networks (class III). (b) Scatter plot of average clustering coefficients for modules identified from the probabilistic functional network (class I), a constructed connectivity network (class II), and random networks (class III).

modules of class I have higher sensitivity in detecting homogeneous components and can predict functions for more hypothetical genes reliably. All these reveal the advantages of our module discovery approach; that is, the module is defined as a robust unit both functionally and topologically so that components within a module are not only tightly interconnected but also functionally consistent.

Figure 11.4 also shows the robustness of identified modules (clusters) compared with random networks. In another study, we tested the robustness of each identified

Table 11.1 Examples of comparisons between modules in class I and modules in class II.

Module	Components in module of class I	Components in module of class II	Unique component in class I
Ubiquitin-dependent protein degradation	YBL041W, YER012W, YER094C, YGL011C, YGR135W, YGR253C, YJL001W, YML092C, YMR314W, YOL038W, YOR157C, **YOR362C**, YPR103W	YBL041W, YER012W, YER094C, YGL011C, YGR135W, YGR253C, YJL001W, YML092C, YMR314W, YOL038W, YOR157C, YPR103W	YOR362C (ubiquitin-dependent protein degradation)
Transcription from Pol III promoter	**YBR154C**, YGL241W, YKL144C, YNR003C, YOL006C, YOR116C, YOR207C, YPR110C, YPR187W, YPR190C	YGL241W, YKL144C, YNR003C YOL006C, YOR116C YOR207C, YPR110C, YPR187W, YPR190C	YBR154C (transcription from Pol III promoter)
mRNA catabolism	YDR432W, YGL049C, **YGR150C**, YGR162W, YGR250C, YIR001C, YNL016W	YDR432W, YGL049C, YGR162W, YGR250C, YIR001C, YNL016W	YGR150C (unannotated protein)
Chromatin modification	YGL165C, YGL166W, YML118W, YMR114C, YNL196C, YNL202W, YNL269W, **YOR064C**	YGL165C, YGL166W, YML118W, YMR114C, YNL196C, YNL202W, YNL269W	YOR064C (chromatin modification)
Adenine catabolism	**YBL039C**, YBR034C, YDR413C, YGL078C, YHR170W, YJL148W, YML093W, YNL141W	YBR034C, YDR413C, YGL078C, YHR170W, YJL148W, YML093W, YNL141W	YBL039C (pyrimidine base biosynthesis)
Establishment of cell polarity (sensu Fungi)	**YCR057C**, YGL019W, YGL207W, YIL035C, YML069W, YMR172W, YNL112W, YOL105C, **YOR039W**, YOR061W, **YOR145C**	YGL019W, YGL207W, YIL035C, YML069W, YMR172W, YNL112W, YOL105C, YOR061W	YCR057C (establishment of cell polarity), YOR039W (establishment of cell polarity), YOR145C (protein complex assembly)

The modules in class I include the components that are functionally consistent but cannot be discovered in modules of class II (shown in bold and the fourth column).

module (cluster) by ranking the distribution of clustering coefficients of identified cluster and 2000 permutated clusters. Each identified cluster shows very strong statistical significance (unpublished results). The number of clusters is increased to 112 when the cluster parameter I, the main factor to control cluster granularity, changes to 4.0. However, the average FSS and C_A values did not change (see Table 11.2), which indicates the robustness of the integrated functional linkage network.

Table 11.2 Average FSS and C_A values of modules identified from the probabilistic functional network (class I), a constructed connectivity network (class II), and random networks (class III) versus parameter I.

	Average value of FSS			Average value of C_A		
	Class I	Class II	Class III	Class I	Class II	Class III
$I = 2.0$	0.44	0.36	0.18	0.65	0.64	0.03
$I = 4.0$	0.44	0.36	0.18	0.65	0.64	0.03

Table 11.3 shows the known and predicted GO indices of proteins in cluster 4 of class I. This cluster is the small nucleolar ribonucleoprotein complex. Our analysis added three unannotated proteins to this cluster, YLR409C, YJL069C, and YKR060W, all of which can be predicted in the small nucleolar ribonucleoprotein complex. Among 86 identified modules in class I, 43 modules contain 98 unannotated proteins with an average of 2.3 unannotated proteins per module, while the average number of components per module that contains unannotated proteins is 9. Thus, identification of functional modules also provides a methodology to assign functions for hypothetical proteins within modules.

Table 11.3 GO indices for proteins in cluster 4 in the small nucleolar ribonucleoprotein complex.

Gene ID	Name	A/P	GO annotation
YLR197W	*SIK1*	A	Biological process of 20S preRNA processing
YLR129W	*DIP2*	A	Biological process of 20S preRNA processing
YPL126W	*NAN1*	A	Biological process of 20S preRNA processing
YPL129W	*TAF14*	A	Biological process of 20S preRNA processing
YOL041C	*NOP12*	A	Biological process of rRNA metabolism.
YJL109C	*UTP10*	A	Biological process of 20S preRNA processing
YDR449C	*UTP6*	A	Biological process of 20S preRNA processing
YER082C	*UTP7*	A	Biological process of processing 20S preRNA
YMR093W	*UTP15*	A	Biological process of 20S preRNA processing
YPL217C	*BMS1*	A	Biological process of 20S preRNA processing
YGL171W	*ROK1*	A	Biological process of 20S preRNA processing
YDL213C	*NOP6*	A	Biological process of rRNA processing
YKL099C	*UTP11*	A	Biological process of processing 20S preRNA
YHR196W	*UTP9*	A	Biological process of 20S preRNA processing
YKR060W	N/A	P	Biological process of rRNA processing (0.98)
YJL069C	N/A	P	Biological process of rRNA processing (0.89)
YLR409C	N/A	P	Biological process of 35S primary transcript processing (0.98)

"A" denotes an annotated protein and "P" denotes the predicted function for an unannotated protein using the global protein function prediction method developed in our previous work [27]. The number in the parenthesis denotes the reliability score ranging from 0 for the least significant to 1 for the most significant.

Detailed information on each identified module is available at http://digbio.missouri.edu/NetCluster/, including (1) the GO biological process indices assigned to each gene and FSS of each module; (2) the web link of each gene to SGD (http://www.yeastgenome.org/); (3) a list of unannotated proteins in each module; and (4) the graph of interaction network.

11.3.3
Module Organization in Yeast Gene Interaction Network

Biological systems are functionally organized in different networks such as metabolic/signaling pathways and regulatory/coexpression network [2,5,6]. Based on the type of particular interactions, the networks can be characterized as static presentations (e.g., protein–protein interaction network) or temporal/transient entities (e.g., gene coexpressed network) [42]. As discussed above, the discovered functional modules could consist of stable interactions or temporal interactions, or both. Among 86 identified modules, we found that 46 modules have only protein–protein interactions, 35 modules have only gene coexpression associations, and very few (only 5) have both. To have a better understanding of the organizing principles of biological systems, we proposed a model of global view of organized modularity, as illustrated in Figure 11.5, to dissect the modularity architecture of the interaction network.

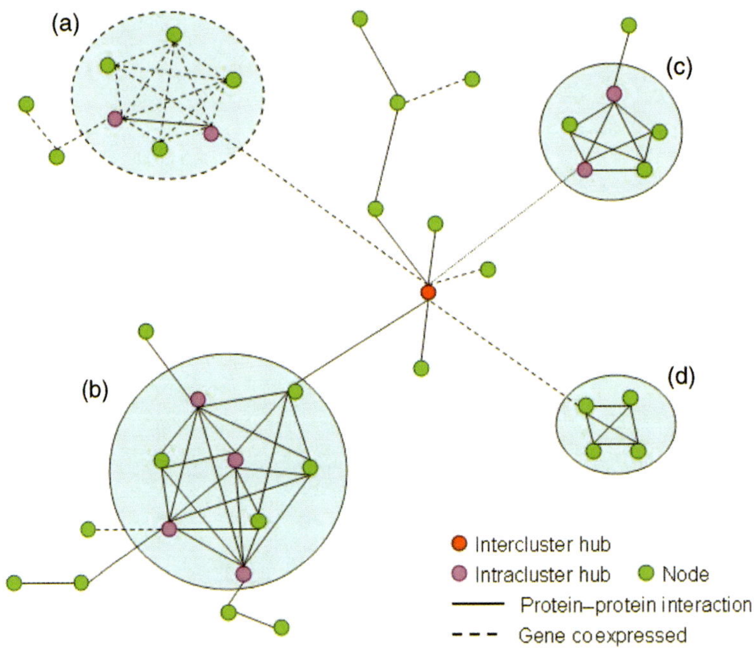

Figure 11.5 Schematic illustration for modularity organization in the probabilistic functional network. Module a is from gene coexpression network and modules b, c, and d are from protein–protein interaction network. The hubs are classified into intra- and intermodule hubs.

In this model, various functional modules that represent the stable molecular machines (complexes) or coexpressed genes (regulons) serve as subnetworks. Proteins (nodes) can be classified according to their roles of intra- and intermodule connections. We define an intramodule hub as a protein (node) inside a module with node degree k greater than 5 and an intermodule hub as a protein (node) outside any module with node degree k greater than 5. The threshold of 5 was used before to define a hub [16]. Our study was inspired by some early work. The dynamic nature of modules has recently been addressed by relating protein interaction networks to gene express profiles [31,32,43]. It was proposed that there are two distinct classes of protein hubs, namely, "party hubs" (where hubs are highly correlated in gene expression with their partners) and "date hubs" (where the correlations of interaction partners depend on timing and location) [31]. In these previous studies, the module decomposition was made from the networks that were constructed based only on the topological properties without the context of quantitative association of functionality. In our study, the modules are identified from a probabilistic functional network, where the modules are characterized as robust units both functionally and topologically. For most of the identified functional modules, all the interactions within a module are either from the static protein–protein interaction network or from temporal gene coexpression network. Moreover, the modules consisting of protein–protein interactions and modules consisting of gene coexpression networks are often connected by the specific intermodule hubs.

We investigated the relationship between protein dispensability and the topological feature of connectivity pattern in a module-organized network (i.e., the nodes

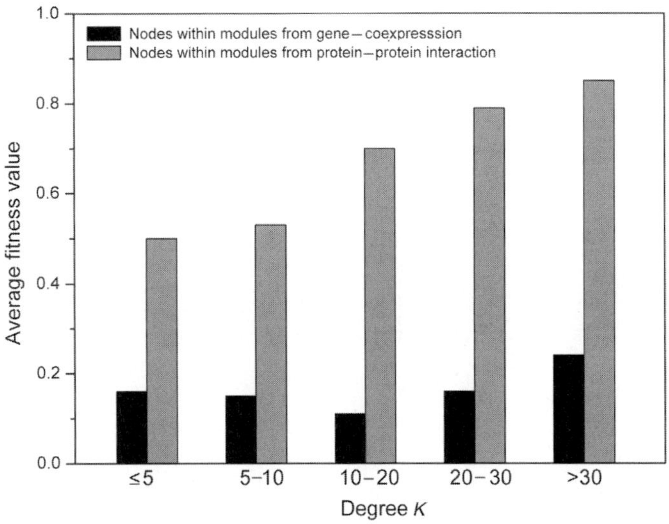

Figure 11.6 Average fitness values of intramodule nodes versus the node degree k. It indicates that the more the connectivity, the higher the value of average fitness (the proteins are more indispensable).

within a module or outside a module). The fitness value calculated from the growth rate of gene deletion mutant was used to quantitatively measure the dispensability of protein. It was already shown that hubs in protein–protein interaction network and gene coexpression network tend to be functional essentiality [3,40]. However, no study linking the intra- and intermodule connectivity to functional essentiality has been conducted yet. Figure 11.6 compares the average fitness values of nodes inside or outside the modules versus the degree *k*. It shows that the more the connections of a node, the higher the fitness value in general. Figure 11.7 shows the average fitness value of intra-and intermodule hubs. The proteins (nodes) inside the functional modules from protein–protein interaction data have the higher average fitness values (i.e., more likely to be essential) than proteins inside the functional modules from gene coexpression network. This is probably because the static functional modules are often used to execute essential tasks of cellular events. Comparisons of intermodule hubs indicate that the more the modules a hub links, the higher the fitness value. Moreover, the intermodule hubs that simultaneously connect various functional modules through both static protein–protein interactions and temporal gene coexpression associations are even more essential to viability (i.e., columns F and G in Figure 11.7). Especially when an intermodule hub connects more than two functional modules (column G), a deletion of the hub will result in lethality (with fitness value as 1). Table 11.4 gives detailed information on all six proteins in column G that connect more than three modules through protein–protein interactions and gene coexpression associations. Interestingly, we found that five out of the six

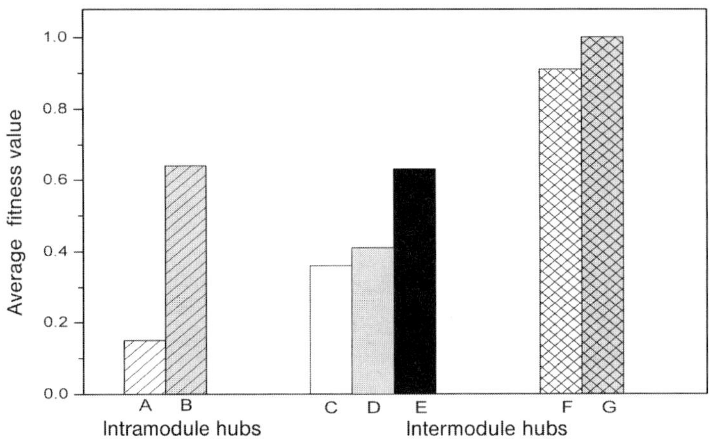

Figure 11.7 Average fitness values of intramodule hubs and intermodule hubs. A: intramodule hubs in the modules made up of gene coexpression associations; B: intramodule hubs in the modules made up of protein–protein interactions; C, D, E: intermodule hubs connecting one, two, and three modules, respectively; F: intermodule hubs that connect modules through both protein–protein interactions and gene coexpression associations; G: intermodule hubs connecting more than two modules through both protein–protein interactions and gene coexpression associations.

Table 11.4 Intermodule hubs that connect more than two modules through protein–protein interactions and/or gene coexpression associations.

Inter-module hub	Gene name	Description	Cellular component	Modules connected through protein–protein interactions	Modules connected through gene coexpression interactions
YMR290C	*HAS1*	ATP-dependent RNA helicase	Nuclear membrane	Regulation of cell growth (37)	Transcription from Pol III promoter (9)
			Nucleolus	Transcription from Pol I promoter (59)	Ribosomal large subunit biogenesis (23)
				Chromatin assembly (60)	
YNL132W	*KRE33*	Essential protein of unknown function	Nucleolus	Protein biosynthesis (11)	rRNA process (2)
				preRNA process (35)	Adenine catabolism (15)
YGR145W	*ENP2*	Essential protein of unknown function	Nucleolus	Regulation of transcription from promoter (7)	Processing of 20S preRNA (4)
				Ribosomal subunit assembly and maintenance (27)	
YOR206W	*NOC2*	Protein belongs to nucleolar complex	Nucleoplasm	rRNA processing (2)	Adenine catabolism (15)
			Nucleolus	Regulation of transcription from promoter (7)	Ribosomal large subunit biogenesis (23)
YLR222C	*UTP13*	U3 snoRNP protein	Nucleolus	Processing of 20S preRNA (4)	Protein biosynthesis (11)
					Chromatin assembly (60)
YCL054W	*SPB1*	AdoMet-dependent methyltrans-ferase	Nucleolus	rRNA processing (2)	rRNA processing (2)
				Ribosomal large subunit biogenesis 23)	Chromatin assembly (60)

The number in the parenthesis denotes the cluster (module) order number and detailed information on each cluster (module) is available at http://digbio.missouri.edu/NetCluster/.

proteins in Table 11.4 (YMR290C, YNL132W, YGR145W, YOR206W, and YLR222C) are also listed in the top 100 hubs in yeast network from the study of Bergmann *et al.* [44] (the data are available at http://barkai-serv.weizmann.ac.il/ComparativeAnalysis/). A possible explanation for the importance of intermodule hubs is that they serve as the connectors for the cross talks among multiple functional modules. The elimination of such connectors will probably have a major impact on the global structure of modularity, as they might play an important role in the coordination of functioning of related modules. Especially for the intermodule hubs that appear in both protein–protein interaction and gene coexpression networks, they might bridge the static functional modules and the temporal functional modules to turn on the complex and dynamic cellular processes or they link together the functional modules in a spatially and temporally ordered way. Most complexes consist of both periodically and constitutively expressed subunits and the formation of complex's activity is controlled by the mechanism of "just-in-time" assembly [34,45]. Moreover, the study of topology of integrated network suggests that an intricate biological process is achieved through coordination of a set of molecular machines [18]. Hence, the intermodule hubs could play an important role in the process of "just-in-time" assembly. For example, in the cell cycle regulation of *Caulobacter crescentus* it is known that CtrA has a top-down control of multiple modules [46]. In a similar fashion, as shown in Table 11.4, the protein KRE33 links the module of preRNA process through physical protein–protein interactions and the module of rRNA process through dynamic association of gene coexpression. KRE33 could be a master regulatory protein coordinating the overall rRNA process. Another example is the protein NOC2 that has two known subcellular localizations – nucleoplasm and nucleolus [47]. We found that NOC2 has linkages with the module that regulates the transcription from the promoter module through the protein–protein interaction with casein kinase II in the nucleoplasm [48] and meanwhile NOC2 is coexpressed with the MAK5 complex in the nucleolus that belongs to the module of ribosomal large subunit biogenesis [49]. This suggests that NOC2 might mediate the complexes that control the intranuclear transport of ribosomal precursors and the assembly of ribosomal complex in a temporal and spatial manner. Overall, the discoveries from network modularity dissection indicate the central importance of the intermodule hubs, shedding some light on the design principles of biological systems. This is supported by an early report that the hubs connecting different modules are more conserved than the hubs whose links are mostly within a single module in the yeast metabolism network [50].

11.4 Discussion

To study biology at the systems level, it is important to understand the structure and dynamics of biomolecular networks since a biological system is much more than a simple assembly of genes/proteins. Characterizing the diagram of interconnections and dynamic patterns of networks can shed some light on the basic cellular mechanisms. Cellular processes are the result of the cascade of events mediated by proteins that act in a cooperative manner. Proteins combine together into macromolecular complexes to execute cellular functions. Genes, which encode proteins that participate in the same pathway or are part of the same protein complex, are often coregulated, and clusters of genes with similar functions often exhibit similar expression patterns under various conditions in microarray experiments [51,52]. The organization of the biomolecular network into functional modules follows the design principles of biological systems that govern multiple types of molecular interactions. Our study provides some quantification for modular architecture and dynamic nature of biological networks. The analysis of the network complexity of the functional ensembles allows us to infer the design principles of biological systems by involving a small number of master regulatory proteins with cross-module hubs coordinating the overall process. Moreover, dissecting the interaction network into modules allows us to discriminate the intra-and intermodule hubs that might help to discover new drug targets essential to the biological systems.

We provide a systematic method for identifying the biological modules based on high-throughput data. A by-product of our method is to assign unannotated genes into some known biological modules and to predict their functions accordingly. The main advantages of our method are that it integrates heterogeneous data spanning different regulation mechanisms and that it developed a statistical framework to quantitatively unify the evidence from diverse sources. Moreover, our method systematically combines functional and topological properties to identify module in a global manner. The framework shows the statistical robustness and can be expanded to other organisms.

We recently learned of one relevant work on network coordinated functionality, which says that the pairwise coordination feature among modules is found in yeast interactome and there is no significant correlation between a protein's traffic (betweenness centrality) and its essentiality [12]. It is worth noting that our work is distinguished from this study in the following aspects. (a) We combined diverse types of data such as protein–protein interactions and gene coexpression profiles to construct an integrated interaction network, while the interactome map is constructed solely on protein–protein interactions (1752 interactions from MIPS) in Valente and Cusick's study. (b) The modules in our study are identified from a probabilistic functional network and are robust both functionally and topologically. However, in Valente and Cusick's study the modular organization of yeast protein–protein network is characterized only from its topological feature in an unsupervised manner. (c) Most of the intermodule hubs defined in our study connect different modules through both protein–protein interactions and coexpression interactions. Moreover, a strong correlation between connectivity and essentiality in gene coexpression networks is also observed [44,53]. Nevertheless, we believe that our work and Valente and Cusick's study complement each other and our work makes contributions to the ongoing effort in understanding the complex global organizations and mechanisms of biological systems.

Acknowledgments

We would like to thank Drs Jeffrey M. Becker, Ying Xu, and Loren Hauser for helpful discussions. Dr Dong Xu's research was supported by USDA/CSREES-2004-25604-14708 and NSF/ITR-IIS-0407204.

References

1 Kitano, H. (2002) Computational systems biology. *Nature*, **420**, 206–210.

2 Hartwell, L.H., Hopfield, J.J., Leibler, S. and Murray, A.W. (1999) From molecular to modular cell biology. *Nature*, **402**, c47–c52.

3 Jeong, H., Mason, S.P., Barabási, A.L. and Oltvai, Z.N. (2001) Lethality and centrality in protein networks. *Nature*, **411**, 41–42.

4 Tanay, A., Sharan, R., Kupiec, M. and Shamir, R. (2004) Revealing modularity and organization in the yeast molecular network by integrated analysis of highly heterogeneous genome-wide data. *Proceedings of the National Academy of Sciences of the United States of America*, **101**, 2981–2986.

5 Rives, A.W. and Galitski, T. (2003) Modular organization of cellular networks. *Proceedings of the National Academy of Sciences of the United States of America*, **100**, 1128–1133.

6 Spirin, V. and Mirny, L. (2003) Protein complexes and functional modules in molecular networks. *Proceedings of the National Academy of Sciences of the United States of America*, **100**, 12123–12128.

7 Gavin, A.C., Aloy, P., Grandi, P., Krause, R., Boesche, M., Marzioch, M., Rau, C. Jensen, L.J., Bastuck, S., Dümpelfeld, B., Edelmann, A., Heurtier, M.A., Hoffman, V., Hoefert, C., Klein, K., Hudak, M., Michon, A.M., Schelder, M., Schirle, M., Remor, M., Rudi, T., Hooper, S., Bauer, A., Bouwmeester, T., Casari, G., Drewes, G., Neubauer, G., Rick, J.M., Kuster, B., Bork, P., Russell, R.B. and Superti-Furga, G. (2006) Proteome survey reveals modularity of the yeast cell machinery. *Nature*, **440**, 631–636.

8 Segal, E., Shapira, M., Regev, A., Pe'er, D., Botstein, D., Koller, D. and Friedman, N. (2003) Module networks: identifying regulatory modules and their condition-specific regulators from gene expression data. *Nature Genetics*, **34**, 166–176.

9 Stephanopoulos, G., Aristidou, A.A. and Nielsen, J. (1998) Metabolic engineering. *Biotechnology and Bioengineering*, **58**, 119–120.

10 Pereira-Leal, J.B., Enright, A.J. and Ouzounis, C.A. (2004) Detection of functional modules from protein interaction networks. *Proteins*, **54**, 49–57.

11 Lubovac, Z., Gamalielsson, J. and Olsson, B.((2006) Combining functional and topological properties to identify core modules in protein interaction networks. *Proteins*, **64**, 948–959.

12 Valente, A.X. and Cusick, M.E. (2006) Yeast protein interactome topology provides framework for coordinated-functionality. *Nucleic Acids Research*, **34**, 2812–2819.

13 Yamada, T., Kanehisa, M. and Goto, S. (2006) Extraction of phylogenetic network modules from the metabolic network. *BMC Bioinformatics*, **7**, 130.

14 Zhang, C., Liu, S. and Zhou, Y. (2006) Fast and accurate method for identifying high-quality protein-interaction modules by clique merging and its application to yeast. *Journal of Proteome Research*, **5**, 801–907.

15 Chen, Y. and Xu, D. (2003) Computation analysis of high-throughput protein–

protein interaction data. *Current Peptide and Protein Science*, **4**, 159–181.

16 Lee, I., Date, S.V., Adai, A.T. and Marcotte, E.M. (2004) A probabilistic functional network of yeast genes. *Science*, **306**, 1555–1558.

17 Jansen, R., Yu, H., Jansen, R., Yu, H., Greenbaum, D., Kluger, Y., Krogan, N.J., Chung, S., Emili, A., Snyder, M., Greenblatt, J.F. and Gerstein, M. (2003) A Bayesian networks approach for predicting protein–protein interactions from genomic data. *Science*, **302**, 449–453.

18 Gunsalus, K.C., Ge, H., Schetter, A.J., Goldberg, D.S., Han, J.D., Hao, T., Berriz, G.F., Bertin, N., Huang, J., Chuang, L.S., Li, N., Mani, R., Hyman, A.A., Sonnichsen, B., Echeverri, C.J., Roth, F.P., Vidal, M. and Piano, F. (2005) Predictive models of molecular machines involved in *Caenorhabditis elegans* early embryogenesis. *Nature*, **436**, 861–865.

19 Rhodes, D.R., Tomlins, S.A., Varambally, S., Mahavisno, V., Barrette, T., Kalyana-Sundaram, S., Ghosh, D., Pandey, A. and Chinnaiyan, A.M. (2005) Probabilistic model of the human protein–protein interaction network. *Nature Biotechnology*, **23**, 951–959.

20 The Gene Ontology Consortium (2000) Gene ontology: tool for the unification of biology. *Nature Genetics*, **25**, 25–29.

21 Uetz, P., Giot, L., Cagney, G., Mansfield, T.A., Judson, R.S., Knight, J.R., Lockshon, D., Narayan, V., Srinivasan, M., Pochart, P., Qureshi-Emili, A., Li, Y., Godwin, B., Conover, D., Kalbfleisch, T., Vijaya-damodar, G., Yang, M., Johnston, M., Fields, S. and Rothberg, J.M. (2000) A comprehensive analysis of protein–protein interactions in *Saccharomyces cerevisiae*. *Nature*, **403**, 623–627.

22 Ito, T., Tashiro, K., Muta, S., Ozawa, R., Chiba, T., Nishizawa, M., Yamamoto, K., Kuhara, S. and Sakaki, Y. (2001) Toward a protein–protein interaction map of the budding yeast: a comprehensive system to examine two-hybrid interactions in all possible combinations between the yeast proteins. *Proceedings of the National Academy of Sciences of the United States of America*, **98**, 4569–4574.

23 Gavin, A., Bosche, M., Krause, R., Grandi, P., Marzioch, M., Bauer, A., Schultz, J., Rick, J., Michon, A. and Cruciat, C. (2002) Functional organization of the yeast proteome by systematic analysis of protein complexes. *Nature*, **415**, 141–147.

24 Ho, Y., Gruhler, A., Heilbut, A., Bader, G.D., Moore, L., Adams, S.L., Millar, A., Taylor, P., Bennett, K., Boutilier, K., Yang, L., Wolting, C., Donaldson, I., Schandorff, S., Shewnarane, J., Vo, M., Taggart, J., Goudreault, M., Muskat, B., Alfarano, C., Dewar, D., Lin, Z., Michalickova, K., Willems, A.R., Sassi, H., Nielsen, P.A., Rasmussen, K.J., Andersen, J.R., Johansen, L.E., Hansen, L.H., Jespersen, H., Podtelejnikov, A., Nielsen, E., Crawford, J., Poulsen, V., Sørensen, B.D., Matthiesen, J. and Hendrickson, R.C. (2002) Systematic identification of protein complexes in *Saccharomyces cerevisiae* by mass spectrometry. *Nature*, **415**, 180–183.

25 Gasch, A.P., Spellman, P.T., Kao, C.M., Carmel-Harel, O., Eisen, M.B., Storz, G., Botstein, D. and Brown, P.O. (2000) Genomic expression programs in the response of yeast cells to environmental changes. *Molecular and Cellular Biology*, **11**, 4241–4257.

26 Chen, Y. and Xu, D. (2005) Understanding protein dispensability through machine-learning analysis of high-throughput data. *Bioinformatics*, **21**, 575–581.

27 Chen, Y. and Xu, D. (2004) Global protein function annotation through mining genome-scale data in yeast *Saccharomyces cerevisiae*. *Nucleic Acids Research*, **32**, 6414–6424.

28 Dongen, S.V. (2000) Performance criteria for graph clustering and Markov cluster experiments. *Technical Report* INS-R0012, National Research Institute for Mathematics and Computer Science in The Netherlands, Amsterdam.

29 Enright, A.J., Kunin, K. and Ouzounis, C.A. (2003) Protein families and TRIBES in genome sequence space. *Nucleic Acids Research*, **31**, 4632–4638.

30 von Mering, C., Zdobnov, E.M., Tsoka, S., Ciccarelli, F.D., Pereira-Leal, J.B., Ouzounis, C.A. and Bork, P. (2003) Genome evolution reveals biochemical networks and functional modules. *Proceedings of the National Academy of Sciences of the United States of America*, **100**, 15428–15433.

31 Han, J.D., Bertin, N., Hao, T., Goldberg, D.S., Berriz, G.F., Zhang, L.V., Dupuy, D., Walhout, A.J., Cusick, M.E., Roth, F.P. and Vidal, M. (2004) Evidence for dynamically organized modularity in the yeast protein–protein interaction network. *Nature*, **430**, 380–385.

32 Yeung, B.G., Phan, H.L. and Payne, G.S. (1999) Adaptor complex-independent clathrin function in yeast. *Molecular Biology of the Cell*, **10**, 3643–3659.

33 Luscombe, N.M., Babu, M.M., Yu, H., Snyder, M., Teichmann, S.A. and Gerstein, M. (2004) Genomic analysis of regulatory network dynamics reveals large topological changes. *Nature*, **431**, 308–312.

34 de Lichtenberg, U., Jensen, L.J., Brunak, S. and Bork, P. (2005) Dynamic complex formation during the yeast cell cycle. *Science*, **307**, 724–727.

35 Cairns, B.R., Lorch, Y., Li, Y., Zhang, M., Lacomis, L., Erdjument-Bromage, H., Tempst, P., Du, J., Laurent, B. and Kornberg, R.D. (1996) RSC an essential, abundant chromatin-remodeling complex. *Cell*, **87**, 1249–1260.

36 Stevens, S.W. and Abelson, J. (1999) Purification of the yeast U4/U6.U5 small nuclear ribonucleoprotein particle and identification of its proteins. *Proceedings of the National Academy of Sciences of the United States of America*, **96**, 7226–7231.

37 Bolser, D., Dafas, P., Harrington, R., Park, J. and Schroeder, M. (2003) Visualisation and graph-theoretic analysis of a large-scale protein structural interactome. *BMC Bioinformatics*, **4** (1), 45.

38 Stuart, J.M., Segal, E., Koller, D. and Kim, S.K. (2003) A gene-coexpression network for global discovery of conserved genetic modules. *Science*, **302**, 249–255.

39 Gagneur, J., Krause, R., Bouwmeester, T. and Casari, G. (2004) Modular decomposition of protein–protein interaction networks. *Genome Biology*, **5**, R57.

40 Maslov, S. and Sneppen, K. (2002) Specificity and stability in topology of protein networks. *Science*, **296**, 910–913.

41 Ravasz, E., Somera, A.L., Mongru, D.A., Oltvai, Z.N. and Barabasi, A.L. (2002) Hierarchical organization of modularity in metabolic networks. *Science*, **297**, 1551–1555.

42 Stelzl, U., Worm, U., Lalowski, M., Haenig, C., Brembeck, F.H., Goehler, H., Stroedicke, M., Zenkner, M., Schoenherr, A., Koeppen, S., Timm, J., Mintzlaff, S., Abraham, C., Bock, N., Kietzmann, S., Goedde, A., Toksoz, E., Droege, A., Krobitsch, S., Korn, B., Birchmeier, W., Lehrach, H. and Wanker, E.E. (2005) A human protein–protein interaction network: a resource for annotating the proteome. *Cell*, **122**, 957–968.

43 Tornow, S. and Mewes, H.W. (2003) Functional modules by relating protein interaction networks and gene expression. *Nucleic Acids Research*, **31**, 6283–6289.

44 Bergmann, S., Ihmels, J. and Barkai, N. (2004) Similarities and differences in genome-wide expression data of six organisms. *PLOS Biology*, **2**, 85–93.

45 Moraru, I.I. and Loew, L.M. (2005) Intracellular signaling: spatial and temporal control. *Physiology*, **20**, 169–179.

46 McAdams, H.H. and Shapiro, L. (2003) A bacterial cell-cycle regulatory network operating in time and space. *Science*, **301**, 1874–1877.

47 Milkereit, P., Gadal, O., Podtelejnikov, A., Trumtel, S., Gas, N., Petfalski, E.,

Tollervey, D., Mann, M., Hurt, E. and Tschochner, H. (2001) Maturation and intranuclear transport of pre-ribosomes requires Noc proteins. *Cell*, **105**, 499–509.

48 Glover, C.V., 3rd (1998) On the physiological role of casein kinase II in *Saccharomyces cerevisiae. Progress in Nucleic Acid Research and Molecular Biology*, **59**, 95–133.

49 Venema, J. and Tollervey, D. (1999) Ribosome synthesis in *Saccharomyces cerevisiae. Annual Review of Genetics*, **33**, 261–311.

50 Guimera, R. and Nunes Amaral, L.A. (2005) Functional cartography of complex metabolic networks. *Nature*, **433**, 895–900.

51 Hughes, T.R., Marton, M.J., Jones, A.R., Roberts, C.J., Stoughton, R., Armour, C.D., Bennett, H.A., Coffey, E., Dai, H., He, Y.D., Kidd, M.J., King, A.M., Meyer, M.R., Slade, D., Lum, P.Y., Stepaniants, S.B., Shoemaker, D.D., Gachotte, D., Chakraburtty, K., Simon, J., Bard, M. and Friend, S.H. (2000) Functional discovery via a compendium of expression profiles. *Cell*, **102**, 109–126.

52 Kim, S.K., Lund, J., Kiraly, M., Duke, K., Jiang, M., Stuart, J.M., Eizinger, A., Wylie, B.N. and Davidson, G.S. (2001) A gene expression map for *Caenorhabditis elegans. Science*, **293**, 2087–2092.

53 Carlson, M.R., Zhang, B., Fang, Z., Mischel, P.S., Horvath, S. and Nelson, S.F. (2006) Gene connectivity functGene connectivity, function, and sequence conservation: predictions from modular yeast co-expression networks. *BMC Genomics*, **7**, 40.

12
Computational Reconstruction of Transcriptional Regulatory Modules of the Yeast Cell Cycle

Wei-Sheng Wu, Wen-Hsiung Li, and Bor-Sen Chen

12.1
Introduction

A transcriptional regulatory module (TRM) is a set of genes that is regulated by a common set of transcription factors (TFs). By organizing the genome into TRMs, a living cell can coordinate the activities of many genes and carry out complex functions. Therefore, identifying TRMs is useful for understanding cellular responses to internal and external signals. The advances in high-throughput genomic technology such as DNA microarray [1,2] and chromatin immunoprecipitation DNA chip (ChIP-chip) [3,4] have greatly facilitated the computational reconstruction of TRMs of a eukaryotic cell.

Genome-wide gene expression analysis has been used to investigate TRMs controlling a variety of cellular processes in yeast [5–9]. Clustering and motif-discovering algorithms have been applied to gene expression data to find sets of coregulated genes and have identified plausible binding motifs of their TFs [7,10,11]. Such approaches have also been expanded to incorporate previous knowledge about the genes, such as cellular functions [12] or promoter sequence motifs [13]. Moreover, some researchers have used model-based approaches such as random Boolean networks [14] and Bayesian networks [15,16] to infer regulatory network architectures. However, this approach provides only indirect evidence of genetic regulatory interactions and does not identify the relevant TFs. On the contrary, the ChIP-chip technique was developed to identify physical interactions between TFs and DNA regions. Using ChIP-chip data, Simon *et al.* [17] investigated how the yeast cell cycle gene-expression program is regulated by each of the nine major transcriptional activators. Lee *et al.* [18] constructed a network of TF–gene interactions and Harbison *et al.* [19] constructed an initial map of yeast's transcriptional regulatory code. However, ChIP-chip data alone cannot tell whether a TF is an activator or a repressor and, most importantly, ChIP-chip data are noisy and, depending on the chosen *P*-value cutoff, include many false positive or false negative TF–DNA binding relationships.

Analysis of Microarray Data: A Network-Based Approach. Edited by F. Emmert-Streib and M. Dehmer

ISBN: 978-3-527-31822-3

Since gene expression and ChIP-chip data provide complementary information, some researchers [20–22] have integrated both types of data in their studies. However, most previous studies except GRAM [21] assumed that a gene is regulated by a TF only if the *P*-value of TF–gene binding in the ChIP-chip data is ≤ 0.001, thus suffering a false negative rate of $\sim 24\%$ in determining TF–gene binding [19].

To reduce the high false negative rate, we develop a method called temporal relationship identification algorithm (TRIA), which uses the information provided by gene expression data to alleviate the effect of using a stringent threshold in determining TF–gene binding. A TF–gene pair is said to have a positively (negatively) temporal relationship if the gene's expression profile is positively (negatively) correlated with the TFs regulatory profile possibly with time lags (see Section 12.2). TRIA identifies TF–gene pairs with temporal relationships. We define that a TF binds to a specific gene if (1) the *P*-value for the TF to bind the gene is ≤ 0.001 in the ChIP-chip data or (2) $0.001 < P \leq 0.01$ and the TF–gene pair has a temporal relationship. That is, we allow the *P*-value cutoff to be relaxed to 0.01 if the TF–gene pair has a temporal relationship. Our approach is different from GRAM [21], which relied on sets of coexpressed gene to relax the stringent *P*-value cutoff.

From the above procedure, we derive a binding score matrix. Then we develop the module finding algorithm (MOFA) that combines this binding score matrix with the gene expression matrix to reconstruct TRMs of the yeast cell cycle (see Section 12.2). For each of the five cell cycle phases (M/G1, G1, S, S/G2, and G2/M), MOFA exhaustively searches for all possible TF combinations and finds their target genes. Once the set of target genes to which a common set of TFs bind is inferred, MOFA identifies a subset of these target genes whose gene expression profiles are highly positively correlated possibly with time lags. That is, the genes of a module not only share a common set of TFs but also have highly positively time-lag correlated expression profiles (see Section 12.2). Our gene module is more general than that of GRAM [21], which only searched coexpressed genes to form a module. MOFA reconstructs 87 TRMs. We then validate the biological relevance of each inferred TRM using existing experimental data, enrichment for genes in the same MIPS functional category [23], known as DNA-binding motifs [7], and so on.

12.2 Methods

12.2.1 Data Sets

We use the ChIP-chip data in [19] and the gene expression data (α factor) of the yeast cell cycle in [7]. Spellman *et al.* [7] used Fourier transform to identify 800 putative cell cycle genes (113 genes in M/G1, 300 in G1, 71 in S, 121 in S/G2, and 195 in G2/M).

By integrating both types of data, MOFA tries to reconstruct TRMs for each of the five cell cycle phases.

12.2.2
Temporal Relationship Identification Algorithm

TRIA is developed to identify TF–gene pairs that have temporal relationships. A cell cycle TF and its binding target are said to have a positively (negatively) temporal relationship if the target gene's expression profile is positively (negatively) correlated with the TFs regulatory profile possibly with time lags.

Let $\vec{y} = (y_1, \ldots, y_N)$ and $\vec{x} = (x_1, \ldots, x_N)$ be the gene expression time profile of gene y and cell cycle TF x, respectively. The regulatory profile $\mathrm{RP}(\vec{x}) = (f(x_1), \ldots, f(x_N))$ of TF x is defined as a sigmoid function:

$$f(x_i) = \frac{1}{1 + \mathrm{e}^{-(x_i - \bar{x})/s)}} \qquad i = 1, 2, \ldots, N,$$

where $\bar{x}$ is the sample mean and s is the sample standard deviation of $\vec{x}$. It is known that TF binding affects gene expression in a nonlinear fashion: it has no effect below some level, and the effect may become saturated above some level. This type of behavior can be modeled using a sigmoid function. Therefore, we define the regulatory profile of a TF as a sigmoid function of its gene expression profile.

Compute the correlation between $\vec{y}$ and $\mathrm{RP}(\vec{x})$ with a lag of k time points:

$$r(k) = \frac{\sum_{i=1}^{N-k} (y_{i+k} - \bar{y})(f(x_i) - \bar{m})}{\sqrt{\sum_{i=1}^{N-k} (y_{i+k} - \bar{y})^2} \sqrt{\sum_{i=1}^{N-k} (f(x_i) - \bar{m})^2}}, \qquad k = 0, 1, \ldots, L,$$

where

$$\bar{y} \triangleq \frac{\sum_{i=1}^{N-k} y_{i+k}}{N-k}, \qquad \bar{m} \triangleq \frac{\sum_{i=1}^{N-k} f(x_i)}{N-k},$$

and L is the maximal time lag of the TFs regulatory profile considered. The value of L is chosen to make the maximal time lag approximately equal to two consecutive cell cycle phases because Simon *et al.* [17] found cases where a cell cycle TF that expresses in one phase of the cell cycle can regulate genes that function in the next phase.

Then we test the null hypothesis H_0: $r(k) = 0$ against the alternative hypothesis H_1: $r(k) \neq 0$ by the bootstrap method (see Appendix 12.A) and get a P-value $p(k)$. The time-lagged correlation (TLC) of $\vec{y}$ and $\mathrm{RP}(\vec{x})$ is defined as $r(j)$ that has the smallest P-value (i.e., $\mathrm{TLC}(\vec{y}, \mathrm{RP}(\vec{x})) = r(j)$ if $p(j) \leq p(k)\ \forall k \neq j$). Note that

$-1 \leq \text{TLC}(\vec{y}, \text{RP}(\vec{x})) \leq 1$. Two possible temporal relationships between $\vec{y}$ and $\text{RP}(\vec{x})$ can be identified by TRIA: $\vec{y}$ and $\text{RP}(\vec{x})$ are (1) positively correlated with a lag of j time points if $\text{TLC}(\vec{y}, \text{RP}(\vec{x})) = r(j) > 0$ and $p(j) \leq p_{\text{Threshold}}$ and (2) negatively correlated with a lag of j time points if $\text{TLC}(\vec{y}, \text{RP}(\vec{x})) = r(j) < 0$ and $p(j) \leq p_{\text{Threshold}}$. The $p_{\text{Threshold}}$ is chosen to ensure that we have at most a 5% false discovery rate (FDR) (see Appendix 12.A).

12.2.3
The Module Finding Algorithm

Before describing MOFA, we define some terms.

Definition 1: Let $E = [e_{ij}]$ be the gene expression matrix whose rows correspond to genes and columns correspond to time points, so that e_{ij} is the expression level of gene i at time point j.

Definition 2: Let $B = [b_{ij}]$ be the binding score matrix whose rows correspond to genes and columns correspond to TFs, so that b_{ij} denotes the binding score of TF j to bind gene i. We set $b_{ij} = 4$ if the P-value for TF j to bind gene i is ≤ 0.001 in the ChIP-chip data and TF j and gene i are found to have a temporal relationship; $b_{ij} = 3$ if $P \leq 0.001$ but no temporal relationship exists; $b_{ij} = 2$ if $0.001 < P \leq 0.01$ and a temporal relationship exists; $b_{ij} = 1$ if $0.001 < P \leq 0.01$ but no temporal relationship exists; and $b_{ij} = 0$ if $P > 0.01$.

Definition 3: Let R be a set of TFs and $C(R, z)$ be the set of target genes to which all the TFs in R bind with a score $\geq z$. In addition, let SP be the set of all genes in a specific cell cycle phase (113 genes in M/G1, 300 in G1, 71 in S, 121 in S/G2, and 195 in G2/M).

Definition 4: The expression coherence score (EC(A)) of a set A is defined as the fraction of gene–gene pairs in A whose TLC is higher than the threshold T [20]. The threshold T was determined to be the 95th percentile TLC value of all pairwise TLC values between 2000 randomly chosen genes in the yeast genome.
Note that $0 \leq \text{EC}(A) \leq 1$.

Remark: EC(A) is a generalization of the expression correlation score used in [13,20]. While expression correlation score can only find coexpressed genes, EC(A) can find genes whose gene expression profiles are highly positively correlated with time lags. Since each gene may have a different response time to the same transcriptional regulatory mechanism in transcribing DNA to RNA, the RNA profiles of coregulated genes may not be coexpressed but rather positively correlated with time lags. Therefore, our measure is more powerful than the expression correlation score to find out the coregulated but not the coexpressed genes.

MOFA performs in two steps (see Figure 12.1). First, for a specific cell cycle phase (M/G1, G1, S, S/G2, or G2/M), it exhaustively searches all possible R's in order to

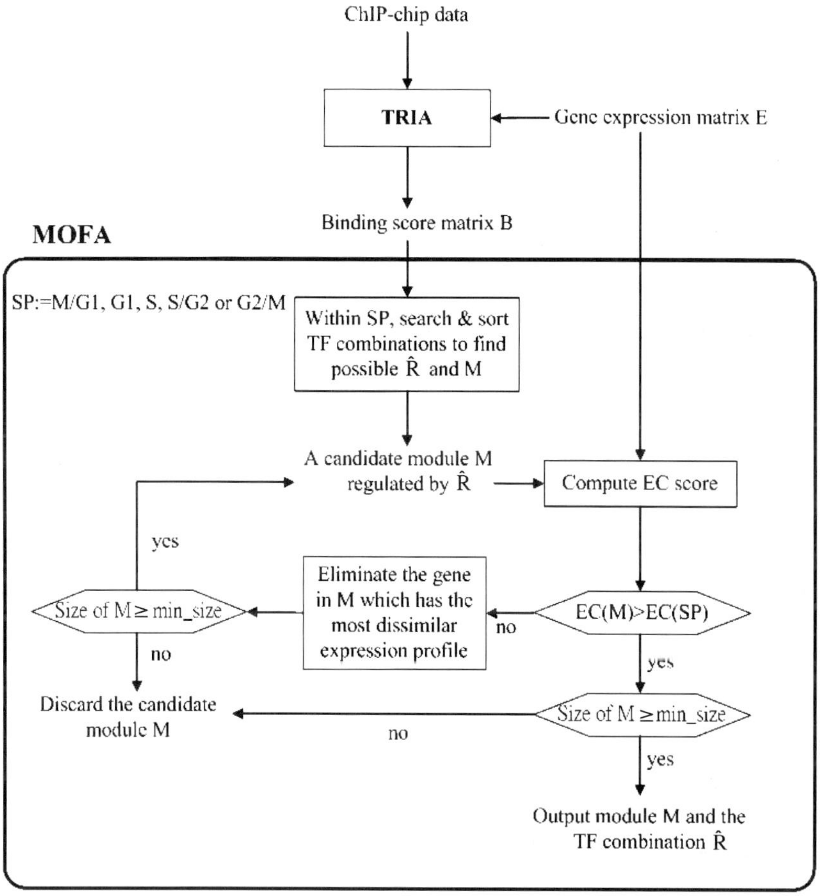

Figure 12.1 Flowchart of MOFA.

find $C(R, z)'s$. A particular R and the corresponding $C(R, z)$ are recorded if $C(R, z)$ contains more than a certain number of genes. MOFA then sorts the recorded R's according to their sizes, denoting the sorting result as $\hat{R}'$s, so that the first $\hat{R}$ is the one with the largest number of members. Second, if $\mathrm{EC}(C(\hat{R}, z)) \leq \mathrm{EC}(\mathrm{SP})$, MOFA iteratively eliminates the genes of the set $C(\hat{R}, z)$ starting from the one with the most dissimilar expression profile until $\mathrm{EC}(\tilde{C}(\hat{R}, z)) > \mathrm{EC}(\mathrm{SP})$, where $\tilde{C}(\hat{R}, z)$ is the set of the remaining genes and SP is the set of all genes in a specific cell cycle phase. That is, MOFA tries to identify a subset of coregulated genes whose gene expression profiles are highly positively correlated possibly with time lags compared to that of the set of all genes in the specific cell cycle phase. Finally, MOFA outputs a module $M(\hat{R}) \triangleq \tilde{C}(\hat{R}, z)$ if $\tilde{C}(\hat{R}, z)$ contains more than a certain number of genes, say 5. The above procedure goes over all $\hat{R}'$s in the specific cell cycle phase. We provide the pseudocode of MOFA in Appendix 12.A.

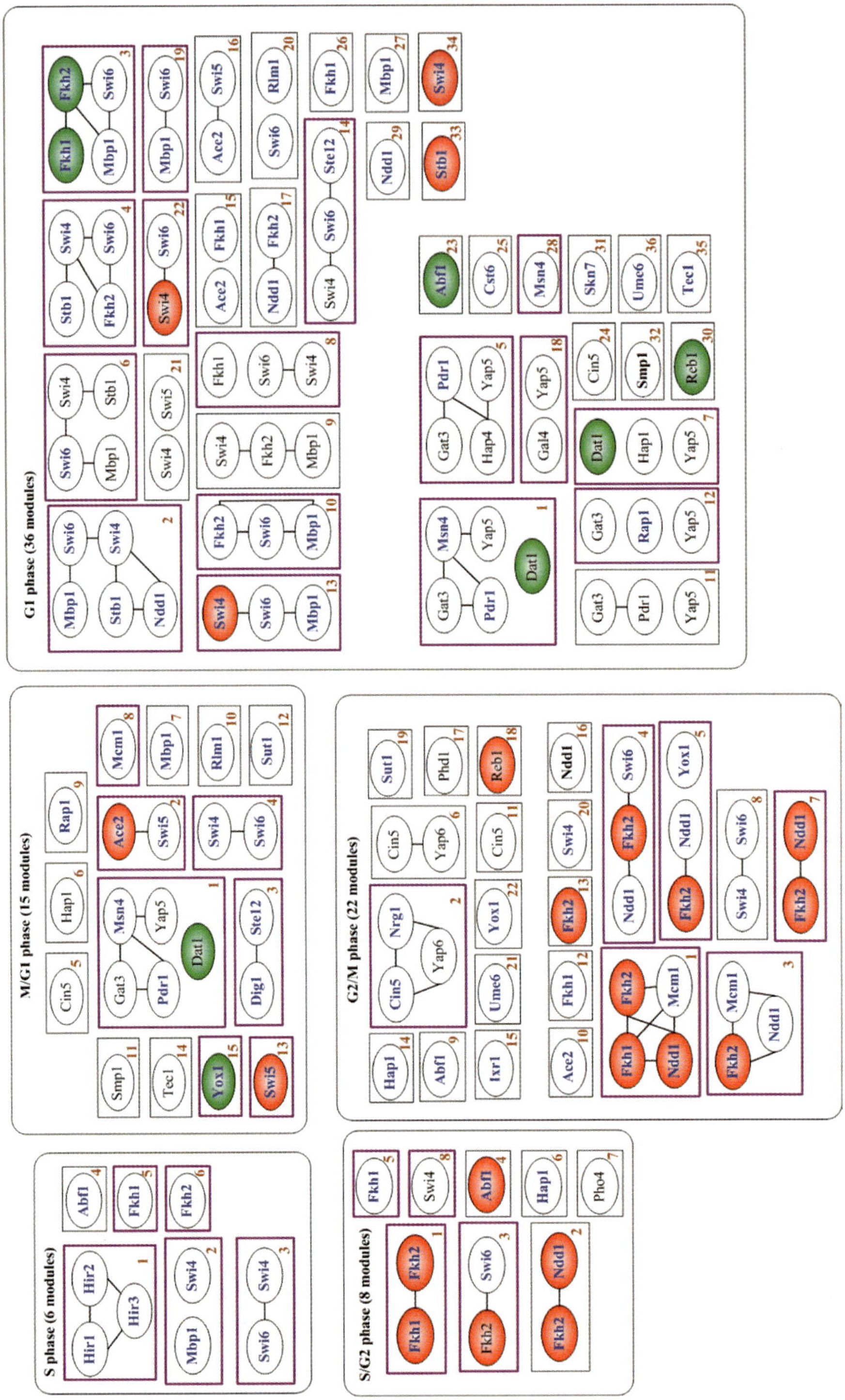
G1 phase (36 modules)
M/G1 phase (15 modules)
G2/M phase (22 modules)
S phase (6 modules)
S/G2 phase (8 modules)

12.3 Results

By integrating the gene expression and ChIP-chip data, MOFA identified 87 TRMs, which together contain 336 distinct genes regulated by 40 distinct TFs (see Figure 12.2 and Supplementary Table 1 at http://www.ee.nthu.edu.tw/bschen/Book-Chapter/MOFA.htm). In the literature [7,23–25], 139 of the 336 genes and 30 of the 40 TFs are known to be involved in the cell cycle.

12.3.1 Validation of the Identified Modules

Analysis of the identified modules suggests that MOFA identifies biologically relevant groups of genes. First, 83 of the 87 modules contain genes that are known to be involved in the cell cycle (see Supplementary Table 1 at http://www.ee.nthu. edu.tw/bschen/BookChapter/MOFA.htm). Second, 51% (44/87) of the identified modules includes groups of genes that function in the same cellular process: each of these modules contains at least one overrepresented MIPS functional category with adjusted P-value < 0.05 (after the Bonferroni correction for multiple tests) using the cumulative hypergeometric distribution (see Supplementary Table 2 at http://www.ee.nthu.edu.tw/bschen/BookChapter/MOFA.htm). Third, the modules are generally accurate in assigning TFs to sets of genes whose functions are consistent with the TFs' known roles. We found that the regulatory functions of 71% (120/169 counting multiplicity) of the TFs are consistent with overrepresented MIPS functional categories of one of their modules, with adjusted P-value < 0.05 (see Figure 12.2). As an example, Dig1 and Ste12 are known to regulate mating and pseudohyphal growth [43] and M/G1 is the critical phase for these processes. All five genes (*FUS1, GPA1, KAR4, SST2,* and *TEC1*) of the {Ste12, Dig1} module are important for mating, pseudohyphal growth, or pheromone response. Fourth, 33% (188/568 counting multiplicity) of the genes are known by previous studies to be regulated by at least one of the TFs that we assigned to the module (see Supplementary Table 1 at http://www.ee.nthu.edu.tw/bschen/BookChapter/MOFA.htm). Fifth, the genes of a module usually have the same binding motifs of the important cell cycle TFs such as SCB (bound by SBF), MCB (bound by MBF), SFF (bound by

◄

Figure 12.2 The 87 TRMs identified in this study. Each rectangle stands for a module and the ovals in a rectangle indicate the TFs that regulate the module. The TF name is colored blue if its function is consistent with overrepresented MIPS functional categories of one of the modules with adjusted P-value < 0.05 (after the Bonferroni correction for multiple tests) using the cumulative hypergeometric distribution, or black if not. Two ovals are connected by a line if previous studies indicated that the two TFs interact with each other physically (forming a protein complex), functionally (in the same cellular process), or statistically (co-occurrence) [19–22,24]. An oval is colored red (green) if the TF is identified as an activator (repressor). The periphery of a rectangle is colored purple if this module implicates important TF combinations or is discussed in the text.

SFF), ECB (bound by Mcm1), and SWI5 (bound by Ace2 and Swi5). We found that in the majority of cases (36/45) in which a module is controlled by at least one of the important cell cycle TFs (SBF, MBF, SFF, Mcm1, Ace2, and Swi5), there always exist genes that have the known binding motifs of the corresponding TFs (see Supplementary Table 1 at http://www.ee.nthu.edu.tw/bschen/BookChapter/MOFA.htm). Finally, in most cases in which a module is controlled by more than one TF, there is evidence that these TFs may interact physically or functionally (see Figure 12.2). About 59% (70/118) of the TF interactions that we identified have been experimentally proven or identified by computational algorithms [19–22,24]. Taken together, these results provide evidence that MOFA identifies not only sets of biologically related genes, but also TFs that individually or cooperatively control these genes.

12.3.2
Identification of Important Cell Cycle TFs and Their Combinations

MOFA identified 40 TFs that regulate genes of the yeast cell cycle, and Figure 12.3 shows the cell cycle phases in which these TFs carry out their regulatory functions. Table 12.1 lists these 40 TFs according to the number of target genes. The nine well-known cell cycle TFs (Ace2, Fkh1, Fkh2, Mbp1, Mcm1, Ndd1, Swi4, Swi5, and Swi6) are ranked within the top 14, suggesting the effectiveness of MOFA to find important cell cycle TFs. Moreover, we found another 21 TFs (Abf1, Cin5, Cst6, Dig1, Gal4, Gat3, Hap4, Hir1, Hir2, Hir3, Ixr1, Msn4, Rap1, Rlm1, Skn7, Stb1, Ste12, Tec1, Ume6, Yap5, and Yox1) that are related to the cell cycle process, consistent with the previous studies [23–25]. The remaining 10 TFs (Dat1, Hap1, Nrg1, Pdr1, Phd1, Pho4, Reb1, Smp1, Sut1, and Yap6) are putative cell cycle related TFs. Among them, Hap1 is more plausible than the others to be related to the cell cycle process, since the number of cell cycle genes that it regulates is much larger than that of the others (see Table 12.1). Actually, it has been shown that Hap1 (also called Ape1 AP endonuclease) regulates *APE1* [44]. Ape1 is a dual-function enzyme and its cell cycle dependent expression might affect both DNA repair and the activity of various transcription factors as a function of the cell cycle [44]. This evidence validates that MOFA has the ability to find novel TFs that may play a role in the cell cycle or are involved in other cellular processes that have cross talk with the cell cycle process.

TF combinations and their target genes that are important for each cell cycle phase are also found. We found that different combinations of a fairly small number of TFs are responsible for regulating a large number of genes in different cell cycle phases. Detailed discussions of the TF combinations and their target genes in each specific cell cycle phase are given below.

12.3.3
The M/G1 Phase

Ace2 and Swi5 have been shown to control certain genes expressed in M/G1 [34]. We successfully found that {Ace2, Swi5} and {Swi5} regulate, respectively, modules 2 and

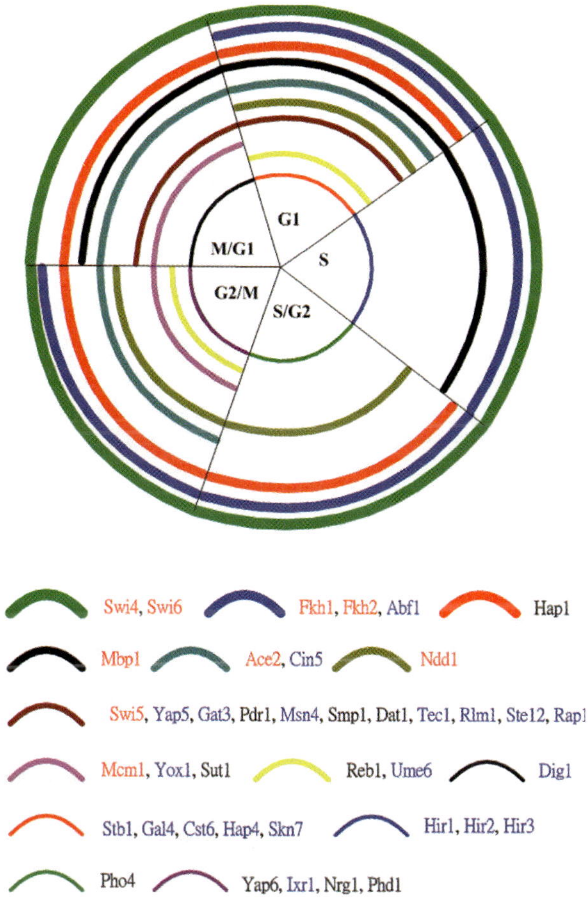

Figure 12.3 The cell cycle phases in which each of the 40 identified TFs carries out its regulatory function. The nine well-known cell cycle TFs are colored red and the other 21 TFs, which are also involved in the cell cycle [23–25], are colored blue.

13 in M/G1. Both Ace2 and Swi5 were found to regulate *EGT2*, whose product is involved in cell wall biogenesis and cytokinesis. Swi5 also regulates *PCL9*, whose product is the only cyclin known to act in M/G1 [45], and *SIC1*, whose product is a cyclin regulator that inhibits the Cdc28–Clb kinase activity. Furthermore, Swi5 regulates several Y′ genes, which are a subgroup of a larger group of subtelomeric genes that share DNA sequence similarity and whose expression peaks in early G1 [7].

It is known that in the absence of Ndd1 and Fkh2, Mcm1 participates in the regulation of genes essential for cellular functions specific to late mitosis and early G1 [42,46]. Indeed, we found that {Mcm1} regulates module 8 in M/G1. It regulates *CDC46*, which encodes a protein involved in prereplication complex formation, and *AGA2*, which is involved in mating. In addition, Yox1 was recently characterized as a

Table 12.1 The number of genes regulated by each of the 40 TFs in each cell cycle phase.

TF	All phases	M/G1	G1	S	S/G2	G2/M
Swi4	78	5	45	9	9	10
Swi6	74	5	50	7	5	7
Fkh2	67		14	5	15	33
Mbp1	57	5	47	5		
Fkh1	53		17	5	18	13
Ndd1	50		12		5	33
Abf1	34		10	5	10	9
Swi5	34	24	10			
Cin5	33	8	7			18
Hap1	30	9	5		5	11
Stb1	24		24			
Mcm1	22	7				15
Yap5	22	5	17			
Ace2	21	6	10			5
Gat3	20	5	15			
Pdr1	17	5	12			
Yox1	16	5				11
Msn4	15	5	10			
Reb1	15		8			7
Smp1	15	8	7			
Dat1	13	5	8			
Sut1	13	5				8
Tec1	13	5	8			
Rlm1	12	7	5			
Ste12	12	5	7			
Ume6	11		6			5
Yap6	11					11
Rap1	10	5	5			
Gal4	6		6			
Hir1	6			6		
Hir2	6			6		
Hir3	6			6		
Cst6	5		5			
Dig1	5	5				
Hap4	5		5			
Ixr1	5					5
Nrg1	5					5
Phd1	5					5
Pho4	5				5	
Skn7	5		5			

The nine well-known cell cycle TFs are bold faced and the other 21 TFs, which are also involved in the cell cycle [23–25], are in italic. The TFs are ordered by the number of their target genes in all phases.

binding partner of Mcm1 in M/G1 [42]. We found that {Yox1}, acting as a repressor, regulates module 15 in M/G1. Three genes *CDC46*, *PIG1*, and *YOR066W* are found to be regulated by both Mcm1 and Yox1, confirming that Yox1 and Mcm1 may coregulate a group of genes.

In addition, some cell wall genes are known to be under the control of the M-phase regulator Mcm1 or the G1-phase regulator SBF. The M/G1 phase is a crucial time for cell wall synthesis because the bud separates from the mother right after the M/G1 phase. We successfully found TF combinations {Mcm1} and {Swi4, Swi6}, whose common target genes include *SWI4*, which encodes a late G1 TF, and *UTR2*, which is involved in cell wall organization and polarized growth. The dual regulation of *SWI4* by Mcm1 and Swi4 has been shown previously [46].

We identified {Dig1, Ste12} to regulate module 3 in M/G1. The genes of this module include *FUS1*, *GPA1*, *KAR4*, *SST2*, and *TEC1*, which are important for mating or pseudohyphal growth. Dig1 and Ste12 are known to regulate mating and pseudohyphal growth [43], supporting the biological relevance of our finding. We also found novel TF combinations. For example, {Dat1, Gat3, Msn4, Pdr1, Yap5} is identified to regulate a group of genes that are similar to subtelomerically encoded proteins.

12.3.4
The G1 Phase

Previous molecular and genetic analysis suggested that SBF and MBF are important activators of genes essential for cellular functions specific to late G1 [17,47]. Our result confirms this model: 10 out of the 36 modules in G1 are regulated by MBF or SBF. SBF regulates *BUD9* and *EXG1* (both of module 8), and *GAS1*, *MNN1*, *OCH1*, and *PSA1* (all of module 22). These genes are involved in the morphological changes associated with cell budding. MBF controls *PDS5*, *RAD51* and *RNR1* (all of module 3), and *DUN1*, *IRR1*, and *RAD27* (all of module 19). These genes are involved in DNA replication and repair. Moreover, the targets of SBF and MBF also include key cell cycle regulators. Both SBF and MBF were found to regulate *CLN1* and *CLB6* (both of module 2), and *PCL1* (of module 6). In addition, SBF regulates *PCL2* (of module 14) and MBF regulates *CLB6* (of module 2).

We found that Stb1 together with SBF (Swi4 + Swi6) or MBF (Mbp1 + Swi6) regulates modules 2, 4, and 6 in G1. It has been known that Stb1 binds to Swi6 *in vitro* and is thought to interact with Swi6, a subunit of both SBF and MBF, to regulate transcription *in vivo* [38]. Also, consistent with our results, Kato *et al.* [22] claimed the presence of the complexes Stb1 + Swi6 + Swi4 and Stb1 + Swi6 + Mbp1. Moreover, we found that {Ste12, Swi4, Swi6} regulates module 14 in G1, which is also consistent with the result of [22].

We found that Fkh1/Fkh2 combines with MBF/SBF to regulate modules 3, 4, 8, and 10 in G1. It is known that Fkh1 and Fkh2 regulate genes expressed in G2/M and also genes expressed in other cell cycle phases [17], supporting our result. We also found some novel TF combinations. For example, {Dat1, Gat3, Msn4, Pdr1, Yap5}, which is also found in M/G1, {Gat3, Hap4, Pdr1, Yap5}, {Dat1, Hap1, Yap5}, {Gat3, Rap1, Yap5}, {Gal4, Yap5}, and {Msn4} are all identified to regulate genes whose products are similar to subtelomerically encoded proteins. All these genes share DNA sequence similarity and are found in Y′ elements, which are located at the ends of chromosomes [7].

12.3.5 The S Phase

We found that {Fkh2} regulates various genes that encode proteins associated with chromatin structure including histone genes *HHF1* and *HHT1* (both of module 6). We found that {Fkh1} regulates *TEL2* (of module 5), a telomere length regulator, and *ARP7* (of module 5), a subunit of the chromatin remodeling Swi/Snf complex. Histone genes can be found in the {Fkh1}, {Fkh2}, {Swi4, Swi6}, and {Mbp1, Swi4} modules, suggesting that SBF, Fkh1, and Fkh2 probably regulate histone genes. Our result is consistent with a few genomic studies [18,48] that indicated the involvement of SBF and Fkh1/Fkh2 in regulating S-phase genes. In addition, we successfully identified {Hir1, Hir2, Hir3} to regulate six histone genes (*HTA1, HTB1, HHT1, HHF1, HHT2, HHF2*) of module 1 in the S phase, supported by existing experimental results [49]. In summary, we suggest that SBF and Fkh1/Fkh2 are activators and Hir1, Hir2, and Hir3 are repressors of histone genes.

12.3.6 The SG2 and G2/M Phases

Simon *et al.* [17] and Lee *et al.* [18] indicated the involvement of SBF and Fkh1/Fkh2 in regulating S/G2 genes. We confirmed that Fkh1, Fkh2, Swi4, and Swi6 are important TFs in this phase since five out of the eight modules in S/G2 are regulated by at least one of these TFs. Fkh2, Swi4, and Swi6 are identified to regulate *SIM1*, which is involved in cell cycle control, and Fkh1 is identified to regulate *CLB4*, which encodes an S/G2 cyclin.

Previous studies have demonstrated that Mcm1 collaborates with Ndd1 and Fkh1/Fkh2 to regulate genes necessary for both entry into and exit from mitosis [28,50]. We successfully identified this TF combination to regulate module 1 in G2/M. Four of the seven genes identified in this module have an SFF (bound by Ndd1 + Fkh1/Fkh2) or ECB (bound by Mcm1) motif (see Supplementary Table 1 at http://www.ee.nthu.edu.tw/bschen/BookChapter/MOFA.htm). The Mcm1 + Ndd1 + Fkh1/Fkh2 protein complex regulates the transcription of *CLB2* (of module 1), whose product is necessary to enter mitosis. Furthermore, SBF and MBF regulate *SWE1* (of module 13 in G1) and *GIN4* (of module 13 in G1). Swe1 is a protein kinase that regulates the G2/M transition by inhibition of Cdc28–Clb2 kinase activity and Gin4 regulates Swe1 [51]. The Mcm1 + Ndd1 + Fkh1/Fkh2 protein complex also sets the stage for exit from mitosis at several levels [17]. First, they regulate two key M/G1 TFs: *SWI5* (of module 3) and *ACE2* (of module 1). Second, they regulate *CDC20* (of module 1), an activator of the anaphase promoting complex (APC). Finally, these activators regulate *SPO12* (of module 3), which encodes a protein that regulates the mitotic exit.

It has been suggested that Fkh2 has a more prominent role than Fkh1 in G2/M transcription [50]. Our analysis agrees with this suggestion since the number of G2/M genes regulated by Fkh2 is much larger than those regulated by Fkh1 (see Table 12.1).

We also found novel TF combinations. For example, we found that SFF instead of combining with Mcm1 can also combine with Swi6 or Yox1 to regulate G2/M genes and {Cin5, Nrg1, Yap6} is identified to regulate a group of genes with unknown functions.

12.4 Discussion

12.4.1 Relationships Between Two TFs of a Module

The relationships between two TFs that regulate the same module fall into three categories. First, both TFs bind DNA in the same promoter region but do not interact with each other. Different TFs may regulate the target gene to execute different functions in different cellular processes. Indeed, we found that TFs in this category usually regulate genes that are required for multiple cellular processes. For example, we found that {Ste12, Swi4, Swi6} regulates module 14 in G1. Since Ste12 and SBF (Swi4 + Swi6) are both DNA-binding TFs and there is no evidence that Ste12 interacts with SBF, the relationship between Ste12 and SBF seems to belong to this category. Ste12 is a regulator of the mating or pseudohyphal growth pathway and SBF is an important regulator in the G1 phase. This indicates that cross talk may exist between these two cellular processes. That is, the TF combination {Ste12, Swi4, Swi6} probably regulates genes needed for the G1 phase and also independently needed for mating, confirming the results of [22]. Second, both TFs bind DNA and interact with each other. For example, we found that {Fkh2, Ndd1, Mcm1} regulates module 3 in G2/M. Both Mcm1 and Fkh2 bind DNA and these two proteins together recruit Ndd1 to form a protein complex to control the transcription of G2/M genes [50]. Third, only one TF binds DNA and the other TF regulates the target genes through binding to the DNA-binding TF. For example, {Mbp1, Swi6} and {Swi4, Swi6} are found to regulate, respectively, modules 19 and 22 in G1. MBF (Mbp1 + Swi6) functions in DNA replication, and SBF (Swi4 + Swi6) predominantly controls the expression of budding and cell wall genes [4]. Since Swi6 is a non-DNA-binding cofactor of Swi4 and Mbp1, the relationship between Swi6 and Swi4/Mbp1 falls into the third category.

12.4.2 Advantages of MOFA

MOFA has two features that make it more powerful than current methods. First, it can reduce false negatives in determining binding events in the ChIP-chip data. Most researchers except for Bar-Joseph *et al.* [21] have chosen a relatively stringent *P*-value threshold (0.001) to determine binding in order to reduce false positives at the expense of false negatives [18–20,22]. In comparison, MOFA allows the *P*-value cutoff to be relaxed to 0.01 if a TF–gene pair has a temporal relationship. (Our approach is different from GRAM [21], which relied on sets of coexpressed gene to relax the stringent *P*-value cutoff.) As an example, consider Swi5, a well-characterized cell cycle TF in M/G1. The {Swi5} module we inferred contains 18 genes that have similar expression patterns (see Supplementary Figure 1 at http://www.ee.nthu.edu.tw/bschen/BookChapter/MOFA.htm). Four of these genes (*YOR264W, PST1, SIC1,* and *YHB1*) would not have been identified as Swi5 targets using the stringent *P*-value threshold (0.001). Previous studies identified these four genes as true targets of Swi5 [7,18]. This attests to the ability of MOFA to lower the rate of false negatives without substantially increasing the rate of false positives. Overall, 87 of the 988 unique TF–gene interactions discovered by MOFA would not have been detected using the current ChIP-chip data with the stringent *P*-value cutoff (0.001). In addition, 312 of the 988 unique TF–gene interactions are supported by gene expression data. That is, each of the 312 TF–gene pairs is identified to have a temporal relationship (see Supplementary Table 1 at http://www.ee.nthu.edu.tw/bschen/BookChapter/MOFA.htm).

Second, MOFA can determine the role of a TF in regulating genes of a module. A TF is said to be an activator (repressor) of a module if the *P*-value of observing TF–gene pairs of the module having a positively (negatively) temporal relationship is $\leq$0.001. The *P*-value is the probability that an observation would be made by chance and is calculated using the cumulative binomial distribution [52]. We found nine activators (Abf1, Ace2, Fkh1, Fkh2, Ndd1, Reb1, Stb1, Swi4, and Swi5) and six repressors (Abf1, Dat1, Fkh1, Fkh2, Reb1, and Yox1), which is consistent with the results of previous studies [26,27,29–33,35–37,39–41]. Interestingly, four TFs (Abf1, Fkh1, Fkh2, and Reb1) are capable of being activators and repressors to regulate different modules. Table 12.2 provides the detailed discussion of how we specify the regulatory roles of TFs and the known experimental evidence that supports our findings.

MOFA is more powerful than GRAM [21] in two ways. First, MOFA has the ability to make a TF to be an activator or/and a repressor (see Table 12.2). On the contrary, GRAM cannot find any repressors or activators that are correlated with its target genes with time lags since GRAM regards a TF to be an activator only when the expression profiles of the TF and genes in the corresponding module are coexpressed. For example, GRAM found only two (Fkh1 and Fkh2) of the nine activators and none of the six repressors that are found by MOFA (see Table 12.2). Second, MOFA is more powerful than GRAM to find out coregulated genes that are not coexpressed. While GRAM assumed that the genes of a module are coexpressed, MOFA allows the genes of a module to be positively correlated with time lags. Since coregulated genes may not be coexpressed [53,54], the relaxation of the coexpressed assumption gives MOFA a better ability to reconstruct gene modules with biological relevance. For example, MOFA identified four genes (*YOR264W, PST1, SIC1,* and *YHB1*) as Swi5 targets (the {Swi5} module in M/G1), which is supported by previous studies [7,18]. However, none of them was found by GRAM.

Table 12.2 Identifying regulatory roles of TFs.

TF	Phase (module number)	Regulatory role	P-value	Evidence from literature
Abf1	S/G2 (4)	Activator	6×10^{-5}	[26]
Abf1	G1 (23)	Repressor	0.001	[27]
Fkh1	S/G2 (1); G2/M (1)	Activator	3×10^{-8}; 1×10^{-7}	[28]
Fkh1	G1 (3)	Repressor	3×10^{-5}	[29]
Fkh2	S/G2 (1) (2) (3); G2/M (1) (3) (4) (5) (7) (13)	Activator	3×10^{-8}; 3×10^{-7}; 3×10^{-5}; 1×10^{-7}; 6×10^{-9}; 3×10^{-5}; 9×10^{-5}; 6×10^{-9}; 3×10^{-5}	[28]
Fkh2	G1 (3)	Repressor	3×10^{-5}	[29]
Reb1	G2/M (18)	Activator	2×10^{-4}	[30]
Reb1	G1 (30)	Repressor	2×10^{-5}	[31–33]
Ace2	M/G1 (2)	Activator	9×10^{-5}	[34,35]
Ndd1	S/G2 (2); G2/M (1) (7)	Activator	3×10^{-5}; 1×10^{-7}; 6×10^{-6}	[36,37]
Stb1	G1 (33)	Activator	2×10^{-11}	[38,39]
Swi4	G1 (13) (22) (34)	Activator	6×10^{-9}; 1×10^{-6}; 2×10^{-4}	[40]
Swi5	M/G1 (13)	Activator	6×10^{-8}	[34,35]
Dat1	M/G1 (1); G1 (1) (7)	Repressor	3×10^{-5}; 3×10^{-7}; 3×10^{-5}	[41]
Yox1	M/G1 (15)	Repressor	3×10^{-7}	[42]

MOFA can determine the regulatory role of a TF in regulating genes of a module. A TF is said to be an activator/repressor of a module if the *P*-value of observing TF–gene pairs of the module having positively/negatively (time-lagged) correlated profiles is ≤ 0.001. The *P*-value is the probability that an observation would be made by chance, and is calculated using the cumulative binomial distribution [52]:

$$P(x \geq n_0) = \sum_{x=n_0}^{N} [N!/(x!(N-x)!)]\, p^x (1-p)^{N-x}$$

where N is the total number of genes in a module, n_0 is the number of genes that have temporal relationships with the TF, and p is the probability of observing an arbitrary gene in the genome that has a temporal relationship with the TF.

12.4.3 Parameter Settings of MOFA

The choices of both the relaxed *P*-value and time-lag parameter have biological meanings. Two previous papers [18,19] used a statistical error model to assign a *P*-value of the binding relationship of a TF–gene pair. They found that if $P \leq 0.001$, the binding relationship of a TF–gene pair is of high confidence and can usually be confirmed by gene-specific PCR. If $P > 0.01$, the binding relationship of a TF–gene pair is of low confidence and cannot be confirmed by gene-specific PCR most of the time. However, if $0.001 < P \leq 0.01$, the binding relationship of a TF–gene pair is ambiguous and can be confirmed by gene-specific PCR in some cases but not in the others. Our aim is to solve this ambiguity. This is why we choose 0.01 to be the relaxed *P*-value. We say that an ambiguous binding relationship of a TF–gene pair is plausible if $0.001 < P < 0.01$ and if this TF–gene pair has a temporal relationship. As to the time-lag parameter, its value is chosen to make the maximal time lag approximately equal to two consecutive cell cycle phases because Simon *et al.* [17] found cases where a cell cycle TF that expresses in one phase of the cell cycle can regulate genes that function in the next phase.

Increasing the value of the relaxed *P*-value or the time-lag parameter may introduce some false positive binding relationships of TF–gene pairs into the binding score matrix. However, decreasing the value of the relaxed *P*-value or the time-lag parameter may fail to rescue some false negative binding relationships of TF–gene pairs. A binding score matrix is used to construct an original TRM, and MOFA refines the TRM by identifying a subset of these coregulated genes in a TRM whose gene expression profiles are highly positively correlated possibly with time lags. MOFA can filter out false positives to some extent because the expression profiles of false positives are unlikely to be similar by chance to those of the highly positively time-lag correlated genes identified by MOFA. As to the false negative problem, MOFA cannot alleviate the harmful effect since these plausible binding relationships of TF–genes pairs are not included in the first place. That is, the false negative problem is a more serious issue than the false positive problem in MOFA. Therefore, if users have no idea about the appropriate values of the relaxed *P*-value and the time-lag parameter, they should first try larger values since MOFA has the ability to reduce this kind of noises.

12.4.4 Refining Clusters from Spellman *et al.*

Spellman *et al.* [7] used a hierarchical clustering algorithm to group together coexpressed genes and searched the promoters of these genes for consensus binding motifs. They tried to use these clusters to understand the transcriptional mechanisms of cell cycle regulation. Their approach has some drawbacks. First, coexpressed genes are not necessarily coregulated. Second, even if the genes in a cluster are coregulated, the relevant TFs still cannot be easily identified by the consensus binding motifs.

MOFA can refine clusters in [7] and provide a better understanding of how the cell regulates the complex expression program of the yeast cell cycle. For example, MOFA reassigned genes of the MCM cluster in [7] to several modules. As shown in Figure 12.4A, these modules differ not only in the set of TFs regulating the modules, but also in the different cell cycle phases to which they belong. Our results confirm previous findings that Mcm1 collaborates with Yox1 to regulate genes in M/G1 (e.g., *YOR066W* and *CDC46*) [42] and collaborates with Ndd1 and Fkh1/Fkh2 to regulate genes in G2/M (e.g., *SPO12* and *KIN3*) [50]. In addition, MOFA provides regulation information of the Y′ cluster in [7]. The Y′ cluster contains genes that share DNA sequence similarity and are found in Y′ elements, which are located at chromosome ends. Spellman *et al.* [7] did not figure out how these genes are regulated. As shown in Figure 12.4B, MOFA reassigned genes of the Y′ cluster to three modules and identified several possible regulators (Dat1, Gal4, Gat3, Hap1, Hap4, Msn4, Pdr1, Rap1, and Yap5), providing information for future experiments.

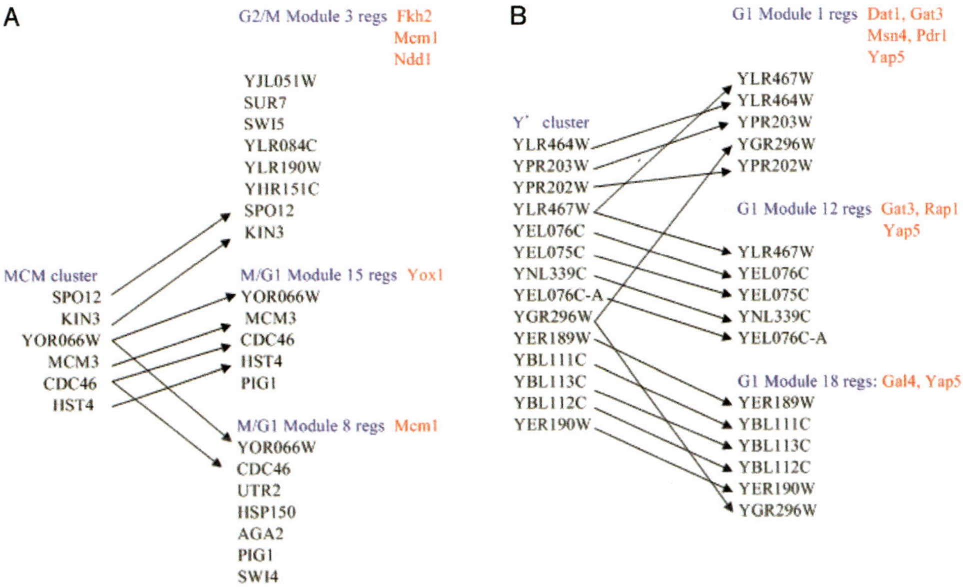

Figure 12.4 Refining clusters from Spellman *et al.* (A) Refining the MCM cluster in [7]. The modules identified by MOFA differ not only in the set of TFs regulating the modules but also in the different cell cycle phases to which they belong, providing a better understanding of how the cell regulates the complex expression program of the yeast cell cycle. Our results confirm previous findings that Mcm1 collaborates with Yox1 to regulate genes in M/G1 (e.g., *YOR066W* and *CDC46*) [42] and collaborates with Ndd1 and Fkh1/Fkh2 to regulate genes in G2/M (e.g., *SPO12* and *KIN3*) [50]. (B) Refining the Y′ cluster in [7]. The Y′ cluster contains genes that share DNA sequence similarity and are found in Y′ elements, which are located at chromosome ends. Spellman *et al.* [7] did not figure out how these genes are regulated. MOFA reassigns genes in the Y′ cluster to three modules and identifies several possible regulators (Dat1, Gal4, Gat3, Hap1, Hap4, Msn4, Pdr1, Rap1, and Yap5), providing information for future experiments.

12.5 Conclusions

We developed a method, called MOFA, for reconstructing TRMs of the yeast cell cycle by integrating gene expression data and ChIP-chip data. MOFA identified 87 TRMs, which together contain 336 distinct genes regulated by 40 TFs. From the literature [7,23–25], 139 of the 336 genes and 30 of the 40 TFs are known to be involved in the cell cycle. The biological relevance of each inferred TRM was validated by using existing experimental data, enrichment for genes in the same MIPS functional category [23], known DNA-binding motifs [7], and so on. Our analysis shows that different combinations of a fairly small number of TFs are responsible for regulating a large number of genes involved in different cell cycle phases and that cross talk may exist between the cell cycle and other cellular processes. Besides, MOFA is capable of finding many novel TF–target gene relationships that could not be identified by using the current ChIP-chip data with the stringent *P*-value cutoff (0.001) or the conventional correlation analysis that only checks the coexpressed relationship. In addition, MOFA can determine the relationships between TFs that regulate the same module and the regulatory roles of these TFs. We found nine activators and six repressors, which is consistent with the results of previous studies [26,27,29–33,35–37,39–41]. Finally, MOFA refines some clusters proposed by previous studies and provides a better understanding of how the complex expression program of the cell cycle is regulated.

We believe that computational analysis of multiple types of data will be a powerful approach for studying complex biological systems when more and more genomic resources such as genome-wide protein activity data and protein–protein interaction data become available.

Appendix 12.A
The Bootstrap Method for Testing the Statistical Significance of the Difference Between *r*(*k*) and 0

We observed *N*–*k* pairs of observations, $Z=\{\mathbf{z}_i\colon i=1,\ldots, N-k \text{ and } \mathbf{z}_i=(f(x_i), y_{i+k})\}$. The correlation coefficient from the sample is calculated as

$$r(k)=\frac{\sum_{i=1}^{N-k}(y_{i+k}-\bar{y})(f(x_i)-\bar{m})}{\sqrt{\sum_{i=1}^{N-k}(y_{i+k}-\bar{y})^2}\sqrt{\sum_{i=1}^{N-k}(f(x_i)-\bar{m})^2}},\quad k=0,1,2,\ldots,$$

where

$$\bar{y}\triangleq\frac{\sum_{i=1}^{N-k}y_{i+k}}{N-k},\qquad \bar{m}\triangleq\frac{\sum_{i=1}^{N-k}f(x_i)}{N-k},$$

and $-1\leq r(k)\leq 1$. We use these observations to test if *r*(*k*) is different from 0 significantly. Suppose that the null hypothesis is H_0: $r(k)=0$ and the alternative hypothesis is H_1: $r(k)\neq 0$. We will apply the bootstrap method to perform this hypothesis testing on the basis of the observations. Keeping the pair relationship of these *N*–*k* pairs to maintain the dependence between $(f(x_i), y_{i+k})$, we sample $\mathbf{z}_i$ with replacement *N*–*k* times to form a bootstrap sample, $Z^*=\{z_i^*: i=1,\ldots, N-k, \text{ and } z_i^* \text{ belongs to } Z\}$. The correlation coefficient from the bootstrap sample Z^* is computed and denoted as $r^*(k)$, $-1\leq r^*(k)\leq 1$. Repeating the resampling procedure *B* times, we will observe $r_1^*(k), r_2^*(k),\ldots, r_B^*(k)$. These bootstrap correlation coefficients are sorted to be $-1\leq r_{(1)}^*(k)\leq r_{(2)}^*(k)\leq\cdots\leq r_{(B)}^*(k)\leq 1$. Then the $(1-\alpha)$ two-sided percentile interval is given by $[r^*_{(B\times\alpha/2)}(k),\ \ r^*_{(B\times(1-\alpha/2))}(k)]$ in this case [55]. If this percentile interval does not contain 0, then the null hypothesis is rejected at the significance level of α. Otherwise, the data fail to reject the null hypothesis at the significance level of α. Since the P-value is the smallest value of α for which the null hypothesis will be rejected based on the observation, the P-value for this test is estimated by the following:

$$\widehat{p}(k) = 2 \times \min\{\widehat{p}_{+}(k), 1-\widehat{p}_{+}(k)\}, \quad \text{where } \widehat{p}_{+}(k) = \sum_{i=1}^{B} \frac{\mathbf{I}\{r_i^*(k) \geq 0\}}{B}$$

where $\mathbf{I}\{\cdot\}$ is the indicator function whose value is one when the event is true and zero otherwise.

False Discovery Rate

False discovery rate control is a statistical method used in multiple hypothesis testing to correct for multiple comparisons. It controls the expected proportion of incorrectly rejected null hypotheses (type I errors) in a list of rejected hypotheses [56]. It is a less conservative comparison procedure with greater power than family-wise error rate (FWER) control, at a cost of increasing the likelihood of obtaining type I errors.

Let $H_1, \ldots, H_m$ be the null hypotheses being tested. The classification of m hypothesis tests can be shown as in the following table.

	Number declared nonsignificant	Number declared significant	Total
Number of true null hypotheses	U	V	m_0
Number of nontrue null hypotheses	T	S	$m - m_0$
Total	$m - R$	R	M

U is the number of true negatives, V is the number of false positives, T is the number of false negatives, S is the number of true positives, m_0 is the number of true null hypotheses, and $m - m_0$ is the number of false null hypotheses.

In m hypothesis tests of which m_0 are true null hypotheses, R is an observable random variable, and S, T, U, and $V$are all unobservable random variables. The false discovery rate is given by $E(V/(V+S)) = E(V/R)$ and one wants to keep this value below a threshold α.

The following procedure ensures that its expected value $E(V/(V+S))$ is less than a given α. This procedure is only valid when the m tests are independent. Let $H_1, \ldots, H_m$ be the null hypotheses and $P_1, \ldots, P_m$ their corresponding P-values. Arrange these values in increasing order and denote them by $P_{(1)}, \ldots, P_{(m)}$. For a given α, find the largest k such that $P_{(k)} \leq (k/m)\alpha$. Then reject (i.e., declare positive) all $H_{(i)}$ for $i = 1, \ldots, k$ [56].

The Pseudocode of MOFA

```
MOFA (E, B, z=2, min_size=5)
{
For SP = M/G1, G1, S, S/G2, or G2/M
{
Search all possible R's in SP
Record a particular R and the corresponding C(R,z) if #(C(R,z)) ≥ min_size
Sort the recorded R's according to their sizes and denote the sorting result as R̂'s
For i=1,2,...
{
If EC(C_i ≜ C(R̂_i, z)) ≤ EC(SP), iteratively eliminate the genes in C_i starting from the one with the most dissimilar expression profile, say h, until
EC(C̃_i ≜ C_i\{h}) > EC(SP)
Output the TF combination R̂_i and module M(R̂_i) ≜ C̃_i if EC(C̃_i) > EC(SP) and #(M(R̂_i)) ≥ min_size,
Mark all genes in M(R̂_i) so that they are not considered for R̂_j ⊂ R̂_i
}
}
}
```

Remark: In this study, we set the binding score $z=2$. This means that a TF is regarded as binding to a gene if (1) the P-value for the TF to bind the gene is ≤ 0.001 in the ChIP-chip data or (2) $0.001 < P \leq 0.01$ and the TF–gene pair have a temporal relationship. Moreover, we require that the number of genes in a module must be ≥ 5. This is the same as that in GRAM [21] for the performance comparison purpose.

References

1 Schena, M., Shalon, D. and Davis, R.W. and Brown, P.O. (1995) Quantitative monitoring of gene expression patterns with a complementary DNA microarray. *Science*, **270**, 467–470.

2 DeRisi, J.L., Iyer, V.R. and Brown, P.O. (1997) Exploring the metabolic and genetic control of gene expression on a genomic scale. *Science*, **278**, 680–686.

3 Ren, B., Robert, F., Wyrick, J.J., Aparicio, O., Jennings, E.G., Simon, I., Zeitlinger, J., Schreiber, J., Hannett, N., Kanin, E., Volkert, T.L., Wilson, C.J., Bell, S.P. and Young, R.A. (2000) Genome-wide location and function of DNA binding proteins. *Science*, **290**, 2306–2309.

4 Iyer, V.R., Horak, C.E., Scafe, C.S., Botstein, D., Snyder, M. and Brown, P.O. (2001) Genomic binding sites of the yeast cell-cycle transcription factors SBF and MBF. *Nature*, **409**, 533–538.

5 Cho, R.J., Campbell, M.J., Winzeler, E.A., Steinmetz, L., Conway, A., Wodicka, L., Wolfsberg, T.G., Gabrielian, A.E.,

Landsman, D., Lockhart, D.J. and Davis, R.W. (1998) A genome-wide transcriptional analysis of the mitotic cell cycle. *Molecular Cell*, **2**, 65–73.

6 Chu, S., DeRisi, J., Eisen, M., Mulholland, J., Botstein, D., Brown, P.O. and Herskowitz, I. (1998) The transcriptional program of sporulation in budding yeast. *Science*, **282**, 699–705.

7 Spellman, P.T., Sherlock, G., Zhang, M.Q., Iyer, V.R., Anders, K., Eisen, M.B., Brown, P.O., Botstein, D. and Futcher, B. (1998) Comprehensive identification of cell cycle-regulated genes of the yeast *Saccharomyces cerevisiae* by microarray hybridization. *Molecular Biology of the Cell*, **9**, 3273–3297.

8 Gasch, A.P., Spellman, P.T., Kao, C.M., Carmel-Harel, O., Eisen, M.B., Storz, G., Botstein, D. and Brown, P.O. (2000) Genomic expression programs in the response of yeast cells to environmental changes. *Molecular Biology of the Cell*, **11**, 4241–4257.

9 Causton, H.C., Ren, B., Koh, S.S., Harbison, C.T., Kanin, E., Jennings, E.G., Lee, T., True, H.L., Lander, E.S. and Young, R.A. (2001) Remodeling of yeast genome expression in response to environmental changes. *Molecular Biology of the Cell*, **12**, 323–337.

10 Eisen, M.B., Spellman, P.T., Brown, P.O. and Botstein, D. (1998) Cluster analysis and display of genome-wide expression patterns. *Proceedings of the National Academy of Sciences of the United States of America*, **95**, 14863–14868.

11 Tavazoie, S., Hughes, J.D., Campbell, M.J., Cho, R.J. and Church, G.M. (1999) Systematic determination of genetic network architecture. *Nature Genetics*, **22**, 281–285.

12 Ihmels, J., Friedlander, G., Bergmann, S., Sarig, O., Ziv, Y. and Barkai, N. (2002) Revealing modular organization in the yeast transcriptional network. *Nature Genetics*, **31**, 370–377.

13 Pilpel, Y., Sudarsanam, P. and Church, G.M. (2001) Identifying regulatory networks by combinatorial analysis of promoter elements. *Nature Genetics*, **29**, 153–159.

14 Liang, S., Fuhrman, S. and Somogyi, R. (1998) REVEAL a general reverse engineering algorithm for inference of genetic network architectures. *Pacific Symposium on Biocomputing*, **3**, 18–29.

15 Friedman, N., Linial, M., Nachman, I. and Pe'er, D. (2000) Using Bayesian networks to analyze expression data. *Journal of Computational Biology*, **7**, 601–620.

16 Segal, E., Shapira, M., Regev, A., Pe'er, D., Botstein, D., Koller, D. and Friedman, N. (2003) Module networks: identifying regulatory modules and their condition-specific regulators from gene expression data. *Nature Genetics*, **34**, 166–176.

17 Simon, I., Barnett, J., Hannett, N., Harbison, C.T., Rinaldi, N.J., Volkert, T.L., Wyrick, J.J., Zeitlinger, J., Gifford, D.K., Jaakkola, T.S. and Young, R.A. (2001) Serial regulation of transcriptional regulators in the yeast cell cycle. *Cell*, **106**, 697–708.

18 Lee, T.I., Rinaldi, N.J., Robert, F., Odom, D.T., Bar-Joseph, Z., Gerber, G.K., Hannett, N.M., Harbison, C.T., Thompson, C.M., Simon, I., Zeitlinger, J., Jennings, E.G., Murray, H.L., Gordon, D.B., Ren, B., Wyrick, J.J., Tagne, J.B., Volkert, T.L., Fraenkel, E., Gifford, D.K. and Young, R.A. (2002) Transcriptional regulatory networks in *Saccharomyces cerevisiae*. *Science*, **298**, 799–804.

19 Harbison, C.T., Gordon, D.B., Lee, T.I., Rinaldi, N.J., Macisaac, K.D., Danford, T.W., Hannett, N.M., Tagne, J.B., Reynolds, D.B., Yoo, J., Jennings, E.G., Zeitlinger, J., Pokholok, D.K., Kellis, M., Rolfe, P.A., Takusagawa, K.T., Lander, E.S., Gifford, D.K., Fraenkel, E. and Young, R.A. (2004) Transcriptional regulatory code of a eukaryotic genome. *Nature*, **431**, 99–104.

20 Banerjee, N. and Zhang, M.Q. (2003) Identifying cooperativity among transcription factors controlling the cell cycle in yeast. *Nucleic Acids Research*, **31**, 7024–7031.

21 Bar-Joseph, Z., Gerber, G.K., Lee, T.I., Rinaldi, N.J., Yoo, J.Y., Robert, F., Gordon, D.B., Fraenkel, E., Jaakkola, T.S., Young, R.A. and Gifford, D.K. (2003) Computational discovery of gene modules and regulatory networks. *Nature Biotechnology*, **21**, 1337–1342.

22 Kato, M., Hata, N., Banerjee, N., Futcher, B. and Zhang, M.Q. (2004) Identifying combinatorial regulation of transcription factors and binding motifs. *Genome Biology*, **5**, R56.

23 Mewes, H.W., Frishman, D., Guldener, U., Mannhaupt, G., Mayer, K., Mokrejs, M., Morgenstern, B., Munsterkotter, M., Rudd, S. and Weil, B. (2002) MIPS: a database for genomes and protein sequences. *Nucleic Acids Research*, **30**, 31–34.

24 Tsai, H.K., Lu, H.S. and Li, W.H. (2005) Statistical methods for identifying yeast cell cycle transcription factors. *Proceedings of the National Academy of Sciences*, **102**, 13532–13537.

25 Yang, Y.L., Suen, J., Brynildsen, M.P., Galbraith, S.J. and Liao, J.C. (2005) Inferring yeast cell cycle regulators and interactions using transcription factor activities. *BMC Genomics*, **6** (1), 90.

26 Miyake, T., Reese, J., Loch, C.M., Auble, D.T. and Li, R. (2004) Genome-wide analysis of ARS (autonomously replicating sequence) binding factor 1 (Abf1p)-mediated transcriptional regulation in *Saccharomyces cerevisiae*. *Journal of Biological Chemistry*, **279**, 34865–34872.

27 Loo, S., Laurenson, P., Foss, M., Dillin, A. and Rine, J. (1995) Roles of ABF1 NPL3, and YCL54 in silencing in *Saccharomyces cerevisiae*. *Genetics*, **141**, 889–902.

28 Kumar, R., Reynolds, D.M., Shevchenko, A., Shevchenko, A., Goldstone, S.D. and Dalton, S. (2000) Forkhead transcription factors Fkh1p and Fkh2p, collaborate with Mcm1p to control transcription required for M-phase. *Current Biology*, **10**, 896–906.

29 Hollenhorst, P.C., Bose, M.E., Mielke, M.R., Muller, U. and Fox, C.A. (2000) Forkhead genes in transcriptional silencing cell morphology and the cell cycle. Overlapping and distinct functions for FKH1 and FKH2 in *Saccharomyces cerevisiae*. *Genetics*, **154**, 1533–1548.

30 Packham, E.A., Graham, I.R. and Chambers, A. (1996) The multifunctional transcription factors Abf1p Rap1p and Reb1p are required for full transcriptional activation of the chromosomal PGK gene in *Saccharomyces cerevisiae*. *Molecular and General Genetics*, **250**, 348–356.

31 Morrow, B.E., Johnson, S.P. and Warner, J.R. (1989) Proteins that bind to the yeast rDNA enhancer. *Journal of Biological Chemistry*, **264**, 9061–9068.

32 Carmen, A.A. and Holland, M.J. (1994) The upstream repression sequence from the yeast enolase gene ENO1 is a complex regulatory element that binds multiple trans-acting factors including REB1. *Journal of Biological Chemistry*, **269**, 9790–9797.

33 Wang, K.L. and Warner, J.R. (1998) Positive and negative autoregulation of REB1 transcription in *Saccharomyces cerevisiae*. *Molecular and Cellular Biology*, **18**, 4368–4376.

34 McBride, H.J., Yu, Y. and Stillman, D.J. (1999) Distinct regions of the Swi5 and Ace2 transcription factors are required for specific gene activation. *Journal of Biological Chemistry*, **274**, 21029–21036.

35 Doolin, M.T., Johnson, A.L., Johnston, L.H. and Butler, G. (2001) Overlapping and distinct roles of the duplicated yeast transcription factors Ace2p and Swi5p. *Molecular Microbiology*, **40**, 422–432.

36 Loy, C.J., Lydall, D. and Surana, U. (1999) NDD1 a high-dosage suppressor of cdc28-1N, is essential for expression of a subset of late-S-phase-specific genes in

Saccharomyces cerevisiae. Molecular and Cellular Biology, **19**, 3312–3327.

37 Darieva, Z., Pic-Taylor, A., Boros, J., Spanos, A., Geymonat, M., Reece, R.J., Sedgwick, S.G., Sharrocks, A.D. and Morgan, B.A. (2003) Cell cycle-regulated transcription through the FHA domain of Fkh2p and the coactivator Ndd1p. *Current Biology*, **13**, 740–745.

38 Ho, Y., Costanzo, M., Moore, L., Kobayashi, R. and Andrews, B.J. (1999) Regulation of transcription at the *Saccharomyces cerevisiae* start transition by Stb1 a Swi6-binding protein. *Molecular and Cellular Biology*, **19**, 5267–5278.

39 Costanzo, M., Schub, O. and Andrews, B. (2003) G1 transcription factors are differentially regulated in *Saccharomyces cerevisiae* by the Swi6-binding protein Stb1. *Molecular and Cellular Biology*, **23**, 5064–5077.

40 Partridge, J.F., Mikesell, G.E. and Breeden, L.L. (1997) Cell cycle-dependent transcription of CLN1 involves swi4 binding to MCB-like elements. *Journal of Biological Chemistry*, **272**, 9071–9077.

41 Reardon, B.J., Winters, R.S., Gordon, D. and Winter, E. (1993) A peptide motif that recognizes A.T tracts in DNA. *Proceedings of the National Academy of Sciences of the United States of America*, **90**, 11327–11331.

42 Pramila, T., Miles, S., GuhaThakurta, D., Jemiolo, D. and Breeden, L.L. (2002) Conserved homeodomain proteins interact with MADS box protein Mcm1 to restrict ECB-dependent transcription to the M/G1 phase of the cell cycle. *Genes and Development*, **16**, 3034–3045.

43 Olson, K.A., Nelson, C., Tai, G., Hung, W., Yong, C., Astell, C. and Sadowski, I. (2000) Two regulators of Ste12p inhibit pheromone-responsive transcription by separate mechanisms. *Molecular and Cellular Biology*, **20**, 4199–4209.

44 Fung, H., Bennett, R.A.O. and Demple, B. (2001) Key role of a downstream specificity protein 1 site in cell cycle-regulated transcription of the AP endonuclease gene APE1/APEX in NIH3T3 cells. *Journal of Biological Chemistry*, **276**, 42011–42017.

45 Aerne, B.L., Johnson, A.L., Toyn, J.H. and Johnston, L.H. (1998) Swi5 controls a novel wave of cyclin synthesis in late mitosis. *Molecular Biology of the Cell*, **9**, 945–956.

46 McInerny, C.J., Partridge, J.F., Mikesell, G.E., Creemer, D.P. and Breeden, L.L. (1997) A novel Mcm1-dependent element in the SWI4 CLN3, CDC6, and CDC47 promoters activates M/G1-specific transcription. *Genes and Development*, **11**, 1277–1288.

47 Koch, C. and Nasmyth, K. (1994) Cell cycle regulated transcription in yeast. *Current Opinion in Cell Biology*, **6**, 451–459.

48 Futcher, B. (2002) Transcriptional regulatory networks and the yeast cell cycle. *Current Opinion in Cell Biology*, **14**, 676–683.

49 Dimova, D., Nackerdien, Z., Furgeson, S., Eguchi, S. and Osley, M.A. (1999) A role for transcriptional repressors in targeting the yeast Swi/Snf complex. *Molecular Cell*, **4**, 75–83.

50 Koranda, M., Schleiffer, A., Endler, L. and Ammerer, G. (2000) Forkhead-like transcription factors recruit Ndd1 to the chromatin of G2/M-specific promoters. *Nature*, **406**, 94–98.

51 Barral, Y., Parra, M., Bidlingmaier, S. and Snyder, M. (1999) Nim1-related kinases coordinate cell cycle progression with the organization of the peripheral cytoskeleton in yeast. *Genes and Development*, **13**, 176–187.

52 Yu, H., Luscombe, N.M., Qian, J. and Gerstein, M. (2003) Genomic analysis of gene expression relationships in transcriptional regulatory networks. *Trends in Genetics*, **19**, 422–427.

53 Liping, J. and Tan, K.L. (2005) Identifying time-lagged gene clusters using gene expression data. *Bioinformatics*, **21**, 509–516.

54 Zhou, X.J., Kao, M.C., Huang, H., Wong, A., Nunez-Iglesias, J., Primig, M., Aparicio, O.M., Finch, C.E., Morgan, T.E. and Wong, W.H.Z. (2005) Functional annotation and network reconstruction through cross-platform integration of microarray data. *Nature Biotechnology*, **23**, 238–243.

55 Efron, B. and Tibshirani, R.J. (1998) *An Introduction to the Bootstrap*, Chapman-Hall/CRC.

56 Benjamini, Y. and Hochberg, Y. (1995) Controlling the false discovery rate: a practical and powerful approach to multiple testing. *Journal of the Royal Statistical Society Series B*, **57**, 289–300.

13
Pathway-Based Methods for Analyzing Microarray Data

Herbert Pang, Inyoung Kim, and Hongyu Zhao

13.1
Introduction

Microarray technologies have flourished over the past decade, and they are widely used in functional genomics studies. Numerous statistical approaches have been developed to analyze microarray data. However, most methods are single-gene based and have not considered the dependencies among genes. In recent years, researchers have started looking at sets of genes instead of one gene at a time. For example, Harris *et al.* [3] considered gene groupings based on gene ontology (GO). Pathways are sets of genes that serve a particular cellular or physiological function. By taking a pathway-based approach in analyzing microarray data, we can make more biological sense of the results. Ranking pathways according to their ability to predict the phenotype of interest, for example, cancer status, can help researchers focus on a more relevant set of genes for further analysis. They can also facilitate researchers to generate biological hypotheses and design experiments to validate them. Moreover, pathway-based analysis can identify pathways with subtle changes of expression levels that may not be detected at the single-gene level [4].

Several research groups have considered enrichment-based approaches to identify significant pathways using various methods, such as Kolmogorov–Smirnov statistic [5], maxmean statistic [6], and random set methods [7]. These pathway-based methods look for sets of genes that are enriched for overexpressed or underexpressed genes. Other model-based methods include global test [8] and ANCOVA [9]. Global test is based on a score test for a random effects model, while the ANCOVA approach is based on a multivariate linear model.

In this chapter, we describe the methods developed by us that are used to analyze gene expression data on the pathway level. The rationale of these methods stems from the fact that the better a pathway is at predicting the phenotype of interest, the more relevant this pathway is to the phenotype. We first describe the method based on Random Forests, a tree-based classification and regression method developed by the late Leo Breiman and Adele Cutler ([10], http://oz.berkeley.edu/users/breiman/Using_random_forests_v4.0.pdf). Researchers have shown that Random Forests

Analysis of Microarray Data: A Network-Based Approach. Edited by F. Emmert-Streib and M. Dehmer

ISBN: 978-3-527-31822-3

has comparable or better performance than other machine learning methods in genomics and proteomics studies (e.g. [11–13]). Secondly, we describe a Bayesian method using hierarchical mixed models for analyzing high-dimensional pathway-based gene expression data [2]. The dependencies among genes within a given pathway are captured by using different kernels.

In addition to identifying significant pathways, both methods can identify important genes that are useful in predicting the outcome of interest, leading to potentially informative biomarkers for disease diagnoses, treatment prognosis, and drug development. These methods may also identify outlying observations in a pathway-based context.

This chapter is organized as follows. In Section 13.2, we describe the Random Forests approach and the Bayesian regression model approach for pathway analysis. In Section 13.3, we apply our methods to several microarray data sets. Section 13.4 concludes this chapter.

13.2 Methods

13.2.1 Random Forests Based Approach

Random Forests [14] is an improved classification and regression tree (CART). In CART, when a classification or regression tree is built, all variables are screened for splitting at each step. The splitting variable and threshold are chosen to achieve the biggest reduction in some node impurity measure, for example, the Gini index. The tree is grown until all the subjects within each terminal node belong to the same class for a discrete outcome or until all the subjects within each terminal node is less than a certain preset number for a continuous outcome. To avoid overfitting the data, the tree is then pruned by cutting back some branches.

13.2.1.1 Random Forests Classification

Random Forests differs from the original CART method by growing many trees and thus the name "forest." Trees within the forest differ owing to two factors. First, instead of considering all the variables for splitting at each node, only a subset of variables are chosen for finding the best split. In the classification case, the default number of variables chosen for every split is $\sqrt{m}$, where m is the total number of variables. Secondly, every tree is built on a bootstrap sample of the original data. In Random Forests, no pruning is involved. See Algorithm 1 in Appendix 13. A for more details.

In the pathway-based analysis of gene expression data, the split variables are the genes within a pathway. The gene expression level of a selected gene is used to split samples in a node. For example, if the split level for a selected gene is 100, all the subjects with values less than or equal to 100 will be in one child node and the other subjects will be in the other child node.

Random Forests uses an internal learning and training set method to assess the prediction accuracy. This results from the use of bootstrap samples. Because not all the original observations are sampled, the out-of-bag (OOB) samples, approximately one third of the original observations, can be used to estimate the prediction accuracy based on the built tree. On the basis of the tree built with the bootstrap sample, each OOB observation can be assigned to a leaf, the terminal node of a tree, according to the values of predictor variables of this observation. For a particular tree, the predictions are made only for the OOB data to get an estimate of the classification accuracy for this tree. The overall prediction accuracy or error is then calculated by averaging over all the trees built. Important pathways can be ranked by the overall prediction accuracies thus obtained. The smaller the OOB error rate for a pathway, the more likely this pathway is associated with the phenotype.

Apart from the OOB error rate, Random Forests also provides a proximity measure to relate different subjects. To obtain this measure, the entire training set and OOB data is run down an unpruned tree. For example, when two observations result in the same terminal node, then the proximity between them is increased by 1. This results in a symmetric proximity matrix of size $n \times n$ and only the off-diagonal values are relevant. The normalized proximity measure is the total count divided by the number of trees. The proximity value between a pair of individuals quantifies whether they behave similarly or not in the context of a given pathway. From these pairwise proximity measures, we can identify outliers that have small proximity measure to most other observations in the same class. An outlier measure for case i in class c can be defined as

$$\mathrm{out}(i) = \frac{1}{\sum_{k \in \mathrm{class}\ c} [\mathrm{proximity}(i,k)]^2}. \tag{13.1}$$

Therefore, if the out(i) value is relatively large for an observation, it is a possible outlier. The normalized measure is defined as $|\mathrm{out}(i) - \mathrm{median}(c)|/\mathrm{deviation}(c)$, where $\mathrm{median}(c) = \mathrm{median}[\mathrm{out}(j)]$ for all j in class c, $\mathrm{deviation}(c) = \sum_{i \in \mathrm{class}\ c} [|\mathrm{out}(i) - \mathrm{median}(c)|]/[\mathrm{no.\ in\ class}\ c]$. A value above 10 is usually considered an outlier [10].

It is often interesting to know which of the features are important in classification. There are two commonly used measures of importance in Random Forests, the mean decrease in accuracy and the Gini index. They give possible ways to quantify which genes are more informative, that is, contribute more to the prediction accuracy than other genes. Because the decrease in the Gini index is not as reliable as the marginal decrease in prediction accuracy [10], we focus on the latter in the following discussion.

To obtain the importance measure for gene G in a particular pathway, the Random Forests algorithm permutes the expression values of gene G in the OOB individuals and uses these permuted expression values to get classifications. Because the permuted values of gene G in the OOB individuals and the outcome of interest become independent after the permutation, a large reduction in prediction accuracy is expected for an informative gene. Margin is defined as (% of votes for true class) − argmax$_c$ (% of votes for other class c). Let us consider the case when there are two classes. The margin is more formally defined as

$$\begin{aligned}\mathrm{mr}_j &= \left(\frac{\sum_{i=1}^{\#\mathrm{trees}} \mathbf{1}_{\{V_j(i)=\mathrm{T\,class}\}}\mathbf{1}_{\{\mathrm{OOB}_j(i)=\mathrm{T}\}}}{\sum_{i=1}^{\#\mathrm{trees}} \mathbf{1}_{\{\mathrm{OOB}_j(i)=\mathrm{T}\}}}\right) \\ &\quad -\left(1-\frac{\sum_{i=1}^{\#\mathrm{trees}} \mathbf{1}_{\{V_j(i)=\mathrm{T\,class}\}}\mathbf{1}_{\{\mathrm{OOB}_j(i)=\mathrm{T}\}}}{\sum_{i=1}^{\#\mathrm{trees}} \mathbf{1}_{\{\mathrm{OOB}_j(i)=\mathrm{T}\}}}\right) \\ &= 2\left(\frac{\sum_{i=1}^{\#\mathrm{trees}} \mathbf{1}_{\{V_j(i)=\mathrm{T\,class}\}}\mathbf{1}_{\{\mathrm{OOB}_j(i)=\mathrm{T}\}}}{\sum_{i=1}^{\#\mathrm{trees}} \mathbf{1}_{\{\mathrm{OOB}_j(i)=\mathrm{T}\}}}\right)-1 \end{aligned} \tag{13.2}$$

where $\mathbf{1}$ is the indicator function, $\mathrm{OOB}_j(i)$ = True when the jth individual is OOB in the ith tree, and $V_j(i)$ = "True class" when a correct vote is given to individual j for the ith tree. The larger the margin, the more confident the Random Forests prediction is. The mean decrease in accuracy MDA(G) for gene G is the reduction of the margin across all individuals divided by the number of cases when gene G is randomly permuted:

$$\mathrm{MDA(G)} = \frac{1}{n_1+n_2}\sum_{j=1}^{n_1+n_2}(\mathrm{mr}_j - \mathrm{mr}_{j(\mathrm{G})}). \tag{13.3}$$

13.2.1.2 Random Forests Regression

Random Forests regression works in a similar manner as Random Forests classification except that the random subset size for splitting is now $m/3$ and the best split is chosen with a different criterion. See Algorithm 2 in Appendix 13.A for more details.

Just like the case for classification, we can rank the pathways based on a Random Forests regression run. The percent variance explained (% Var Explained), which is defined as 1 − (mean squared error)/[variance (response)], is used in place of the OOB error rate in classification. The mean squared error (MSE) is the sum of squared residuals divided by the sample size, that is, $\sum_{\mathrm{subjects}}$ (predicted value − observed value)$^2/n$. It indicates how well the set of gene expressions in a particular pathway can explain the variation in the response of interest. The percent variance explained is viewed as a pseudo r-square. It is pseudo in the sense that it can take negative values when the ratio of MSE and the variance of the response is greater than 1. A pathway that helps explain the variation in the response variable is an indication of the informativeness of the pathway relative to other pathways with smaller percent variance explained. In the regression case, one cannot output the outlier plot under the current R implementation, as the definition of the outlying measure for different classes no longer applies. Despite this, the proximity measure between two observations can be obtained in the same way as classification.

As in the classification case, two importance measures can be obtained for regression as well. They are the mean decrease in accuracy and mean decrease in MSE. The set of informative genes found using these measures may be important biomarkers for the response of interest.

13.2.2
Regression Model Based Approach

13.2.2.1 Bayesian Hierarchical Model

A Bayesian hierarchical model relating a continuous outcome Y with clinical covariates and gene expression levels can be defined as

$$\begin{aligned}[y|\mathbf{x}, r(\mathbf{z})] &\sim N\{\mathrm{x}^{\mathrm{T}}\beta + r(\mathbf{z}), \sigma^2\};\\ r(\mathbf{z}) &\sim \mathrm{GP}\{0, \tau K(\mathbf{z}, \mathbf{z}')\},\end{aligned}$$

where n is the sample size, $\mathbf{X}$ is a $q \times n$ matrix, $\mathbf{Z}$ is a $p \times n$ matrix, q is the number of covariates (small), p is the number of genes in each pathway (can be larger than n), $\mathbf{r} = \{r(\mathbf{z}_1), r(\mathbf{z}_2), \ldots, r(\mathbf{z}_n)\}$ has a Gaussian process distribution, $\mathbf{K}$ is the $n \times n$ matrix with the ijth component $K(\mathbf{z}_i, \mathbf{z}_j)$, $i, j = 1, \ldots, n$. We consider the following kernels for $K(\cdot, \cdot)$:

- The dth order polynomial kernel, $K(\mathbf{z}, \mathbf{z}') = (\mathbf{z}\mathbf{z}')^d$, $d = 1, 2$.
- Gaussian kernel, $K(\mathbf{z}, \mathbf{z}') = \exp(||\mathbf{z}-\mathbf{z}'||^2/\rho)$, where $||\mathbf{z}-\mathbf{z}'|| = \sum_{k=1}^{p}(z_k - z'_k)^2$.
- Neural network kernel, $K(\mathbf{z}, \mathbf{z}') = \tanh(\mathbf{z} \cdot \mathbf{z}')$.

These kernels impose stronger pathway effect similarities between individuals with similar expression profiles, and they differ in the exact form of such dependence.

Let θ denote the collection of the parameters. Under the Gaussian kernel, $\theta = (\beta, \sigma^2, r, \tau, \rho)$ and for other kernels, $\theta = (\beta, \sigma^2, r, \tau)$. For a given set of prior distributions of the parameters, we can obtain their posterior distributions on the basis of the observed data. In the following discussion, we consider the following prior distributions: $\beta \sim N(0, \Sigma_\beta)$, $\sigma^2 \sim \mathrm{IG}(a_\sigma{}^2, b_\sigma{}^2)$, $\mathbf{r} \sim N(0, \tau\mathbf{K})$, $\tau \sim \mathrm{IG}(a_\tau, b_\tau)$, and $\rho \sim$ Unit $[C_1, C_2]$, where $\Sigma_\beta = \sigma_\beta^2 I$ with a fixed large value σ_β^2, a fixed positive small value C_1, and large value C_2 with the assumption $C_2 \sim O\{E(||\mathbf{z}-\mathbf{z}'||^2)\}$.

13.2.2.2 A Bayesian MCMC Approach

The conditional distribution of (y, $\mathbf{r}$) and the marginal distribution of y are

$$\begin{aligned}[y|x, \mathbf{r}, \theta][\mathbf{r}|\theta] \sim \frac{1}{(2\pi)^n(\sigma^2)^{n/2}\tau^{n/2}|K|} &\exp\left\{-\frac{(\mathrm{y}-\mathrm{x}^{\mathrm{T}}\beta-\mathbf{r})^{\mathrm{T}}(\mathrm{y}-\mathrm{x}^{\mathrm{T}}\beta-\mathbf{r})}{2\sigma^2}\right\}\\ &\exp\left\{-\frac{r^{\mathrm{T}}K^{-1}r}{2\tau}\right\};\end{aligned}$$

$$[y|x, \theta] \sim N(X^{\mathrm{T}}\beta, \Sigma).$$

The complete conditional distributions of θ under the Gaussian kernel are proportional to

$$[\beta|\text{rest}] \sim N\{\sigma^{-2}(\Sigma_\beta^{-1}+\sigma^{-2}\mathbf{X}\mathbf{X}^T)^{-1}\mathbf{X}(y-r),(\Sigma_\beta^{-1}+\sigma^{-2}\mathbf{X}\mathbf{X}^T)^{-1}\};$$

$$[\sigma^2|\text{rest}] \sim \text{IG}\left[\frac{n}{2}+a_{\sigma^2},\frac{\{y-(\mathbf{X}^T\beta+\mathbf{r})\}^T\{y-(\mathbf{X}^T\beta+\mathbf{r})\}}{2}+b_{\sigma^2}\right];$$

$$[\mathbf{r}|\text{rest}] \sim N[\sigma^{-2}\{(\tau\mathbf{K})^{-1}+\sigma^{-2}I\}^{-1}(y-\mathbf{X}^T\beta),\{(\tau\mathbf{K})^{-1}+\sigma^{-2}I\}^{-1}];$$

$$[\tau|\text{rest}] \sim \text{IG}\left[\frac{n}{2}+a_\tau,\frac{\mathbf{r}^T\mathbf{K}(\rho)^{-1}\mathbf{r}}{2}+b_\tau\right];$$

$$[\rho|\text{rest}] \propto \frac{1}{(2\pi)^{n/2}|\tau\mathbf{K}(\rho)|^{1/2}}\exp\left\{-\frac{\mathbf{r}^T\mathbf{K}(\rho)^{-1}\mathbf{r}}{2\tau}\right\}.$$

Samples from the joint posterior distribution [θ|Y, **X**, **Z**] are drawn using the Markov Chain Monte Carlo (MCMC) techniques (see Kim *et al.* [2]).

13.3
Real Data Analysis

In this section, we demonstrate the usefulness of the pathway-based analyses through their applications to five microarray data sets summarized in Table 13.1.

The `Breast` data set has three tumor classes of 49 breast cancer patients [15]. Tumors are classified into luminal, basal, and apocrine classes. The `canine` data set was generated from investigative toxicology studies designed to identify the molecular pathogenesis of drug-induced vascular injury in coronary arteries of dogs treated with adenosine receptor agonist CI-947 [16]. Each animal was assigned a lesion score ranging from 0 to 21 on the basis of the number of lesions observed. In addition, a recovery group consisting of three treated (10 mg/kg) and four control dogs was also included in the data set. The canine genes were mapped to human orthologs for pathway analysis. The corresponding human orthologs for dogs were generated by matching the gene sequences using BLAST. See Pang *et al.* [1] for more details. `Diabetes` [4] data set contains 17 samples with normal glucose tolerance and 18 samples with type II diabetes mellitus. A comparison of female and male

Table 13.1 Data sets used in this chapter.

Data set	Reference	*n*	Genes	Response type	Classes	Continuous
Breast	[15]	49	22215	Three tumor types	3	No
Canine	[16]	31	12473	Lesion score	n/a	Yes
Diabetes	[4]	35	22215	Glucose level	n/a	Yes
Gender	Broad Institute	32	22283	Male versus female cells	2	No
p53	Broad Institute	50	12625	p53+/p53 mutant	2	No

lymphoblastoid cell lines was the subject of interest for the `Gender` data set. The `p53` data set contains 50 NCI-60 cell lines with patients carrying mutations in the p53 gene.

13.3.1 Pathways and Gene Sets

A total of 441 gene sets plus 3 gene set files from Broad Institute were used for the analysis. Some of these pathways are wired diagrams of genes and molecules from KEGG and BioCarta. Others are manually curated. The distribution is as follows:

- A total of 129 pathways were taken from KEGG, a pathway database with the majority responsible for metabolism, degradation, and biosynthesis. There are also a few signal-processing pathways and others related to human diseases. KEGG has a wide variety of organisms ranging from human, mouse, and rat to bacteria like *Escherichia coli* with direct links to the genes or gene products involved in the biochemical reactions. The pathways are divided into the following categories: (1) metabolism, (2) genetic information processing, (3) environmental information processing, (4) cellular processes, (5) human diseases, and (6) drug development.
- We considered 312 BioCarta pathways, which is more than twice the number of pathways compared with KEGG. Most of these pathways are related to signal transduction for human and mouse with a smaller set of metabolic pathways.
- Two gene sets are manually curated "leukocyte adhesion" and "eicosenoid metabolism."
- Three gene set files were obtained from the Broad Institute (http://www.broad.mit.edu/).

13.3.2 Data Analysis Using Random Forests

In this section, we consider Random Forests based approach and explain how to choose the number of trees and detect outliers based on classification and regression.

13.3.2.1 Canine Data Set

Classification We used 441 gene sets whose sizes vary between 3 and 151 for Random Forests classification and sought to distinguish between dogs with lesion and those without. Although lesion score is continuous, we can also perform classification on dogs with or without lesion, that is, positive lesion score or zero score, respectively.

Choosing the Number of Trees As the number of trees may affect the classification error, we ran 500 (default), 5000, 50 000, and 100 000 trees. We found that the classification error became stable at around 5000 for most pathways. Some examples of OOB error rate plot versus the number of trees are given in Figure 13.1. These are

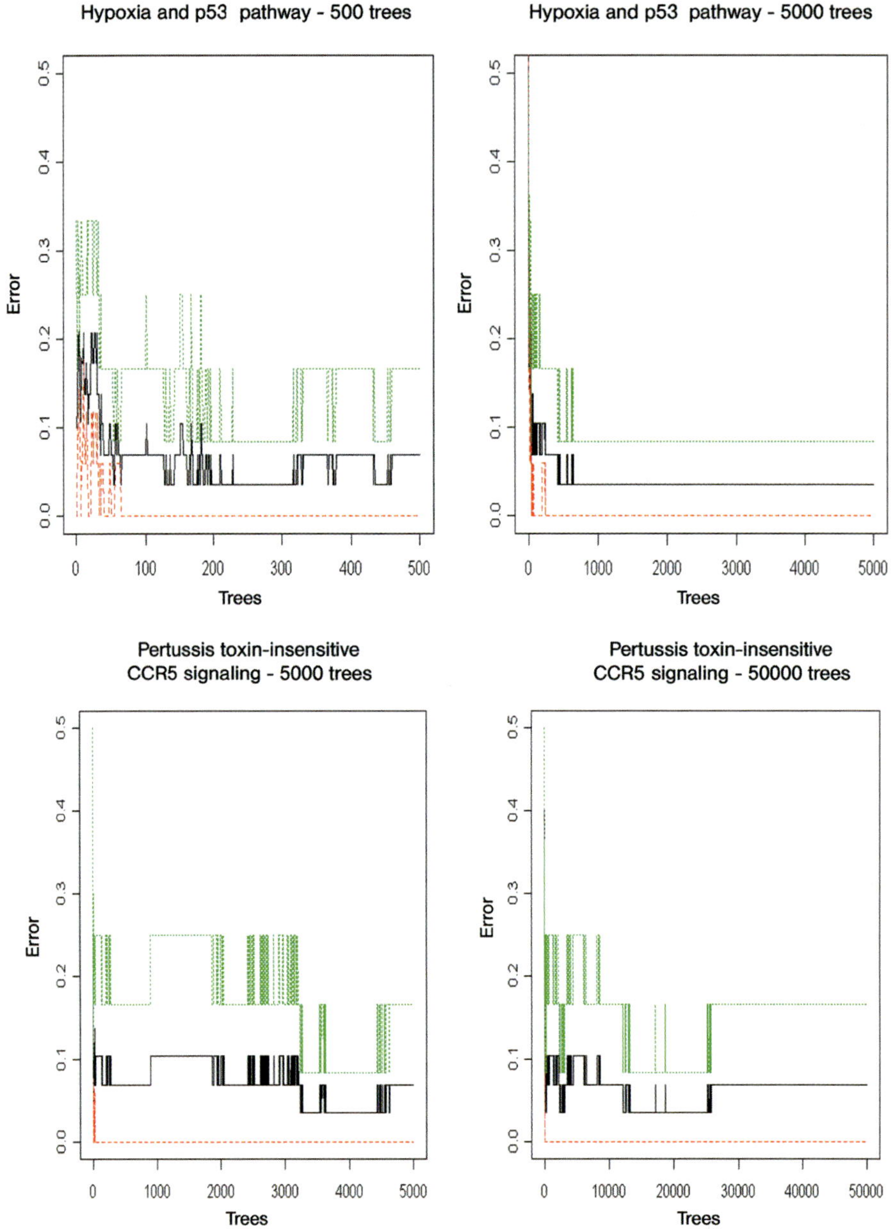

Figure 13.1 Error rate plots versus the number of trees for different pathways.

plots of the estimated error rate against the number of trees for the hypoxia and p53 in the cardiovascular system and the pertussis toxin insensitive CCR5 signaling pathway. With about 1000 trees, we can see that the OOB error has stabilized. For some pathways, there are more fluctuations, such as the pertussis toxin insensitive CCR5 signaling pathway. Therefore, we decided to use 50 000 trees. Solid lines (black) represent the actual classification error rate, and dotted lines (green) and dash lines (red) are the upper and lower confidence bound, respectively.

Detection of Outliers As outliers can highly affect the accuracy of classification, we should remove them before identifying important pathways using the OOB error rate. The best OOB error rate achieved was 9.68% using all the 31 dogs in the study, which means the best we could do is to have 28 dogs correctly classified (see Table 13.2). The top pathways from the table were chosen for outlier detections. From the outlier plot (Figure 13.2), we found that animals #10 and #19 were frequently outliers. Animal #10 was an outlier in the circadian rhythm, low density protein (LDL) pathway, and Msp-Ron receptor signaling pathway, while animal #19 was an outlier in leukocyte adhesion, LDL pathway, and Msp-Ron receptor signaling pathway. No animal was considered as an outlier in the CTCF pathway. The detection of animals #10 and #19 as outliers seems biologically plausible as those two dogs had lesion score either higher/lower from the other three dogs sampled at the same time point and dosage group, and thus they were classified in the other lesion group. Using 50 000 trees and removing animals #10 and #19, the classification error rates dropped from 12.9 to 3.4% for both LDL pathway and Msp-Ron receptor signaling pathway and even more for hypoxia and p53 in the cardiovascular system pathway. Our analysis indicates that these pathways were the most informative in classifying between the lesion and nonlesion groups (Table 13.3).

Table 13.2 Canine data set: pathways ranked by OOB error rates of less than or equal to 12.9% with outliers.

Pathway	OOB	Number of genes
CTCF: first multivalent nuclear factor	9.68%	18
Circadian rhythm	9.68%	9
Leukocyte adhesion	9.68%	59
Ascorbate and aldarate metabolism	12.90%	9
CDK regulation of DNA replication	12.90%	10
Circadian rhythms (BioCarta)	12.90%	4
Double stranded RNA induced gene expression	12.90%	6
Eukaryotic protein translation	12.90%	18
Granzyme A mediated apoptosis pathway	12.90%	5
Internal ribosome entry pathway	12.90%	11
LDL pathway during atherogenesis	12.90%	4
Monocyte and its surface molecules	12.90%	7
Msp-Ron receptor signaling pathway	12.90%	6
PTEN dependent cell cycle arrest and apoptosis	12.90%	17
Regulation of eIF4e and p70 S6 Kinase	12.90%	27

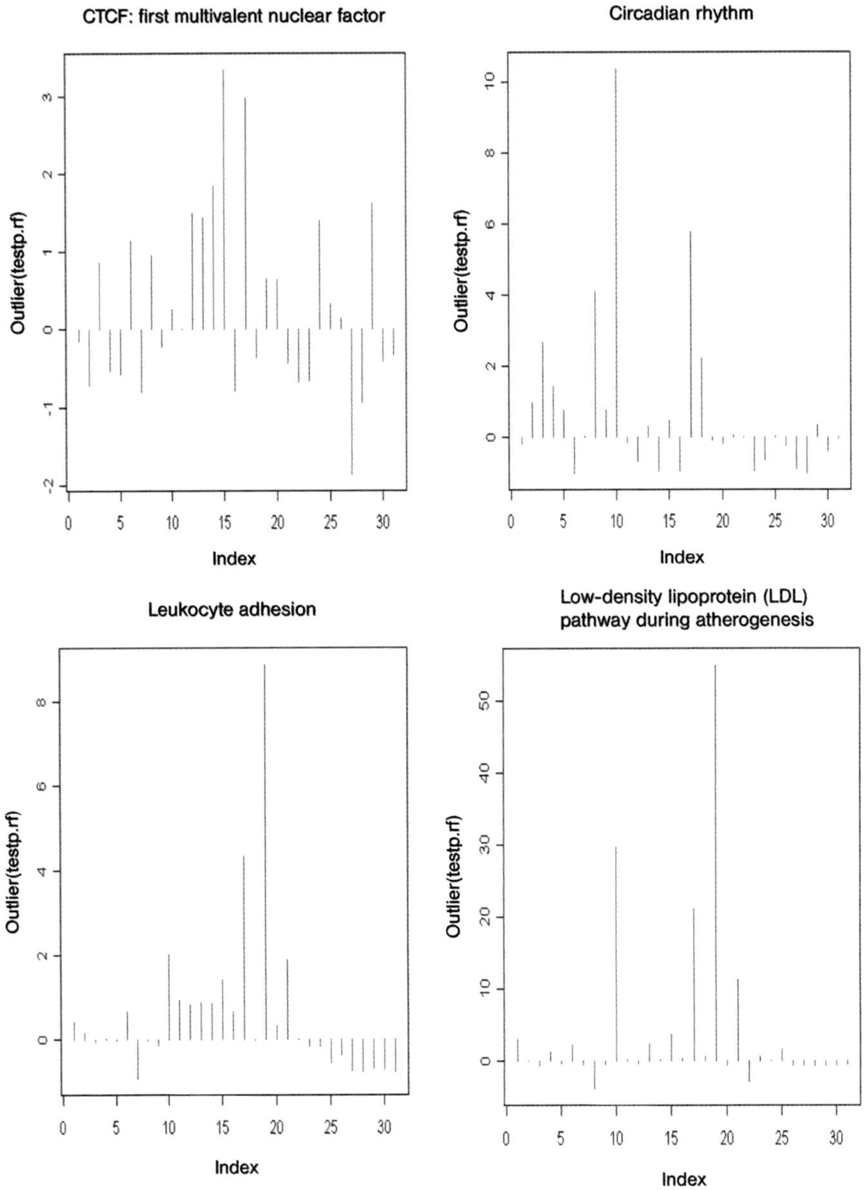

Figure 13.2 These four pathways are the top ranked pathways with all the 31 dogs and have the lowest OOB error rate.

Table 13.3 Canine data set: pathways ranked by OOB error rates of less than or equal to 6.9% with outliers.

Pathway	OOB	Number of genes
LDL pathway during atherogenesis	3.40%	4
Msp-Ron receptor signaling pathway	3.40%	6
Hypoxia and p53 in the cardiovascular system	3.40%	14
Role of Ran in mitotic spindle regulation	6.90%	8
Granzyme A mediated apoptosis pathway	6.90%	5
CTCF first multivalent nuclear factor	6.90%	18
CDK regulation of DNA replication	6.90%	10
Sumoylation by RanBP2 reg. trans. repression	6.90%	8
Circadian rhythm	6.90%	9
Nitric oxide signaling pathway	6.90%	14
Aminosugars metabolism	6.90%	19
Wnt signaling pathway	6.90%	68
Aminoacyl–tRNA biosynthesis	6.90%	19
Pertussis toxin insensitive CCR5 signaling in macrophage	6.90%	15
Leukocyte adhesion	6.90%	59

These analyses yield biologically relevant results. As shown in Table 13.3, the most significant pathways in the data set include (1) LDL pathway during atherogenesis, (2) Msp-Ron receptor signaling pathway, and (3) hypoxia and p53 in the cardiovascular system. The first two pathways are associated with inflammatory response and the last one is related to hypoxic stress and induction of apoptosis [17]. These are clearly related to vascular injury. For example, pathway 1 contains merely four genes but they have high classification power. It is a collection of signaling low-density lipoprotein, colony stimulating factor, cytokines, and macrophage in arterial blood and the neighboring liver and endothelial cells. Pathway 2 is a collection of macrophage-stimulating proteins in liver that act through the transmembrane receptor kinase to play a part in inflammation and tissue injury, and pathway 3 is a collection of genes induced by hypoxic stress causing p53 apoptotic activity in a different way compared to p53 induction by DNA damage [18].

In the set of pathways with OOB error 6.9%, a few of them are related to apoptosis, immune response, cardiovascular function, and sheer stress. They are the granzyme A mediated apoptosis pathway (apoptosis, immune surveillance), circadian rhythm (cardiovascular function), and nitric oxide (NO) pathway (NO plays an important role in shear stress) [19]. Apart from identifying which pathways are important in classification, we can also look at which animals are anomalies. For every pathway, we can find out which dogs are misclassified. For the pathway of interest, the researcher can make use of the proximity measure that can inform us which subjects/dogs are more similar to each other. This is particularly useful in seeing, for example, how a set of recovery dogs are similar to other dogs measured at different time points. The multidimensional scaling (MDS) plot (Figure 13.3) projects a high-dimensional proximity matrix to a 2D surface showing similarities

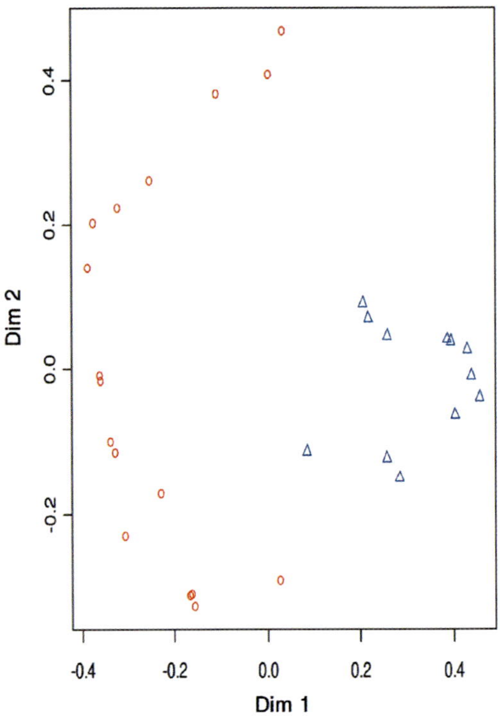

Figure 13.3 An MDS plot for hypoxia and p53 in the cardiovascular system pathway.

among dogs and their respective classes. In this plot, the lesion (triangles) and the nonlesion (circles) groups are well separated.

In the top ranked pathways by OOB error, we can look at the importance measure of each gene in these pathways. For the top pathways with OOB error $\leq$6.9%, Figure 13.4 plots all the genes that have positive "decrease in accuracy" importance measure. In Figure 13.4, the darker a genes is shaded, the more informative it is. Leukocyte adhesion, LDL pathway, and MSP and pertussis toxin pathway, all share the component CCL2, chemokine (C–C motif) ligand 2 also known as MCP-1, which is related to vascular injury and one of the strongly upregulated chemokines in endothelial microvascular injury [20–22]. A gene like CSNK1A1 is shared between circadian rhythm, hypoxia and p53 pathway, and WNT signaling pathway [23]. PRKCA2 and PRKCA were also found in three pathways, WNT signaling, NO signaling, and pertussis toxin pathways. It is interesting to note that PRKCA, protein kinase Ca, forms a complex with active RhoA, which is involved in a signaling pathway in the endothelial cell and contributes to vascular inflammation [24]. Among all the top pathways with OOB error $\leq$6.9%, there are three pathways that do not have any overlaps with the others: aminosugars metabolism, aminoacyl–tRNA biosynthesis, and granzyme A mediated apoptosis pathway. As seen from Figures 13.4 and 13.5, there are 11 important genes that are colored red and connected to at least two pathways, while 24 important genes are unique to one particular pathway.

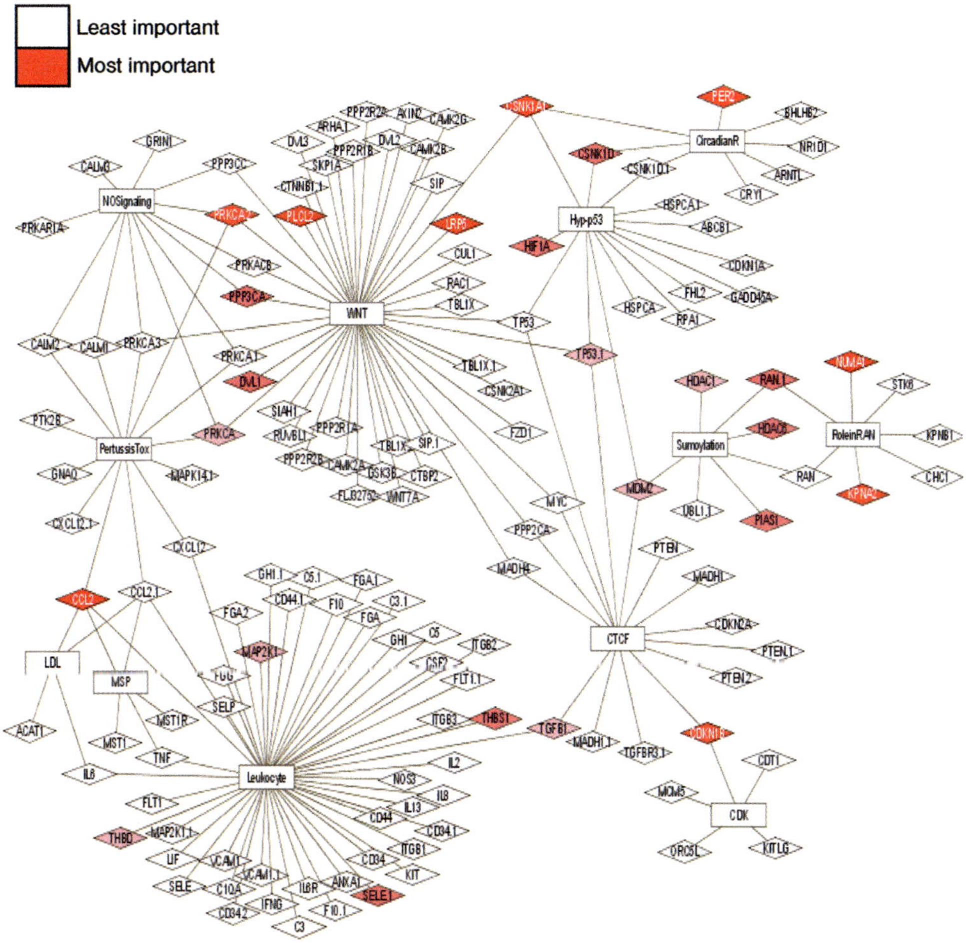

Figure 13.4 Overlapping genes with positive importance measure of top ranked pathways for classification (pathways are rectangular-shaped nodes).

As there is a close tie between cytokines and inflammatory diseases, the only important gene in the IFN alpha signaling pathway, STAT2, was picked up probably because the STAT family members are phosphorylated in response to cytokines and growth factors using Entrez Gene [25]. SELE, E-selectin, plays the role of an adhesion ligand for recruiting leukocytes during inflammation [26] and is a good biomaker for endothelial function and inflammatory response [27,28]. DNCL1 or DYNLL1, which mediates DNA damage induced p53 nuclear accumulation, is no doubt closely tied to the hypoxia and p53 in cardiovascular system pathway [29].

Regression In the regression setting, we used pathway-based Random Forests regression to predict the lesion score. In this analysis, we also excluded dogs

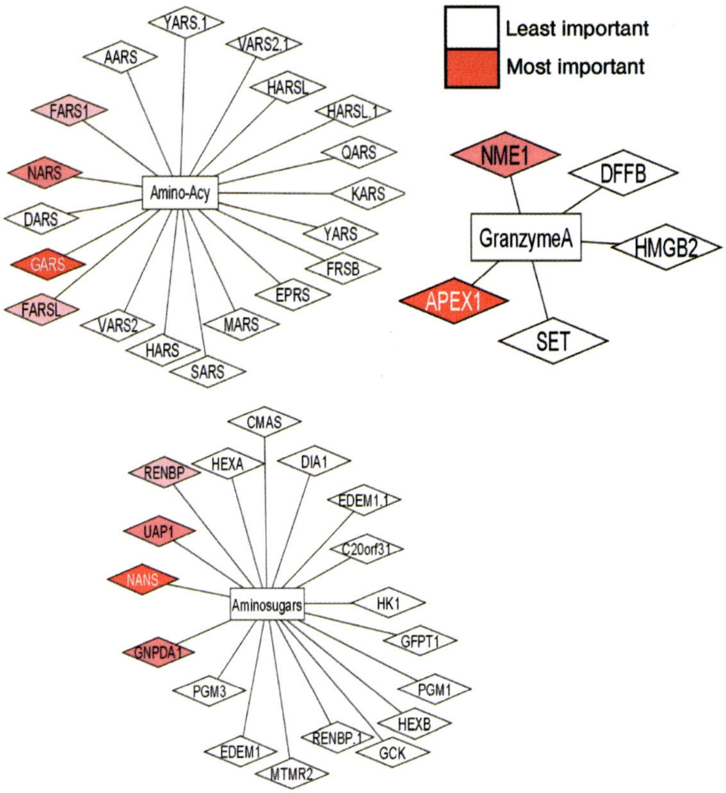

Figure 13.5 Other top ranked pathways for classification with positive importance measure genes (pathways are rectangular-shaped nodes).

#10 and #19 that were found as outliers in our Random Forests classification analysis. The same 441 gene sets were used. As in the classification case, the number of trees could affect the MSE, which is used to calculate the percent variance explained. Once again we saw that the MSE became stable after the number of trees reached 50 000 for most pathways. Some examples of the MSE plot versus the number of trees are shown in Figure 13.6. For example, for pathway hypoxia and p53 in the cardiovascular system, the MSE plot against the number of trees no longer fluctuates after around 1000 trees. A similar pattern can be seen for other pathways, such as the pertussis toxin insensitive CCR5 signaling pathway, which require more trees to get a stable MSE. Therefore, 50 000 trees were used for the regression case.

The top pathways ranked by percent variance explained are listed in Table 13.4. Out of 441, 144 were found to have a positive percent variance explained. It was not surprising to see that the majority of the top 15 pathways found by classification were also highly ranked in the regression case. Our analysis shows that the following pathways had percent variance explained of >20%: (1) one-carbon pool by folate, (2)

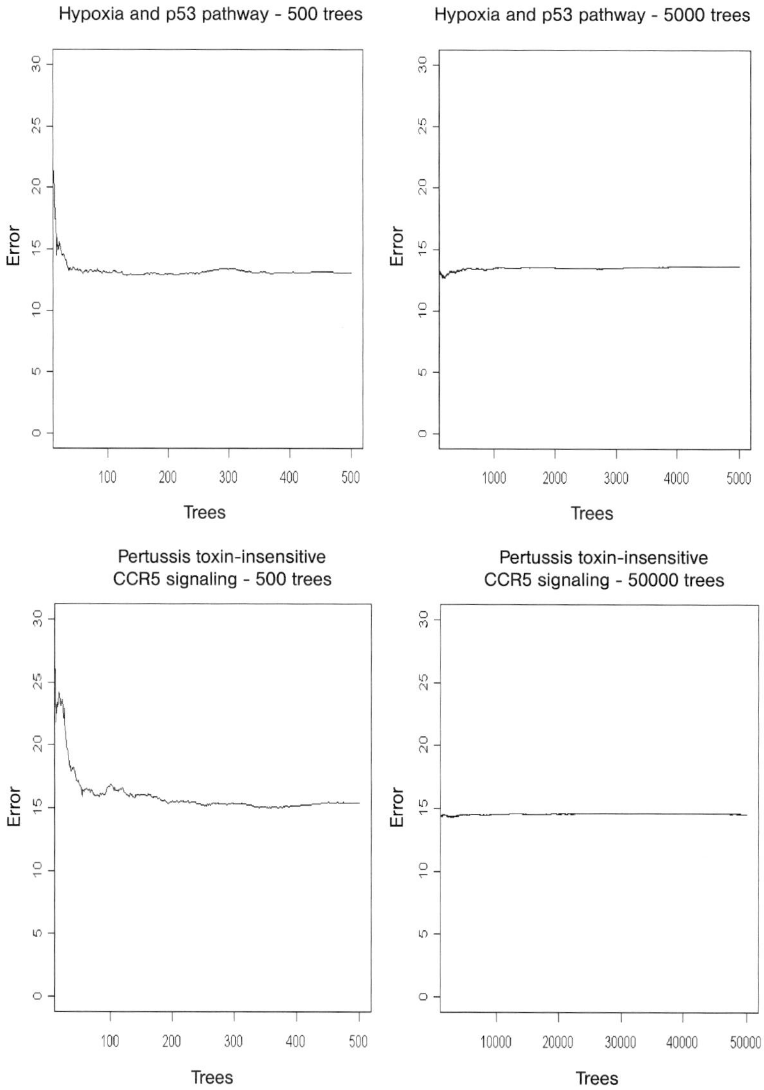

Figure 13.6 MSE plot versus the number of trees for Random Forests regression.

antisense pathway, (3) aminoacyl–tRNA biosynthesis, (4) hypoxia and p53 in the cardiovascular system, (5) adhesion molecules on lymphocyte, and (6) neutrophil and its surface molecules. All were useful in explaining the variation in the lesion scores among the animals.

One-carbon pool by folate, the antisense pathway, and aminoacyl–tRNA biosynthesis may not seem directly related to inflammatory or the response to vascular pathology, but as we shall see in the next paragraph, some of the genes within them explained why they were picked up and biologically plausible. As described in the

Table 13.4 Canine data set: pathways ranked by percent variance explained of more than 16.0% without outliers.

Pathway	% Var explained	Number of genes
One-carbon pool by folate	26.4	11
Antisense pathway	26.0	4
Aminoacyl–tRNA biosynthesis	22.0	19
Hypoxia and p53 in the cardiovascular system	21.4	14
Adhesion molecules on lymphocyte	21.4	6
Neutrophil and its surface molecules	21.0	5
Monocyte and its surface molecules	19.7	7
LIS1 gene in neuronal migration and development	19.1	8
IFN alpha signaling pathway	18.4	1
Role of Ran in mitotic spindle regulation	18.0	8
Wnt signaling pathway	17.2	68
Granzyme A mediated apoptosis pathway	17.0	5
Circadian rhythm	16.6	9
Dicer pathway	16.1	2
Pertussis toxin insensitive CCR5 signaling in macrophage	16.0	15
Msp-Ron receptor signaling pathway	16.0	6

previous section, hypoxia and p53 in the cardiovascular system is a pathway related to hypoxic stress and apoptosis. As for (5) adhesion molecules on lymphocyte and (6) neutrophil and its surface molecules, these two pathways differ only by one gene and consist of molecules interacting with endothelial cells that are responsible for sending inflammatory signals and triggering immune response.

Although an outlier plot cannot be plotted, an MDS plot can be used to see which of the animals are closer to each other in a particular pathway. As for this pathway, we can see three distinct groups when the proximity matrix is projected onto a 2D plane (see Figure 13.7). When we applied the importance measures to identify important genes, the majority of the genes were found to have references that support their relation with vascular injury. The most informative gene GARS in aminoacyl–tRNA pathway has been shown to be a target of autoantibodies in the human autoimmune diseases by Entrez Gene [25]. For the antisense pathway, it contains three genes, ADAR, SFPQ, and MATR3, which are particularly important in explaining the variance among the lesion scores. ADAR, adenosine deminase, is a marker for T cell activation and is related to the production of neutrophils, which have close interaction with endothelial cells [30]. MTHFD2, methylenetetrahydrofolate dehydrogenase, within the one-carbon pool by folate pathway, has been shown to be upregulated in vascular endothelial cells when treated with a chemical [31].

Genes that are found to be informative in the same pathway in both classification and regression have some common genes. Unlike the case for classification, besides the three pathways, (1) monocyte and its surface molecules, (2) adhesion molecules on lymphocyte, and (3) neutrophil and its surface molecules that actually differ in not more than a couple of genes, the top 15 pathways do not have many top

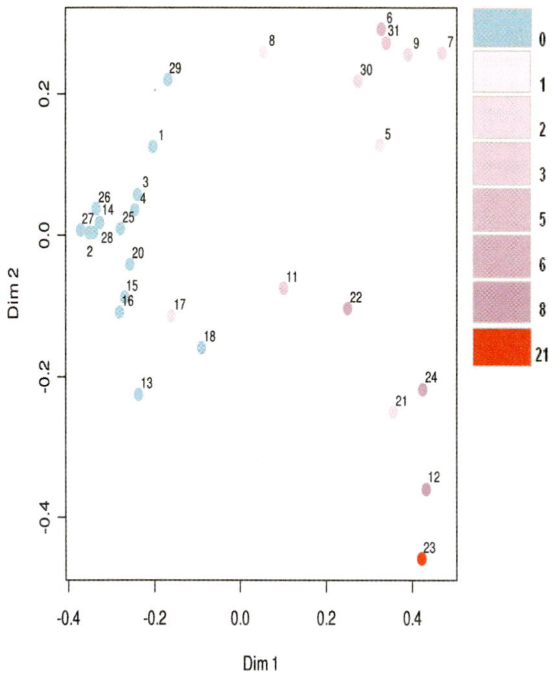

Figure 13.7 MDS plot for the proximity matrix of the 29 canine cases for "one-carbon pool by folate" pathway.

important genes in common. Two other examples are PRKCA2, which is shared between the WNT signaling pathway and the pertussis toxin insensitive CCR5 signaling pathway, and the CCL2 for MSP-RON and pertussis toxin insensitive CCR5 signaling pathway.

13.3.2.2 **Breast Cancer Data Set**

We next turn to a study of breast cancer with three tumor subtypes (Table 13.5). We used Random Forests classification to see which pathways is informative in classifying the patients into the correct subgroups. The CARM1 and regulation of estrogen receptor seem reasonable because the estrogen receptor plays an important role in normal breast development and is expressed in common cancer subtypes [32]. In addition, the regulation of BAD phosphorylation pathway contains a widely expressed BCL-2 family member and has been studied for epidermal growth factor receptor (EGFR) targeted therapy for breast cancer [33]. A recent paper in *Cancer Research* [34] has identified GATA3 as a potential prognostic marker for breast cancer. Researchers have demonstrated that tumors have abnormal bioenergetics. Subjects with cancer can show a systematic loss of energy involving the interaction of tumor glycolysis and gluconeogenesis [35]. This may be the reason why glycolysis and gluconeogenesis pathway also has a low error rate.

Table 13.5 Breast data set: pathways ranked by OOB error rates of less than or equal to 12.3%.

Pathway	OOB	Number of genes
CARM1 and regulation of the estrogen receptor	8.16	24
Fructose and mannose metabolism	8.16	39
GATA3 participate in activating the Th2 cytokine gene expression	8.16	21
Glycolysis – gluconeogenesis	8.16	68
Regulation of BAD phosphorylation	8.16	24
Carbon fixation	10.20	25
Downregulated of MTA-3 in ER-negative breast tumors	10.20	19
Estrogen-responsive protein Efp controls cell cycle and breast tumors growth	10.20	10
JAK-stat signaling pathway	10.20	71
mCalpain and friends in cell motility	10.20	33
Pentose phosphate pathway	10.20	22
Valine leucine and isoleucine degradation	10.20	46
Limonene and pinene degradation	12.24	26
Map kinase inactivation of SMRT corepressor	12.24	16
Phenylalanine metabolism	12.24	22
Role of ERBB2 in signal transduction and oncology	12.24	29
Sulfur metabolism	12.24	9
Trefoil factors initiate mucosal healing	12.24	33
Tryptophan metabolism	12.24	94
Tyrosine metabolism	12.24	37

13.3.2.3 **Diabetes Data Set**

This data set contains the glucose level as the continuous outcome. We used Random Forests regression to see which pathways can predict an individual's glucose level (see Table 13.6). It is not surprising to see that the OXPHOS HG-U133A and oxidative phosphorylation gene sets came in fifth and sixth, respectively, in ranking, which is consistent with previous findings in Mootha *et al.* [4]. ATP synthesis, a complex within oxidative phosphorylation pathway, is ranked fourth. ATP is necessary for the formation of aminoacyl–tRNA complex, amino acid + ATP $\rightarrow$ aminoacyl–tRNA + AMP + inorganic pyrophosphate [36]. Therefore, it is quite reasonable to observe aminoacyl–tRNA biosynthesis at the top of the list, as all these pathways are somehow interconnected. Aminoacyl–tRNA biosynthesis gene set contains small metabolic and biosynthesis modules involving tRNAs. It also has genes from the highly ranked alanine and aspartate metabolism pathway. Reductive carboxylate cycle pathway is connected to alanine and aspartate metabolism, and both are highly ranked in the table. The less well annotated gene sets "c11 U133 probes" contain a bunch of genes ALP, CEACAM1, EIF2B1, HIRA, MME, TAP1, and TSN that are related to diabetes and glucose. For example, diabetic rats show an increased activity of ALP [37], and elevations in ALP level have been shown to correlate with insulin resistance [38]. Negative mutation in mice of CEACAM1 results in glucose intolerance [39]. Also, in SHP-1-deficient cells or mice, tyrosine phosphorylation of CEACAM1, a regulator of insulin clearance, is markedly

Table 13.6 Diabetes data set: pathways ranked by percent variance explained of more than 8.5%.

Pathway	% Var explained	Number of genes
MAP00970 Aminoacyl–tRNA biosynthesis	20.4	27
Reductive carboxylate cycle (CO_2 fixation)	17.7	12
MAP00020 citrate cycle TCA cycle	17.4	26
ATP synthesis	16.2	49
OXPHOS HG-U133A probes	16.0	121
Oxidative phosphorylation	14.1	113
Alanine and aspartate metabolism	14.0	18
MAP00720 reductive carboxylate cycle CO_2 fixation	13.9	11
c11U133 probes	13.1	131
c17 U133 probes	13.1	116
TCA HG-U133A probes	12.5	30
c7 U133 probes	12.4	218
MAP00061 fatty acid biosynthesis path 1	10.8	4
MAP00190 oxidative phosphorylation	10.5	58
c16 U133 probes	9.6	138
Aminoacyl–tRNA biosynthesis	9.4	27
Glyoxylate and dicarboxylate metabolism	9.1	13
Integrin mediated cell adhesion	8.9	91
MAP00630 glyoxylate and dicarboxylate metabolism	8.7	13
MAP00252 alanine and aspartate metabolism	8.6	21

increased [40,41]. There is a reduced expression of TAP1 for diabetic mice [42] and the activity of EIF2B1 is markedly lower in nonexercised diabetic mice compared to nondiabetic mice after exercise [43]. "c17 U133 probes" consists of genes related to human insulin signaling, for example, ARF6, CAP1, IKBKB, MAP3K4, MAP2K6, and SGK [44], and several glucose-related genes.

13.3.2.4 Comparison with Other Machine Learning Approaches

We compared Random Forests with other classification tools using their implementations in *R*. The classifiers chosen were (1) linear discriminant analysis `lda`, (2) neural network, which consists of an input-trained hidden layer of nonlinear functions, `nnet` with three units in the hidden layer, (3) bagging, bootstrap aggregation of trees also developed by Breiman, `bagging` in the `ipred` package, (4) support vector machine (SVM), a method to fit maximum-margin hyperplanes, `svm` with C-classification type and the radial basis kernel, (5) *k*-nearest neighbor `ipredknn` by considering both 1 and 3 neighbors, and (6) naive Bayes, which calculates the conditional a posteriori probability from predictor variables based on Bayes' rule, `naiveBayes`.

To compare different machine learning approaches, we used a fivefold cross validation (CV) and average 632plus [45] to assess the error rate. Average 632plus is a bootstrapping method that uses a weighted average of the resubstitution error and the "leave-one-out" bootstrap error. The weight is determined by the amount of overfitting. Cross validation is described in Appendix 13.A.

Table 13.7 Average fivefold cross-validation error rates of the top 10 pathways with lowest error rates from each of the seven classification methods.

Method	Canine data set	Gender data set	p53 data set
Random Forests	5.52%	18.8%	16.4%
SVM	3.79%	26.3%	19.2%
LDA	6.90%	20.9%	20.2%
k-NN ($k=1$)	7.24%	14.4%	21.2%
k-NN ($k=3$)	7.24%	15.3%	20.4%
Naive Bayes	8.28%	26.3%	23.2%
Neural network	24.8%	31.9%	27.4%

Table 13.8 Average 632+ error rates of the top 10 pathways with lowest error rates from each of the eight classification methods.

Method	Canine data set	Gender data set	p53 data set
Random Forests	5.52%	20.6%	16.0%
SVM	6.38%	27.9%	20.1%
Bagging[a)]	24.5%	27.5%	13.2%
LDA	9.04%	24.3%	24.9%
k-NN ($k=1$)	9.22%	13.7%	19.9%
k-NN ($k=3$)	9.61%	15.2%	21.9%
Naive Bayes	9.93%	31.4%	22.0%
Neural network	33.0%	NA	NA

a) OOB error rates.

Since there is no consensus as to whether k-fold cross validation or average 632plus is better of the two [46–48], we decided to use both. To compare the classification methods, we considered the three data sets and two methods to estimate the error rates. From Tables 13.7 and 13.8, we see that Random Forests performs similar to SVM for the canine data set and better than SVM in the other two data sets. Random Forests only falls to k nearest neighbor classifiers for the gender data set and bagging for the p53 data set. Surprisingly, naive approaches, such as linear discriminant analysis and naive Bayes classifiers, performed well compared to other methods. Random Forests has its positivities over the other classification methods, as it can handle more than two classes as demonstrated in the applications section in addition to its ability to perform regression. For other applications of our pathway-based Random Forests approach, see Pang *et al.* [1].

13.3.3 Data Analysis Using Bayesian Approach

We apply the Bayesian approach to a microarray expression data on type II diabetes [4]. In this data, there were 17 samples with normal glucose tolerance and

18 samples with type II diabetes mellitus. There were 22 283 genes and 277 pathways that consisted of 128 KEGG pathways and 149 curated pathways. The 149 curated pathways were constructed from the literature by Mootha and colleagues.

In our study, we let Y be the log transformation of glucose level, X be the age, and Z be the $p \times n$ gene expression levels within each pathway, where n is 35 and p is the number of genes in a specific pathway and varies across pathways ($4 \leq p \leq 200$). The goal of our study is to identify pathway effects on the glucose level related to diabetes after adjusting for the age effect and also to rank genes within a significant pathway. With the Gaussian kernel, the null hypothesis H_0: $\{r(\mathbf{z})$ is a point mass at zero$\} \cup \{r(\mathbf{z})$ has a constant covariance matrix as a function of $\mathbf{z}\}$ is equivalent to $H_0 : \tau/\rho = 0$ or $\rho = 0$. For the polynomial and neural network kernels, we test H_0: $\tau = 0$. We adopted the following simulation procedure to assess statistical significance: we first estimate τ/ρ, ρ, τ using the observed data by fitting the semiparametric model. We then calculate the residuals $\hat{e}_{0i}$ from $y_i = x_i\beta + e_{0i}$ and permute the residuals as $\hat{e}^*_{0i}$. The simulated outcomes are created as $y_i^* = x_i^T\hat{\beta} + e^*_{0i}$. On the basis of y^*, $\mathbf{x}$, and $\mathbf{z}$, we fit the semiparametric model and then estimate $\hat{\tau}^*/\hat{\rho}^*$, $\hat{\rho}^*$, and $\hat{\tau}^*$. For each pathway, we calculate the proportion of times that $\hat{\tau}^*/\hat{\rho}^* > \hat{\tau}/\hat{\rho}$, $\hat{\rho}^* > \hat{\rho}$, $\hat{\tau}^* > \hat{\tau}$, where $\hat{\tau}/\hat{\rho}$, $\hat{\rho}$, $\hat{\tau}$ are the estimated values from the observed data by fitting the Bayesian hierarchical model. The P-value is estimated by this percentage, and significant pathways can be selected on the basis of these estimated P-values.

On the basis of our Bayesian approach, we found 58, 57, 55, and 23 significant pathways using the polynomial with $d = 1, 2$, for Gaussian and neural network kernels, respectively. We draw the Venn diagrams on the overlap patterns among the pathways identified by different kernels in Figure 13.8. One of the seven pathways, common

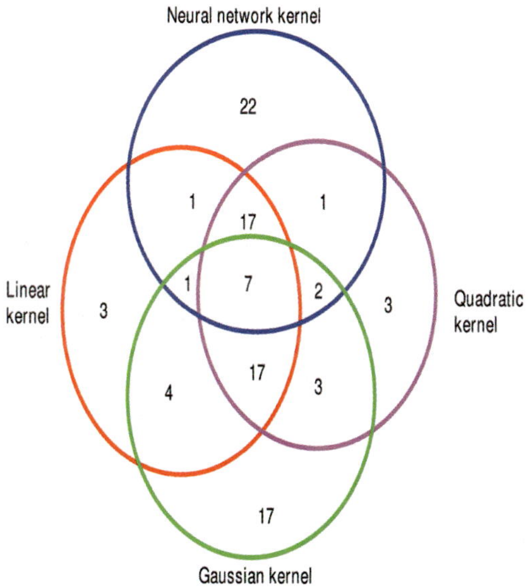

Figure 13.8 Overlap pathways among top 50 pathways selected by Bayesian MCMC approach with different kernels.

Table 13.9 Leave-one-out cross-validation prediction rates using several prediction methods.

Method	Pathway (no. of genes)						
	P4 (18)	P36 (116)	P133 (58)	P140 (21)	P209 (21)	P229 (113)	P254 (25)
LDA	0.5714	0.7142	0.4857	0.5714	0.6286	0.4857	0.5714
SVML	0.6000	0.6571	0.4857	0.7428	0.6000	0.6285	0.6000
SVMP	0.5714	0.5714	0.4857	0.5714	0.6000	0.5714	0.6000
KNN	0.6857	0.6000	0.5142	0.6571	0.6286	0.4517	0.5714
NNet	0.5428	0.6857	0.4000	0.6285	0.6571	0.5428	0.7143

LDA = linear discrimination analysis; SVML = support vector machine with linear kernel; SVMP = support vector machine with polynomial kernel; KNN = k nearest neighbor classifier ($k = 1$); NNet = neural network classifier.

among all kernels, is "oxidative phosphorylation," named pathway 229, known to be related to diabetes [4,49,50]. It is a process of cellular respiration in humans (or in general eukaryotes). This pathway contains coregulated genes across different tissues and is related to insulin/glucose disposal. It consists of ATP synthesis, a pathway involved in energy transfer. Pathway 36, named "$c17_U133$_probes," is the most significant pathway. However, its function is not well annotated. Both of these pathways were in the top 10 ranked pathways in the Random Forests regression analysis described in Section 13.3.2.3.

We further calculated the prediction accuracy based on leave-one-out cross validation for each of these seven pathways. The prediction accuracy is given in Table 13.9 on the basis of several classification methods including discriminant analysis, support vector machine, k-nearest neighbor classifier, and neural network classifier. We found that "$c17_U133$_probes" gives the best prediction accuracy among all the pathways considered. Using "$c17_U133$_probes," we estimated $\hat{\mathbf{r}}$ and then plotted $\hat{\mathbf{r}}$ versus 35 samples consisting of 17 normal and 18 type II diabetes samples for "oxidative phosphorylation" and "$c17_U133$_probes." In Figure 13.9, it can be seen that the two groups (17 normal and 18 diabetes) are discriminated very well. We note that there are no common genes between these two pathways even if two plots are very similar to each other.

To rank important genes within a pathway, we can fit the following model

$$Y = \mathbf{X}^T\beta + \mathbf{r}(\mathbf{Z}_{-(g)}) + e,$$

where $\mathbf{Z}_{-(g)}$ is the gene expression data for a pathway without gene g.

We bootstrap samples B times by keeping the two groups (normal and diabetes), clinical covariates, and the order of the genes. The bootstrap approach was used because each group has small size, and asymptotic results may not be reliable. We estimate $\tau^b_{(-g)}, \rho^b_{(-g)}$, and $(\tau/\rho)^b_{(-g)}$, $b = 1,\ldots, B$ using the bootstrap samples, and calculate the absolute differences $|\tau^b_{(-g)} - \tau|, |\rho^b_{(-g)} - \rho|$, and $|(\frac{\tau}{\rho})^b_{(-g)} - (\frac{\tau}{\rho})|$. The top genes with large mean absolute differences, $1/B\sum_{b=1}^{B}|\tau^b_{(-g)} - \tau|$, are considered important in a pathway. The top 20 genes are shown in Table 13.10. Genes MEF2C,

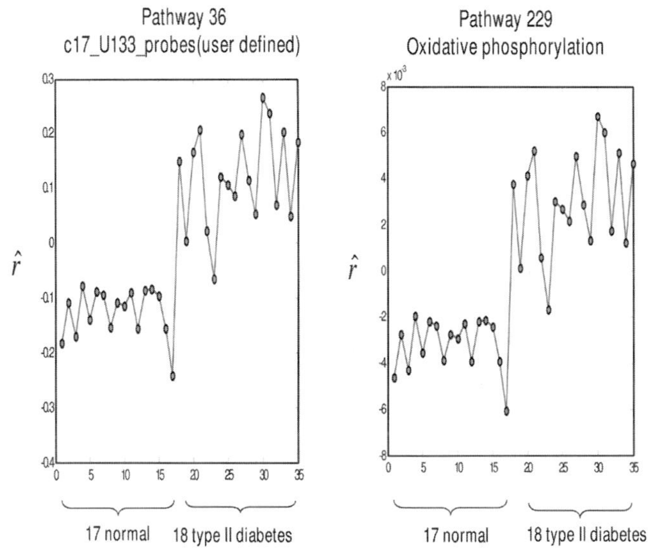

Figure 13.9 Plot between $\hat{r}(Z)$ and 35 samples with 17 normal and 18 diabetes using pathway 36 and pathway 229.

Table 13.10 Top genes ranked by the mean of absolute values of the difference $1/B\sum_{b=1}^{B}\left|\hat{\tau}-\hat{\tau}_{(-g)}^{b}\right|$.

Gene ID	Gene name	$1/B\sum_{b=1}^{B}\left\|\hat{\tau}_{(-g)}^{b}-\hat{\tau}\right\|$	$1/B\sum_{b=1}^{B}\left\|\hat{\rho}_{(-g)}^{b}-\hat{\rho}\right\|$	$1/B\sum_{b=1}^{B}\left\|(\hat{\tau}/\hat{\rho})_{(-g)^{b}}-(\tau/\rho)\right\|$
202305_s_at	FEZ2	13.7285	1.2464	58.5893
203138_at	HAT1	6.7900	0.6121	58.4642
215000_s_at	FEZ2	6.1778	1.1308	64.1316
215489_x_at	HOMER3	5.9030	0.7320	61.5046
220966_x_at	ARPC5L	3.6316	1.1279	66.3728
202569_s_at	MARK3	2.7196	1.1278	67.1803
208903_at	RPS28	2.6710	0.3435	61.7000
207968_s_at	MEF2C	2.0194	1.1255	67.7968
217150_s_at	NF2	1.4378	0.0131	37.3021
217852_s_at	SBNO1	1.3753	0.2019	62.5390
207741_x_a	TPS1	1.1823	0.0669	51.2447
205963_s_at	DNAJA3	1.1215	0.2183	64.2275
38918_at	SOX13	0.9110	0.1569	63.4542
217161_x_at	AGC1	0.8227	0.0512	52.5880
217780_at	PTD008	0.8119	0.1889	65.0175
208904_s_at	RPS28	0.7673	0.2034	65.5661
213798_s_at	CAP1	0.6807	0.1830	65.5825
202804_at	ABCC1	0.6268	0.1871	65.9588
215148_s_at	RAB3D	0.5916	0.1680	65.7499
218188_s_at	RPS5P1	0.5128	0.0235	46.0110

NR4A1, SOX1, and TPS1 are known to be related to glucose [51–54], and CAP1 is known to be related to human insulin signaling [44].

13.4 Conclusions and Discussion

In this chapter, we have described two pathway-based approaches for analyzing microarray data; one based on Random Forests and the other based on a Bayesian semiparametric model. The first approach is helpful in identifying pathways that are useful in classifying a categorical outcome or predicting a continuous measure and isolating important genes. Under the Bayesian hierarchical framework, we estimate the nonparametric pathway effect of gene expressions in relation to a continuous clinical outcome. The modeling of the dependencies among genes within a pathway is done through different kernels: Gaussian, polynomial, and neural network kernels. These two methods have a number of distinct features to help both biologists and bioinformaticians understand biological systems better. Random Forests provides us with a proximity matrix defining distance among subjects and allows the detection of outliers. These can identify subjects who behave differently from others in a particular pathway, and such detections are often useful in biomedical studies. The proximity matrix can also be visualized on a multidimensional scaling plot. Moreover, we can make use of the importance measure from Random Forests to identify genes that are more informative in pathway-based prediction. The bootstrap procedure described serves a similar purpose in our Bayesian model. This can help identify important genes of interest for future studies.

As we know, genes are not independent of each other but rather work in pathways. Pathway-based analysis allows a researcher to make better use of the biological information available by combining microarray data with information from externally available databases such as KEGG and BioCarta. We have demonstrated that the proposed pathway-based methods may be useful in identifying genes that are good classifiers or predictors. Finding important pathways allows researchers to focus on a subset of genes that explain the response of interest. We were able to rank pathways and genes that are related to various diseases and pathologies of interest for a few microarray data sets. In comparison with other machine learning methods, Random Forests remains within the top two in terms of prediction accuracy for the data sets examined here and simulated data sets shown in Pang *et al.* [1]. In summary, our simulation studies have demonstrated that Random Forests has comparable or better performance than SVM. As for our second approach, simulation study suggests our Bayesian MCMC approach has reasonable type I error, power, and mean square error.

Although it is difficult to conclude which pathway method is superior, we have shown the benefits of our pathway-based methods through several microarray data sets analyzed here. In reality, many clinical outcomes are measured as continuous measures, such as glucose level or white blood cells count. The Random Forests and Bayesian approaches described are one of the firsts that make use of a continuous measure for ranking pathways. Finally, we note that our approaches consider each pathway separately. It is known that pathways are not independent of each other, in fact pathways are often defined subjectively because some share the same genes and they interact with each other to serve a particular function. Future developments are needed to thoroughly consider interactions among pathways.

Appendix 13.A Algorithm of Random Forests

Algorithm 1: Random Forests classification

```
  0.  Input: pathway with m genes and n subjects
  1.  For I = 1 to t do
  2.  Sample with replacement from the original data
  3.  Approximately 1/3 of the original data not in 2. are called
Out-of-Bag (OOB) samples
  4.  Construct a classification tree using the data from 2
        4a.  Randomly choose √m genes to do split
        4b.  Perform the split using the gene that gives the
biggest decrease in Gini impurity criterion
        4c.  Repeat step 4a/b until each terminal node contains
subject(s) of the same class
  5.  Run the OOB subjects from 3. down the tree
  6.  Subject i from 3. gets a vote for the class of the terminal
node it ends up in end For
  7.  Aggregate the votes across t for each subject and divide
by the number of times subject is OOB
  8.  Each subject will be classified as the class they have the
highest proportion from 8
  9.  OOB error rate can be calculated by summing the off
diagonal values of a confusion table (true class vs. classified
class) and divide by the number of subjects
end Algorithm
```

Algorithm 2: Random Forests regression

The algorithm for Random Forests regression is essentially the same as Algorithm 1 but differs from it from step 4:

```
  4.  Construct a regression tree using the data from 2
        4a.  Randomly choose m/3 genes to do split
        4b.  Perform the split using the gene that minimizes the
sum of the squared deviations from the mean in the separate
nodes
```

```
    4c. Repeat step 4a/b until each terminal node contains ≤5
subjects
  5. Run the OOB subjects from 3. down the tree
  6. Subject i from 3. gets a predicted value for the terminal
node it ends up in end For
  7. Aggregate the predicted values across t for each subject
and divide by the number of times each subject is OOB
  8. We can calculate the differences between the fitted value
in 7. and the observed value
  9. The percent variance explained,
1 −[ MSE/Var(response)] , is then calculated
end Algorithm
```

Ten times fivefold CV

```
  1. Split the data set into k subsets, in this case k = 5
  2. One of the k subsets will be used as a test set and the
remaining k − 1 subsets will be used as the training set
  3. Use the test set to measure the error rate of the trained
classifier
  4. Repeat so that every k-th subset will be used as a test set
once
  5. Summarize the result by averaging the error rate
  6. Repeat, steps 1 to 5, 10 times.
```

Acknowledgments

This study was supported in part by the National Science Foundation grant DMS0241160, NIDA grant P30 DA018343, and NINDS grant U24 NS051869. Part of this work was supported through a grant from Pfizer. We also thank OUP for letting us reproduce some of the figures and tables published in *Bioinformatics*.

References

1 Pang, H., Lin, A., Holford, M., Enerson, B., Lu, B., Lawton, M., Floyd, E. and Zhao, H. (2006) Pathway analysis using random forests classification and regression. *Bioinformatics (Oxford, England)*, **22**, 2028–2036.

2 Kim, I., Pang, H. and Zhao, H. (2007) Bayesian semiparametric approach for gene pathway analysis. *Technical Report.*

3 Harris, M. and Gene Ontology Consortium (2004) The Gene Ontology (GO) database and informatics resource. *Nucleic Acids Research,* **32**, D258–D261.

4 Mootha, V., Lindgren, C., Eriksson, K., Subramanian, A., Sihag, S., Lehar, J., Puigserver, P., Carsson, E., Ridderstrale, M., Laurila, E., Houstis, N., Daly, M., Patterson, P., Mesirov, J., Golub, T., Tamayo, P., Spiegelman, B., Lander, E.,

Hirschhorn, J., Altshuler, D. and Groop, L. (2003) PGC-1alpha-responsive genes involved in oxidative phosphorylation are coordinately downregulated in human diabetes. *Nature Genetics*, **34** (3), 267–273.

5 Subramanian, A., Tamayoa, P., Mootha, V., Mukherjee, S., Ebert, B., Gillette, M., Paulovich, A., Pomeroy, S., Golub, T., Lander, E. and Mesurov, J. (2005) Gene set enrichment analysis: a knowledge-based approach for interpreting genome-wide expression profiles. *Proceedings of the National Academy of Sciences of the United States of America*, **102** (43), 15545–15550.

6 Efron, B. and Tibshirani, R. (2006) Testing the significance of sets of genes. *Technical Report.*

7 Newton, M., Quintana, F., den Boon, A., Sengupta, S. and Ahlquist, P. (2007) Random-set methods identify distinct aspects of the enrichment signal in gene-set analysis. *Annals of Applied Statistics*, **1**, 85–106.

8 Goeman, J., van de Geer, S., de Kort, F. and van Houwelingen, H. (2004) A global test for groups of genes: testing association with a clinical outcome. *Bioinformatics (Oxford England)*, **20** (1), 93–99.

9 Mansmann, U. and Meister, R. (2005) Testing differential gene expression in functional groups. Goeman's global test versus an ANCOVA approach. *Methods of Information in Medicine*, **44** (3), 449–453.

10 Breiman, L. (2003) Manual on Setting Up, Using, and Understanding Random Forests, V4.0http://oz.berkeley.edu/users/breiman/Using_random_forests_V4.0.pdf.

11 Svetnik, V., Liaw, A., Tong, C., Culberson, J., Sheridan, R. and Feuston, B. (2003) Random forest: a classification and regression tool for compound classification and QSAR modeling. *Journal of Chemical Information and Computer Sciences*, **43** (6), 1947–1958.

12 Wu, B., Abbott, T., Fishman, D., McMurray, W., Mor, G., Stone, K., Ward, D., Williams, K. and Zhao, H. (2003) Comparison of statistical methods for classification of ovarian cancer using mass spectrometry data. *Bioinformatics (Oxford England)*, **19** (13), 1636–1643.

13 Qi, Y., Bar-Joseph, Z. and Klein-Seetharaman, J. (2006) Evaluation of different biological data and computational classification methods for use in protein interaction prediction. *Proteins*, **63** (3), 490–500.

14 Breiman, L. (2001) Random forests. *Machine Learning*, **45**, 5–32.

15 Farmer, P., Bonnefoi, H., Becette, V., Tubiana-Hulin, M., Fumoleau, P., Larsimont, D., Macgrogan, G., Bergh, J., Cameron, D., Goldstein, D., Duss, S., Nicoulaz, A., Brisken, C., Fiche, M., Delorenzi, M. and Iggo, R. (2005) Identification of molecular apocrine breast tumours by microarray analysis. *Oncogene*, **24** (29), 4660–4671.

16 Enerson, B., Lin, A., Lu, B., Zhao, H., Lawton, M. and Floyd, E. (2006) Acute drug-induced vascular injury in beagle dogs: pathology and correlating genomic expression. *Toxicologic Pathology*, **34** (1), 27–32.

17 Appella, E. and Anderson, C. (2001) Post-translational modifications and activation of p53 by genotoxic stresses. *European Journal of Biochemistry*, **268** (10), 2764–2772.

18 Iida, T., Mine, S., Fujimoto, H., Suzuki, K., Minami, Y. and Tanaka, Y. (2002) HIF-1-induced apoptosis of endothelial cells. *Genes to Cells*, **7**, 143–149.

19 Desai, A., Miller, M., Huang, X. and Warren, J. (2003) Nitric oxide modulates MCP-1 expression in endothelial cells: implications for the pathogenesis of pulmonary granulomatous vasculitis. *Inflammation*, **27** (4), 213–223.

20 Charo, I. and Taubman, M. (2004) Chemokines in the pathogenesis of vascular disease. *Circulation Research*, **95**, 858.

21 Panzer, U., Steinmetz, O., Reinking, R., Meyer, T., Fehr, S., Schneider, A., Zahner,

G., Wolf, G., Helmchen, U., Schaerli, P., Stahl, R. and Thaiss, F. (2006) Compartment-specific expression and function of the chemokine IP-10/CXCL10 in a model of renal endothelial microvascular injury. *Journal of the American Society of Nephrology*, **17**, 454–464.

22 Rothenbacher, D., Muller-Scholze, S., Herder, C., Koenig, W. and Kolb, H. (2005) Differential expression of chemokines risk of stable coronary heart disease, and correlation with established cardiovascular risk markers. *Arteriosclerosis, Thrombosis, and Vascular Biology*, **26**, 26–194.

23 Zhao, Y., Qin, S., Atangan, L., Molina, Y., Okawa, Y., Arpawong, H., Ghosn, C., Xiao, J., Vuligonda, V., Brown, G. and Chandraratna, R. (2004) Casein kinase 1alpha interacts with retinoid X receptor and interferes with agonist-induced apoptosis. *Journal of Biochemistry*, **279** (29), 30844–30849.

24 Bolick, D., Srinivasan, S., Whetzel, A., Fuller, L. and Hedrick, C. (2005) 12/15-Lipoxygenase regulates intercellular adhesion molecule-1 expression and monocyte adhesion to endothelium through activation of RhoA and nuclear factor-κB. *Arteriosclerosis Thrombosis, and Vascular Biology*, **25**, 2301.

25 Maglott, D., Ostell, J., Pruitt, K. and Tatusova, T. (2005) Entrez Gene: gene-centered information at NCBI. *Nucleic Acids Research*. **33**, D54–D58 (database issue).

26 Harari, O., McHale, J., Marshall, D., Ahmed, S., Brown, D., Askenase, P. and Haskard, D. (1999) Endothelial cell E- and P-selectin up-regulation in murine contact sensitivity is prolonged by distinct mechanisms occurring in sequence. *Journal of Immunology*, **163** (12), 6860–6866.

27 Hetzel, J., Balletshofer, B., Rittig, K., Walcher, D., Kratzer, W., Hombach, V., Haring, H., Koenig, W. and Marx, N. (2005) Rapid effects of rosiglitazone treatment on endothelial function and inflammatory biomarkers. *Arteriosclerosis, Thrombosis, and Vascular Biology*, **25** (9), 1804–1809.

28 Eriksson, E., Karlof, E., Lundmark, K., Rotzius, P., Hedin, U. and Xie, X. (2005) Powerful inflammatory properties of large vein endothelium *in vivo*. *Arteriosclerosis Thrombosis, and Vascular Biology*, **25** (4), 723–728.

29 Lo, K., Kan, H., Chan, L., Xu, W., Wang, K., Wu, Z., Sheng, M. and Zhang, M. (2005) The 8-kDa dynein light chain binds to p53-binding protein 1 and mediates DNA damage-induced p53 nuclear accumulation. *Journal of Biological Chemistry*, **280** (9), 8172–8179.

30 Erkilic, K., Evereklioglu, C., Cekmen, M., Ozkiris, A., Duygulu, F. and Dogan, H. (2003) Adenosine deaminase enzyme activity is increased and negatively correlates with catalase superoxide dismutase and glutathione peroxidase in patients with Behcet's disease: original contributions/clinical and laboratory investigations. *Mediators of Inflammation*, **12** (2), 107–116.

31 Sato, N., Kokame, K., Shimokado, K., Kato, H. and Miyata, T. (1998) Changes of gene expression by lysophosphatidylcholine in vascular endothelial cells: 12 up-regulated distinct genes including 5 cell growth-related 3 thrombosis-related, and 4 others. *Journal of Biochemistry*, **123** (6), 1119–1126.

32 Shao, W. and Brown, M. (2004) Advances in estrogen receptor biology: prospects for improvements in targeted breast cancer therapy. *Breast Cancer Research*, **6** (1), 39–52.

33 Motoyama, A. and Hynes, N. (2003) BAD: a good therapeutic target? *Breast Cancer Research*, **5** (1), 27–30.

34 Mehra, R., Varambally, S., Ding, L., Shen, R., Sabel, M., Ghosh, D., Chinnaiyan, A. and Kleer, C. (2005) Identification of GATA3 as a breast cancer prognostic marker by global gene expression meta-analysis. *Cancer Research*, **65** (24), 11259–11264.

35 Perumal, S., Shanthi, P. and Sachdanandam, P. (2005) Therapeutic effect of tamoxifen and energy-modulating vitamins on carbohydrate-metabolizing enzymes in breast cancer. *Cancer Chemotherapy and Pharmacology*, **56** (1), 105–114.

36 Alarcon, C., Wicksteed, B., Prenki, M., Corkey, B. and Rhodes, C. (2002) Succinate is a preferential metabolic stimulus-coupling signal for glucose-induced proinsulin biosynthesis translation. *Diabetes*, **51**, 2496–2504.

37 Gopalakrishnan, V., Arunakaran, J., Aruldhas, M. and Srinivasan, N. (2006) Effects of streptozotocin-induced diabetes mellitus on some bone turnover markers in the vertebrae of ovary-intact and ovariectomized adult rats. *Biochemistry and Cell Biology*, **84** (5), 728–736.

38 McKenzie, J., Fisher, B., Jaap, A., Stanley, A., Paterson, K. and Sattar, N. (2006) Effects of HRT on liver enzyme levels in women with type 2 diabetes: a randomized placebo-controlled trial. *Clinical Endocrinology*, **65** (1), 40–44.

39 Park, S., Cho, Y., Kim, H., Hong, E., Higashimori, T., Lee, S., Goldberg, I., Shulman, G., Najjar, S. and Kim, J. (2006) Mechanism of glucose intolerance in mice with dominant negative mutation of CEACAM1. *American Journal of Physiology. Endocrinology and Metabolism*, **291** (3), E517–E524.

40 Dubois, M., Bergeron, S., Kim, H., Dombrowski, L., Perreault, M., Fournes, B., Faure, R., Olivier, M., Beauchemin, N., Shulman, G., Siminovitch, K., Kim, J. and Marette, A. (2006) The SHP-1 protein tyrosine phosphatase negatively modulates glucose homeostasis. *Nature Medicine*, **12** (5), 549–556.

41 Poy, M., Yang, Y., Rezaei, K., Fernstrom, M., Lee, A., Kido, Y., Erickson, S. and Najjar, S. (2002) CEACAM1 regulates insulin clearance in liver. *Nature Genetics*, **30** (3), 270–276.

42 Yan, G., Fu, Y. and Faustman, D. (1997) Reduced expression of Tap1 and Lmp2 antigen-processing genes in the non-obese diabetic (NOD) mouse due to a mutation in their shared bidirectional promoter. *Journal of Immunology*, **159** (6), 3068–3080.

43 Kostyak, J., Kimball, S., Jefferson, L. and Farrell, P. (2001) Severe diabetes inhibits resistance exercise-induced increase in eukaryotic initiation factor 2B activity. *Journal of Applied Physiology*, **91**, 79–84.

44 Dahlquist, K., Salomonis, N., Vranizan, K., Lawlor, S. and Conklin, B. (2002) GenMAPP a new tool for viewing and analyzing microarray data on biological pathways. *Nature Genetics*, **31**, 19–20.

45 Efron, B. and Tibshirani, R. (1997) Improvements on cross-validation: the.632+ bootstrap estimator. *Journal of the American Statistical Association*, **92** (438), 548–560.

46 Braga-Neto, U. and Dougherty, E. (2004) Is cross-validation valid for small-sample microarray classification? *Bioinformatics (Oxford, England)*, **20** (3), 374–380.

47 Fu, W., Carroll, R. and Wang, S. (2005) Estimating misclassification error with small samples via bootstrap cross-validation. *Bioinformatics (Oxford, England)*, **21** (9), 1979–1986.

48 Molinaro, A., Simon, R. and Pfeiffer, R. (2005) Prediction error estimation: a comparison of resampling methods. *Bioinformatics (Oxford, England)*, **21** (15), 3301–3307.

49 Mootha, V., Handschin, C., Arlow, D., Xie, X., Pierre, J., Sihag, S., Yang, W., Altshuler, D., Puigserver, P., Patterson, N., Willy, P., Schulman, I., Heyman, R., Lander, E. and Spiegelman, B. (2004) Errα and Gabpa/b specify PGC-1α-dependent oxidative phosphorylation gene expression that is altered in diabetic muscle. *Proceedings of the National Academy of Sciences of United States of America*, **101**, 6570–6575.

50 Misu, H., Takamura, T., Matsuzawa, N., Shimizu, A., Ota, T., Sakurai, M., Ando, H., Arai, K., Yamashita, T., Honda, M., Yamashita, T. and Kaneko, S. (2007)

Genes involved in oxidative phosphorylation are coordinately upregulated with fasting hyperglycaemia in livers of patients with type 2 diabetes. *Diabetologia*, **50**, 268–277.

51 Razeghi, P., Young, M., Cockrill, T., Frazier, O. and Taegtmeyer, H. (2002) Downregulation of myocardial myocyte enhancer factor 2C and myocyte enhancer factor 2C regulated gene expression in diabetic patients with nonischemic heart failure. *Journal of the American Heart Association*, **106**, 407–411.

52 Voisine, P., Ruel, M., Khan, T., Bianchi, C., Xu, S., Kohane, I., Libermann, T., Otu, H., Saltiel, A. and Sellke, F. (2004) Differences in gene expression profiles of diabetic and nondiabetic patients undergoing cardiopulmonary bypass and cardioplegic arrest. *Journal of the American Heart Association*, **110**, 280–286.

53 Zhang, D., Zhou, Z., Li, L., Weng, J., Huang, G., Jing, P., Zhang, C., Peng, J. and Xiu, L. (2005) Islet autoimmunity and genetic mutations in Chinese subjects initially thought to have Type 1B diabetes. *Diabetic Medicine*, **23**, 67–71.

54 Chen, Q., Behar, K.L., Xu, T., Fan, C. and Haddad, G. (2003) Expression of *Drosophila* trehalose-phosphate synthase in HEK-293 cells increases hypoxia tolerance. *Journal of Biological Chemistry*, **278**, 49113–49118.

14
The Most Probable Genetic Interaction Networks Inferred from Gene Expression Patterns

Timothy R. Lezon, Jayanth R. Banavar, Marek Cieplak, Nina V. Fedoroff, and Amos Maritan

14.1
Introduction

An important challenge in molecular biology lies in inferring the underlying gene interaction network from high-throughput gene expression data. A key problem here is that the system is underdetermined, because the number of genes sampled in a microarray experiment is invariably much larger than the number of samples; as a consequence, myriad networks can reproduce the observed data with fidelity. In order to select between these networks, a common approach has been to impose additional constraints based on assumptions pertaining to the form of interactions, the form of data, or the topology of the interaction network [1–14]. The difficulty, of course, is that these assumptions, while plausible in some instances, neither can be tested nor are guaranteed to be correct.

To minimize the effects of unwarranted assumptions on network inference, we have developed [15] a new approach to gene network inference from gene expression data that relies on Boltzmann's concept of entropy maximization. The strength of our approach is that it transforms the tables of genetic expression data into a system of interacting genes without incorporating external assumptions about the network structure. Our method provides a network interpretation of the expression data in which the genes are the nodes and the links of various strengths and signs describe the nature of the effective gene–gene interactions that come from an implicit mediation by proteins. It also provides a framework in which three-gene and higher order interactions can be assessed in a perturbative fashion. More important, the entropy maximization approach identifies the most probable genetic network consistent with correlations derived from the experimental data and relies minimally on missing information [16,17].

As an illustration of our approach, we consider the microarray data obtained by Klevecz *et al.* [18] for genes involved in metabolic oscillations in *Saccharomyces cerevisiae*. The pairwise couplings that we find between genes exactly account for

Analysis of Microarray Data: A Network-Based Approach. Edited by F. Emmert-Streib and M. Dehmer

ISBN: 978-3-527-31822-3

the expression data, and we find that the entropy-based identification of the hubs is consistent with the current understanding of the signaling pathways in the yeast cells.

14.2 Entropy Maximization

Starting with Boltzmann's pioneering work [16], the concept of entropy has provided a simple framework for understanding phenomena as complex as equilibrium thermodynamics and nonequilibrium global climate [19–23]. The basic idea is that the most likely situation ought to be observed macroscopically when the number of ways of realizing it microscopically is overwhelmingly large compared to other macrostates. For the particular case of microarray experiments, the macrostates refer to the observed expression level measurements provided by the microarray, whereas the microstates refer to the set of molecular states within the cell that correspond to the observed expression levels.

Entropy is conceptually linked to the amount of disorder or uncertainty in a system: the higher a system's entropy, the less certain we can be about its exact state. This connection is formalized in Shannon's theory of information [17]. The amount of information about a system that is provided by the outcome of a measurement of its state is equal to $-\ln(\rho(s))$, where $\rho(s)$ is the probability that the system is in the measured state s. The expectation value of the information contained in an arbitrary measurement of a system's state is just the weighted average of the information provided by all possible outcomes of the measurement, defined as the entropy $S = -\sum_i \rho(i) \ln \rho(i)$. For a discrete system with n possible states, entropy is maximized when all states occur with equal probability, $\rho(i) = 1/n \; \forall i$. Conversely, a system about which a set of n outcomes of a measurement provides the most information is the system for which each outcome is equally likely. Thus, if we are given n observed states, the system that most likely generated this set of states is the one that maximizes the entropy.

When reconstructing the workings of a system from a set of observed states, it is not sufficient to assume that any system that produces all observed states with equal probability will correctly account for the observed data. In general, many putative systems will be able to produce all of the observed states with equal probability, forcing us to choose from amongst the available maximally entropic solutions. More important, by ignoring information contained in the data, we artificially inflate the entropy of the system. For example, a coin known never to come up either heads or tails on two consecutive tosses has an entropy much lower than a completely unbiased coin. The unbiased coin has 2^N possible outcomes for a series of N tosses, whereas the biased coin has just two outcomes showing that observed constraints on the data must be taken into account when inferring a probability distribution through entropy maximization. Beyond these observed constraints, however, we are still maximally uncertain about the nature of the system. Therefore, as pointed out by Jaynes [19,20], when inferring an unknown probability distribution from data, the least-biased estimate is the one that maximizes the system entropy subject to the constraints of the data.

A single DNA microarray chip provides information about genetic expression in a culture at a single point in time. Using P chips allows us to record the expression levels of G genes at P different times and thus catch temporal variations in genes. The formalism also applies if we consider P different stimuli or P different tissues (e.g., cancer tissues) of the same organism when seeking non-temporal patterns. Typically, $P < G$. The expression levels can be represented as vectors $\vec{x}^{(k)}$ where the component $x_i^{(k)}$ is the expression level of the ith gene at the kth time. A series of P measurements, then, has associated with it P distinct state vectors $\vec{x}^{(1)}, \ldots, \vec{x}^{(P)}$.

We ask what is the probability distribution function, $\rho(\vec{x})$, which is consistent with the partial knowledge contained in the P measurements of the gene network. As argued by Jaynes [19,20], the best guess is the one that maximizes the entropy but, at the same time, is constrained by available information, such as the mean expression value of a gene and two-gene covariances. The system entropy is defined as

$$S = -\sum_{\vec{x}} \rho(\vec{x}) \ln \rho(\vec{x}) \tag{14.1}$$

and the constraints are provided by the equations

$$\sum_{\vec{x}} \rho(\vec{x}) = 1, \tag{14.2}$$

$$\langle x_i \rangle = \sum_{\vec{x}} \rho(\vec{x}) x_i = \frac{1}{P} \sum_{k=1}^{P} x_i^k, \tag{14.3}$$

and

$$\langle x_i x_j \rangle = \sum_{\vec{x}} \rho(\vec{x}) x_i x_j = \frac{1}{P} \sum_{k=1}^{P} x_i^k x_j^k. \tag{14.4}$$

Equation 14.2 provides the normalization condition that the probabilities of all observable states sum to 1. Equations 14.3 and 14.4 ensure that the distribution $\rho(\vec{x})$ preserves the mean expression level of each gene as well as the correlations between genes. It should be noted that, in principle, we can incorporate additional conditions, which relate to higher order, on the entropy, for example, triple point and correlation measures. However, the associated mathematics then becomes much more complicated due to the resulting non-Gaussian nature of the probability distribution function. Instead, we truncate the list of conditions at the two-point level to get the leading, or "unperturbed," form of $\rho(\vec{x})$ and then use it to estimate the higher order solution in a perturbative fashion. The constraints in Equations 14.3 and 14.4 can be used to define the covariance matrix C as

$$C_{ij} = \langle x_i x_j \rangle - \langle x_i \rangle \langle x_j \rangle. \tag{14.5}$$

The entropy maximization is carried out by introducing the Lagrange multipliers ν, μ_i, and M_{ij} and maximizing

$$\begin{aligned}\zeta &= S - \nu \sum_{\vec{x}} \rho(\vec{x}) - \sum_{i=1}^{G} \mu_i \sum_{\vec{x}} \rho(\vec{x}) x_i - \frac{1}{2} \sum_{i,j=1}^{G} M_{ij} \sum_{\vec{x}} \rho(\vec{x}) x_i x_j \\ &= -\sum_{\vec{x}} \rho(\vec{x}) \left[\ln\rho(\vec{x}) + \nu + \vec{\mu} \cdot \vec{x} + \frac{1}{2} \vec{x} M \vec{x} \right]. \end{aligned} \tag{14.6}$$

We set $\partial\zeta/\partial\rho(\vec{x}) = 0$ and get

$$\begin{aligned}\rho(\vec{x}) &= e^{-1-\nu-\vec{\mu}\cdot\vec{x}-(1/2)\vec{x}M\vec{x}} \\ &= \frac{1}{Z} e^{-(1/2)\vec{y}M\vec{y}}, \end{aligned} \tag{14.7}$$

where $\vec{y} = \vec{x} + M^{-1}\vec{\mu}$ is the expression level vector that is shifted by a constant and the quantity $Z = e^{-(1/2)\vec{\mu}M^{-1}\vec{\mu}} e^{1+\nu}$ serves to normalize the probability. We use a simplified notation in which a vector written to the left of a matrix is meant to be in a transposed form.

The matrix element M_{ij} has the natural interpretation of the interaction between genes i and j. To calculate the elements of M, we note that the states $\vec{x}$ span a continuous space, and we replace the summations in the constraints (Equations 14.2–14.4) with integrals. Thus, the normalization condition of Equation 14.2 becomes

$$1 = \int d^G x \rho(\vec{x}) = \frac{1}{Z} \int d^G y \, e^{-(1/2)\vec{y}M\vec{y}}, \tag{14.8}$$

and Equation 14.3 can similarly be rewritten as

$$\begin{aligned}\langle x_i \rangle &= \int d^G x \rho(\vec{x}) x_i \\ &= \frac{1}{Z} \int d^G y \, e^{-(1/2)\vec{y}M\vec{y}} \left(y_i - \sum_j M_{ij}^{-1} \mu_j \right) \\ &= -\sum_j M_{ij}^{-1} \mu_j. \end{aligned} \tag{14.9}$$

Note that Equation 14.9 and the previous definition $\vec{y} = \vec{x} + M^{-1}\vec{\mu}$ imply that $\vec{y} = \vec{x} - \langle \vec{x} \rangle$.. Thus, when the data are shifted so that $\langle x_i \rangle = 0 \; \forall i$, the elements of M are the coupling constants that link the genes directly.

Applying the continuum treatment to Equation 14.4, we find

$$\begin{aligned}\langle x_i x_j \rangle &= \int d^G x \rho(\vec{x}) x_i x_j \\ &= \frac{1}{Z} \int d^G y \, e^{-(1/2)\vec{y}M\vec{y}} [\langle x_i \rangle \langle x_j \rangle + y_i \langle x_j \rangle + y_j \langle x_i \rangle + y_i y_j] \\ &= \langle x_i \rangle \langle x_i \rangle + \frac{1}{Z} \int d^G y \, e^{(1/2)\vec{y}M\vec{y}} y_i y_j. \end{aligned} \tag{14.10}$$

The integral in Equation 14.10 can be solved by defining the generating function

$$\Upsilon(\vec{J}) = \int \mathrm{d}^G y\, \mathrm{e}^{-(1/2)\vec{y}M\vec{y}+\vec{J}\cdot\vec{y}}. \tag{14.11}$$

The source field $\vec{J}$ in Equation 14.11 allows us to write the integral in Equation 14.10 in terms of derivatives of $\Upsilon(\vec{J})$ taken with respect to $\vec{J}$ and then setting $\vec{J}=0$:

$$\left.\frac{\partial^2\Upsilon(\vec{J})}{\partial J_i \partial J_j}\right|_{\vec{J}=0} = \int \mathrm{d}^G y\, \mathrm{e}^{-(1/2)\vec{y}M\vec{y}} y_i y_j. \tag{14.12}$$

By making the substitution $\vec{z}=\vec{y}-M^{-1}\vec{J}$, Equation 14.11 can be rewritten as

$$\Upsilon(\vec{J}) = \mathrm{e}^{(1/2)\vec{J}M^{-1}\vec{J}} \int \mathrm{d}^G z\, \mathrm{e}^{-(1/2)\vec{z}M\vec{z}} = \mathrm{e}^{(1/2)\vec{J}M^{-1}\vec{J}}\Upsilon(\vec{J})|_{\vec{J}=0}, \tag{14.13}$$

yielding

$$\begin{aligned}\frac{1}{Z}\int \mathrm{d}^G y\, \mathrm{e}^{(1/2)\vec{y}M\vec{y}} y_i y_j &= \left.\frac{1}{\Upsilon(\vec{J})}\frac{\partial^2}{\partial J_i \partial J_j}\Upsilon(\vec{J})\right|_{\vec{J}=0} \\ &= M_{ij}^{-1}.\end{aligned} \tag{14.14}$$

Substituting back into Equation 14.10, we find that $M^{-1}=C$, and the interaction matrix is simply the inverse of the covariance matrix.

It should be noted that the form of M is independent of the mean expression levels $\langle x_i\rangle$, because neither the covariance matrix C nor its inverse is affected by changes in the mean expression levels. Furthermore, this is a robust result for linear systems and can be derived in several ways. An alternative way of arriving at this result without invoking the maximization of entropy follows from the assumptions that $\ln \rho(\vec{x})$ peaks at $\vec{x}^{(0)}$, is normalizable, and is a smooth function that can be expressed in a Taylor expansion up to quadratic order:

$$\ln \rho(\vec{x}) = \ln \rho(\vec{x}^{(0)}) - \frac{1}{2}\sum_{i,j}(x_i - x_i^{(0)})M_{ij}(x_j - x_j^{(0)}) + \cdots, \tag{14.15}$$

where the neglected terms are of cubic order in $(x_i - x_i^{(0)})$ and $-M$, the matrix of the second derivative of $\ln\rho(\vec{x})$ with respect to $\vec{x}$, is negative definite. Note that $\vec{x}^{(0)} = \langle\vec{x}\rangle$.. Within this Gaussian approximation, we again obtain the result that M is the inverse of C. Not surprisingly, the same result is arrived at in the graphical Gaussian model, in which expression level data are assumed to be drawn from a Gaussian distribution [13].

If $G > P$, the problem is underdetermined and C is singular. This is usually the case, as microarray experiments typically generate data for thousands of genes at only a few tens of time points. To overcome the problem of ill-conditioned matrices, we use the spectral decomposition of C,

$$C_{ij} = \sum_{\omega=1}^{G} \lambda_\omega v_i^{(\omega)} v_j^{(\omega)}, \tag{14.16}$$

where λ_ω is the ωth eigenvalue of C and $\vec{v}^{(\omega)}$ is its corresponding eigenvector. The form of C (Equation 14.5) ensures that only $P-1$ eigenvalues of C are nonzero. Adopting the convention that the lowest indices refer to the nonzero eigenvalues, we have $\lambda_\omega \neq 0$ for $\omega = 1, \ldots, (P-1)$ and $\lambda_\omega \neq 0$ for $P \leq \omega \leq G$. We construct M as the inverse of C in the nonzero eigenspace:

$$M_{ij} = \sum_{\omega=1}^{P-1} \frac{1}{\lambda_\omega} v_i^{(\omega)} v_j^{(\omega)}, \tag{14.17}$$

where the summation is over only the nonzero eigenvalues. This partial inversion is robust against adding small noise to the raw data.

It should be noted that the eigenvectors with large eigenvalues contribute the most to C, but have little effect on M. The gross features of the data are captured in these eigenvectors, and therefore such general features indicate little about the nature of couplings between genes. However, the eigenvectors with small eigenvalues dominate the calculation of M. These eigenvectors correspond to the residual fluctuations in expression levels that remain when the common, large-scale fluctuations are removed.

The elements of the matrix M are, by definition, the effective pairwise gene interactions that reproduce the gene profile covariances exactly while maximizing the entropy of the system. The strength and the sign of the interaction represent the mutual influence of the expression levels of a pair of genes on each other. This is necessarily indirect, since gene interactions are mediated by proteins. The magnitude of the element M_{ij} is a measure of the strength of the net interaction between genes i and j. The sign of the interaction indicates the nature of the coupling: a negative coupling between genes indicates that a change in the expression level of one gene is accompanied by a similar change in the expression level of the other gene. Conversely, a positive coupling indicates that a change in the expression level of one gene is accompanied by an opposite change in the other. The diagonal element M_{ii} provides a measure of the influence that gene i has on the whole network. Nodes with large diagonal values have strong couplings with several other nodes, whereas nodes with smaller diagonal elements generally have couplings of lesser magnitude. The gene couplings integrate all of the influences that are not considered as part of the network. It should be noted, however, that the nature of the correlation between the expression profiles of the two genes cannot be deduced directly from their interaction strength.

14.3 Recovering the Data

We now continue with a further theoretical analysis of the entropy maximization method and consider how to use the interaction matrix M to recover expression levels of all genes in a temporal state based on the knowledge of the expression levels only of any $P-1$ genes in this state. The covariance matrix C (Equation 14.5) for G genes with expression levels that are known for P distinct states has $P-1$ nonzero eigenvalues λ_ω: $\omega=1,\ldots,(P-1)$. The vector $\vec{x}$ of expression levels of G genes in any state that is consistent with C can be written as a linear combination of the corresponding $P-1$ eigenvectors $\vec{v}^{(\omega)}$. Given partial information about $\vec{x}$, that is, the expression levels of a subset of q genes, we can find the state $\vec{x}$ in the nonzero eigenspace that maximizes $\rho(\vec{x})$. Defining $\{q\}$ as the set of q genes whose expression levels are initially known, we can define the vectors

$$\vec{f}_i=\hat{e}_i-\sum_{\omega=P}^{G}\vec{v}^{(\omega)}(\vec{v}^{(\omega)}\cdot\hat{e}_i)\quad\forall i\in\{q\}\tag{14.18}$$

orthogonal to the space of zero eigenvalues of C. The basis vector $\hat{e}_i$ is defined as $\hat{e}_i=(\delta_{1i},\ldots,\delta_{Gi})$, and thus $\vec{f}_i$ is the part of $\hat{e}_i$ within the nonzero eigenspace of C. Our task is to maximize $\rho(\vec{x})$ subject to the constraints

$$\vec{f}_i\cdot\vec{x}=x_i\quad\forall i\in\{q\}\tag{14.19}$$

and

$$\vec{x}\cdot\vec{v}^{(\omega)}=0\quad\forall\omega\geq P.\tag{14.20}$$

The first constraint comes directly from the data, whereas the second comes from the requirement that $\vec{x}$ is orthogonal to the zero eigenspace of C. Both constraints can be imposed via Lagrange parameters as before, but for reasons that will soon become clear, we impose the constraint of Equation 14.20 quadratically. Introducing the Lagrange parameters μ_i and ν_i, Equation 14.7 can be rewritten as

$$\begin{aligned}\rho(\vec{x})&=\frac{1}{Z}\exp\left\{-\frac{1}{2}\vec{x}M\vec{x}-\frac{1}{2}\sum_{\omega=P}^{G}\mu_\omega(\vec{x}\cdot\vec{v}^{(\omega)})^2+\sum_{i\in\{q\}}\nu_i\vec{f}_i\cdot\vec{x}\right\}\\&=\frac{1}{Z}\exp\left\{-\frac{1}{2}\vec{x}M'\vec{x}+\vec{F}\cdot\vec{x}\right\},\end{aligned}\tag{14.21}$$

where

$$\begin{aligned}M'&=M+\sum_{\omega=P}^{G}\mu_\omega\vec{v}^{(\omega)}\otimes\vec{v}^{(\omega)},\\\vec{F}&=\sum_{i\in\{q\}}\nu_i\vec{f}_i,\end{aligned}\tag{14.22}$$

and the matrix element $(\vec{v}\otimes\vec{v})_{ij} = v_i v_j$. The quadratic term in Equation 14.21 allows us to define the matrix M' that satisfies Equation 14.20 and is invertible so long as $\mu_\omega \neq 0\ \forall\omega$. The probability $\rho(\vec{x})$ is maximized when $\partial\rho(\vec{x})/\partial x_\alpha = 0\ \forall\alpha = 1, \ldots, G$, yielding

$$\begin{aligned}\vec{x} &= M'^{-1}\vec{F} \\ &= \sum_{\omega=1}^{P-1}\lambda_\omega \vec{v}^{(\omega)}(\vec{v}^{(\omega)}\cdot\vec{F}).\end{aligned} \tag{14.23}$$

Using Equation 14.23 in Equation 14.19 gives

$$x_i = \sum_{\omega=1}^{P-1}\lambda_\omega(\vec{f}_i\cdot\vec{v}^{(\omega)})(\vec{v}^{(\omega)}\cdot\vec{F}), \tag{14.24}$$

and using Equation 14.22, we find

$$x_i = \sum_{j\in\{q\}}\sum_{\omega=1}^{P-1}\lambda_\omega(\vec{f}_i\cdot\vec{v}^{(\omega)})(\vec{v}^{(\omega)}\cdot\vec{f}_j)v_j \tag{14.25}$$

or

$$v_i = \sum_{j\in\{q\}} A_{ij}^{-1}x_j, \tag{14.26}$$

where

$$A_{ij} = \sum_{\omega=1}^{P-1}\lambda_\omega(\vec{f}_i\cdot\vec{v}^{(\omega)})(\vec{v}^{(\omega)}\cdot\vec{f}_j). \tag{14.27}$$

An exact solution for $\vec{x}$ exists as long as the $q \times q$ matrix A is nonsingular, which is indeed the case for $q<q$, as $\vec{x}$ is confined to a space of $P-1$ dimensions, specifying P or more components of $\vec{x}$ overconstrains the problem. Using Equation 14.23 and $\vec{v}^{(\omega)}\cdot\vec{f}_i = v_i^{(\omega)}$ when $\omega< P$, the components of $\vec{x}$ can be expressed in terms of the genes in $\{q\}$ as

$$x_\alpha = \sum_{\omega=1}^{P-1}\lambda_\omega v_\alpha^{(\omega)}\sum_{ij\in\{q\}} A_{ij}^{-1}x_j v_i^{(\omega)}. \tag{14.28}$$

The response of gene α to a change in gene $Q\in\{q\}$ is

$$\frac{dx_\alpha}{dx_Q} = \sum_{\omega=1}^{P-1}\lambda_\omega v_\alpha^{(\omega)}\sum_{j\in\{q\}} A_{jQ}^{-1}v_j^{(\omega)}. \tag{14.29}$$

In the special case where Q is the only gene with an initially defined value (i.e., $\{q\} = Q$), Equation 14.27 gives $A_{ij} = A_{ij}^{-1} = \delta_{iQ}\delta_{jQ}$. Inserting into Equation 14.29, we find $dx_\alpha/dx_Q = \sum_{\omega<P} \lambda_\omega v_\alpha^{(\omega)} v_Q^{(\omega)} = C_{\alpha Q}$, the covariance between genes α and Q. The interaction network not only maximizes the entropy of the system, but also reproduces the experimentally measured covariances exactly.

14.4 Integrating Over Interactions

Each gene probed by the microarray constitutes a degree of freedom of the system, but it is not necessary to consider all these genes to infer a network of effective interactions. One can choose to select as the relevant degrees of freedom any subset of genes for which there are expression data, in which case the effect of each gene that is not explicitly considered in the network is integrated into the interactions between the remaining genes. In other words, the method considers exactly the direct and indirect effects of all the genes and cell products not explicitly considered as part of the network. As the number of genes in the network is reduced, the elements of M more closely correspond to the correlations between expression profiles until, when only two genes are considered, their effective interaction directly relates to the observed correlation between their profiles. This is illustrated in Figure 14.1.

To guarantee self-consistency, the method must allow us to integrate an existing interaction matrix over individual genes, thereby calculating new effective interactions that recover the remaining gene profiles without explicit consideration of the

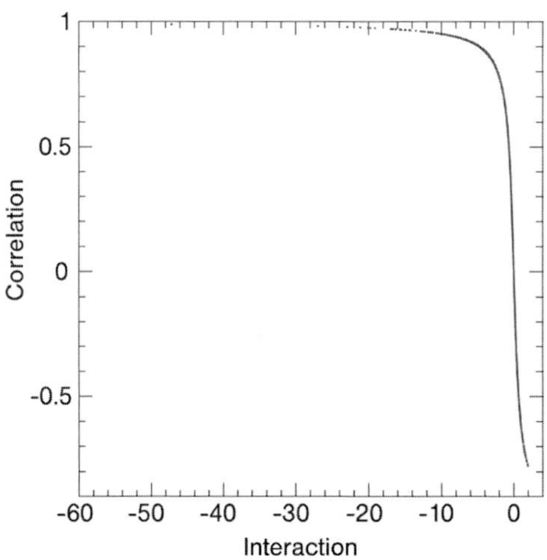

Figure 14.1 A cross-plot of correlations between pairs of genes and their effective pairwise interaction when only the two genes are retained as the system's relevant degrees of freedom. The one-to-one correspondence confirms that their effective interaction captures the correlation of their expression profiles.

integrated genes. For example, one might start with the probability distribution of Equation 14.7. Many of the genes described in M may not be of interest, and one might instead wish to determine the form of the interactions if several of the genes are integrated over. Let $\{q\}$ be the set of genes that we wish to retain. We aim to find the form of the interaction matrix R given by

$$\frac{1}{Z}\int\prod_{k\notin\{q\}}\mathrm{d}x_k\,\mathrm{e}^{-(1/2)\vec{x}M\vec{x}}=\frac{1}{Z'}\exp\left\{-\frac{1}{2}\sum_{i,j\in\{q\}}x_iR_{ij}x_j\right\},\tag{14.30}$$

where Z' is some normalization factor. This can be readily solved by introducing the matrix $\overline{M}$ with elements $\overline{M}_{ij}=M_{ij}$, if neither i nor j is in $\{q\}$ and $\overline{M}_{ij}=0$ otherwise, as well as the vector $\vec{J}$ with elements $J_i=\sum_{j\in\{q\}}x_jM_{ji}$. Then the integral in Equation 14.30 can be written as

$$\int\prod_{k\notin\{q\}}\mathrm{d}x_k\,\mathrm{e}^{-(1/2)\vec{x}M\vec{x}}=\exp\left\{-\frac{1}{2}\sum_{i,j\in\{q\}}x_iM_{ij}x_j\right\}\int\prod_{k\notin\{q\}}\mathrm{d}x_k\,\mathrm{e}^{-(1/2)\vec{x}\overline{M}\vec{x}-\vec{J}\cdot\vec{x}}.\tag{14.31}$$

Solving and equating to the right-hand side of Equation 14.30, we find that the "dressed" interactions between the genes in the set $\{q\}$ are given by

$$R_{ij}=M_{ij}-\sum_{\alpha\beta}M_{i\alpha}\overline{M}^{-1}_{\alpha\beta}M_{j\beta}.\tag{14.32}$$

However, we will generally be working in only a subspace of the gene states, and we must restrict ourselves to this subspace when calculating effective interactions by integrating over genes.

To find the matrix of dressed interactions, R, when working in a limited subspace of gene expression, we begin by writing a generating function, $\Upsilon(\vec{J})$, similar to that in Equation 14.11, but confined to only the nonzero gene expression eigenspace. We can then find the elements of the matrix R^{-1} through differentiation of $\Upsilon(\vec{J})$, and the dressed interactions as the inverse of R^{-1}. Let the nonzero eigenspace be spanned by the $P-1$ eigenvectors corresponding to $\lambda_1,\ldots,\lambda_{P-1}$, let μ_i be the Lagrange multipliers, and let

$$\Upsilon(\vec{J})=\int d^Gx\prod_{\omega\geq P}\delta\left(\vec{x}\cdot\vec{v}^{(\omega)}\right)\exp\left\{-\frac{1}{2}\vec{x}M\vec{x}+\vec{J}\cdot\vec{x}-\frac{1}{2}\sum_{\omega\geq P}\mu_\omega\left(\vec{x}\cdot\vec{v}^{(\omega)}\right)^2\right\}\tag{14.33}$$

be our generating function. The product of delta functions is taken over the eigenvectors with zero eigenvalue and it ensures that the integral is confined to the nonzero eigenspace. Owing to this restriction, the last term in the exponent does not affect the integral. However, by setting $\mu_\omega>0\ \forall\omega$, the quadratic form of this last term allows us to construct the invertible matrix

$$M'_{ij} = M_{ij} + \sum_{\omega \geq P} \mu_\omega v_i^{(\omega)} v_j^{(\omega)} \tag{14.34}$$

with the property $M'\vec{v}^{(\omega)} = (1/\lambda'_\omega)\vec{v}^{(\omega)}$, where $1/\lambda'_\omega = 1/\lambda_\omega$ if $\omega < P$ and $1/\lambda'_\omega = \mu_\omega$ if $P \leq \omega \leq G$. We can then write Equation 14.33 using the delta function representation $\delta(x) = (1/2\pi)\int \mathrm{d}z\, e^{-\mathrm{i}zx}$:

$$\begin{aligned}
\Upsilon(\vec{J}) &= \int \mathrm{d}^G x \int \prod_{\omega \geq P} \frac{\mathrm{d}z_\omega}{2\pi} \exp\left\{-\frac{1}{2}\vec{x}M'\vec{x} + \vec{J}\cdot\vec{x} + \mathrm{i}\sum_{\omega \geq P} \vec{x}\cdot\vec{v}^{(\omega)} z_\omega\right\} \\
&= (2\pi)^{G/2-(G-P+1)}(\det M')^{-1/2} \\
&\quad \int \prod_{\omega \geq P} \mathrm{d}z_\omega \exp\left\{\frac{1}{2}\left(\vec{J} + \mathrm{i}\sum_{\omega \geq P} z_\omega \vec{v}^{(\omega)}\right) M'^{-1} \left(\vec{J} + \mathrm{i}\sum_{\omega \geq P} z_\omega \vec{v}^{(\omega)}\right)\right\} \\
&= (2\pi)^{-G/2+P-1} \prod_{\omega=1}^{G} (\lambda'_\omega)^{1/2} \exp\left\{\frac{1}{2}\sum_{\omega=1}^{P-1} \lambda_\omega (\vec{J}\cdot\vec{v}^{(\omega)})^2\right\} \\
&\quad \int \prod_{\omega \geq P} \mathrm{d}z_\omega \exp\left\{\frac{1}{2}\sum_{\omega \geq P} \frac{(\vec{J}\cdot\vec{v}^{(\omega)} + \mathrm{i}z_\omega)^2}{\mu_\omega}\right\},
\end{aligned} \tag{14.35}$$

where the identity $(M'_{ij})^{-1} = \sum_{\omega=1}^{G} \lambda'_\omega v_i^{(\omega)} v_j^{(\omega)}$ was used in the last step. By changing the variable of integration to $z'_\omega = z_\omega - \mathrm{i}\vec{J}\cdot\vec{v}^{(\omega)}$, we find

$$\Upsilon(\vec{J}) = (2\pi)^{(P-1)/2} \prod_{\omega < P} (\lambda_\omega)^{1/2} \exp\left\{\frac{1}{2}\vec{J}R^{-1}\vec{J}\right\}, \tag{14.36}$$

where

$$R_{ij}^{-1} = \sum_{\omega < P} \lambda_\omega v_i^{(\omega)} v_j^{(\omega)}, \qquad i, j \in \{q\}. \tag{14.37}$$

The elements of R^{-1} are then the covariances between the genes in $\{q\}$, and the interactions between genes in $\{q\}$ are found by inverting this covariance matrix, exactly as we had found earlier.

14.5 Higher Order Interactions

The mathematics of our method is greatly simplified by assuming that the interactions between genes are primarily pairwise and the simple result is that the effective couplings between genes can be derived from the matrix of their covariances. Nevertheless, the genetic network need not strictly be Gaussian, and higher order interactions are measures of non-Gaussian effects. Our approach can be generalized to systematically handle three-body and higher order interactions in a perturbative fashion since the nonlinear case with the three-gene interaction is not amenable to exact analytic calculation. We use standard perturbation theory, which assumes that the characteristic three-gene correlation function is much smaller than the pairwise correlation. We begin by considering a system for which the data have been adjusted such that $\langle x_i \rangle = 0 \; \forall i$ and defining the unperturbed probability distribution function

$$\rho_0(\vec{x}) = \frac{1}{Z_0(\vec{h})} \exp\left\{ -\frac{1}{2} \vec{x} M \vec{x} + \vec{h} \cdot \vec{x} \right\}, \tag{14.38}$$

where $M^{-1} \equiv C$ (see Equation 14.16) and the unperturbed partition function is given by

$$Z_0(\vec{h}) = \int d^G x \exp\left\{ -\frac{1}{2} \vec{x} M \vec{x} + \vec{h} \cdot \vec{x} \right\} = Z_0(0) e^{(1/2)\vec{h} M^{-1} \vec{h}}. \tag{14.39}$$

As before, we will express quantities of interest in terms of derivatives with respect to the components of the field term, $\vec{h}$, and then set $\vec{h} = 0$ at the end of the calculations.

When the three-gene term is included, we find the perturbed probability distribution function

$$\rho(\vec{x}) = \frac{1}{Z} \exp\left\{ -\frac{1}{2} \sum_{ij} x_i M_{ij} x_j + \sum_i h_i x_i + \frac{1}{6} \sum_{ijk} \Gamma_{ijk} x_i x_j x_k \right\}, \tag{14.40}$$

where Z is the perturbed partition function and can be expressed in terms of the unperturbed distribution. Defining the unperturbed expectation value of an arbitrary observable $\cdot$ by $\langle \cdot \rangle_0 = \int d^G x \rho_0(\vec{x}) \cdot \leq$, we have

$$\begin{aligned} Z &= \int d^G x \exp\left\{ -\frac{1}{2} \sum_{ij} x_i M_{ij} x_j + \sum_i h_i x_i + \frac{1}{6} \sum_{ijk} \Gamma_{ijk} x_i x_j x_k \right\} \\ &= Z_0 \left\langle \exp\left\{ \frac{1}{6} \sum_{ijk} \Gamma_{ijk} x_i x_j x_k \right\} \right\rangle_0 \\ &= Z_0 \left(1 + \frac{1}{6} \sum_{ijk} \Gamma_{ijk} \langle x_i x_j x_k \rangle_0 + O(\Gamma^2) \right), \end{aligned} \tag{14.41}$$

where the exponent in Equation 14.41 has been expanded and terms beyond the leading order dropped. We can find the form of the three-gene interactions through the three-body correlation,

$$\langle x_\alpha x_\beta x_\gamma \rangle = \frac{\partial^3 \ln Z}{\partial h_\alpha \partial h_\beta \partial h_\gamma} + \langle x_\alpha \rangle \langle x_\beta x_\gamma \rangle + \langle x_\beta \rangle \langle x_\alpha x_\gamma \rangle + \langle x_\gamma \rangle \langle x_\alpha x_\beta \rangle - 2 \langle x_\alpha \rangle \langle x_\beta \rangle \langle x_\gamma \rangle. \tag{14.42}$$

Because all the data have zero mean, only the first term is nonzero, leaving

$$\begin{aligned}\langle x_\alpha x_\beta x_\gamma\rangle &= \frac{\partial^3 \ln Z}{\partial h_\alpha \partial h_\beta \partial h_\gamma}\\ &= \frac{\partial^3 \ln Z}{\partial h_\alpha \partial h_\beta \partial h_\gamma} + \frac{\partial^3}{\partial h_\alpha \partial h_\beta \partial h_\gamma}\frac{1}{6}\sum_{ijk}\Gamma_{ijk}\langle x_i x_j x_k\rangle_0\\ &= \frac{1}{6}\sum_{ijk}\Gamma_{ijk}\frac{\partial^3\langle x_i x_j x_k\rangle_0}{\partial h_\alpha \partial h_\beta \partial h_\gamma}.\end{aligned} \tag{14.43}$$

Differentiating and dropping terms containing $\langle x\rangle_0$ or $\langle x^3\rangle_0$, we find

$$\begin{aligned}\frac{\partial^3\langle x_i x_j x_k\rangle_0}{\partial h_\alpha \partial h_\beta \partial h_\gamma} &= \langle x_i x_j x_k x_a x_\beta x_\gamma\rangle_0 - \langle x_a x_\beta\rangle_0\langle x_i x_j x_k x_\gamma\rangle_0\\ &\quad - \langle x_a x_\gamma\rangle_0\langle x_i x_j x_k x_\beta\rangle_0 - \langle x_\beta x_\gamma\rangle_0\langle x_i x_j x_k x_\alpha\rangle_0.\end{aligned} \tag{14.44}$$

This can be simplified using Wick's theorem, which states that the many-body correlation averaged over Gaussian distributions can be written as the sum of all distinct products of pair correlations. A four-gene average $\langle x_1 x_2 x_3 x_4\rangle_0$ can then be written as

$$\langle x_1 x_2 x_3 x_4\rangle_0|_{\vec{h}=0} = \langle x_1 x_2\rangle_0\langle x_3 x_4\rangle_0|_{\vec{h}=0} + \langle x_1 x_3\rangle_0\langle x_2 x_4\rangle_0|_{\vec{h}=0} + \langle x_1 x_4\rangle_0\langle x_2 x_3\rangle_0|_{\vec{h}=0}. \tag{14.45}$$

Similarly, the six-gene correlation $\langle x_1 x_2 x_3 x_4 x_5 x_6\rangle_0$ can be expanded into a sum of 15 sets of products of the three-pair correlations. Expanding Equation 14.44 in this way and using the identity $\langle x_i x_j\rangle_0|_{\vec{h}=0} = M_{ij}^{-1}$, we find

$$\begin{aligned}\frac{\partial^3\langle x_i x_j x_k\rangle_0}{\partial h_\alpha \partial h_\beta \partial h_\gamma}\Big|_{\vec{h}=0} &= M_{\alpha i}^{-1}M_{\beta j}^{-1}M_{\gamma k}^{-1} + M_{\alpha i}^{-1}M_{\beta k}^{-1}M_{\gamma j}^{-1} + M_{\alpha j}^{-1}M_{\beta i}^{-1}M_{\gamma k}^{-1}\\ &\quad + M_{\alpha j}^{-1}M_{\beta k}^{-1}M_{\gamma i}^{-1} + M_{\alpha k}^{-1}M_{\beta i}^{-1}M_{\gamma j}^{-1} + M_{\alpha k}^{-1}M_{\beta j}^{-1}M_{\gamma i}^{-1},\end{aligned} \tag{14.46}$$

and the three-gene interaction (Equation 14.43) is

$$\Gamma_{ijk} = \sum_{\alpha\beta\gamma=1}^{G}\langle x_\alpha x_\beta x_\gamma\rangle M_{i\alpha}M_{j\beta}M_{k\gamma}. \tag{14.47}$$

The distribution of the strengths of the perturbatively obtained three-gene interactions in a yeast gene network is shown in Figure 14.2. Although the mean value of $\langle x_i x_j x_k\rangle^2$ for $i \neq j$, $i \neq k$, and $j \neq k$ is found to be an order of magnitude smaller than $|\langle x_i x_j\rangle^3|$ for $i \neq j$ in the data considered, we find that there are some quite large three-gene correlations and indeed a few of the computed three-gene interactions turn out to have a large magnitude. It is therefore possible that many body effects may in fact play an important role in determining the behavior of the cell. A discussion of the

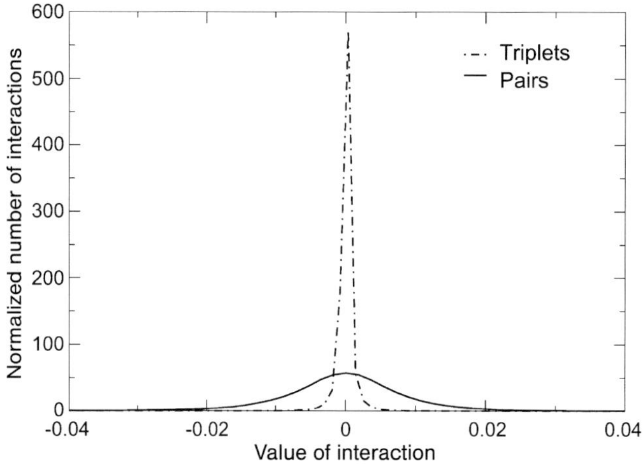

Figure 14.2 The distribution of the relative strengths of pairwise and three-gene interactions among the 582 genes in the 40-minute data set. Each curve is normalized to unit area.

biological significance of the genes that are involved in prominent three-gene interactions appears at the end of the next section.

14.6 Network Analysis

14.6.1 Metabolic Oscillations in Yeast

Our method can be validated by applying it to a well-understood system, so we use it to determine the network of effective interactions between genes in the well-studied yeast *S. cerevisiae* undergoing metabolic oscillations. Yeast cells under nutrient-limiting conditions oscillate between states of fermentation and oxidative reduction. In the presence of sugar, the enzymes required for utilization of other carbon sources are repressed. This is termed catabolite or glucose repression and results not only in the downregulation of genes encoding proteins of the mitochondrial tricarboxylic acid cycle, galactose utilization, and gluconeogenesis, but also activates expression of genes encoding glycolytic enzymes, hexose transporters, and ribosomal protein genes. Conversely, when a yeast culture growing on a glucose-containing medium depletes it of glucose, it undergoes what is termed as the diauxic shift and begins upregulating genes encoding enzymes involved in respiration, as well as other mitochondrial functions, and downregulating genes involved in other cellular functions, such as protein synthesis. When other nutrients are limiting, cells adjust their growth rate to the available nutrient supply by activating the metabolic pathways that allow them to compensate for the least abundant ones [24]. The observed stable cycling between oxidofermentative and oxidative metabolism arises under diverse conditions in response to a limited nutrient supply [24,25]. Thus, the precise state of the underlying genetic network is exquisitely tuned to the availability of a variety of nutrients.

The signaling pathways orchestrate a change in the metabolic mode. It has been established [26] that there are at least two such pathways: one involving the protein kinase A (PKA) and another involving a protein known as the target of rapamycin (TOR). The two pathways overlap partially. By tinkering with the nutrient supply rates one can generate metabolic oscillation of various periods that are stable against small variations in the temperature. These oscillations appear to be coupled to periodic variations in the number of the yeast cells in the culture [24] indicating the existence of some overlap between the genes involved in the signaling and the genes involved in the cell division process.

14.6.2
Polishing of the Data and Selection of Subsets of Genes

The task at hand is to identify the crucial elements of the genetic network and to elucidate relationships among them. We consider microarray data for *S. cerevisiae* undergoing metabolic oscillations corresponding to a period of about 40 minutes. In an experiment performed by Klevecz *et al.* [18], metabolic oscillations are induced in yeast grown under nutrient-limiting conditions and 32 genome-wide expression level measurements are taken from the culture at $\sim$4-minute intervals over three oscillatory periods. We use these expression level data to find the network of effective pairwise interactions between the monitored genes.

The temporal variance of the expression levels of individual genes varies by several orders of magnitude across the genome, skewing the covariance matrix C toward those genes with the largest profile amplitudes. For this reason, some reduction to a "common denominator" is necessary; and prior to analysis, each row of the data table is polished by subtracting its mean value and normalizing it to unit variance. In this way, we can avoid biasing the calculations toward genes with high expression levels and make sure that the genetic influences are assessed on the basis of the shape, and not the magnitude, of the expression profile. An example of how the polished data look is shown in Figure 14.3 for seven genes, which turned out to be the hubs in the network.

A side effect of normalizing the profiles by their variances is the amplification of experimental noise in the small-amplitude profiles. We reduce the effects of amplified noise by including or excluding each gene based on the standard deviation σ of its raw expression profile. We define $\langle\sigma\rangle = (1/G)\sum_{i=1}^{G}\sigma_i$ as the mean value of the standard deviation, taken over all expression profiles, and we define $\sigma_\sigma^2 = (1/G)\sum_{i=1}^{G}(\langle\sigma_i^2\rangle - \langle\sigma_i^2\rangle)$ as the variance of expression profile standard deviations. Our criterion is to include genes with a profile standard deviation greater than $\langle\sigma\rangle + \kappa\sigma$, where κ is a multiplicative parameter that determines the cutoff. The resulting networks are robust to small variations in κ. If κ varies from 0.9 to 1.1, for example, the topology of the strongly interacting network component

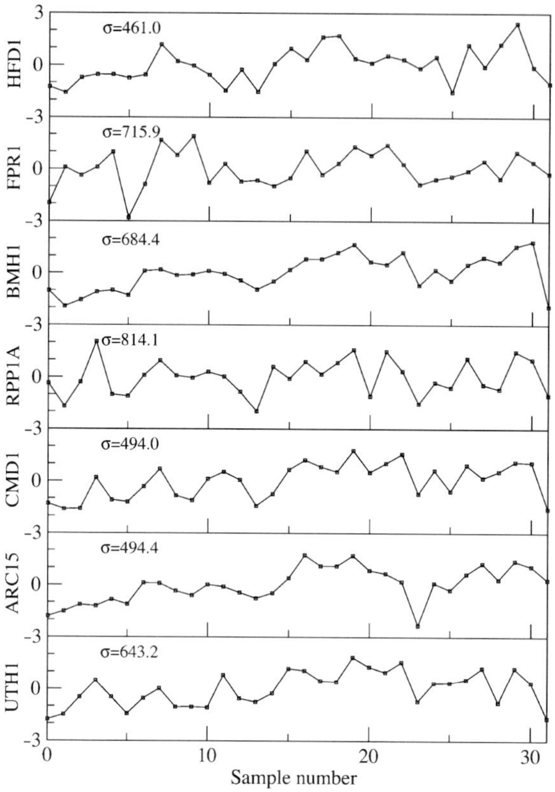

Figure 14.3 Examples of the polished data for seven hub genes in the 582-gene network. Each normalized profile shown has a mean expression level of zero and a variance of one. The standard deviation σ of each gene's raw expression profile is provided. Although the expression levels are sampled over three metabolic cycles, this periodicity is not obvious from the profiles of the hub genes.

remains largely unchanged. The raw data contain measurements for 4670 genes (other genes had expression level below the experimental resolution), and in our initial analysis we use the cutoff value $\kappa = 1$, leaving $G = 582$ genes that vary the most in the data, as shown in Figure 14.4. This set of genes will be denoted as "level 1" description and we would discuss later on the effects of expanding it to "level 2" description in which the cutoff is lowered to deal with a set of 1008 genes.

14.6.3
The Nature of the Network

Figure 14.5 shows the strongest 110 pairwise interactions inferred by entropy maximization for the system of 582 genes, whereas Figure 14.6 is a graph of the

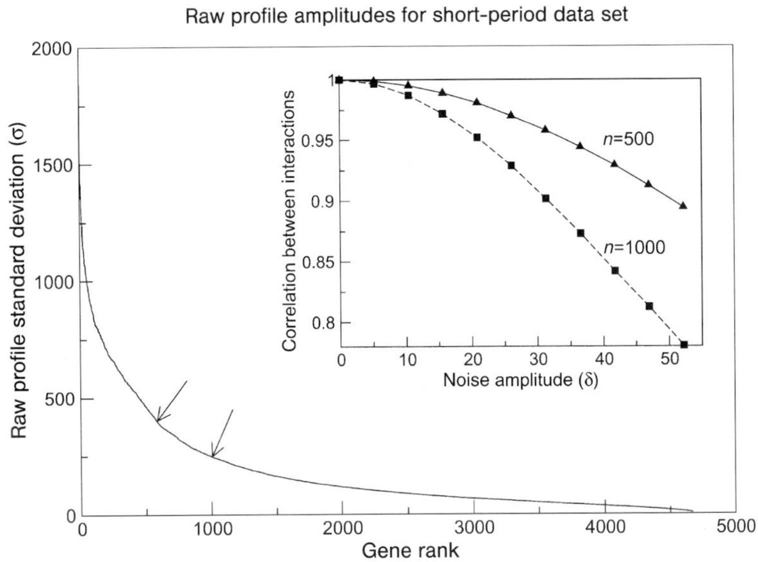

Figure 14.4 A plot of the rank-ordered standard deviations (σ) of raw expression profiles for the 4670 genes in the short-period data set. Arrows indicate cutoff values for 582 and 1008 genes. The inset shows the Spearman correlation between the calculated interactions and those that result when noise of a fixed amplitude is added to the raw profiles. Random noise from a Gaussian distribution of width δ is added to each of the raw expression profiles, and the network for the noise-enhanced data is calculated and compared with the network from the raw data using the Spearman correlation. When the 500 genes with largest profile amplitudes are retained (solid line), noise of $\delta = 52.2$, about 10 times the estimated background level, does not significantly change the network. The network is more sensitive to noise when 1000 genes are retained (dashed line), but the correlation is still 0.9 between the original network and that with noise added at six times the estimated background.

110 weakest interactions in the same system. The solid (dotted) lines indicate positive (negative) M_{ij}. There is a qualitative difference between the subgraph composed of only the largest magnitude interactions and the subgraph of only the smallest magnitude interactions. For $P \ll G$, the strongly coupled component forms a cluster of highly, and sometimes hierarchically, interconnected nodes, whereas the weakly coupled component forms a discontinuous network mainly composed of isolated pairwise interactions. In the strong coupling subnetwork, only the single pair of genes (Dal4 and Gap1) is not connected to the interconnected main cluster. This apparent disconnect results from our arbitrary selection of displaying only the strongest 110 interactions in the network. Indeed, the 111th strongest interaction is between Gap1 and Arc15. It should be noted that irrespective of the size of the subset of the full genome considered, the deduced interactions arise from the influence of all variables. Hence, even if a gene is not explicitly considered in the network calculation, its effect is nonetheless integrated into the interactions among the remaining genes. The analysis yields a measure of the

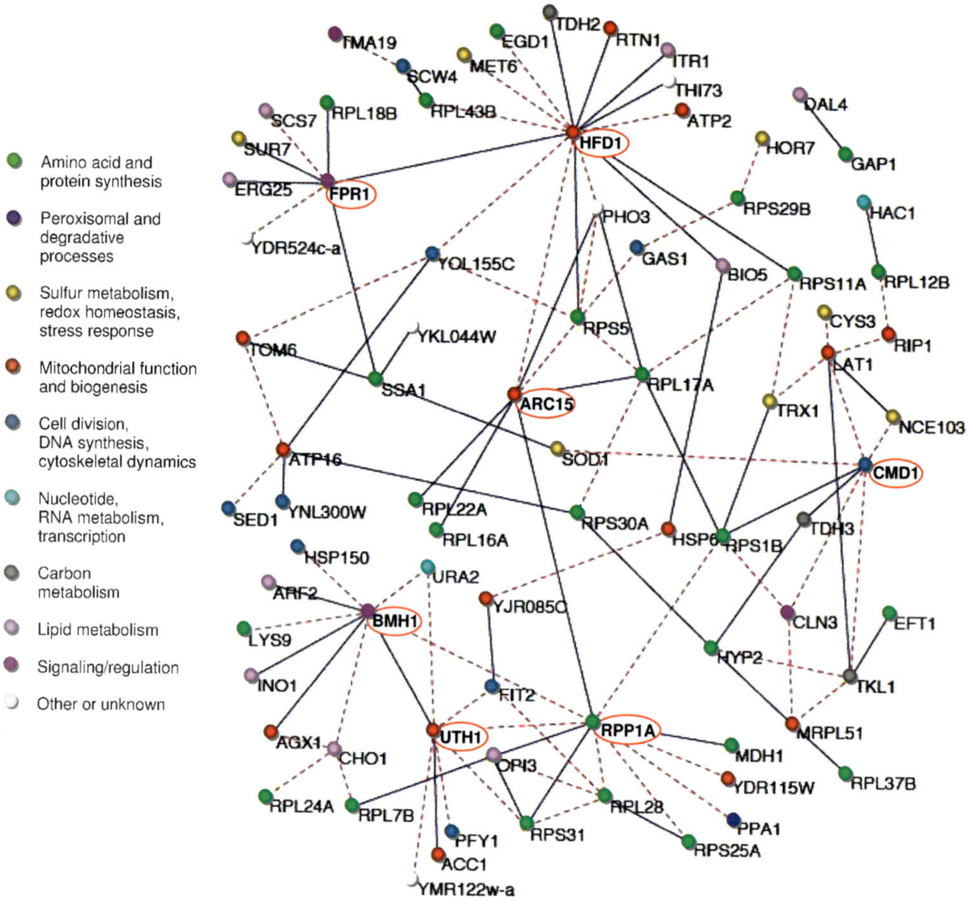

Figure 14.5 The network of the strongest 110 pairwise interactions inferred by entropy maximization using the 582 genes showing the most marked fluctuations in transcript levels in the data set from yeast chemostat cultures showing 40-minute metabolic oscillations [18]. Nodes are identified by gene names and color-coded to indicate the cell process in which they participate (there is some ambiguity in assigning genes to categories). The solid blue lines denote positive couplings and the dashed red lines denote negative couplings. The identity of the hubs circled in red is discussed in the text [27].

magnitude and the sign of the pairwise interactions most likely to give rise to the observed data.

The network nodes have been color coded in the figure to illustrate the connectivity of genes encoding proteins involved in different metabolic functions. Moreover, the nodes in this subset vary markedly in their connectivity. Strikingly, the most highly connected genes (hubs) encode proteins involved in critical processes that tune cell growth and division to the nutrient supply. We selected the hubs with more than six edges for a detailed analysis. These seven hub genes encode proteins

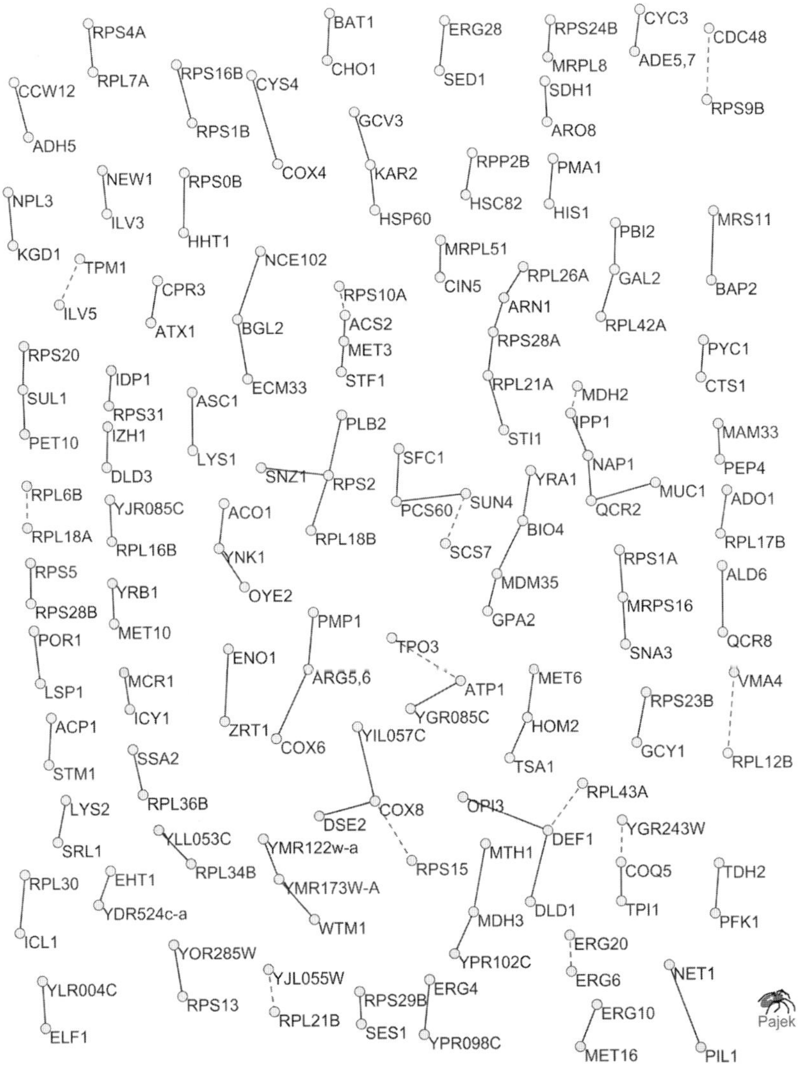

Figure 14.6 A graph of the 110 weakest interactions in the network inferred from analysis of the top 582 genes in the short-period data set. The disjointed nature of the graph implies randomness and is topologically distinct from the highly connected graphs formed by genes with strongest interactions (Figure 14.5). The solid blue lines denote positive couplings and the dashed red lines denote negative couplings [27].

involved in TOR signaling (Fpr1, Bmh1, and Uth1), the outer mitochondrial membrane proteins Hfd1 and Arc15, the ribosomal stalk protein Rpp1A, and calmodulin (Cmd1). Moreover, the processes identified by the hub genes are themselves highly interconnected.

14.6.4 The Biological Interpretation of the Network

We now focus on the functions of the genes that are hubs in Figure 14.5. The summary of the major functions is shown in the "level 1" portion of Figure 14.7.

The Fpr1 (aka Fkbp12) protein is a key component of the critical TOR nutrient signaling pathway. Mutations in the Fpr1, Tor1, and Tor2 genes confer rapamycin resistance; the Fpr1 protein, a peptidyl-prolyl trans-isomerase, binds rapamycin and the Fpr1–rapamycin complex physically interacts with the TOR proteins [26]. Tor1 and Tor2 are protein kinases that control cell growth in response to nutrients and both are components of larger complexes. The Tor1 complex regulates translation initiation, transcription, mRNA stability, nutrient uptake, ribosome biogenesis, and retrograde signaling from the mitochondrion to the nucleus, as well as autophagy and entrance into the cell cycle, while the Tor2 complex controls the polarization of the actin cytoskeleton required for establishing cell polarity [26]. Fpr1 also negatively regulates calcineurin, interacts with the high mobility group protein Hmo1 that stabilizes chromatin structure, and mediates aspartokinase feedback inhibition. Thus, the Fpr1 gene in the network identifies a key cellular

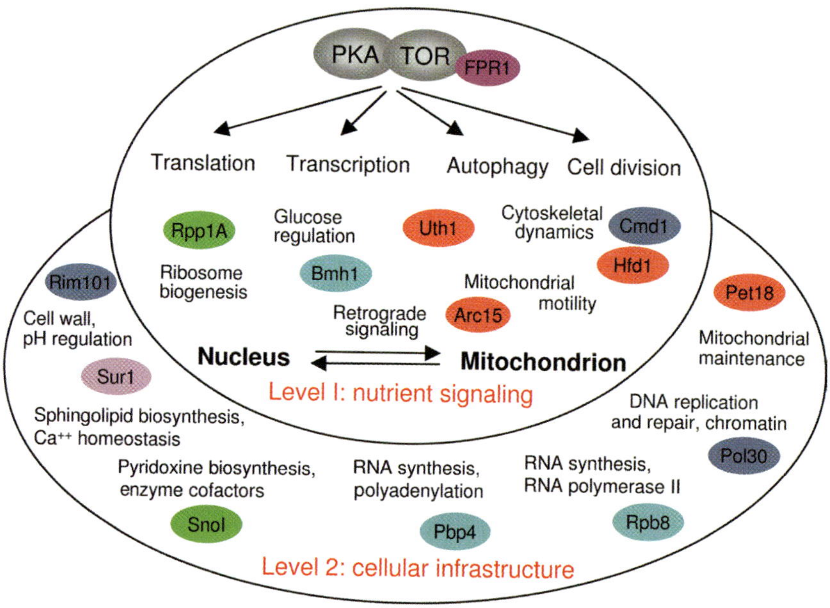

Figure 14.7 A diagrammatic representation of the cellular processes identified by the network hubs among the 582 (level 1) and the 1008–2000 (level 2) genes exhibiting the most marked fluctuations in transcript levels during 40-minute metabolic oscillations. PKA and TOR represent the protein kinase A and target of rapamycin nutrient signaling pathways; other ovals contain the designations of hub genes identified as described in the text and color coded by cell process as in Figure 14.5

hub that integrates essential cytoskeletal processes involved in cell division with cell growth and physiology. The interaction between the Fpr1 gene and the Ssa1 gene is the strongest in the represented network. Ssa1 is a chaperone protein that interacts with Cdc25, the guanine nucleotide exchange factor for Ras, to control cAMP production and the activation of PKA [28]. PKA is the central kinase in the PKA pathway that overlaps the TOR pathway in mediating glucose and nutrient signaling [29].

Bmh1 is one of the two redundant yeast 14-3-3 proteins (Bmh1 and Bmh2) that are central regulators in retrograde signaling from the mitochondrion to the nucleus in response to nutrient availability [30]. The interface between carbon and nitrogen metabolism in yeast is mediated by several genes, which encode proteins in the tricarboxylic acid cycle, such as aconitase, citrate synthetase, and isocitrate dehydrogenase. Expression of these genes is regulated by a complex of the transcription factors Rtg1 and Rtg3. Bmh1 and Bmh2 are involved in the cytoplasmic sequestration of the Rtg1/3 complex. Activation of retrograde signaling by nutrient limitation, mitochondrial dysfunction, or rapamycin inhibition of TOR kinases stimulates nuclear accumulation of the complex. These proteins function similarly to constrain the Msn2 and Msn4 transcription factors for stress-inducible genes to the cytoplasm [26]. Bmh1 and Bmh2 have also been implicated in the regulation of glucose repression, which is mediated by a type 1 protein phosphatase complex [31], and in cytoplasmic sequestration of the Rim15 kinase, which integrates signals from several pathways, including PKA and TOR, to regulate entry into G0 in response to carbon and nitrogen availability [32]. Thus, the Bmh1 hub in the network represents an intersection between carbon and nitrogen nutrient sensing through the TOR pathway and retrograde signaling from mitochondrion to the nucleus.

Rpp1A is a component of the ribosome stalk, a lateral protuberance on the large ribosomal subunit comprised of five acidic proteins [33]. The stalk proteins are the only ribosomal proteins found free in the cytoplasm; their stability is regulated by phosphorylation and interactions among them [34]. The stalk proteins interact with translation elongation factor EF-2 and may regulate ribosome function [33]. The genes encoding the stalk proteins, as well as all other ribosomal protein genes, are subject to regulation through the overlapping TOR and PKA pathways [29]. In addition, ribosomal protein mRNA stability is subject to TOR regulation [35]. Hence, the Rpp1A hub represents the connection between nutrient supply, ribosome biogenesis, and protein synthesis.

Uth1 is present both in the cell wall and in the mitochondrial outer membrane and is involved in mitochondrial biogenesis [36]. Cells lacking the Uth1 gene are more tolerant to carbon and nitrogen starvation than the wild type. Rapamycin, which binds to the Fpr1 protein, induces autophagy, mimicking the nitrogen starvation. Rapamycin-treated cells rapidly degrade mitochondria; mutants lacking the Uth1 gene are rapamycin resistant. Thus, the Uth1 protein is a critical link between stress, nutrient supply, and mitochondrial fate through the TOR signaling pathway.

Cmd1 is the yeast calmodulin and is essential for organization of the actin cytoskeleton, endocytosis, nuclear division, and bud emergence. It is a component of the core protein complex of the spindle pole body, the yeast microtubule organizing center [37]. Cmd1 interacts physically with the Arc35 subunit of the Arp2/3 complex in endocytosis, a process that involves the actin cytoskeleton [38]. Parenthetically, the Cln3 gene linked to the Cmd1 hub encodes a key cyclin whose increased expression in response to glucose gates the transition from G1 to S phase prior to cell division [39]. Arc15 is a subunit of the Arp2/3 complex, which assembles and maintains actin-based cellular structures [40]. The Arp2/3 complex coordinates signals to the actin cytoskeleton and is localized primarily in actin patches in areas of polarized growth and septation [40]. Mitochondria show a pattern of movement that resembles that of chromosomes during cell division, moving to and being retained at the poles, resulting in equal distribution between mother and daughter cells. Both Arp2 and Arc15 are associated with mitochondria and are required for mitochondrial motility [41]. The Cmd1 and Arc15 hubs therefore identify connections between cytoskeletal organization, cell division, and mitochondrial movement during cell division.

Hfd1, an outer mitochondrial membrane and lipid particle protein, is the yeast homologue of the fatty aldehyde dehydrogenase gene whose deficiency in humans underlies the Sjogren-Larsson syndrome, a severe metabolic disorder associated with ichthyosis, mental retardation, and spastic diplegia or tetraplegia [42,43]. Since the hfd1 mutant of yeast shows a defect in spindle pole body organization [44], the highly interconnected Hfd1 node in the network may also be involved in the connection between the cytoskeletal dynamics and mitochondrial localization.

We conclude that the network hubs identify essential cellular processes that adjust the metabolic state and pace cell division to the carbon and nitrogen supply through the pervasive TOR and PKA signaling pathways. Bmh1, Fpr1, and the mitochondrial protein Uth1 interconnect the TOR pathway with the metabolic and physical state of the mitochondrion, as well as with the retrograde signaling system that adjusts expression of nuclear genes encoding mitochondrial proteins in response to changes in carbon and nitrogen nutrient supplies. Rpp1A may be an important translational regulatory protein and transcription and stability of both its mRNA and those of other ribosomal proteins are regulated through the TOR signaling pathway [29]. Cmd1 and the mitochondrial Arc15 and Hfd1 proteins are involved in the actin cytoskeletal dynamics, essential for endocytosis, cell division, and mitochondrial motility; these interconnect with the TOR signaling pathway through the Fpr1 protein. The Fpr1 gene shows a strong linkage to the Ssa1 gene encoding a key regulator of the overlapping PKA nutrient-signaling pathway. The signaling pathways identified in the analysis here function primarily through mechanisms that alter protein activity or abundance; their representation in the networks inferred from transcript profiling implies that at least one component of the pathway is subject to regulation by a mechanism that adjusts transcription frequency or transcript stability (or both). We conclude that the application of our entropy maximization technique to the 582 genes exhibiting the greatest variance in the short-period data set accurately identifies genes encoding key protein participants in intracellular communication pathways known to be crucial to adjusting both cellular metabolic activity and the cell cycle to the available supply of carbon and nitrogen nutrients.

14.6.5
The Larger Subset of Genes

One may ask how the network structure changes when more genes are included in the analysis, that is, when one considers level 2 of the description – see Figure 14.7.

Since we initially rank genes on the basis of the magnitude of variation in transcript abundance, expanding the gene set brings in genes whose expression levels show smaller fluctuations. The subsequent profile normalization, which eliminates the differences among them, affects the membership of the subset of strongest interactions. When the number of genes is expanded to 1008, hubs representing such fundamental cellular processes as pH regulation and cell wall biosynthesis, DNA replication, pyridoxine biosynthesis, mitochondrial organization, and biogenesis are added to the network, although all of the original seven hubs are still represented among the genes showing the strongest interactions. Rim101 is a transcription factor that is proteolytically processed in response to alkaline pH and acts as both a transcriptional activator and a repressor; it also functions in cell wall assembly. Pol30 encodes proliferating cell nuclear antigen (PCNA), a key organizer of proteins involved in DNA replication, repair, and modification, as well as in transcriptional silencing and epigenetic inheritance. Sno1 encodes a subunit of a glutaminase required for pyridoxine biosynthesis and its transcription is nutrient regulated through the TOR pathway. Pet18 is a complex locus whose loss renders the cells respiration deficient and unable to grow on nonfermentable carbon sources, although the precise function of the Pet18 protein is still not known. Further expansion of the gene set to 1500 adds Pbp4 gene, which encodes a protein involved in the regulation of polyA tail synthesis in mRNA biogenesis, as a hub; while inclusion of 2000 genes adds two additional hubs corresponding to genes, which encode an essential RNA polymerase II subunit (Rpb8) and Sur1, an enzyme in the sphingolipid biosynthetic pathway. The Sur1 gene is required for calcium homeostasis and has been implicated in TOR signaling, as well. Thus, the subset of genes that shows the most marked fluctuations encodes proteins involved in adjusting the cell's immediate metabolic state to the nutrient supply, while genes encoding proteins involved in more fundamental cellular processes show less marked fluctuations in the course of the metabolic adjustments.

14.6.6
Metabolic Oscillations with Longer Periods

A fundamental question posed by the differences in the periodicities of observed metabolic and transcript fluctuations is whether there is a unitary underlying clocklike mechanism that coordinates a variety of cellular functions or whether there exist a variety of states of the genetic and cellular network that support oscillatory behaviors on different time scales. The availability of transcriptional data for both short- and long-period yeast metabolic oscillations provides an opportunity to ask this question directly. As reported in the original publications, correlation clustering yields superficially similar results [18,45], identifying groups of genes that encode proteins involved in amino acid and protein synthesis, RNA metabolism, sulfur metabolism, DNA replication, and mitosis, as well as in mitochondrial structure and function. However, although the categories are same and roughly equally represented in both data sets, there is little overlap in the genes represented in each category. Moreover, pairs of genes whose expression patterns are highly correlated in one data set are generally not correlated in the other (see Figure 14.8).

The genetic network inferred from the long-period data set [45] using the entropy maximization method described here is also quite different [15] from that extracted from short-period data set. However, the Rpp1A hub is common to both networks; this and several additional ribosomal protein gene hubs in the long-period network are all regulated through the TOR signaling pathway [46]. Moreover, mitochondrial protein genes, albeit different ones, constitute hubs in both short- and long-period networks. We conclude that although some of the same signaling pathways are involved, rather different states of the gene network support the observed short- and

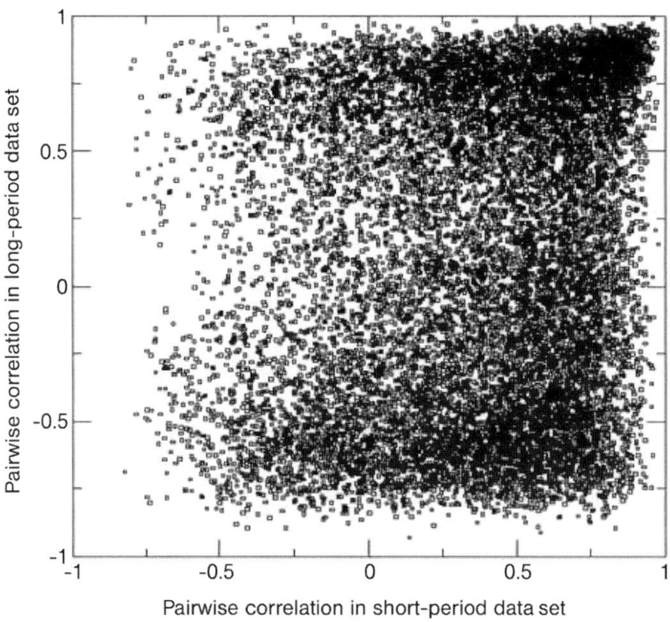

Figure 14.8 The correlation coefficient between pairs of profiles in the short-period data set are plotted against the pairwise correlation coefficient for the same pair in the long-period data set. The data shown are a random sample of 1000 pairwise correlations from among the 582^2 possible correlations between gene pairs in the short-period data set.

long-period metabolic oscillations. This prediction has been subsequently supported by Murray *et al.* [47].

14.6.7
Three-Gene Interactions

A study of the strongest three-gene interactions also identified genes that encode proteins likely to be important in regulating the metabolic activity. The Pnc1 gene, the most highly interconnected hub in the triplet network and involved in 74 of the top 100 three-gene interactions, encodes a nicotinamide deaminase that plays a major role in yeast lifespan extension in response to caloric restriction, precisely the conditions of the experiments from which the data set was derived [48]. The second most highly interconnected gene, participating in 66 of the strongest 100 three-gene interactions, is the Tma19 gene – the yeast homologue of the well-studied mammalian translationally controlled tumor protein (TCTP) gene – a calcium-binding protein that interacts with microtubules, regulates translation, and exerts an apoptotic effect. The yeast Tma19 protein, which interacts with microtubules, exhibits redox-dependent translocation to mitochondria under stress conditions and influences lifespan, maybe, of a similar multifunctional protein [49].

14.7
Conclusion

We have established a method for calculating effective interactions between genes based only on their expression data and entropy maximization. At the pairwise level, these interactions are found as the elements of the inverse of the matrix of gene expression covariances. We have shown analytically that the calculated interactions can reproduce their generating data from partial information, that the method is self-consistent insofar as the number of effective genes can be reduced by straightforwardly integrating over a subset of genes, and that the entropy maximization can be applied to calculate higher order interactions. Finally, we have validated our method by demonstrating that it produces biologically relevant results when applied to microarray data.

Our network inference technique has several advantages over other methods. First, the estimation of maximum entropy does not place any unwarranted assumptions upon the structure of the network. Any discrepancy between the inferred network and the microscopic cellular activity, therefore, results not from a deficiency in the model, but from differences between true cellular activity and that which is measured by the microarray. Given completely noiseless microarray data, the technique produces a network that is entirely faithful to the biochemistry of the cell. A second strength of our technique is that it takes a systems approach to biological networks. The observed gene profiles result within the context of all of the other genes and cell products and are thus the emergent properties of the system of interacting genes. Our method does not merely calculate the interactions of isolated pairs of genes, but also considers the behavior of the entire system of genes as input into the network calculation. Finally, the entropy maximization technique is not limited to pairwise interactions between genes. Unlike other methods, ours can explicitly address three-gene and higher order interactions. These advantages, together with the encouraging results from its application to yeast data, suggest that entropy maximization has great potential and value in microarray analysis.

Acknowledgment

Support from grant N N202 0852 33 from the Ministry of Science in Poland is acknowledged.

References

1 Liang, S.S., Fuhrman, S. and Somogyi, R. (1998) Reveal, a general reverse engineering algorithm for inference of genetic network architectures. *Pacific Symposium on Biocomputing*, **3**, 18–29.

2 Akutsu, T., Miyano, S. and Kuhara, S. (2000) Algorithms for identifying Boolean networks and related biological networks based on matrix multiplication and fingerprint function. *Journal of Computational Biology*, **7**, 331–343.

3 Butte, A. J. and Kohane, I. S. (2000) Mutual information relevance networks: functional genomic clustering using pairwise entropy measurements. *Pacific Symposium on Biocomputing*, **3**, 418–429.

4 Chen, K.C., Csikasz-Nagy, A., Gyorffy, B., Val, J., Novak, B. and Tyson, J.J. (2000) Kinetic analysis of a molecular model of the budding yeast cell cycle. *Molecular Biology of the Cell*, **11**, 369–391.

5 Friedman, N., Linial, M., Nachman, I. and Pe'er, D. (2000) Using Bayesian networks to analyze expression data. *Journal of Computational Biology*, **7**, 601–620.

6 Ideker, T., Thorsson, V., Ranish, J.A., Christmas, R., Buhler, J., Eng, J.K., Bumgarner, R., Goodlett, D.R., Aebersold, R. and Hood, L. (2001) Integrated genomic and proteomic analyses of a systematically perturbed metabolic network. *Science*, **292**, 929–934.

7 Shmulevich, I., Dougherty, E.R., Kim, S. and Zhang, W. (2002) Gene perturbation and intervention in probabilistic Boolean networks. *Bioinformatics*, **18**, 261–274.

8 Toh, H. and Horimoto, K. (2002) Inference of a genetic network by a combined approach of cluster analysis and graphical Gaussian modeling. *Bioinformatics*, **18**, 287–297.

9 Yeung, M.K., Tegner, J. and Collins, J.J. (2002) Reverse engineering gene networks: integrating genetic perturbations with dynamical modeling. *Proceedings of the National Academy of Sciences of the United States of America*, **99**, 6163–6168.

10 Gardner, T.S., di Bernardo, D., Lorenz, D. and Collins, J.J. (2003) Inferring genetic networks and identifying compound mode of action via expression profiling. *Science*, **301**, 102–105.

11 Tegnér, J., Yeung, M.K., Hasty, J. and Collins, J.J. (2003) Reverse engineering gene networks: integrating genetic perturbations with dynamical modeling. *Proceedings of the National Academy of Sciences of the United States of America*, **100**, 5944–5949.

12 Friedman, N. (2004) Inferring cellular networks using probabilistic graphical models. *Science*, **303**, 799–805.

13 Schafer, J. and Strimmer, K. (2005) An empirical Bayes approach to inferring large-scale gene association networks. *Bioinformatics*, **21**, 754–764.

14 Li, S., Wu, L. and Zhang, Z. (2006) Constructing biological networks through combined literature mining and microarray analysis: a LMMA approach. *Bioinformatics*, **22**, 2143–2150.

15 Lezon, T.R., Banavar, J.R., Cieplak, M., Maritan, A. and Fedoroff, N.V. (2006) Using the principle of entropy maximization to infer genetic interaction networks from gene expression patterns. *Proceedings of the National Academy of Sciences of the United States of America*, **103**, 19033–19038.

16 Boltzmann, L. (1964) *Lectures on Gas Theory*, Cambridge University Press, London.

17 Shannon, C.E. (1948) A mathematical theory of communication. *Bell System Technical Journal*, **27**, 379–423.

18 Klevecz, R.R., Bolen, J., Forrest, G. and Murray, D.B. (2004) A genomewide oscillation in transcription gates DNA replication and cell cycle. *Proceedings of the National Academy of Sciences of the United States of America*, **101**, 1200–1205.

19 Jaynes, E.T. (1957) Information theory and statistical mechanics. *Physical Review*, **106**, 620–630.

20 Jaynes, E.T. (1957) Information theory and statistical mechanics, II. *Physical Review*, **108**.

21 Dewar, R. (2003) Information theory explanation of the fluctuation theorem, maximum entropy production and self-organized criticality in non-equilibrium stationary states. *Journal of Physics A*, **36**, 631–641.

22 Dewar, R.C. (2005) Maximum entropy production and the fluctuation theorem. *Journal of Physics A*, **38**, L371–381.

23 Whitfield, J. (2005) Order out of chaos. *Nature*, **436**, 905–907.

24 Saldanha, A.J., Brauer, M.J. and Botstein, D. (2004) Nutritional homeostasis in batch and steady-state culture of yeast. *Molecular Biology of the Cell*, **15**, 4089–4104.

25 Richard, P. (2003) The rhythm of yeast. *FEMS Microbiology Reviews*, **27**, 547–557.

26 Rohde, J.R. and Cardenas, M.E. (2004) Nutrient signaling through TOR kinases controls gene expression and cellular differentiation in fungi. *Current Topics in Microbiology and Immunology*, **273**, 53–72.

27 Batagelj, V. and Mrvar, A.Pajek – program for large network analysis.http://vlado.fmf.unilj.si/pub/networks/pajek/.

28 Geymonat, M., Wang, L., Garreau, H. and Jacquet, M. (1998) Ssa1p chaperone interacts with the guanine nucleotide exchange factor of ras Cdc25p and controls the cAMP pathway in *Saccharomyces cerevisiae*. *Molecular Microbiology*, **30**, 855–864.

29 Chen, J.C. and Powers, T. (2006) Coordinate regulation of multiple and distinct biosynthetic pathways by TOR and PKA kinases in *S. cerevisiae*. *Current Genetics*, 1–13.

30 Butow, R.A. and Avadhani, N.G. (2004) Mitochondrial signaling: the retrograde response. *Molecules and Cells*, **14**, 1.

31 Dombek, K.M., Kacherovsky, N. and Young, E.T. (2004) The Reg1-interacting proteins Bmh1, Bmh2, Ssb1, and Ssb2, have roles in maintaining glucose repression in *Saccharomyces cerevisiae*. *The Journal of Biological Chemistry*, **279**, 39165–39174.

32 Wanke, V., Pedruzzi, I., Cameroni, E., Dubouloz, F. and De Virgilio, C. (2005) Regulation of G(0) entry by the Pho80–Pho85 cyclin–CDK complex. *EMBO Journal*, **24**, 4271–4278.

33 Lalioti, V.S., Perez-Fernandez, J., Remacha, M. and Ballesta, J.P. (2002) Characterization of interaction sites in the *Saccharomyces cerevisiae* ribosomal stalk components. *Molecular Microbiology*, **46**, 719–729.

34 Nusspaumer, G., Remacha, M. and Ballesta, J.P. (2000) Phosphorylation and N-terminal region of yeast ribosomal protein P1 mediate its degradation, which is prevented by protein P2. *EMBO Journal*, **19**, 6075–6084.

35 Albig, A.R. and Decker, C.J. (2001) The target of rapamycin signaling pathway regulates mRNA turnover in the yeast *Saccharomyces cerevisiae. Molecular Biology of the Cell*, **12**, 3428–3438.

36 Camougrand, N.M., Mouassite, M., Velours, G.M. and Guerin, M.G. (2000) The 'SUN' family: UTH1 an ageing gene, is also involved in the regulation of mitochondria biogenesis in *Saccharomyces cerevisiae. Archives of Biochemistry and Biophysics*, **375**, 154–160.

37 Muller, E.G., Snydsman, B.E., Novik, I., Hailey, D.W., Gestaut, D.R., Niemann, C.A., O'Toole, E.T. Jr, Giddings, T.H., Sundin, B.A. and Davis, T.N. (2005) The organization of the core proteins of the yeast spindle pole body. *Molecular Biology of the Cell*, **16**, 3341–3352.

38 Schaerer-Brodbeck, C. and Riezman, H. (2000) Functional interactions between the p35 subunit of the Arp2/3 complex and calmodulin in yeast. *Molecular Biology of the Cell*, **11**, 1113–1127.

39 Newcomb, L.L., Diderich, J.A., Slattery, M.G. and Heideman, W. (2003) Glucose regulation of *Saccharomyces cerevisiae* cell cycle genes. *Eukaryotic Cell*, **2**, 143–149.

40 Machesky, L.M. and Gould, K.L. (1999) The Arp2/3 complex: a multifunctional actin organizer. *Current Opinion in Chemical Biology*, **11**, 117–121.

41 Boldogh, I.R., Yang, H.C., Nowakowski, W.D., Karmon, S.L., Hays, L.G., Yates, J.R. III and Pon, L.A. (2001) Arp2/3 complex and actin dynamics are required for actin-based mitochondrial motility in yeast. *Proceedings of the National Academy of Sciences of the United States of America*, **98**, 3162–3167.

42 Rizzo, W.B. (1993) Sjogren-Larsson syndrome. *Seminars in Dermatology*, **12**, 210–218.

43 Zahedi, R.P., Sickmann, A., Boehm, A.M., Winkler, C., Zufall, N., Schonfisch, B., Guiard, B., Pfanner, N. and Meisinger, C. (2006) Proteomic analysis of the yeast mitochondrial outer membrane reveals accumulation of a subclass of pre-proteins. *Molecular Biology of the Cell*, **17**, 1436–1450.

44 Okamoto, S. and Iino, T. (1982) Genetic block of outer plaque morphogenesis at the second meiotic division in an hfd1-1 mutant of *Saccharomyces cerevisiae. Journal of General Microbiology*, **128**, 1309–1317.

45 Tu, B.P., Kudlicki, A., Rowicka, M. and McKnight, S.L. (2005) Logic of the yeast metabolic cycle: temporal compartmentalization of cellular processes. *Science*, **310**, 1152–1158.

46 Rohde, J., Heitman, J. and Cardenas, M.E. (2001) The TOR kinases link nutrient sensing to cell growth. *The Journal of Biological Chemistry*, **276**, 9583.

47 Murray, D.B., Beckmann, M. and Kitano, H. (2007) Regulation of yeast oscillatory dynamics. *Proceedings of the National Academy of Sciences of the United States of America*, **104**, 2241–2246.

48 Anderson, R.M., Bitterman, K.J., Wood, J.G., Medvedik, O. and Sinclair, D.A. (2003) Nicotinamide and PNC1 govern lifespan extension by calorie restriction in *Saccharomyces cerevisiae. Nature*, **423**, 181–185.

49 Rinnerthaler, M., Jarolim, S., Heeren, G., Palle, E., Perju, S., Klinger, H., Bogengruber, E., Madeo, F., Braun, R.J. and Breitenbach-Koller, L. (2006) MMI1 (YKL056c TMA19), the yeast orthologue of the translationally controlled tumor protein (TCTP) has apoptotic functions and interacts with both microtubules and mitochondria. *Biochimica et Biophysica Acta*, **1757**, 631–638.

Index

Analysis of Microarray Data: A Network-Based Approach. Edited by F. Emmert-Streib and M. Dehmer

ISBN: 978-3-527-31822-3